A Bibliography of the Philosophy of Science, 1945-1981

A Bibliography of the Philosophy of Science, 1945-1981

Compiled by
Richard J. Blackwell

G
P

Greenwood Press
Westport, Connecticut • London, England

Library of Congress Cataloging in Publication Data

Blackwell, Richard J., 1929-
  A bibliography of the philosophy of science, 1945-1981.

  Bibliography: p.
  Includes index.
  1. Science—Philosophy—Bibliography.   I. Title.
Z7405.P74B57 1983   [Q175]   016.501   83-5671
ISBN 0-313-23124-9 (lib. bdg.)

Library of Congress Catalog Card Number: 83-5671
ISBN: 0-313-23124-9

First published in 1983

Greenwood Press
A division of Congressional Information Service, Inc.
88 Post Road West, Westport, Connecticut 06881

Printed in the United States of America

10 9 8 7 6 5 4 3 2 1

For Rosemary

Contents

Acknowledgments   ix

Introduction   xi

List of Abbreviations   xvii

BIBLIOGRAPHY 1945-1981

1. Bibliographies   1
2. Topics Related to the Disciplines   6
   2.1 Nature of Philosophy of Science   6
   2.2 Recent History of Philosophy of Science   14
   2.3 Relations Between the Disciplines   23
   2.4 Unity of Science   33
3. General Works in Philosophy of Science   38
   3.1 General Studies   38
   3.2 Collected Papers and Symposia   68
   3.3 Textbooks   84
4. Aspects of Scientific Method   87
   4.1 Scientific Method—General   87
   4.2 Evidence, Observation, Experiments   102
   4.3 Scientific Discovery   111
   4.4 Models and Analogies   121
   4.5 Simplicity   127
   4.6 Meaning of Scientific Terms   131
   4.7 Scientific Laws   149
   4.8 Theories   158

	4.9	Research Programs	170
	4.10	Explanation and Prediction	173
	4.11	Deduction and Proof in Science	191
	4.12	Induction	193
	4.13	Probability	221
	4.14	Verification and Confirmation	248
	4.15	Falsification, Duhem's Thesis, Adhocness	265
	4.16	Popperianism	269
	4.17	Operationalism	285
	4.18	Measurement	287
5.		Philosophical Issues Concerning Science	294
	5.1	Empiricism and the A Priori	294
	5.2	Causality	300
	5.3	Mechanism and Teleology	308
	5.4	Reduction and Emergence	316
	5.5	Realism and Instrumentalism	322
	5.6	Scientific Change and Growth	334
	5.7	Objectivity and Rationality	352
	5.8	Truth in Science	360
	5.9	Determinism and Indeterminism	364
6.		Special Topics in the Philosophy of the Physical Sciences	368
	6.1	Physical Sciences—General	368
	6.2	Physics—General	371
	6.3	Matter	385
	6.4	Cosmology	389
	6.5	Classical Mechanics	395
	6.6	Quantum Mechanics	400
	6.7	Relativity Theory	435
	6.8	Space, Time, Space-Time	454
7.		Special Topics in the Philosophy of the Biological Sciences	478
	7.1	Biological Sciences—General	478
	7.2	Biological Species	496
	7.3	Evolution	499
	7.4	Genetics	514

APPENDIX 1.   BOSTON STUDIES IN THE PHILOSOPHY OF
                          SCIENCE   517
APPENDIX 2.   SYNTHESE LIBRARY   522
APPENDIX 3.   THE UNIVERSITY OF WESTERN ONTARIO
                          SERIES IN PHILOSOPHY OF SCIENCE   531
APPENDIX 4.   TITLES OF PERIODICALS CITED   533

Index of Personal Names   539

Acknowledgments

It is a pleasant duty to express appreciation for being granted access to the library research collections, for the use of the physical facilities, and for the helpful and friendly assistance of the professional librarians at the following institutions: Saint Louis University, Washington University (St. Louis), University of Illinois (Champaign-Urbana), University of Chicago, Duke University, Yale University, Cornell University, and the University of Pennsylvania.

Many individuals have helped in many ways, for which we are thankful. Special appreciations must be extended to my colleague, Dr. Vernon J. Bourke, for originally suggesting this project and for his technical advice; to Mr. William R. Rehg, S. J., for his help as research assistant; to Ms. Linda Lashley, for seeing the text through the word processor; and especially to Mr. Paul A. Kobasa, former Acquisitions Editor at Greenwood Press, for his cooperation and advice from start to finish.

We are grateful to Saint Louis University for providing sabbatical leave time for work on this book.

Introduction

A bibliography is first and foremost a research tool, and as a result its value is primarily, if not exclusively, an instrumental one. This pragmatic consideration has held the central place in the construction of this bibliography. Our purpose is to provide assistance to both the neophyte and the expert in finding one's way around within the vast contemporary literature on the philosophy of science, and insofar as the users of this volume are so assisted, we will be satisfied. Although there already have been published numerous specialized bibliographies in this field (see Chapter 1), there exists to date no printed, volume length, general bibliography of contemporary philosophy of science. The closest alternative is Jean-Dominique Robert's *Philosophie et science. Philosophy and Science. Eléments de bibliographie. Elements of Bibliography* (Paris: Beauchesne, 1968) (see entry 35), but the primary focus of that listing is on the relations between philosophy and science rather than philosophy of science as such. Hopefully this volume will also fill a gap in previously available research tools.

The Status of the Contemporary Literature

Philosophy of science in its broadest sense is a critical reflection on the conceptual content, the methodologies, and the cultural implications of the various sciences. Its history dates back to the very beginnings of Western science among the Greeks. Over the centuries its fortunes have waxed and waned as the sciences themselves have grown and changed. Not unexpectedly, it has experienced vigorous revivals during and immediately after the creative periods of such figures as Galileo, Newton, Darwin, and Einstein. Since the beginning of the twentieth century there has been a continuously increasing interest in the issues traditionally discussed by philosophers of science. The resulting literature in this field has recently become truly enormous in size and variation.

Moreover, since the end of World War II, several special developments have redefined the character of the philosophy of science. This in turn has prompted us to adopt as chronological limitations for this bibliography the period from 1945 through to the end of 1981. The central change was the emergence of the philosophy of science in these years not only as a separate and distinctive branch of philosophy but more significantly as a role model which has had considerable impact on the more traditional parts of philosophy. It came to supply much of the terminology, the methodology, the problem structures, the conceptual categories, and the standards of evaluation for philosophy in general. And in one preeminent case, Kuhn's notion of cognitive paradigms, philosophy of science has had a major impact far beyond philosophy into most of contemporary intellectual culture. This new role as a model has made the literature in philosophy of science since 1945 especially important. Whether this discipline still enjoys such a status within philosophy in the 1980s is debatable and beyond our competence or intention to judge here. If the future proves that this status has been lost, then the period covered in this bibliography will be in a sense the "classical era" in modern philosophy of science.

One of the consequences of the emergence of the philosophy of science as a paradigm for other disciplines was the progressive professionalization of the field since the end of World War II. A number of national and international scholarly associations devoted to it have come into existence during this period. Many universities have established academic departments, degree programs, and special professorships in philosophy of science. In the 1980s numerous scholarly journals devoted primarily or exclusively to philosophy of science are being published on a regular basis. Most of these first appeared in the 1960s and 1970s, with only a few pre-dating 1945, and even then only by a short duration (*Philosophy of Science* first appeared in 1934; *Synthese* in 1936; *Erkenntnis* was published from 1930 to 1940, and then resumed in 1975). This bibliography is focused on the literature produced during this recent professionalized state of the philosophy of science.

During this period the character of the literature has also undergone significant changes, which would be more evident if we had listed the entries chronologically as a whole rather than alphabetically by topic. As is well known to specialists in the field, interest levels, and hence publication rates, have varied over the years in roughly the following key areas: the empirical basis of science and the meaning of its terms, models of explanation-prediction, the structure of theories, scientific revolutions, the theory-ladenness of observations, scientific change and progress, the rationality and objectivity of science. Problems in relativity theory and quantum mechanics have attracted an almost constant interest, with issues in the philosophy of biology becoming more prominent in the past two decades. All of the topics listed in the table of contents have been actively discussed throughout the period but with uneven emphasis.

A wholly chronological listing would have revealed additional characteristics of some interest. Over the years the sheer quantity of publication and the range of journals in which it has appeared have increased steadily. At the same time the focus of attention has narrowed in individual studies, subject matters under discussion are more delimited, and the tools of philosophical investigation are much more specialized and technical. The more comprehensive and foundational type of study of science, somewhat prominent in the earlier part of the period, is less in vogue today. Also early in this time period logical empiricism served more or less as a common (although not universally accepted) epistemological framework for discussions among philosophers of science. By the midpoint this had disappeared, and significantly there has emerged no replacement for it to serve the role of a shared framework. This fact, coupled with a seeming increase in the verbal contentiousness endemic to philosophers, has not helped to maintain unity within the field. Further, value questions regarding the sciences, once shunned in almost exclusive preference for conceptual and methodological issues by orthodox philosophers of science, have for several reasons come to attract increasing attention in recent years. If this continues and strengthens in the years to come, this will mean some basic changes in the character of the field. Lastly, the boundary lines between philosophy of science and other closely related disciplines have always been hazy at best. But in the past decade or so these lines have become increasingly ill-defined, especially in relation to logic, linguistics, ideologies, and the history of science, as philosophers of science have sought new approaches to their discipline. In particular the newer "historical way" of pursuing philosophy of science is quite prominent today, even though philosophers and historians of science still experience mutual professional suspicions as strongly as ever.

These brief observations, offered after an extensive and systematic survey of the contemporary literature in the philosophy of science, are intended as more descriptive than evaluative. Whether these developments as a whole bode well or ill for the future of this discipline remains to be seen. What does seem to be clear is that the increase in prominence, professionalization, and specialization experienced by the philosophy of science in the past four decades is not an unmixed set of blessings and is not necessarily a permanent condition.

The Structure and Use of the Bibliography

Most of the material in this bibliography is self-explanatory. Its main guide is, of course, the table of contents. The chapter titles and their subdivisions are not taken from any other bibliographic source, nor do they represent any a priori model of what the philosophy of science is or ought to be. It is not our intention to define the nature and limits of the discipline. Rather, the designations for the chapters and their subdivisions in this bibliography grew out of a close inductive survey of the contents of the major journals in the field along with pragmatic considerations of what classifications of the literature would be helpful for the researcher.

Special difficulties were encountered concerning how and where to draw lines of distinction between philosophy of science and other related fields, and were resolved on the basis of what could reasonably be included in a one-volume collection, what areas already have good bibliographic tools, and what would be most helpful. Equally competent experts will undoubtedly remain in disagreement on which specific topics are clearly to be included, which are on the fringes, and which are clearly to be excluded on such standards. We have decided to focus on the literature indicated by the chapter titles. The closest cognate areas that we have thus excluded are: philosophy of logic, mathematics, and language; philosophy of technology and of the social sciences, including history; the history and sociology of science; and value questions raised by the sciences. It is not that these areas are any less important; rather they deserve separate bibliographic treatment, which is already available in some cases, for example, the *Isis* "Critical Bibliographies" for history of science (see entry 32). Even so, decisions on inclusion, as well as on the appropriate choice of subdivision for some listings, were often difficult. In a few cases these decisions were also complicated by the fact that titles of entries can be misleading.

There is unavoidably some degree of overlapping among the classifications used in this bibliography, while for reasons of economy we have avoided duplicate listings. As a result the reader is advised to peruse related subdivisions for fuller bibliographic information. Also, although we have had in hand in our research most of the items listed, judgments of classifications are not infallible.

A one-volume bibliography of a literature as large as the philosophy of science cannot be critical. We have not attempted to include comments or summaries of content. Thus inclusion of books and articles in this compilation is not a judgment of scholarly quality.

A special feature of this bibliography is the somewhat frequent use of a system of subentries, indicated by lowercase letters under the entry numbers. This is to help unify the literature for the reader. Two types of items are included in these subentries: (a) journal articles that respond to previous articles, sometimes in different journals; and (b) book review essays, which are listed together with the book itself. Routine book reviews and notices have not been included in the bibliography.

In each subdivision of the bibliography entries are listed alphabetically by author or editor. There are only two exceptions to this: (a) Chapter 1: "Bibliographies," where entries are listed alphabetically by subject matter, and (b) section 3.2 "Collected Papers and Symposia," where some entries are listed alphabetically by title for convenience and ease of reference (for example, it seemed more helpful to list all nine volumes of the *Minnesota Studies in the Philosophy of Science* series together (entries 1052-1060).

Leaving journal publications aside, much of the literature in the philosophy of science appears in volumes consisting of independent papers, often on quite disparate topics. Many of these anthologies are mostly or exclusively collections of reprints. But many others are composed of original and basic research papers which have not been published previously (for example, the *Minnesota Studies* series, the relevant volumes edited by Paul A. Schilpp, the Pittsburgh series edited by R. G. Colodny, the *Proceedings* over the years of the Philosophy of Science Association and of the International Congress for Logic, Methodology and Philosophy of Science—all listed in section 3.2 "Collected Papers and Symposia"). In compiling this bibliography a special effort has been made to give a separate listing of such original research papers to make them more accessible to researchers.

A number of conventions have been adopted for consistency and to conserve space. Lengthy titles of frequently quoted serials have been abbreviated. The reader should consult the List of Abbreviations which follows this Introduction for identification of the symbols adopted. The abbreviations "BSPS," "SL," and "WOSPS," which are very frequently used, refer respectively to *Boston Studies in the Philosophy of Science, Synthese Library,* and *The University of Western Ontario Series in Philosophy of Science.* For the convenience of the reader we have included in Appendices 1, 2, and 3, respectively, a listing of the volumes in each of these series in the numbered sequence assigned by their publisher and have cross-listed the first two of these appendices. Thus "BSPS 23, SL 66" refers to volume 23 in the *Boston Studies* series (Appendix 1) and to volume 66 in the *Synthese Library* series (Appendix 2) which happen to be the same book.

Another convention relates to personal names. We have placed all name prefixes, such as "de," "di," "von," "van," after the family name, even though this is sometimes artificial or unfamiliar. Thus "Bas C. van Fraassen" is listed as "Fraassen, Bas C. van." Compound family names, occurring mostly in Spanish, and hyphenated family names have been alphabetized by the first letter. Names beginning with "Mac" or "Mc" have been placed before other "M" entries. The listing and alphabetizing of personal names is difficult to carry through with full clarity and consistency, and the reader should consult the various possible locations for complex names.

The Index of Personal Names includes all such names printed in the body of the bibliography and not just the names of authors and editors. In this index the underlining of a number indicates that the name appears in the title of a book or article.

In the literature listed in the bibliography Russian names are sometimes romanized in slightly different ways. All these variant spellings are preserved in the entries exactly as they appear in the original sources. However, in the Index of Personal Names we have included all the variants under the most commonly used spelling.

This bibliography does not list doctoral dissertations in the philosophy of science. For that information consult *Comprehensive Dissertation Index, 1861-1972* (Ann Arbor, Mich.: Xerox University Microfilms) 37 Vols., and the companion annual supplementary volumes which have appeared since 1973. This *Index* in turn serves as a table of contents for *Dissertation Abstracts* and *Dissertation Abstracts International.*

List of Abbreviations

AJAPS:	*Annals of the Japan Association for Philosophy of Science (Kagaku Kisoron Gakkai)*
Am. Phil. Quart.:	*American Philosophical Quarterly*
BJPS:	*British Journal for the Philosophy of Science*
BSPS:	*Boston Studies in the Philosophy of Science* (see Appendix 1)
DZP:	*Deutsche Zeitschrift für Philosophie*
FP:	*Foundations of Physics*
IPQ:	*International Philosophical Quarterly*
J. Phil.:	*The Journal of Philosophy*
LMPS-n:	*Logic, Methodology and Philosophy of Science: Proceedings of the nth International Congress for Logic, Methodology and Philosophy of Science* (see entries 1037-1041)
MSPS:	*Minnesota Studies in the Philosophy of Science* (see entries 1052-1060)
Phil. Quart.:	*The Philosophical Quarterly* (Scotland)
Phil. Review:	*The Philosophical Review*
Phil. Studies:	*Philosophical Studies* (Holland)
PN:	*Philosophia Naturalis*
Poznan Studies:	*Poznań Studies in the Philosophy of the Sciences and the Humanities*
PPR:	*Philosophy and Phenomenological Research*

<u>Proc. & Add. Am. Phil Assn.</u>:	*Proceedings and Addresses of the American Philosophical Association*
<u>Proc. Am. Cath. Phil. Assn.</u>:	*Proceedings of the American Catholic Philosophical Association*
<u>Proc. Arist. Soc.</u>:	*Proceedings of the Aristotelian Society*
PS:	*Philosophy of Science*
<u>PSA 19xx</u>:	*PSA 19xx: Proceedings of the 19xx Biennial Meeting of the Philosophy of Science Association* (see entries 1088-1093)
<u>Revue Phil.</u>:	*Revue Philosophique de la France et de l'Etranger*
RIP:	*Revue Internationale de Philosophie*
RMM:	*Revue de Metaphysique et de Morale*
SHPS:	*Studies in History and Philosophy of Science*
SL:	*Synthese Library: Studies in Epistemology, Logic, Methodology, and Philosophy of Science* (see Appendix 2)
WOSPS:	*The University of Western Ontario Series in Philosophy of Science* (see Appendix 3)
ZAW:	*Zeitschrift für Allgemeine Wissenschaftstheorie*
ZPF:	*Zeitschrift für Philosophische Forschung*

BIBLIOGRAPHY
1945-1981

1.
Bibliographies

1. "Bibliography of Joseph Agassi." BSPS 65 (1981) 502-11.

2. "Bibliografia prac Kazimierz Ajdukiewicz." Studia Logica. 16 (1965) 39-44.

3. "Scientific Works of Yehoshua Bar-Hillel." BSPS 43 (1976) xix-xxviii.

4. Gochet, Paul. "Philosophy of Science in Belgium: A Selective Bibliography." ZAW 6 (1975) 182-86.

5. "Bibliography: Works by Gustav Bergmann," in The Ontological Turn: Studies in the Philosophy of Gustav Bergmann. M. S. Gram and E. D. Klemke, eds. Iowa City: University of Iowa Press, 1974. 301-10.

6. Staal, J. F. "Bibliography of E. W. Beth." Synthese. 16 (1966) 90-106.

7. Butts, Robert E., and John Galinaitis. "Veröffentlickungen kanadischer Wissenschaftstheoretiker." ZAW 5 (1974) 390-406.

8. Benson, Arthur J., "Bibliography of the Writings of Rudolf Carnap," in The Philosophy of Rudolf Carnap. P.A. Schilpp, ed. LaSalle, Illinois: Open Court, 1963. 1015-70.

9. Ross, Alan, and Danny Steinberg. "Bibliography." [on decision theory vis-a-vis science and values] WOSPS 1 (1973) 191-213.

10. Schroeder-Heister, Peter. "Bibliographie Hugo Dingler (1881-1954)." ZPF 35 (1981) 283-98.

11. Shields, Margaret C. "Bibliography of the Writings of Albert Einstein to May, 1951," in Albert Einstein: Philosopher-Scientist. P.A. Schilpp, ed. New York: Tudor, 1949, 1951. 689-760.

12. Feigl, Herbert. "Selected Bibliography of Logical Empiricism." RIP 4 (1950) 95-102.

13. Dürr, Karl. "Der logische Positivismus." Bibliographische Einführungen in das Studium der Philosophie. 11. Bern: A. Francke AG Verlag, 1948. 24 pp.

14. "Writings by James K. Feibleman 1922-1976." Tulane Studies in Philosophy, 25 (1976) 107-18.

15. Hintikka, Jaakko. "Philosophy of Science in Finland." ZAW 1 (1970) 119-32.

16. Lalande, André. "Principal Publications on the Philosophy of the Sciences Brought out in France since 1900," in Philosophic Thought in France and the United States, 2nd ed. Marvin Farber, ed. Albany: State University of New York Press, 1968. 169-80.

17. "A Bibliography of Philipp Frank: Selected Writings on the Philosophy of Science." BSPS 2 (1965) xxxi-xxxiv.

18. Hartmann, Max. "Deutsche Philosophisch-Biologische Veröffent-licken vom Kriegsende bis Ende 1948." PN 1 (1950) 285-98.

19. Zecha, Gerhard. "Bibliographie: Veröffentlichungen osterrei-chischer Wissenschaftstheoretiker." ZAW 1 (1970) 311-21.

20. "Publications de Ferdinand Gonseth." Dialectica. 14 (1960) 267-74.

21. "Norwood Russell Hanson: Publications." BSPS 3 (1967) xliii-xlix.

22. "Publications (1934-1969) by Carl G. Hempel." SL 24 (1970) 266-70.

23. Bona, E., and J. Farkas. "Bibliographie: Veroeffentlichungen ungarischer Wissenschaftstheoretiker." ZAW 4 (1973) 188-93.

24. Slaght, Ralph L. "Induction, Acceptance, and Rational Belief: A Selected Bibliography." SL 26 (1970) 186-227.

25. "Bibliography." [of Jørgen Jørgensen] Danish Yearbook of Philo-sophy. 1 (1964) 183-96.

26. "Nachträge und Ergänzungen zur Bibliographie der Schriften von Bela Juhos." ZAW 2 (1971) 338-39.

27. Frey, Gerhard. "Nachträge und Ergänzungen zur Bibliographie der Schriften von Victor Kraft." ZAW 6 (1975) 179-81.

28. Risse, Wilhelm. Bibliographia Logica. I: 1472-1800; II: 1801-1969. Hildesheim: Georg Olms, 1965, 1973. 293, 494 pp.

29. Beth, E. W. "Symbolische Logik und Grundlegung der exakten Wissenschaften." Bibliographische Einführungen in das Studium der Philosophie. 3. Bern: A. Francke A.G. Verlag, 1948. 28 pp.

30. Laudan, Laurens. "Theories of Scientific Method from Plato to Mach: A Bibliographical Review." History of Science. 7 (1968) 1-63.

31. "Arne Naess: Selected List of His Philosophical Writings in the English and German Languages. 1936-1970." Synthese. 23 (1971) 348-52.

32. "Critical Bibliography of the History of Science and its Cultural Influences." Isis. 1913 - .

   In 1981 the 106th "Critical Bibliography" was published. Each bibliography contains a sub-section on philosophy of science.

33. The Philosopher's Index: An International Index to Philosophical Periodicals and Books. Richard H. Lineback, ed. Bowling Green, Ohio: Philosophy Documentation Center, 1967-.

   A quarterly index, by subject and author, with abstracts, of more than 300 philosophy and interdisciplinary journals. Annual cumulative volume. Two retrospective indexes have also been published: Retrospective Index to U.S. Publications from 1940, and Retrospective Index to non-U.S. English Language Publications from 1940.

34. Dialog. Philosophy Documentation Center, Bowling Green State University, Bowling Green, Ohio, 43403.

   A computerized bibliographic search service providing custom bibliographies, with abstracts, on philosophical topics.

35. Robert, Jean-Dominique. Philosophie et science. Philosophy and Science. Eléments de bibliographie. Elements of Bibliography. (Bibliothèque Archives de Philosophie, nouv. ser. 8). Paris: Beauchesne, 1968. 384 pp.

   Supplemented for 1968 and 1969 in Archives de philosophie. 33 (1970) 111-27, 295-324.

36. Filiasi, Carcano. "Rassegne di filosofia della scienza." Rassegne de filosofia. 1 (1952) 14-26.

37. Pitt, Joseph, and Donna Pitt. "General Bibliography." SL 52 (1973) 246-82.

Approx. 1100 items on philosophy of science in general, with special emphasis on conceptual change.

38. Feigl, Herbert, and Charles Morris. "Bibliography and Index," in International Encyclopedia of Unified Science, Vol. II, no 10. Chicago: University of Chicago Press, 1969. 108 pp.

39. Guérard des Lauriers, M. L. "Bulletin de philosophie des sciences." Revue des sciences philosophiques et théologiques. 35 (1951) 431-69.

40. "Bibliographie: Kongressberichte and Serien zur Wissenschaftstheorie." ZAW 3 (1972) 392-400; 6 (1975) 383-410.

41. Harré, Rom, et al. A Selective Bibliography of Philosophy of Science. 2nd ed. Oxford: Sub-Faculty of Philosophy, 1977. 69 pp.

42. "General Bibliography," in Philosophy of Science: The Delaware Seminar. Bernard Baumrin, ed. New York: Interscience Publishers, 1963. Vol. I, 353-65. Vol II, 525-46.

43. Ofierska, Urszula. "Recent Polish Methodology of Empirical Sciences: A Bibliographical Note." Organon. 7 (1970) 295-308.

44. Niklas, Urszula. "Polish Writings on Methodology (1970-71): A Bibliographical Note." Organon. 10 (1974) 267-74.

45. Hansen, Troels, Eggers. "Bibliography of the Writings of Karl Popper." The Philosophy of Karl Popper. P. A. Schilpp, ed. LaSalle, Illinois: Open Court, 1974. 1201-87.

46. Scheibe, Erhard. "Bibliographie zu Grundlagenfragen der Quantenmechanik." PN 10 (1967-68) 249-90.

47. Nilson, Donald Richard. "Bibliography on the History and Philosophy of Quantum Physics." SL 78 (1976) 457-520.

48. "Publications of W. V. Quine." SL 21 (1975) 353-73. Reprinted in Southwestern Journal of Philosophy. 9 (1978) 171-87.

49. "Bibliography of Writings of Hans Reichenbach," in Hans Reichenbach: Selected Writings, 1909-1953. 2 Vols. M. Reichenbach and R. S. Cohen, eds. Dordrecht: Reidel, 1978. Vol I, 481-97; Vol II, 413-29. [Same listings printed in each volume.]

50. Denonn, Lester E. "Bibliography of the Writings of Bertrand Russell to 1962," in The Philosophy of Bertrand Russell. P. A. Schilpp, ed., Fourth edition. New York: Harper & Row, 1963. 743-828.

Expansion and revision of the bibliography in the first edition of this volume.

51. "General Bibliography for Wilfrid Sellars (through 1975)," in The Synoptic Vision: Essays on the Philosophy of Wilfrid Sellars.

Notre Dame, Indiana: University of Notre Dame Press, 1977. 189-202.

52. Nyberg, Tauno. "Bibliography of the Philosophical Writings of
Erik Stenius: 1935-1971." Acta Philosophica Fennica. 25 (1972)
257-62.

53. "Bibliography of Patrick Suppes," in Patrick Suppes. Radu J.
Bogdan, ed. Dordrect: Reidel, 1979. 235-58.

54. Lauener, Henri. "Veröffentlichungen schweizerischer Wissenschaftstheoretiker." ZAW 2 (1971) 340-51; 3 (1972) 176-89.

55. "Bibliographie" [on logic and methodology of the sciences in the
USSR]. RIP 25 (1971) 602-25.

2.
Topics Related to the Disciplines

2.1 Nature of Philosophy of Science

56. Adams, Jr., John Stokes. "Contemporary Philosophy and Philosophy of Science." PS 18 (1951) 218-22.

57. Agassi, Joseph. "Sociologism in Philosophy of Science." Metaphilosophy. 3 (1972) 103-22. Reprinted in BSPS 65 (1981) 85-103.

58. Amsterdamski, Stefan. "Science as Object of Philosophical Reflection." Organon. 9 (1973) 35-60.

59. Ardley, Gavin. "Philosophy of Science: Dualist or Integral?" Philosophical Studies (Ireland). 16 (1967) 230-51.

60. Baum, Robert J. "Can Government Support of Philosophy of Science Research be Justified?" PSA 1976. I, 289-312.

61. Benjamin, A. Cornelius. "Is the Philosophy of Science Scientific?" PS 27 (1960) 351-58.

62. Bernays, P. "Von der Syntax der Sprache zur Philosophie der Wissenschaften." Dialectica. 11 (1957) 233-46.

63. Beth, E. W. "The Present Analysis of Science." Philosophy Today. 1 (1957) 159-62.

64. _____. "Carnap's Views on the Advantages of Constructed Systems over Natural Languages in the Philosophy of Science," in The Philosophy of Rudolph Carnap. P. A. Schilpp, ed. LaSalle, IL: Open Court, 1963. 469-502.

65. Blackwell, Richard J. "A New Direction in the Philosophy of Science." Modern Schoolman. 59 (1981) 55-59.

66. Braun, E., and H. Radermacher, hrsg. Wissenschaftstheoretisches Lexikon. Graz: Styria Verlag, 1978. 713 pp.

67. Brown, James Robert. "History and the Norms of Science." PSA 1980. I, 236-48.

68. Buchdahl, Gerd. "Inductivist versus Deductivist Approaches in the Philosophy of Science as Illustrated by Some Controversies Between Whewell and Mill." Monist. 55 (1971) 343-67.

69. Bunge, Mario. "On Method in the Philosophy of Science." Archives de Philosophie. 34 (1971) 551-74.

70. Burian, Richard M. "More Than a Marriage of Convenience: On the Inextricability of History and Philosophy of Science." PS 44 (1977) 1-42.

71. Bushkovitch, A. V. "Philosophy of Science as a Model for All Philosophy." PS 37 (1970) 307-11.

72. Coffey, Brian. "The Philosophy of Science and the Scientific Outlook." Modern Schoolman. 26 (1948) 23-35; 331-36.

73. Cohen, I. Bernard. "History and the Philosopher of Science," in The Structure of Scientific Theories. F. Suppe, ed. Urbana, Illinois: University of Illinois Press, 1974, 308-49.

   a. "History and Philosophy of Science: A Reply to Cohen," by Peter Achinstein. 350-60.

74. Compton, John. "Reinventing the Philosophy of Nature." Review of Metaphysics. 33 (1979) 3-28.

   a. "Compton on the Philosophy of Nature," by Ernan McMullin. 29-58.

75. Connell, Richard J. "The Character of Natural Philosophy." New Scholasticism. 51 (1977) 277-302.

76. Dieks, D. "De wankele basis van de wetenschapsfilosophie." Algemeen Nederlands Tijdschrift voor Wijsbegeerte. 72 (1980) 69-84.

77. Dingler, Hugo. "Analyse oder Synthese in der Philosophie der Wissenschaften?" Methodos. 6 (1954) 165-94.

78. Feigl, Herbert. "Beyond Peaceful Coexistence." MSPS 5 (1970) 3-11.

79. Feyerabend, Paul K. "Philosophy of Science: A Subject with a Great Past." MSPS 5 (1970) 172-83.

80. Fiedler, F. "Der Wissenschaft als Gegenstand der Wissenschaft." DZP 16 (1968) 558-70.

81. Finocchiaro, Maurice A. "The Uses of History in the Interpretation of Science." Review of Metaphysics. 31 (1977) 93-107.

82. _____. "On the Importance of Philosophy for History of
Science: Studies in the Logic of Erudition." Synthese. 42 (1979)
411-41.

83. Flach, Werner. Thesen zum Begriff der Wissenschaftstheorie.
Bonn: Bouvier, 1979. 106 pp.

84. _____. "Kritische Erwägungen zum Logikkonzept der analytischen Wissenschaftstheorie." Philosophisches Jahrbuch. 87 (1980)
142-49.

85. Gagnebin, H.-S. "Sur l'idée de dialectique dans la philosophie
des sciences contemporaine." Dialectica. 1 (1947) 72-75.

86. Gagnon, Maurice. "La philosophie de la nature est-elle encore
possible?" Dialogue. 20 (1981) 415-29.

87. Giere, Ronald N. "The Structure, Growth and Application of
Scientific Knowledge: Reflections on Relevance and the Future of
Philosophy of Science." BSPS 8, SL 39 (1971) 539-51.

88. Gonseth, Ferdinand. "Sur les buts et la méthode de la philosophie des sciences." Synthese. 5 (1947) 381-89.

89. Grene, Marjorie. "Philosophy of Medicine: Prologomena to a
Philosophy of Science." PSA 1976. II, 77-93.

90. Grewendorf, Günther. "Nicht-empirische Argumente. Zur Problematik ihrer wissenschaftstheoretischen Untersuchung." ZAW 9 (1978)
21-40.

91. Grünbaum, Adolf. "The Special Theory of Relativity as a Case
Study of the Importance of the Philosophy of Science for the History
of Science," in Philosophy of Science: The Delaware Seminar, II.
New York: Interscience Publishers, 1963. 171-204. Also in Annali
di Mathematica, Pura e Applicata. 57 (1962) 257-82.

92. Gutting, Gary. "Continental Philosophy of Science," in Current
Research in Philosophy of Science. P. D. Asquith and H. E. Kyburg,
Jr., eds. East Lansing, MI: Philosophy of Science Assn., 1979.
94-117.

93. Guzzo, Augusto. "Une philosophie de la nature est-elle encore
possible?" RIP 10 (1956) 131-43.

94. Hanson, Norwood Russell. "On Philosophy of Science." Granta
(1954) 14-16.

95. _____. "The Irrelevance of History of Science to the
Philosophy of Science." J. Phil. 59 (1962) 574-86.

96. Heelan, Patrick A. "Continental Philosophy and the Philosophy of Science," in Current Research in Philosophy of Science. P. D. Asquith and H. E. Kyburg, Jr., eds. East Lansing, MI: Philosophy of Science Assn., 1979. 84-93.

97. Hesse, Mary. "Hermeticism and Historiography: An Apology for the Internal History of Science." MSPS 5 (1970) 134-60.

98. Hintikka, Jaakko. "On the Ingredients of an Aristotelian Science." Nous. 6 (1972) 55-69.

   a. "Comments on Hintikka's Paper 'On the Ingredients of an Aristotelian Science'," by Dorothea Frede. Synthese. 28 (1974) 74-90.

   b. "Reply to Dorothea Frede," by Jaakko Hintikka. 91-96.

99. Horn, J.C. "Ist Natur mehr als Naturwissenschaft erfasst?" Wiener Jahrbuch fur Philosophie. 9 (1976) 104-13.

100. Hübner, Kurt. "Was heisst und zu welchem Ende studiert man Naturphilosophie?" PN 7 (1961-62) 129-43.

101. Kamlah, Andreas. "Wie arbeitet die analytische Wissenschaftstheorie?" ZAW 11 (1980) 23-44.

   a. "Eine Anmerkung zu Andreas Kamlahs Darstellung der 'normativ-analytischen' Wissenschaftstheorie," by Ulrich Charpa. 12 (1981) 135-37.

102. Kitchener, Richard F. "Genetic Epistemology, Normative Epistemology, and Psychologism." Synthese. 45 (1980) 257-80.

103. König, Gert. Was heisst Wissenschaftstheorie? Düsseldorf: Philosophia Verlag, 1971. 59 pp.

104. Kopnin, P. V., and V. A. Lektorsky. "Gnoseological Aspects of Present-Day Science." LMPS-4 (1971) 323-32.

105. Kosing, A. "Wissenschaftstheorie in der Sicht der marxistischen Philosophie." DZP 15 (1967) 759-71.

106. Kmita, Jerzy. "The Methodology of Science as a Theoretical Discipline." Soviet Studies in Philosophy. 12 (1974) 38-51. [English reprint from Voprosy filosofii. (1973, No. 5)].

107. Kuhn, Thomas S. "The History of Science," in International Encyclopedia of the Social Sciences. New York: Crowell Collier and Macmillan, 1968. Vol. 14, 74-83. Reprinted in T. S. Kuhn, The Essential Tension. Chicago: University of Chicago Press, 1977. 105-26.

108. Kyburg, Jr., Henry E. "The Application of Formal Methods in the Philosophy of Science," in Current Research in Philosophy of

Science. P. D. Asquith and H. E. Kyburg, Jr., eds. East Lansing,
MI: Philosophy of Science Assn., 1979. 28-39.

109. Laszlo, E., and H. Margenau. "The Emergence of Integrative
Concepts in Contemporary Science." PS 39 (1972) 252-59.

110. Laudan, Larry. "Historical Methodologies: An Overview and
Manifesto," in *Current Research in Philosophy of Science*. P. D.
Asquith and H. E. Kyburg, Jr., eds. East Lansing, MI: Philosophy of
Science Assn., 1979. 40-54.

111. Lazerowitz, Morris. "Moore and Philosophical Analysis."
Philosophy. 33 (1958) 193-220.

    a. "Philosophy of Science and Analysis: A Reply to M. Lazero-
witz," by Wolfgang Yourgrau. 35 (1960) 147-51.

    b. "Hume and Philosophical Analysis: A Reply to Professor
Lazerowitz," by Richard Wasserstrom. 151-53.

112. Leclerc, Ivor. "The Necessity Today of the Philosophy of
Nature." *Process Studies*. 3 (1973) 158-68.

113. Le Grand, Y. "Physiologie du cerveau et philosophie des
sciences." *Revue Phil*. 160 (1970) 135-43.

114. Lorenz, Kuno. "The Concept of Science: Some Remarks on the
Methodological Issue 'Construction' versus 'Description' in the
Philosophy of Science." SL 133 (1979) 177-90.

    a. "Transcendentalism and Protoscience: Comment on Lorenz,"
by Rüdiger Bubner. 191-95.

115. MacKinnon, Edward, "Analysis and the Philosophy of Science."
IPQ 7 (1967) 213-50.

116. McMullin, Ernan. "Philosophies of Nature." *New Scholasticism*.
43 (1969) 29-74.

117. _____. "The History and Philosophy of Science: A Taxonomy."
MSPS 5 (1970) 12-67.

118. _____. "Two Faces of Science." *Review of Metaphysics*.
27 (1974) 655-76.

119. _____. "History and Philosophy of Science: A Marriage
of Convenience?" BSPS 32 (1976) 585-602.

120. _____. "The Ambiguity of 'Historicism'," in *Current Research in Philosophy of Science*. P. D. Asquith and H. E. Kyburg,
Jr., eds. East Lansing, MI: Philosophy of Science Assn., 1979.
55-83.

121. McNicholl, Ambrose. "Contemporary Challenge to the Traditional
Ideal of Science." *The Thomist*. 24 (1961) 583-604. Reprinted in

The Dignity of Science. James A. Weisheipl, ed. Baltimore: The Thomist Press, 1961. 447-68.

122. Madden, Edward H. "Ontologizing Science." Proc. Am. Cath. Phil. Assn. 54 (1980) 164-73.

123. Malisoff, William Marias. "On the Possible Philosophies of Science." PS 12 (1945) 231-36.

124. Martin, Michael. "Value Judgements and the Acceptance of Hypotheses in Science and Science Education." Philosophic Exchange. 1, No. 4 (1973) 83-100.

    a. "Values in Science and Science Education," by Alex C. Michalos. 103-06.

    b. "Some Problems in Communication between Philosopher and Scientist," by K. Thomas Finley. 109-12.

125. _____. "The Relevance of Philosophy of Science for Science Education." BSPS 32 (1976) 293-300.

126. Maull, Nancy L. "Reconstructed Science as Philosophical Evidence." PSA 1976. I, 119-29.

127. Merrill, G. H. "Moderate Historicism and the Empirical Sense of 'Good Science'." PSA 1980. I, 223-35.

128. Metz, A. "De quelques conditions d'une philosophie des sciences." Archives de Philosophie. 29 (1966) 368-96.

129. Mitroff, Ian I. "The Methodology of Methodology: An Essay on the Nature of a Feeling Science." Theory and Decision. 2 (1972) 274-90.

130. _____. "Integrating the Philosophy and the Social Psychology of Science, or a Plague on Two Houses Divided." BSPS 32 (1976) 529-48.

131. Naess, Arne. "Plea for Pluralism in Philosophy and Physics," in Physics, Logic and History. W. Yourgrau and A. D. Breck, eds. New York: Plenum Press, 1970. 129-39.

132. Novik, I. B. "Some Aspects of the Interrelation of Philosophy and Natural Science." Soviet Studies in Philosophy, 8 (1969-70) 295-310. [English reprint from Voprosy filosofii. (1969, No. 9)].

133. Oeser, Erhard. "Analytische Wissenschaftstheorie and kritische Rekonstruktion." Wiener Jahrbuch für Philosophie. 4 (1971) 73-95.

134. Palter, Robert. "Philosophy of Science, History of Science, and Science Education." BSPS 32 (1976) 313-21.

135. Politis, Constantine. "Limitations of Formalization." PS 32 (1965) 356-60.

136. Popper, Karl R. "The Nature of Philosophical Problems and Their Roots in Science." BJPS 3 (1952) 124-56. Reprinted in Conjectures and Refutations. New York: Basic Books; London: Routledge & Kegan, Paul, 1962; New York: Harper & Row, 1968. 66-96.

137. Radnitzky, Gérard. "Du positivisme logique au rationalisme critique en passant par la théorie critique." Archives de Philosophie. 44 (1981) 99-116.

138. _____. "Entre Wittgenstein et Popper: Philosophie analytique et théorie de la science." Archives de Philosophie. 45 (1982) 3-62.

139. Radnitzky, Gerard, Håkan Törnebohm, and Göran Wallén. "Wissenschaftstheorie als Forschungswissenschaft." ZAW 2 (1971) 115-19.

140. Rantala, Veikko. "The Old and the New Logic of Metascience." Synthese. 39 (1978) 233-48.

141. Rogers, Robert. "Mathematical and Philosophical Analyses." PS 31 (1964) 255-64.

142. Sagal, Paul T. "On Science." Journal of Value Inquiry. 14 (1980) 301-07.

143. Salmon, Wesley C. "Informal Analytic Approaches to the Philosophy of Science," in Current Research in Philosophy of Science. P. D. Asquith and H. E. Kyburg, eds. East Lansing, MI: Philosophy of Science Association, 1979. 3-15.

144. Senior, James K. "The Vernacular of the Laboratory." PS 25 (1958) 163-68.

145. Shea, William R. "Beyond Logical Empiricism." Dialogue. 10 (1971) 223-42.

146. Sikora, S. J., Joseph J. The Scientific Knowledge of Physical Nature: An Essay on the Distinction between the Philosophy of Nature and Physical Science. Bruges-Paris: Desclée de Brouwer, 1966. 165 pp.

147. Sontag, Frederik. "Philosophy of Science and the Revival of Classical Ontology." J. Phil. 53 (1956) 597-607.

    a. "Philosophy of Science and the Revival of Classical Ontology: A Reply," by Richard F. Grabau. 54 (1957) 131-37.

    b. "Ontology and the Philosophy of Science: A Reply," by F. Sontag, 55 (1958) 337-39.

148. Strawson, P. F. "Carnap's Views on Constructed Systems versus Natural Languages in Analytic Philosophy," in The Philosophy of Rudolph Carnap. P. A. Schilpp, ed. La Salle, IL: Open Court, 1963. 503-18.

149. Suppes, Patrick. "Some Remarks on Problems and Methods in the Philosophy of Science." PS 21 (1954) 242-48.

150. _____. "The Role of Formal Methods in the Philosophy of Science," in Current Research in Philosophy of Science. P. D. Asquith and H. E. Kyburg, Jr., eds. East Lansing, MI: Philosophy of Science Assn., 1979. 16-27.

151. Szaniawski, Klemens. "Information and Decision as Tools of Philosophy of Science." Danish Yearbook of Philosophy. 10 (1973) 47-59.

152. Thackray, Arnold. "Science: Has its Present Past a Future?" MSPS 5 (1970) 112-27.

   a. "Comment" by Laurens Laudan. 127-32.

   b. "Reply" by Arnold Thackray. 132-33.

153. Walcott, Gregory Dexter. "An Inquirying Mind." PS 23 (1956) 315-24.

154. Wartofsky, Marx W. "How to Begin Again: Medical Therapies for the Philosophy of Science." PSA 1976. II, 109-22.

155. _____. "The Relation Between Philosophy of Science and History of Science." BSPS 39 (1976) 717-38.

156. Westland, Gordon. "Psychology and the Philosophy of Science." Mind. 81 (1972) 533-42.

157. Whitbeck, Caroline. "The Relevance of Philosophy of Medicine for the Philosophy of Science." PSA 1976. II, 123-35.

158. Whitrow, G. J. "The Study of the Philosophy of Science." BJPS 7 (1956) 189-205.

159. Wittenberg, A. "May Philosophy of Science Preach Empiricism and Practice Apriorism?" Dialectica. 16 (1962) 15-24.

160. Wohlrapp, Harald. "Analytischer versus konstruktiver Wissens- chaftsbegriff." ZAW 6 (1975) 252-75.

161. Worrall, John. "Thomas Young and the 'Refutation' of Newtonian Optics: A Case-Study in the Interaction of Philosophy of Science and History of Science," in Method and Appraisal in the Physical Sciences. C. Howson, ed. Cambridge: Cambridge University Press, 1976. 107-79.

162. Wykstra, Stephen J. "Toward a Historical Meta-Method for Assessing Normative Methodologies: Rationability, Serendipity, and the Robinson Crusoe Fallacy." PSA 1980. I, 211-22.

2.2 Recent History of Philosophy of Science

163. Agassi, Joseph. "The Present State of the Philosophy of
Science." Philosophica. 15 (1975) 5-20. Reprinted in BSPS 65
(1981) 18-32.

164. Agazzi, Evandro. "Recent Developments of the Philosophy of
Science in Italy." ZAW 3 (1972) 359-71.

165. Ambacher, Michael. "La Philosophie des Sciences de Gaston
Bachelard." Dialogue. 2 (1963) 13-24.

166. Andersson, Gunnar, and Gérard Radnitzky. "Le progrès de la
connaissance: où en sont les théories de la science?" Archives de
Philosophie. 39 (1976) 619-28.

167. Ayer, A. J. "The Vienna Circle." Midwest Studies in Philosophy.
6 (1981) 173-87.

168. Bergmann, Peter G. "Ernst Mach and Contemporary Physics."
BSPS 6 (1970) 69-78.

169. Beth, Evert W. "Critical Epochs in the Development of the
Theory of Science." BJPS 1 (1950) 27-42.

170. Bhattacharya, Nikhil. "John Dewey's Philosophy of Science."
Philosophical Forum. 7 (1976) 105-25.

171. Bird, Otto. "Peirce's Theory of Methodology." PS 26 (1959)
187-200.

172. Blackmore, John T. Ernst Mach: His Life, Work, and Influence.
Berkeley: University of California Press, 1972. xx + 414 pp.

    a. "Essay Review" by Wolfram Swoboda. SHPS 5 (1974) 187-201.

173. Blackwell, Richard J. "The Inductivist Model of Science: A
Study in Nineteenth Century Philosophy of Science." Modern Schoolman.
51 (1974) 197-212.

174. Blake, Ralph M., Ducasse, Curt J., and Madden, Edward H.
Theories of Scientific Method: The Renaissance Through the Nineteenth Century. Seattle: University of Washington Press, 1960. iv
+ 346 pp.

175. Bohnert, Herbert G. "Carnap's Logicism." SL 73 (1975) 183-216.

176. Bóna, E., and J. Farkas. "Die Lage der Wissenschaftstheorie
in Ungarn." ZAW 4 (1973) 133-46.

177. Bradie, Michael P. "The Development of Russell's Structural
Postulates." PS 44 (1977) 441-63.

178. Bradley, J. Mach's Philosophy of Science. London: Athlone Press, 1971. 226 pp.

179. Brüning, Walther. "Der Gesetzesbegriff im Positivismus der Wiener Schule." Beihefte zur Zeitschrift für Philosophische Forschung. 10 (1954) 101 pp.

180. Buchdahl, Gerd. Methaphysics and the Philosophy of Science. The Classical Origins: Descartes to Kant. Oxford: Blackwell; Cambridge, MA: MIT Press, 1969. xii + 714 pp.

    a. "Essay Review" by Robert McRae. SHPS 3 (1972) 89-99.

181. _____. "Hegel's Philosophy of Nature and the Structure of Science." Ratio 15 (1973) 1-27.

182. Butts, Robert E. "Philosophy of Science in Canada." ZAW 5 (1974) 341-58.

183. Canguilhem, G. "Note sur la situation faite en France à la philosophie biologique." RMM 52 (1947) 322-32.

184. Čapek, Milič. "The Development of Reichenbach's Epistemology." Review of Metaphysics. 11 (1957) 42-67.

185. _____. "Ernst Mach's Biological Theory of Knowledge." Synthese. 18 (1968) 171-91. Reprinted in BSPS 5 (1969) 400-20.

186. Caton, Hiram. "Carnap's First Philosophy." Review of Metaphysics. 28 (1975) 623-59.

187. Clark, Joseph T. "Remarks on the Role of Quantity, Quality, and Relations in the History of Logic, Methodology, and Philosophy of Science." LMPS-1 (1960) 604-12.

188. Clavelin, Maurice. "Les deux positivismes du Cercle de Vienne." Archives de Philosophie. 43 (1980) 33-56.

189. Coffa, J. Alberto. "Carnap's Sprachanschauung Circa 1932." PSA 1976. II, 205-41.

190. Cohen, Robert S. "Ernst Mach: Physics, Perception and the Philosophy of Science." Synthese. 18 (1968) 132-70. Reprinted in BSPS 6 (1970) 126-64.

191. Cohen, Robert S., and Raymond J. Seeger, eds. Ernst Mach: Physicist and Philosopher. BSPS 6. SL 27. Dordrecht: Reidel, 1970. 295 pp.

192. Delaney, C. F. "Peirce on Induction and the Uniformity of Nature." Philosophical Forum. 4 (1973) 436-48.

193. Dima, Teodor. "The Philosophy of Science in Romania." ZAW 6 (1975) 355-68.

194. Dingle, Herbert. The Sources of Eddington's Philosophy.
Cambridge: Cambridge University Press, 1954. 64 pp.

195. Ellegård, Alvar. "The Darwinian Theory and Nineteenth-Century
Philosophies of Science." Journal of the History of Ideas. 18
(1957) 362-93.

196. Escobar, Edmundo F., y M. G. Gorostieta. La teoría de la
ciencia en México. Mexico: 1967. 91 pp.

197. Esposito, Joseph L. "Reichenbach's Philosophy of Nature."
SHPS 10 (1979) 189-200.

198. Favrholdt, David. "Niels Bohr and Danish Philosophy." Danish
Yearbook of Philosophy. 13 (1976) 206-20.

199. Feigl, Herbert. "Some Major Issues and Developments in the
Philosophy of Science of Logical Empiricism." MSPS 1 (1956) 3-37.

200. _____. "Logical Positivism after Thirty-Five Years."
Philosophy Today. 8 (1964) 228-45.

201. Feuer, Lewis S. "Dialectical Materialism and Soviet Science."
PS 16 (1949) 105-24.

202. Fraassen, Bas C. van. "A Re-examination of Aristotle's Philosophy of Science." Dialogue. 19 (1980) 20-45.

203. Frank, Philipp. "The Importance of Ernst Mach's Philosophy of
Science for Our Times." BSPS 6 (1970) 219-34.

204. Gavin, William J. "William James' Philosophy of Science."
New Scholasticism. 52 (1978) 413-20.

205. Geldsetzer, Lutz. "Neueste Tendenzen und Problemstellungen
der Philosophie und Wissenschaftstheorie in der Bundesrepublik
Deutschland aus der Sicht sowjetischer Beobachter." ZAW 10 (1979)
394-404.

206. Gochet, Paul. "Recent Trends in Philosophy of Science in
Belguim." ZAW 6 (1975) 145-63.

207. Gould, James A. "The Origins of Poincaré's Conventionalism."
RIP 15 (1961) 115-18.

208. Graham, Loren R. Science and Philosophy in the Soviet Union.
New York: Alfred A. Knopf, 1972. xii + 584 + xvi pp.

209. Gribanov, D. P. "The Philosophical Views of Albert Einstein,"
Soviet Studies in Philosophy. 18 (1979) 72-94. [English reprint
from Voprosy filosofii. (1979, No. 2)].

210. Griffin, N. "Einstein's Philosophy of Science." Scientia.
106 (1971) 25-37.

211. Gunter, Pete A. Y. "Bergson's Philosophical Method and Its
Application to the Sciences." Southern Journal of Philosophy. 16
(1978) 167-82.

212. Hall, Thomas S. "On Biological Analogs of Newtonian Paradigms."
PS 35 (1968) 6-27.

213. Hausman, Alan, and Fred Wilson. Carnap and Goodman: Two
Formalists. Iowa Publications in Philosophy, Vol. 3. Iowa City,
Iowa: University of Iowa Press, 1968. viii + 225 pp.

214. Hempel, Carl G. "Implications of Carnap's Work for the Philosophy of Science," in The Philosophy of Rudolf Carnap. P. A. Schlipp,
ed. La Salle, Illinois: Open Court, 1963. 685-709.

215. Hiebert, Erwin N. "The Genesis of Mach's Early Views on
Atomism." BSPS 6 (1970) 79-106.

216. _____. "Mach's Philosophical Use of the History of
Science." MSPS 5 (1970) 184-203.

217. Hintikka, Jaakko. "Philosophy of Science (Wissenschaftstheorie)
in Finland." ZAW 1 (1970) 119-32.

218. Holton, Gerald J. "Mach, Einstein, and the Search for Reality."
Daedalus 97 (1968) 636-73. Reprinted in BSPS 6 (1970) 165-99.

219. Howson, Colin, and John Worrall. "The Contemporary State of
Philosophy of Science in Britain." ZAW 5 (1974) 363-74.

220. Hübner, Kurt. "Duhems historistische Wissenschaftstheorie und
ihre gegenwärtige Weiterentwicklung." PN 13 (1971-72) 81-97.

221. Juhos, Béla. "Formen des Positivismus." ZAW 2 (1971) 27-62.

222. Kaminski, Stanislaw. "The Development of Logic and the Philosophy of Science in Poland after the Second World War." ZAW 8
(1977) 163-71.

223. Kisiel, Theodore, and Galen Johnson. "New Philosophies of
Science in the USA: A Selective Survey." ZAW (1974) 138-91.

224. Kockelmans, Joseph J., ed. Philosophy of Science: The Historical Background. New York: Free Press, 1968. xiii + 496 pp.

225. Kopnine, P. V. "La logique de la science et les voies de son
développement." RIP 25 (1971) 431-55.

226. Krauth, Lothar. Die Philosophie Carnaps. Wien and New York:
Springer-Verlag, 1970. 234 pp.

227. Kraft, Victor. Der Wiener Kreis. Wien: Springer-Verlag,
1950. vi + 179 pp. English tr. by A. Pap: The Vienna Circle: The
Origins of Neo-Positivism. New York: Philosophical Library, 1953.

a. Review essay by Ivan Boh. <u>New Scholasticism</u> 44 (1970) 611-19.

228. Kropp, Gerhard. "Die philosophischen Gedanken Max Plancks." ZPF 6 (1951) 434-58.

229. Kursanov, G. A. "Philipp Frank and His Philosophy of Science." <u>Daedalus</u>. 91 (1962) 617-41.

230. Kwiatkowski, Tadeusz. "Le caractère et la structure de la science selon Tadeusz Czeżowski." <u>Organon</u>. 8 (1971) 231-56.

231. Ladrière, Jean. "Le Congrès international de philosophie des sciences." <u>Revue Philosophique de Louvain</u>. 48 (1950) 102-19.

232. Laudan, Larry. "Towards a Reassessment of Comte's 'Methode Positive'." PS 38 (1971) 35-53.

233. _____. "Peirce and the Trivialization of the Self-Correcting Thesis." in <u>Foundations of Scientific Method: The Nineteenth Century</u>. R. N. Giere and R. S. Westfall, eds. Bloomington: Indiana University Press, 1973. 275-306.

234. Lauener, Henri. "Wissenschaftstheorie in der Schweiz." ZAW 2 (1971) 291-317.

235. Lawrence, Nathaniel. "Whitehead's Method of Extensive Abstruction." PS 17 (1950) 142-63.

236. Lenzen, Victor F. "Einstein's Theory of Knowledge," in <u>Albert Einstein: Philosopher-Scientist</u>. P. A. Schilpp, ed. New York: Tudor, 1951. 355-84.

237. MacKinnon, Edward. "Motion, Mechanics and Theology." <u>Thought</u>. 36 (1961) 344-70.

238. _____. "The New Materialism." <u>Heythrop Journal</u>. 8 (1967) 5-26.

239. McGuinness, B. F. "Philosophy of Science in the <u>Tractatus</u>." RIP 23 (1969) 155-66.

240. McKinney, John C. "George H. Mead and the Philosophy of Science." PS 22 (1955) 264-71.

241. McMullin, Ernan. "The Analytic Approach to Philosophy." Proc. Am. Cath. Phil. Assn. 34 (1960) 50-79.

242. _____. "Recent Work in Philosophy of Science." <u>New Scholasticism</u>. 40 (1966) 478-518.

243. _____. "Le déclin du fondationnalisme." <u>Revue Philosophique de Louvain</u>. 74 (1976) 235-55.

244. Malherbe, J.-F. "Le scientisme du Cercle de Vienne." Revue Philosophique de Louvain. 72 (1974) 562-73.

245. Margenau, Henry. "Present Status and Needs of the Philosophy of Science." Proceedings of the American Philosophical Society. 99 (1955) 334-37.

246. Mays, W. "History and Philosophy of Science in British Commonwealth Universities." BJPS 11 (1960) 192-211.

247. _____. "Carnap on Logic and Language." Proc. Arist. Soc. 62 n.s. (1961-62) 21-38.

248. _____. Whitehead's Philosophy of Science and Metaphysics. The Hague: Martinus Nijhoff, 1977. 144 pp.

249. Menger, Karl. "Mathematical Implications of Mach's Ideas: Positivistic Geometry, The Clarification of Functional Connections." BSPS 6 (1970) 107-25.

250. Mercier, André. "L'evolution de la pensée de xxe siècle en physique et le déclin de l'intuition sensible." RIP 17 (1963) 171-89.

251. Mikulak, Maxim W. "Soviet Philosophic-Cosmological Thought." PS 25 (1958) 35-50.

252. Mirski, E.M. "Wissenschaftswissenschaft in der UdSSR (Geschichte, Probleme, Perspektiven)." ZAW 3 (1972) 127-44.

253. Mises, Richard von. "Ernst Mach and the Empiricist Conception of Science." BSPS 6 (1970) 245-70.

254. Mittelstrass, Jürgen. "The Galilean Revolution: The Historical Fate of a Methodological Insight." SHPS 2 (1972) 297-328.

255. Morris, Charles. "On the History of the International Encyclopedia of Unified Science." Synthese. 12 (1960) 517-21.

256. Müller-Markus, Siegfried. "Zur Diskussion der Relativitätstheorie in der Sowjetwissenschaft." PN 6 (1960-61) 327-48.

257. _____. "Die Organisation der sowjetischen Philosophie der Physik seit December 1960." Studies in Soviet Thought. 2 (1962) 49-63.

258. _____. "Zur sowjetischen Philosophie der Physik in Jahre 1962." Studies in Soviet Thought, 3 (1963) 33-52.

259. _____. "Niels Bohr in the Darkness and Light of Soviet Philosophy." Inquiry. 9 (1966) 73-93.

260. Nagai, Hiroshi. "Some Aspects of the Philosophy of Science in Japan." AJAPS 1, No. 1 (1956) 63-90.

261. _____. "Philosophy of Science in Japan 1966-1970." AJAPS 4, No. 1 (1971) 68-70.

262. _____. "Recent Trends in Japanese Research on the Philosophy of Science." ZAW 2 (1971) 101-14.

263. Nagel, Ernest. "Russell's Philosophy of Science," in The Philosophy of Bertrand Russell. P. A. Schilpp, ed. Evanston and Chicago: Northwestern University Press, 1944. 317-50.

264. _____. "Dewey's Theory of Natural Science," in John Dewey: Philosopher of Science and Freedom. S. Hook, ed. New York: Dial Press, 1950. 231-48.

265. Nelson, Benjamin. "The Early Modern Revolution in Science and Philosophy." BSPS 3 (1967) 1-40.

266. Neurath, Otto. "After Six Years." Synthese. 5 (1946) 77-82.

267. Nordenstam, Tore, and Hans Skjervheim. "Philosophy of Science in Norway." ZAW 4 (1973) 147-64.

268. Northrop, Filmer S. C. "Einstein's Conception of Science," in Albert Einstein: Philosopher-Scienctist. P. A. Schilpp, ed. New York: Tudor, 1951. 385-408.

269. Norton, Bryan G. "On the Metatheoretical Nature of Carnap's Philosophy." PS 44 (1977) 65-85.

270. Oeser, Erhard. "Victor Krafts konstruktiver Empirismus und seine Bedeutung für die gegenwärtige Wissenschaftstheorie." Wiener Jahrbuch für Philosophie. 8 (1975) 85-93.

271. Ohe, S. "Philosophy of Science in Japan 1956-1965." AJAPS 3, No. 1 (1966) 31-32.

272. Palter, Robert M. Whitehead's Philosophy of Science. Chicago: University of Chicago Press, 1960. 248 pp.

    a. Review essay by Adolf Grünbaum. Phil. Review. 71 (1962) 218-29.

273. _____. "Whitehead and the Philosophy of Science." International Studies in Philosophy. 12/1 (1980) 81-86.

274. Paris, Carlos. "Las grandes sistematizaciones de la Filsofía de la Ciencia y el ideal de una filosofía científica." Pensamiento. 29 (1973) 263-85.

275. Patin, Henry A. "Pragmatism, Intuitionism, and Formalism." PS 24 (1957) 243-52.

276. Peursen, C. A. van, and R. J. A. van Dijk. "Wissenschaftstheorie in den Niederlanden." ZAW 3 (1972) 372-79.

277. Pitt, Joseph C., ed. The Philosophy of Wilfrid Sellars:
Queries and Extensions. Philosophical Studies in Philosophy, 12.
Dordrecht: Reidel, 1978. x + 304 pp.

278. _____. Pictures, Images, and Conceptual Change: An
Analysis of Wilfrid Seller's Philosophy of Science. SL 151. Dordrecht: Reidel, 1981.

279. Plamondon, Ann L. Whitehead's Organic Philosophy of Science.
Albany, NY: State University of New York Press, 1979. x + 174 pp.

280. Poirer, R. "Henri Poincaré et le problème de la valeur de la
science." Revue Phil. 144 (1954) 485-513.

281. _____. "L'épistémologie de Pierre Duhem et sa valeur
actuelle." Les Etudes Philosophiques. 22 (1967) 399-420.

282. Radnitzky, Gérard. Contemporary Schools of Metascience. Vol.
I: Anglo-Saxon Schools of Metascience. Vol II: Continental Schools
of Metascience. Göteborg, Sweden: Akademiförlaget, 1968. Vol. I:
200 pp. Vol II: 199 pp. 2nd, Revised edition: New York: Humanities
Press, 1970.

    a. Review essay, "Positivism versus the Hermeneutic-Dialectic
    School," by Alex C. Michalos. Theoria. 35 (1969) 267-78.

    b. Review essay by Nils Roll-Hansen. Methodology and Science.
    3 (1970) 66-77.

    c. Review essays by Andrew McLaughlin and John B. O'Malley.
    Philosophy Forum. 11 (1972) Supplement. 13-19; 19-24.
    "Author's Response," 25-35.

    d. Review essay by Dietrick Böhler. Philosophische Rundschau.
    19 (1972) 165-92.

    e. Review essay by Robert C. Neville. IPQ 12 (1972) 131-36.

    f. Review essay, "Considerazioni sul concetto e l'uso di
    'metascienza'," by Rosaria Egidi. Proteus. 11-12 (1973)
    171-92.

283. _____. "Ways of Looking at Science: On a Synoptic Study
of Contemporary Schools of 'Metascience'." Scientia. 104 (1969)
49-57. French tr.: Supplement, 28-35.

284. _____. "Analytic Philosophy as the Confrontation Between
Wittgensteinians and Popper." BSPS 67 (1981) 239-86.

285. Rescher, Nicholas. Peirce's Philosophy of Science: Critical
Studies in His Theory of Induction and Scientific Method. Notre
Dame, IN: University of Notre Dame Press, 1977. x + 125 pp.

286. Roldán, S. J., Alejandro. "La filosofia de las ciencias en la
Argentina." Pensamiento. 9 (1953) 355-68.

287. Runggaldier, Edmund. "Carnap's Early Conventionalism." IPQ 19 (1979) 73-84.

288. Russo, F. "L'épistémologie de Mario Bunge." Archives de Philosophie. 36 (1973) 373-93.

289. Rutte, Heiner. "Moritz Schlick, der Positivismus und der Neupositivismus." ZPF 30 (1976) 246-68.

290. Salmon, Wesley C. "The Philosophy of Hans Reichenbach." Synthese. 34 (1977) 5-88. Reprinted in SL 132 (1979) 1-85.

291. Simon, Yves. "Maritain's Philosophy of the Sciences." Thomist. 5 (1943) 85-102. French tr.: Revue Philosophique de Louvain. 70 (1972) 220-36.

292. Smart, J. J. C. "Quine's Philosophy of Science." Synthese. 19 (1968) 3-13. Reprinted in SL 21 (1975) 3-13.

293. Somenzi, Vittorio. "La filosofia e la metodologia della scienza oggi in Italia." Man and World. 2 (1969) 285-95.

294. Stock, Wolfgang G. "Die Bedeutung Ludwig Flecks für die Theorie der Wissenschaftsgeschichte." Grazer Philosophische Studien. 10 (1980) 105-18.

295. Suppe, Frederick. "The Search for Philosophic Understanding of Scientific Theories," and "Afterword-1977," in The Structure of Scientific Theories. F. Suppe, ed. Urbana, Illinois: University of Illinois Press, 1977. 1-241; 615-730.

296. Sweigart, John. "Carnap's Ontology and Challenges." PS 30 (1963) 71-80.

297. Takeda, K. "An Appraisal of Feibleman's Conception of the Philosophy of Science." Studium Generale. 24 (1971) 673-77.

298. Teranaka, Heiji. "Philosophy of Science in Japan, 1971-1975." AJAPS 5, No. 2 (1977) 49-53.

299. Tonini, V. "Nouvelles tendances réalistes dans l'interprétation des théories physiques modernes." RMM 67 (1962) 152-62. English tr.: Philosophy Today. 7 (1963) 62-69.

300. Toulmin, Stephen. "From Form to Function: Philosophy and History of Science in the 1950's and Now." Daedalus. 106 (1977) 143-62.

301. Ushenko, Andrew Paul. "Einstein's Influence on Contemporary Philosophy," in Albert Einstein: Philosopher-Scientist. P. A. Schilpp, ed. New York: Tudor, 1951. 607-46.

302. Walentynowicz, Bohdan. "Wissenschaft der Wissenschaft in Polen." ZAW 3 (1972) 145-55.

303. Wetter, Gustav. "Dialectical Materialism and Natural Science." Philosophy Today. 2 (1958) 196-206.

304. Witt-Hansen, Johannes. Exposition and Critique of the Conceptions of Eddington Concerning the Philosophy of Physical Science. Copenhagen: G. E. C. Gads forlag, 1958. 125 pp.

305. _____. "Jorgen Jorgensen and the Grammar of Science." Danish Yearbook of Philosophy. 1 (1964) 159-72.

306. _____. "Philosophy of Science (Wissenschaftstheorie) in Denmark." ZAW 1 (1970) 264-83.

307. Yaker, Henri Marc. "Medieval Thought, Modern Physics, and the Physical World." PS 18 (1951) 144-53.

308. Yolton, John W. The Philosophy of Science of A.S. Eddington. The Hague: M. Nijhoff, 1960. 151 pp.

309. Zahar, Elie. "Positivismus und Konventionalismus." ZAW 11 (1980) 292-301.

310. Zaslawsky, Denis. "La philosophie des sciences (Wissenschaftstheorie) en France (1950-1971)." ZAW 2 (1971) 318-25.

311. Zecha, Gerhard. "Die geganwärtige Situation der Wissenschaftstheorie in Oesterreich." ZAW 1 (1970) 284-92.

312. Zeman, Vladimir. "The Philosophy of Science in Eastern Europe: A Concise Survey." ZAW 1 (1970) 133-41.

2.3 Relations Between the Disciplines

313. Agassi, Joseph. "Epistemology as an Aid to Science." BJPS 10 (1959) 135-46. Reprinted in BSPS 65 (1981) 45-54.

314. _____. "Art and Science." Scientia. 114 (1979) 127-40. Italian tr.: 141-52.

315. Agazzi, Evandro. "What Have the History and Philosophy of Science to do for One Another?" SL 146 (1981) 214-48.

    a. "Comment," by I. A. Markova. 249-52.

316. Ardley, G. W. R. "Prolegomena to Any Natural Science Which Can be Called Philosophical." Modern Schoolman. 32 (1955) 101-14.

317. Ayala, Francisco J. "The Autonomy of Biology as a Natural Science," in Biology, History, and Natural Philosophy. A. D. Breck and W. Yourgrau, eds. New York: Plenum, 1972. 1-16.

318. Baekers, S. F. "Fenomenologie en moderne wetenschapsfilosofie." Algemeen Nederlands Tijdschrift voor Wijsbegeerte. 69 (1971) 260-68.

319. Barber, Bernard. "On the Relations Between Philosophy of Science and Sociology of Science," in Current Research in Philosophy of Science. P. D. Asquith and H. E. Kyburg, Jr., eds. East Lansing, MI: Philosophy of Science Assn., 1979. 129-37.

320. Barbour, Ian G., ed. Science and Religion: New Perspectives on the Dialogue. New York: Harper & Row, 1968. 323 pp.

321. _____. Issues in Science and Religion. Englewood Cliffs, NJ: Prentice-Hall, 1966. Reprint: New York: Harper & Row, 1971. 470 pp.

322. Bertholet, Edmond. "Vers un organon de la science et de la philosophie." Dialectica. 24 (1970) 29-39.

323. Beth, E. W. "Science and Classification." Synthese. 11 (1959) 231-44.

324. Blair, G. A. "Science, Sufficient Ground, and the Possibility of Metaphysics." Dialectica. 14 (1960) 53-79.

325. Blanché, Robert. "Sur les rapports présents de la physique et de la philosophie." Les Etudes Philosophiques. 18 (1963) 21-34.

326. Bochner, Salomon. "Aristotle's Physics and Today's Physics." IPQ 4 (1964) 217-44.

327. Bonnet de Viller, A. "Science et philosophie." Dialectica. 19 (1965) 91-135.

328. Bonsack, F. "La métaphysique et la science." Dialectica. 14 (1960) 157-66.

329. Brüning, Walther. "Naturwissenschaft und Naturphilosophie." PN 3 (1955) 261-82.

330. Buchdahl, Gerd. "Philosophische Grundlagen einer historischen Bewertung der Wissenschaft." Wiener Jahrbuch für Philosophie. 12 (1979) 16-42.

331. Butts, Robert E. "Methodology and the Functional Identity of Science and Philosophy." SL 146 (1981) 253-70.

332. Caldin, E. F. "Science and Philosophy: Implications or Presuppositions?" BJPS 1 (1950) 196-210.

333. Cantore, E. "Some Reflections on Man's Unending Quest for Understanding." Dialectica. 22 (1968) 132-66.

334. _____. "The Humanistic Significance of Science." PS 38 (1971) 395-412.

335. Čapek, Milič. "Sur quelques résistances philosophiques à la physique du vingtieme siècle." Dialectica. 28 (1974) 211-22.

336. Cartwright, Nancy Delaney. "How Do We Apply Science?" BSPS 32 (1976) 713-20.

337. Chauchard, P. "Valeur et limites de l'apport scientifique à la philosophie." Dialectica. 13 (1959) 123-43.

338. Clark, S. J., Joseph T. "The History of Science and the Enterprise of Philosophy: A Prelude to Partnership." Proc. Am. Cath. Phil. Assn. 38 (1964) 23-35.

339. _____. "Science and Some Other Components of Intellectual Culture," in Mind and Cosmos. Robert G. Colodny, ed. Pittsburgh: University of Pittsburgh Press, 1966. 292-307.

340. Cohen, Robert S. "Marxism and Scientific Philosophy." Review of Metaphysics. 4 (1951) 445-58.

341. Connolly, F. G. "Science vs. Philosophy." Modern Schoolman. 29 (1952) 197-210.

342. Coulson, C. A. Science and Religion: A Changing Relationship. Cambridge: Cambridge University Press, 1955. 36 pp.

343. _____. Science and Christian Belief. London: Oxford University Press, 1955. x + 127 pp.

344. _____. Science and the Idea of God. Cambridge: Cambridge University Press, 1958. vi + 51 pp.

345. Cowan, Thomas A. "The Historian and the Philosophy of Science." Isis. 38 (1947-48) 11-18.

346. Creed, Walter. "Philosophy of Science and Theory of Literary Criticism: Some Common Problems." PSA 1980. I, 131-40.

347. Delaney, C. F. "Science and Philosophy." Proc. Am. Cath. Phil. Assn. 54 (1980) 92-100.

348. Dingle, Herbert. Science and Literary Criticism. Edinburgh: Thomas Nelson & Sons, 1949. 184 pp.

349. Elkana, Y., ed. The Interaction Between Science and Philosophy. Atlantic Highlands, NJ: Humanities Press, 1974. xvii + 481 pp.

350. Elsasser, Walter M. "The Transition from Theoretical Physics into Theoretical Biology," in Biology, History, and Natural Philosophy. A. D. Breck and W. Yourgrau, eds. New York: Plenum, 1972. 135-63.

351. Ennis, Robert H. "Research in Philosophy of Science Bearing on Science Education," in Current Research in Philosophy of Science. P. D. Asquith and H. E. Kyburg, Jr., eds. East Lansing, MI: Philosophy of Science Assn., 1979. 138-70.

352. Firestone, Joseph M. "Remarks on Concept Formation: Theory Building and Theory Testing." PS 38 (1971) 570-604.

353. Foss, Laurence. "Art as Cognitive: Beyond Scientific Realism." PS 38 (1971) 234-50.

  a. "Cognitive Aspects of Art and Science," by Ronald C. Hoy. 40 (1973) 294-97.

  b. "Does Don Juan Really Fly?", by Laurence Foss, 298-316.

354. Frank, Philipp. "The Peace of Logic and Metaphysics in the Advancement of Modern Science." PS 15 (1948) 275-86.

355. _____. "The Origin of the Separation Between Science and Philosophy." Proceedings of the Academy of Arts and Sciences. 80 (1952) 115-39.

356. Freundlich, Yehudah. "Methodologies of Science as Tools for Historical Research." SHPS 11 (1980) 257-66.

357. Frey, Gerhard. "Methodological Problems of Interdisciplinary Discussions." Ratio 15 (1973) 161-82.

358. Galtung, Johan. "Notes on the Differences Between Physical and Social Sciences." Inquiry. 1 (1958) 7-34.

359. Gex, M. "La philosophie d'inspiration scientifique." Dialectica. 13 (1959) 144-59.

  a. "Note sur la philosophie d'inspiration scientifique," by J.-Claude Piguet. 191-207.

  b. "Défense de la philosophie d'inspiration scientifique: Résponse à la 'Note sur la philosophie d'inspiration scienti-fique' de M. J.-Claude Piguet," by M. Gex. 14 (1960) 21-31.

  c. "Résponse à la Note de M. J.-Claude Piguet," by R. Ruyer, 32-36.

360. Gilson, Etienne. "Science, Philosophy, and Religious Wisdom." Proc. Am. Cath. Phil. Assn. 26 (1952) 5-13.

361. Glas, Eduard. Chemistry and Physiology in their Historical and Philosophical Relations. Delft University Press, 1979. xiii + 179 pp.

362. Glass, Bentley. "The Relation of the Physical Sciences to Biology -- Indeterminancy and Causality," in Philosophy of Science: The Delaware Seminar, I. New York: Interscience Publishers, 1963. 223-49.

363. Gonseth, F. "De l'unité du savoir." Dialectica. 4 (1950) 148-57.

364. _____. "Motivation et structure d'une philosophie ouverte."
Dialectica. 6 (1952) 9-29.

365. Grassi, E., and F. Gonseth. "Les sciences et la philosophie."
Dialectica. 2 (1948) 25-44.

366. Grünbaum, Adolf. "The Relevance of Philosophy to the History
of the Special Theory of Relativity." J. Phil. 59 (1962) 561-74.

367. Hainard, R. "Science et art." Dialectica. 14 (1960) 188-96.

368. Hanson, Norwood Russell. "Scientists and Logicians: A Confrontation." Science. 138 (1962) 1311-14.

369. _____. "The Contributions of Other Disciplines to Nineteenth Century Physics." Scientia. 59 (1965) 149-57.

370. Harris, Errol E. Revelation through Reason. London: Allen
and Unwin, 1959. 123 pp.

371. Hartman, Robert S. "The Logical Difference Between Philosophy
and Science." PPR 23 (1963) 353-79.

372. Hartung, Frank E. "On the Contribution of Sociology to the
Physical Sciences." PS 15 (1948) 109-15.

373. Heim, Karl. Christian Faith and Natural Science. Nevill
Horton Smith, tr. London: S.C.M. Press, 1953. 256 pp.

    a. Review essay, "Science and Religion," by Herbert Dingle.
    BJPS 4 (1953) 235-46.

374. _____. The Transformation of the Scientific World View.
W.A. Whitehouse, tr. London: S.C.M. Press, 1953. 262 pp.

375. Henle, Robert J. "Science and the Humanities." Thought. 35
(1960) 513-36.

376. Herzfeld, Karl. "Philosophy and Experimental Physics." Proc.
Am. Cath. Phil. Assn. 26 (1952) 54-61.

377. Hübner, Kurt. "Der systematische Zusammenhang von Natur-und
Geschictswissenschaften." Tijdschrift voor Filosofie. 40 (1978)
183-201. English tr.: Epistemologia. 1 (1978) 93-112.

378. Isaye, S. J., R. P. Gaston. "Métaphysique réflexive et philo-
phie de la nature." RIP 10 (1956) 174-202.

379. Itō, Schuntaro. "Biologische Erkenntnis und Moderne Physik."
PS 25 (1958) 195-97.

380. Jones, W. T. The Sciences and the Humanities. Berkeley:
University of California Press, 1965. 291 pp.

381. Kapp, Reginald O. Facts and Faith: The Dual Nature of Reality.
Oxford University Press, 1955. 63 pp.

382. Kedrow, B. M. "Ueber die Klassifizierung der Wissenschaften."
DZP 4 (1956) 178-205.

383. Kelly, Joseph P. "Science Ventures into Philosophy." Thought
24 (1949) 598-616.

384. Knorr-Cetina, Karin D. "Social and Scientific Method, or What
do we Make of the Distinction between the Natural and the Social
Sciences." Philosophy of the Social Sciences. 11 (1981) 335-59.

385. Kohak, Erazim V. "Physics, Meta-Physics, and Metaphysics."
Metaphilosophy. 5 (1974) 18-35.

386. Kokoszyńska, M. "On the Difference between Deductive and
Non-Deductive Sciences." SL 87 (1977) 201-32. [First published in
Fragmenty Filozoficzne. 3 (1967).]

387. Körner, Stephan. "Philosophy, Science and Commonsense."
AJAPS 2, No. 2 (1962) 60-65.

388. _____. "Some Relations Between Philosophical and Scientific
Theories." BJPS 17 (1967) 265-78.

389. _____. "On a Difference between the Natural Sciences and
History," in Biology, History, and Natural Philosophy. A. D. Breck
and W. Yourgrau, eds. New York: Plenum, 1972. 243-50.

390. Krüger, L. "Does a Science Need Knowledge of Its History?"
Acta Philosophica Fennica. 30, 2-4 (1978) 51-61.

391. Kuhn, Thomas S. "Comment on the Relations of Science and
Art." Comparative Studies in Society and History. 11 (1969) 403-12.
Reprinted in T. S. Kuhn, The Essential Tension. Chicago: University
of Chicago Press, 1977. 340-51.

392. _____. "The Relations between History and the History of
Science." Daedalus 100 (1971) 271-304. Reprinted in T. S. Kuhn,
The Essential Tension. Chicago: University of Chicago Press, 1977.
127-61.

393. _____. "The Relations between the History and the Philosophy of Science," in T. S. Kuhn, The Essential Tension. Chicago:
University of Chicago Press, 1977. 3-20.

394. _____. "History of Science," in Current Research in
Philosophy of Science. P. D. Asquith and H. E. Kyburg, Jr., eds.
East Lansing, MI: Philosophy of Science Assn., 1979. 121-28.

395. Ladrière, Jean. "Philosophy and Science." Philosophical
Studies (Ireland). 8 (1958) 3-23.

396. Laer, P. Henry van. Philosophy of Science: Part Two: A
Study of the Division and Nature of Various Groups of Sciences.
Duquesne Studies, Philosophical Series, 14. Pittsburgh: Duquesne
University Press, 1962. xiii + 342 pp.

397. Lenzen, V.F. "Science and Philosophy." PPR 8 (1948) 448-55.

398. Lucca, John De. "Science and Philosophy." Philosophical
Quarterly (India). 37 (1964) 69-78.

399. MacKinnon, Edward. "Aristotelianism and Modern Physics."
Proc. Am. Cath. Phil. Assn. 38 (1964) 102-09.

400. McKeon, Richard. "Philosophy and the Development of Scientific
Methods." Journal of the History of Ideas. 27 (1966) 3-22.

401. McMullin, Ernan. "Medieval and Modern Science: Continuity or
Discontinuity?" IPQ 5 (1965) 103-29.

402. McMurrin, Sterling M. "Philosophy, Science, and Education,"
in Philosophy of Science: The Delaware Seminar, II. New York:
Interscience Publishers, 1963. 449-62.

403. McNicholl, O.P., A. J. "Science and Philosophy." Thomist. 8
(1945) 68-130.

404. Macifie, A. L. "On the Break between the Natural and the
Human Sciences." Phil. Quart. 1 (1951) 140-51.

405. Manier, Edward. "History, Philosophy and Sociology of Biology:
A Family Romance." SHPS 11 (1980) 1-24.

406. Mascall, E. L. Christian Theology and Natural Science: Some
Questions on their Relations. London: Longmans, 1956. New York:
Ronald Press, 1957. xxi + 328 pp.

407. Maziarz, Edward A. "From Meta-Science to Meta-Theology."
Proc. Am. Cath. Phil. Assn. 44 (1970) 122-29.

408. Mehlberg, Henry. "Can Science Absorb Philosophy?" RIP 13
(1959) 61-87.

409. Mercier, A. "Science et philosophie." Dialectica. 11 (1957)
276-95.

410. Merlan, Philip. "Metaphysics and Science -- Some Remarks."
J. Phil. 56 (1959) 612-19.

411. Miles, T. R. Religion and the Scientific Outlook. London:
Allen and Unwin, 1959. 224 pp.

412. Mirsky, E. M. "Philosophy of Science, History of Science, and
Science of Science." SL 146 (1981) 295-300.

413. Mitchell, E. T. "Metaphysics and Science." PS 13 (1946)
274-80.

414. Murdoch, John E. "Utility versus Truth: At Least One Reflection on the Importance of the Philosophy of Science for the History
of Science." SL 146 (1981) 311-19.

415. Noren, Stephen J. "The Conflict Between Science and Common
Sense and Why it is Inevitable." Southern Journal of Philosophy.
13 (1975) 331-46.

416. Northrop, F. S. C. The Logic of the Sciences and the Humanities.
New York: Macmillan, 1947. 397 pp.

417. _____. "The Implications of Traditional Modern Physics
for Modern Philosophy." RIP 3 (1949) 176-202.

418. _____. "The Relation between the Natural and the Normative Sciences," in Philosophy of Science: The Delaware Seminar, I.
New York: Interscience Publishers, 1963. 3-19.

419. Olmstead, Paul S. "Some Thoughts on 'What the Natural Scientist Needs from the Social Scientist'." PS 15 (1948) 85-86.

420. Orens, Irving P. "Physical Science and the Social Sciences."
PS 15 (1948) 90-95.

421. Pannenberg, Wolfhart. Theology and the Philosophy of Science.
Francis McDonagh, tr. London: Darton, Longman and Todd, 1976.
458 pp.

        a. Review essay by Donald Wiebe. Philosophical Studies.
        (Ireland) 26 (1977) 210-18.

        b. Review essay by Helmut Kuhn. Philosophische Rundschau.
        25 (1978) 264-77.

422. Pantin, C. F. A. The Relations Between the Sciences. London:
Cambridge University Press, 1968. x + 206 pp.

423. Passmore, J. A. "Philosophy and Scientific Method." Proc.
Arist. Soc. 49 n.s. (1948-49) 17-32.

424. Peacocke, A. R., ed. The Sciences and Theology in the Twentieth-
Century. London: Loutledge & Kegan Paul, 1981. 320 pp.

425. Perelman, Ch. "Sciences et philosophie." RIP 17 (1963)
133-40. English tr.: Philosophy Today. 9 (1965) 273-77.

426. Phelps, Everett R. "What the Physical Scientists Need from
the Social Scientists." PS 15 (1948) 87-89.

427. Piaget, Jean. "Du rapport des sciences avec la philosophie."
Synthese. 6 (1947) 130-50.

428. Pilet, P.-E., and D. Zaslawsky. "Confrontation dialectique
entre le biologiste et le philosophe: expérimentation et épistémologie." RIP 32 (1978) 100-19.

429. Preston, Malcolm G. "Concerning an Essential Condition of
Cooperative Work." PS 15 (1948) 96-99.

430. Quay, S. J., Paul M. "A Distinction in Search of a Difference:
The Psycho-Social Distinction between Science and Theology." Modern
Schoolman. 51 (1974) 345-59.

431. Robert, J. -D. "Approaches méthodologiques des problèmes
posés par la distinction et les rapports de droit entre disciplines
scientifiques et disciplines philosophiques." Sciences Ecclésiasti-
ques. 19 (1967) 169-213.

432. _____. "Le Problème des 'limites' respectives de la
philosophie et de la science devant la montée actuelle des sciences
de l'homme." Science et Esprit. 20 (1968) 195-222, 409-31.

433. Rochhausen, Rudolf, ed. Die Klassification der Wissenschaften
als philosophisches Problem. Berlin: Deutscher Verlag der Wissenschaften, 1968. 158 pp.

434. Rodni, N I. "Histoire de la science, logique de la science et
logique du developpement de la science." RIP 25 (1971) 456-66.

435. Rose, Lynn E. "The Domination of Astronomy Over Other Disciplines." BSPS 32 (1976) 469-76.

436. Rousseau, G. S. "Literature and Science: The State of the
Field." Isis. 69 (1978) 583-91.

437. Ruyer, Raymond. "La philosophie de la nature et le mythe."
RIP 10 (1956) 166-73.

438. _____. "La science et la philosophie considérées comme
des traductions." Les Etudes philosophiques. 18 (1963) 13-20.

439. _____. "Les rapports de la science et de la philosophie
et l'étroitesse de la conscience." RIP 17 (1963) 141-54.

440. Schaerer, René. "Oeuvre de science et oeuvre d'art." Studia
Philosophica. 5 (1945) 68-90.

441. Schlesinger, George. Religion and Scientific Method. Philosophical Studies Series in Philosophy, 10. Dordrecht: Reidel,
1977. vi + 201 pp.

442. Sciacca, Michele Federico. "Moment scientifique et moment
metaphysique." RIP 8 (1954) 218-35.

443. Scriven, Michael. "A Possible Distinction between Traditional
Scientific Disciplines and the Study of Human Behavior." MSPS 1
(1956) 330-39.

444. Sellars, Roy Wood. "Do the Natural Sceinces Have Need for the
Social Sciences?" PS 15 (1948) 104-08.

445. Sikora, Joseph J. "The Speculative Value of Physical Science."
Thought. 35 (1960) 494-512.

446. _____. "Sources of Disagreement between Philosophers and
Scientists," Modern Schoolman. 40 (1963) 263-74.

447. Singerman, Ora. "The Relation between Philosophy and Science --
A Comparison between the Positions of Bergson and Whitehead." [In
Hebrew] Iyyun. 19 (1968) 65-91.

448. Sommerville, John. "Soviet Science and Dialectical Materialism." PS 12 (1945) 23-29.

449. Spinney, G. H. "On the Contrast Between Scientific and Philosophic Hypotheses." Philosophy. 30 (1955) 15-32.

450. Spirito, Ugo. "The Limits of Science." Phil. Quart. 2
(1952) 208-17.

451. Stern, H. S. "Implications of the Methodology of the Physical
Sciences for the Social Sciences." Dialectica. 16 (1962) 255-74.

452. Swinton, W. E. "Historical Interrelations of Geology and
Other Sciences." Journal of the History of Ideas. 36 (1975) 729-38.

453. Tassi, Aldo. "The New Philosophy of Science and Philosophy."
Proc. Am. Cath. Phil. Assn. 54 (1980) 174-80.

454. Tatarkiewicz, Wladyslaw. "Nomological and Typological Sciences."
J. Phil. 57 (1960) 234-41.

455. Taton, René. "Historical Observations Concerning the Relationship between Biology and Mathematics," in Biology, History, and Natural
Philosophy. A. D. Breck and W. Yourgrau, eds. New York: Plenum,
1972. 171-80.

456. Taylor, H. Austin. "A Scientist Questions the Philosophers."
Thought 35 (1960) 252-68.

457. Thompson, W. R. "The Unity of the Organism." Modern Schoolman.
24 (1947) 125-57.

458. Titze, H. "Philosophie und Wissenschaft." PN 16 (1976-77)
318-33.

459. Tonini, Valerio. "Science et philosophie: la fondation d'une
épistémologie ouverte." Dialectica. 26 (1972) 93-102.

460. Topitsch, E. "Das Verhältnis zwischen Sozial-und Naturwissenschaften." Dialectica. 16 (1962) 211-31.

461. Veatch, Henry. "Concerning the Distinction between Descriptive and Normative Science." PPR 6 (1945) 284-306.

462. Vetterling, Mary K. "On a Supposed Methodological Difference between the Natural and Social Sciences." PS 40 (1973) 292-93.

463. Waelhens, A. de. "Savoir scientifique et savoir phénoménologique." Archives de Philosophie. 27 (1964) 439-53.

464. Wallace, O. P., William A. "Thomism and Modern Science: Relationships Past, Present, and Future." Thomist. 32 (1968) 67-83.

465. Wein, Hermann. "Heutiges Verhältnis und Missverhältnis von Philosophie und Naturwissenschaft." PN 1 (1950) 56-75, 189-222.

466. Werner, Charles G. "Science and Philosophy." Southern Journal of Philosophy. 2 (1964) 8-13.

467. Whiteley, C. H. "Metaphysics and Science." Phil. Quart. 9 (1959) 244-49.

468. Zaslawski, Denis. "Quelques rapports méthodologiques entre les sciences biologiques et la philosophie." Dialectica. 29 (1975) 223-36.

2.4 Unity of Science

469. Agassi, Joseph. "Unity and Diversity in Science." BSPS 4 (1969) 463-522.

470. Akchurin, J. A. The Unity of Natural Scientific Knowledge. [In Russian] Moscow, 1974.

471. Bohr, Niels. "Analysis and Synthesis in Science," in International Encyclopedia of Unified Science. Vol. I, No. 1. O. Neurath, R. Carnap, and C. W. Morris, eds. Chicago: University of Chicago Press, 1938-55. 28.

472. Carnap, Rudolf. "Logical Foundations of the Unity of Science," in International Encyclopedia of Unified Science. Vol. I, No. 1. O. Neurath, R. Carnap, and C. W. Morris, eds. Chicago: University of Chicago Press, 1938-55. 42-62.

473. Causey, Robert L. "Unified Theories and Unified Science." BSPS 32 (1976) 3-13.

474. _____. Unity of Science. SL 109 Dordrecht: Reidel, 1977. 185 pp.

   a. Review essay by J. J. C. Smart. Synthese. 41 (1979). 451-59.

b. Review essay by Peter Achinstein. <u>Nous</u>. 15 (1981) 67-75.

475. Churchman, C. West, and Russell L. Ackoff. "Varieties of Unification." PS 13 (1946) 287-300.

476. Dewey, John. "Unity of Science as a Social Problem," in <u>International Encyclopedia of Unified Science</u>. Vol. I, No. 1, O. Neurath, R. Carnap, and C. W. Morris, eds. Chicago: University of Chicago Press, 1938-55. 29-38.

477. Emmet, Dorothy M. "Philosophy and 'The Unity of Knowledge'." <u>Synthese</u>. 5 (1946) 134-7.

478. Feigl, Herbert. "Physicalism, Unity of Science and the Foundations of Psychology," in <u>The Philosophy of Rudolf Carnap</u>. P. A. Schilpp, ed. La Salle, IL: Open Court, 1963. 227-67.

479. Feuer, Lewis S. "Mechanism, Physicalism, and the Unity of Science." PPR 9 (1949) 627-43.

480. Fiedler, F. "Die Einheit der Wissenschaft und die neopositivistische Theorie der 'Einheitswissenschaft'." DZP 15 (1967) 838-49.

481. _____. <u>Einheitswissenschaft oder Einheit der Wissenschaft?</u> Berlin: Dietz Verlag, 1971. 283 pp.

482. Fodor, J. A. "Special Sciences (or: The Disunity of Science as a Working Hypothesis)." <u>Synthese</u>. 28 (1974) 97-116.

483. Frank, Philipp. "The Institute for the Unity of Science, its Background and its Purpose." <u>Synthese</u>. 5 (1947) 160-67.

484. _____. "Ernst Mach and the Unity of Science." BSPS 6 (1970) 235-44.

485. Frazer, William R. "Some Indications of Unity Among the Sciences." PS 22 (1955) 135-39.

486. Gomperz, Heinrich. "Unified Science and Value." <u>Erkenntnis</u>. 9 (1975) 5-10.

487. Gott, V. S. "The Material Unity of the World and the Unity of Scientific Knowledge." <u>Soviet Studies in Philosophy</u>. 17 (1978) 4-21. [English reprint from <u>Voprosy filosofii</u>. (1977, No. 12)].

488. Jørgensen, Jørgen. "Empiricism and Unity of Science." <u>Journal of Unified Science</u>. 9 (1941) 181-88. Reprinted in <u>Danish Yearbook of Philosophy</u>. 6 (1969) 108-14.

489. Kallen, Horace M. "The Meanings of 'Unity' Among the Sciences, Once More." PPR 6 (1946) 493-96.

a. "The Orchestration of the Sciences by the Encyclopedism of Logical Empiricism," by Otto Neurath. 496-508.

b. "The Significance of the Unity of Science Movement," by
Charles Morris. 508-15.

c. "Reply" by Horace M. Kallen, 512-26.

d. "For the Discussion: Just Annotations, Not a Reply," by
Otto Neurath. 526-28.

e. "An Annotation to the Annotation," by H. M. Kallen.
528-29.

490. Kedrov, B. M. "Marx and the Unity of Science -- Natural and
Social." Soviet Studies in Philosophy. 7, No. 2 (1968) 3-14.
[English reprint from Voprosy filosofii. (1968, No. 5)].

491. Kelle, V. Zh. and E. S. Markarian. "The Methodology of Social
Knowledge and the Problem of the Integration of the Sciences."
WOSPS 10 (1977) 259-68.

492. Kraft, V. "Die Einheit der Wissenschaften." Studium Generale.
9 (1956) 333-39.

493. Lamb, Matthew. "Towards a Synthetization of the Sciences."
PS 32 (1965) 182-91.

494. MacKinnon, Edward. "Niels Bohr on the Unity of Science."
PSA 1980. II, 224-44.

495. McCauley, Robert N. "Hypothetical Identities and Ontological
Economizing: Comments on Causey's Program for the Unity of Science."
PS 48 (1981) 218-27.

a. "Reduction and Ontological Unification: Reply to McCauley,"
by Robert L. Causey. 228-31.

496. Massey, Gerald J. "Reflections on the Unity of Science."
AJAPS 4, No. 3 (1973) 47-56.

497. Maugé, Francis. La synthèse totale des sciences, ses conditions et son principe. Paris: Hermann, 1955. 192 pp.

498. Maull, Nancy L. "Unifying Science Without Reduction." SHPS 8
(1977) 143-62.

499. Morris, Charles. "The Science of Man and Unified Science."
Proceedings of the American Academy of Arts and Sciences. 80 (1951)
37-44.

500. Müller, Henning. "Einheit als Ziel der Naturwissenschaft."
PN 8 (1964) 418-30.

501. Neurath, Otto. "Unified Science as Encyclopedic Integration,"
in International Encyclopedia of Unified Science. Vol. I, No. 1.
O. Neurath, R. Carnap, and C. W. Morris, eds. Chicago: University
of Chicago Press, 1938-55. 1-27.

502. Nickles, Thomas. "Theory Generalization, Problem Reduction, and the Unity of Science." BSPS 32 (1976) 33-75.

503. Oppenheim, Paul, and Hilary Putnam. "Unity of Science as a Working Hypothesis." MSPS 2 (1958) 3-36.

504. Ovchinnikov, N. F., and I A. Akchurin. "Concerning Unity of Knowledge in Physics." LMPS-4 (1971) 583-92.

505. Polikarov, Azarya. "The Proliferation and Synthesis of Physical Theories." Epistemologia. 4 (1981) 433-56.

506. Raymond, Arnold. "De l'unité et de la méthode dans les sciences." Synthese. 5 (1947) 365-80; 475-85.

507. Russell, Bertrand. "On the Importance of Logical Form," in International Encyclopedia of Unified Science. Vol. I, No. 1. O Neurath, R. Carnap, and C. W. Morris, eds. Chicago: University of Chicago Press, 1938-55. 39-41.

508. Ruytinx, Jacques. "Le développement du problème de l'unité de la science dans l'empiricisme logique." Synthese. 9 (n.d.) 26-41.

509. _____. La Problématique philosophique de l'unité de la science. Paris: Société d'Édition "Les Belles Lettres," 1962. viii + 368 pp.

   a. "Étude Critique," by Jerzy A. Wojciechowski. Dialogue. 2 (1963) 346-58.

510. _____. "The Unity of Science: Present State of the Problem." RIP 21 (1967) 183-98.

511. Sachs, Mendel. "Philosophical Implications of Unity in the Contemporary Arts and Sciences." PPR 34 (1974) 489-503.

512. Sastri, P. S. "Can Unified Sciences Replace Logic? Ontology and Unified Science." Philosophical Quarterly (India). 26 (1954) 253-62.

513. Schaffner, Kenneth F. "The Unity of Science and Theory Construction in Molecular Biology." BSPS 11, SL 58 (1974) 497-533.

514. Sklar, Lawrence. "The Evolution of the Problem of the Unity of Science." BSPS 11, SL 58 (1974) 535-45.

515. Snyder, D. Paul. Toward One Science: The Convergence of Traditions. New York: St. Martin's Press, 1978. xiii + 224 pp.

516. Ströker, Elisabeth. "Die Einheit der Naturwissenschaften. Bermerkungen zu einer fragwürdigen Idee." Philosophische Perspektiven. 3 (1971) 176-93.

517. Suppes, Patrick. "The Plurality of Science." PSA 1978. II, 3-16.

518. Székely, D. L. "A Preliminary Report on the Theory of Unification of Sciences and Its Concept Transforming Automaton." Notre
Dame Journal of Formal Logic. 3 (1962) 234-42.

519. _____. "A New Image of Unified Science." Logique et
analyse. 12 (1969) 382-92.

520. _____. "The Principles of the Theory of the Unification
of Sciences." Notre Dame Journal of Formal Logic. 10 (1969) 181-213.

521. Weaver, Warren. "The Emerging Unity of Science." AJAPS 2,
No. 2 (1962) 44-59.

522. Weizsäcker, C. F. von. "Die Einheit der Physik als konstruktive Aufgabe." PN 9 (1965-66) 247-65.

523. _____. "The Unity of Physics." BSPS 5 (1969) 460-73.

      a. "Supplementary Comments to Weizsäcker's Paper," by Francis
      J. Zucker, 474-82,

524. Wilkinson, John. "The Concept of Information and the Unity of
Science." PS 28 (1961) 406-13.

3. General Works in Philosophy of Science

3.1 General Studies

525. Achinstein, Peter. <u>Concepts of Science: A Philosophical Analysis</u>. Baltimore: Johns Hopkins Press, 1968. 266 pp.

    a. Review essay, "On Achinstein's <u>Concepts of Science</u>," by Fred Wilson, PS 38 (1971) 442-52.

526. _____. <u>An Essay in the Philosophy of Science</u>. Oxford: Clarendon Press, 1971. xii + 168 pp.

527. Addis, Laird. <u>The Logic of Society: A Philosophical Study</u>. Minneapolis: University of Minnesota Press, 1975. 226 pp.

528. Agassi, Joseph. "Towards an Historiography of Science." <u>History and Theory</u>. Vol. II, Beiheft 2 (1962) 117 pp.

529. _____. "Questions of Science and Metaphysics." <u>Philosophical Forum</u>. 5 (1974) 529-56.

530. _____. "Between Science and Technology." PS 47 (1980) 82-99.

531. _____. The Legitimation of Science." <u>Dialogos</u>. 35 (1980) 27-35. Reprinted in BSPS 65 (1981) 77-84.

532. Agazzi, E. "Science and Metaphysics Confronting Nature." <u>Dialectics and Humanism</u>. 4, No. 3 (1977) 127-36.

    a. "Comments," by S. A. Matczak. 137-40.

533. Alexander, Peter. <u>A Preface to the Logic of Science</u>. London: Sheed and Ward, 1963. viii + 144 pp.

534. Aliotta, Antonio. La reazione idealistica contro la scienza.
Naples: Libreria Scientifica Editrice, 1970. 573 pp.

535. Amsterdamski, Stefan. Between Experience and Metaphysics.
Philosophical Problems of the Evolution of Science. BSPS 35. SL
77. Dordrecht: Reidel, 1975. 193 pp.

536. Arthur, R. "The Empiricist Account of Scientific Knowledge --
A Polemical Evaluation." Poznan Studies. 3 (1977) 125-41.

537. Asti Vera, Armando. Fundamentos de la filosofía de la ciencia.
Buenos Aires: Editorial Nova, 1967. 121 pp.

538. Aune, Bruce. Knowledge, Mind and Nature. New York: Random
House, 1967. 246 pp.

539. Avaliani, Sergei S. Ocherki filosofii estestvoznaniia.
(Essays on the Philosophy of Science) Tbilisn: Metsniereba, 1968.
311 pp.

540. Bachelard, Gaston. Le rationalisme appliqué. Paris: Presses
Universitaires de France, 1949. 216 pp.

541. _____. L'activité rationaliste de la physique contemporaine. Paris: Presses Universitaires de France, 1951. 225 pp.

542. _____. L'engagement rationaliste. Paris: PUF, 1972.
190 pp.

543. _____. La philosophie du non: essai d'une philosophie
du nouvel esprit scientifique. 6 ed. Paris: PUF, 1973. 145 pp.
English tr.: The Philosophy of No: A Philosophy of the New Scientific Mind. New York: Orion, 1968. xii + 123 pp.

544. Baldini, Massimo. Epistemologia e storia della scienza.
Florence: Citta di Vita, 1974. 91 pp.

545. Barraud, H.-J. Science et philosophie. Essai. Louvain:
Editions Nauwelaerts, 1968. 400 pp.

546. _____. La science et le materialisme: essai de philosophie réaliste. Paris: Éditions M. Rivière, 1973. 240 pp.

547. Bateson, Gregory. Mind and Nature: A Necessary Unity. New
York: Dutton, 1979. xii + 238 pp.

548. Bavink, B. Ergebnisse und Probleme der Naturwissenschaften.
Neunte Auflage. Zurich: Hirzel, 1948. viii + 814 pp.

549. Bayertz, Kurt. Wissenschaft als historischer Prozess: die
antipositivistische Wende in der Wissenschaftstheorie. München: W.
Fink, 1980. 260 pp.

550. Becker, Georges. La mycologie et ses corollaires: Une philosophie des sciences naturelles. Paris: Maloine-Doin éditeurs,

1974. 242 pp.

551. Benjamin, A. Cornelius. "On Defining 'Science'." Scientific Monthly. 68 (1949) 192-98.

552. _____. "Science and Its Presuppositions." Scientific Monthly. 33 (1951) 150-53.

553. Bennett, J. G. The Dramatic Universe. Volume I: The Foundations of Natural Philosophy. London: Hodder and Stoughton, 1956. 534 pp.

554. Bennett, H. Stanley. "The Scope and Limitations of Science." Zygon. 3 (1968) 343-53.

555. Benvenuto, Edoardo. Materialismo e pensiero scientifico. Milan: Tamburini, 1974. viii + 317 pp.

556. Bergmann, Gustav. Philosophy of Science. Madison: University of Wisconsin Press, 1957. xiv + 181 pp.

     a. Review essay by Sidney Morgenbesser. J. Phil. 55 (1958) 169-76.

557. _____. Meaning and Existence. Madison: University of Wisconsin Press, 1960. xi + 274 pp.

558. Beth, Evert Willem. Naturphilosophie. Gorinchem, 1948. 230 pp.

559. _____. "Towards an Up-to-Date Philosophy of the Natural Sciences." Methodos. 1 (1949) 178-85.

560. _____. "Fundamental Features of Contemporary Theory of Science." BJPS 1 (1951) 291-302.

561. _____. Science, A Road to Wisdom. Dordrecht: Reidel, 1968. xiii + 127 pp.

562. Blackmore, John T. "A New Conception of Epistemology and Its Relation to the Methodology and Philosophy of Science." Methodology and Science. 14 (1981) 95-126.

563. Blackwell, Richard J. "The Adaptation Theory of Science." IPQ 13 (1973) 319-34.

564. Blanché, Robert. La science physique et la réalité. Réalisme, positivisme, mathématisme. Paris: Presses Universitaires de France, 1948. 213 pp.

565. _____. Contemporary Science and Rationalism. Edinburgh: Oliver & Boyd, 1968. ix + 92 pp.

566. Bloomfield, Leonard. "Linguistic Aspects of Science," in International Encyclopedia of Unified Science. Vol. I, No. 4. O.

Neurath, R. Carnap, and C. W. Morris, eds. Chicago: University of Chicago Press, 1938-55. 215-277.

567. Bloor, David. Knowledge and Social Imagery. London: Routledge & Kegan Paul, 1976. xi + 156 pp.

    a. "How Strong is Dr. Bloor's 'Strong Programme'?" by Gad Freudenthal. SHPS 10 (1979) 67-83.

568. Bohm, David. Wholeness and the Implicate Order. London: Routledge and Kegan Paul, 1980. xv + 224 pp.

569. Böhm, Walter. Die metaphysischen Grundlagen der Naturwissenschaft und Mathematik. Wien: Herder, 1965. 194 pp.

570. Boirel, René. Science et technique. Neuchatel, Suisse: Éditions du Griffon, 1955. 116 pp.

571. Bornemisza, Stephen Thyssen. The Unified System Concept of Nature. New York: Vantage, Press, 1955. viii + 131 pp.

572. Brillouin, Léon. Vie, matière et observation. Paris: A. Michel, 1959. 245 pp.

573. _____. Scientific Uncertainty and Information. New York: Academic Press, 1964. xiv + 164 pp.

574. Broad, C. D. Scientific Thought. Paterson, NJ: Littlefield, Adams, 1959. 555 pp.

575. Bronowski, J. The Common Sense of Science. London: William Heinemann; 1951. Cambridge, MA: Harvard University Press, 1958. 154 pp.

576. _____. Nature and Knowledge: The Philosophy of Contemporary Science. Eugene: Oregon State System of Higher Education, 1969. 95 pp.

577. _____. The Origins of Knowledge and Imagination. New Haven: Yale University Press, 1978. xiii + 146 pp.

578. Brown, R. Hanbury. "The Nature of Science." Zygon. 14 (1979) 201-15.

579. Buchdahl, Gerd. Metaphysics and the Philosophy of Science. Oxford: Blackwell, 1969. xii + 714 pp.

    a. Review essay by W. von Leyden. Philosophy. 46 (1971) 38-42.

580. Bunge, Mario. Scientific Research I: The Search for System. II: The Search for Truth. Berlin-Heidelberg-New York: Springer-Verlag, 1967. 536, 374 pp.

a. "Changing our Background-Knowledge: Review Essay," by Joseph Agassi. <u>Synthese</u>. 19 (1969) 453-64.

581. _____. <u>Method, Model, and Matter</u>. SL 44. Dordrecht: Reidel, 1973. 196 pp.

582. _____. "Les présupposés et les produits métaphysiques de la science et de la technique contemporaines." <u>Dialogue</u>. 13 (1974) 443-54.

583. _____. "The Limits of Science." <u>Epistemologia</u>. 1 (1978) 11-32.

584. Burgers, J. M. <u>Experience and Conceptual Activity</u>. Cambridge, MA: MIT Press, 1965. 277 pp.

585. Caldin, E. F. <u>The Power and Limits of Science</u>. London: Chapman & Hall, 1949. ix + 196 pp.

586. Callot, Emile. <u>La philosophie de la science et de la nature: Essais dialectiques et critiques sur la forme et le contenu de la connaissance de la réalité sensible</u>. Paris: Éditions Orphrys, 1979. 267 pp.

587. Canguilhem, Georges. <u>Etudes d'histoire et de philosophie des sciences</u>. Paris: J. Vrin, 1968. 395 pp.

588. Cantore, Enrico. "Science as a Dialogical Humanizing Process: Highlights of a Vocation." <u>Dialectica</u>. 25 (1971) 293-316.

589. _____. <u>Scientific Man: The Humanistic Significance of Science</u>. New York: ISH Publications, 1977. 487 pp.

590. Carmichael, Peter A. "The Metaphysical Matrix of Science." PS 20 (1953) 208-16.

591. Carnap, Rudolf. <u>Philosophical Foundations of Physics: An Introduction to the Philosophy of Science</u>. New York: Basic Books, 1966. x + 300 pp. German tr.: <u>Einführung in die Philosophie der Naturwissenschaft</u>. München: Nymphenburger Verlagshandlung, 1969. 296 pp.

a. Review essay by Eike v. Savigny. <u>Philosophische Rundschau</u>. 19 (1972) 192-98.

592. _____. <u>The Logical Structure of the World and Pseudo-problems in Philosophy</u>. Berkeley: University of California Press, 1967. 390 pp.

593. Carrelli, Antonio. <u>Limiti e possibilità della sceinza</u>. Bari: G. Laterza, 1947. 133 pp.

594. Catel, Werner. <u>Grundlagen und Grenzen des naturwissenschaftlichen Weltbildes</u>. Stuttgart: Ferdinand Enke Verlag, 1948.

595. Cavaillès, Jean. Sur la logique et la théorie de la science.
Troisième édition. Paris: Vrin, 1976. 79 pp.

596. Caws, Peter. "A Reappraisal of the Conceptual Scheme of
Science." PS 24 (1957) 221-34.

597. Chalmers, A. F. What Is This Thing Called Science? The University of Queensland/The Humanities Press, 1976. xii + 157 pp.

    a. Review essay by James H. Fetzer. Erkenntnis. 14 (1979)
    393-404.

598. Chaudhury, Pravas Jivan. "Science and Commonsense." Philosophical Quarterly (India). 26 (1953) 153-76.

599. _____ . The Philosophy of Science. Calcutta: Progressive
Publications, 1955. 184 pp.

600. _____ . "Science and Epistemology." Philosophical Quarterly
(India). 30 (1957) 237-42.

601. Chwistek, Leon. The Limits of Science. Outline of Logic and
of the Methodology of the Exact Sciences. New York: Harcourt,
Brace, 1948. lvii + 347 pp.

602. Clifford, William Kingdom. The Common Sense of the Exact
Sciences. Preface by Karl Pearson, ed. Introduction by James R.
Newman, ed. Preface by Bertrand Russell. New York: Knopf, 1946.
Lxvi + 249 pp. Reprint: New York: Dover, 1955.

603. Cohen, I. B. Science, Servant of Man. Boston: Little, Brown
and Co., 1948.

604. Collingwood, R. G. The Idea of Nature. Oxford: Clarendon
Press, 1945. viii + 183 pp.

605. Compton, J. J. "Understanding Science." Dialectica. 16
(1962) 155-76.

606. Conant, James B. On Understanding Science. New Haven: Yale
University Press, 1947. 157 pp.

607. _____ . Science and Common Sense. New Haven: Yale University Press, 1951. xii + 372 pp.

608. _____ . Modern Science and Modern Man. New York: Columbia
University Press, 1952. 111 pp.

609. Conger, George P. Epitomizations: A Study in the Philosophy
of the Sciences. Minneapolis: Burgess, 1949. vi + 878 pp.

610. Conradt, Rüdiger. "Grundzüge einer naturwissenschaftlichen
Erkenntnistheorie." PN 12 (1970) 3-46.

611. Cooper, Neil. "The Aims of Science." Phil. Quart. 14 (1964) 328-33.

612. Cornforth, Maurice. Science and Idealism. New York: International Publishers, 1947. 267 pp.

    a. Review essay: "Materialism, Idealism and Science," by Lewis S. Feuer. PS 15 (1948) 71-75.

613. Dascal, Marcelo, and Asher Idan. "Procedures in Scientific Research and in Language Understanding." ZAW 12 (1981) 226-49.

614. Davies, J. T. The Scientific Approach. London and New York: Academic Press, 1973. xi + 185 pp.

615. Destouches, Jean Louis. "Sciences, modèles, objectivité." Epistemologia. 2 (1979) Fascicolo Speciale. 141-54.

616. Delattre, Pierre. Système, structure, fonction, évolution: Essai d'analyse épistémologie. Paris: Maloine-Doin, 1971. 185 pp.

617. _____. "Concepts de formalisation et concepts d'exploration." Scientia. 109 (1974) 427-58. English tr.: 459-81.

618. Desanti, Jean T. La philosophie silencieuse, ou Critique des philosophies de la science. Paris: Seuil, 1975. 283 pp.

619. Dewey, John. "Common Sense and Science: Their Respective Frames of Reference." J. Phil. 45 (1948) 197-207.

620. Diéguez, Manuel de. Science et nescience. Paris: Gallimard, 1970. 548 pp.

621. Diemer, Alwin. "Zur Grundlegung eines allgemeinen Wissenschaftsbegriffes." ZAW 1 (1970) 209-27.

622. Diez Blanco, A., and P. Gomez Bosque. "Die Mathematisierung der Wissenschaft." PN 11 (1969) 3-74.

623. Dingle, Herbert. The Scientific Adventure. Essays in the History and Philosophy of Science. London: Sir Isaac Pittman & Sons, 1952. x + 372 pp.

    a. Review essay by F. Kröner. Dialectica. 8 (1954) 258-67.

624. _____. Science at the Crossroads. London: Martin Brian and O'Keefe, 1972. 256 pp.

625. Dobrov, Gennadii M. Aktuelle Probleme der Wissenschaftswissenschaft. Berlin: Dietz, 1970. 81 pp.

626. Drieschner, Michael. "Objekte der Naturwissenschaft." Neue Hefte fur Philosophie. 6/7 (1974) 104-28.

627. Duhem, Pierre. The Aim and Structure of Physical Theory. Phillip P. Wiener, tr. Foreward by Prince Louis de Broglie. Princeton: Princeton University Press, 1954. xxii + 344 pp.

    a. Review essay, "Duhem versus Galileo," by J. Agassi. BJPS 8 (1957) 237-48.

628. Dumitriu, A. "La structure axiomatique de la science moderne." Scientia. 105 (1970) 737-56. English tr.: Supplement, 221-36.

629. Egidi, Rosaria. Studi di logica e filosofia della scienza. Roma: Bulzoni, 1971. 245 pp.

630. Eigen, Manfred, and Ruthild Winkler. Laws of the Game: How the Principles of Nature Govern Chance. New York: Knopf, 1980. xiv + 347 pp. English tr. of Das Spiel.

631. Elkana, Yehuda. "Introduction: Culture, Cultural System and Science." BSPS 39 (1976) 99-108.

632. Esposito, Joseph L. "Some Grounds for a Moral Criticism of Science." Southern Journal of Philosophy. 13 (1975) 47-54.

633. Essler, Wilhelm K. Wissenschaftstheorie. 4 Vols. Freiburg/ München: Verlag Karl Alber, 1970-79. 162, 164, 175, 259 pp.

634. Faggiani, D. "Dimensioni fisiche e legittimità semantica." Scientia. 108 (1973) 301-11. English tr.: 312-22.

635. _____. "Ordine, linguaggi e modelli delle scienze empiriche." Scientia. 109 (1974) 679-700. English tr.: 701-20.

636. Feibleman, James. "A Set of Postulates and a Definition for Science." PS 15 (1948) 36-38.

637. _____. "The Scientific Philosophy." PS 28 (1961) 238-59.

638. Fetzer, J. A. Scientific Knowledge: Causation, Explanation, and Corroboration. BSPS 69 Dordrecht: Reidel, 1981. 339 pp.

639. Fevrier, Paulette. "Ce qui fait la science en tant que science." Epistemologia. 2 (1979) Fascicolo speciale. 107-18.

640. Feyerabend, Paul. "'Science'. The Myth and Its Role in Society." Inquiry. 18 (1975) 167-81.

    a. "Why Not Science for Anarchists Too? A Reply to Feyerabend," by Arne Naess, 183-94.

641. _____. Science in a Free Society. London: New Left Books, 1978. 221 pp.

642. _____. Der wissenschaftstheoretische Realismus und die Autorität der Wissenschaft. Braunschweig: Vieweg, 1978. viii + 368 pp.

643. _____. "In Defence of Aristotle: Comments on the Condition of Content Increase." BSPS 58, SL 125 (1978) 143-80.

644. _____. Erkenntnis für freie Menschen. Frankfurt a. M.: Suhrkamp, 1979. 272 pp.

645. _____. Philosophical Papers. Cambridge: Cambridge University Press, 1981.

646. _____. Problems of Empiricism. Cambridge: Cambridge University Press, 1981. xii + 255 pp.

647. Fitzgerald, John J. "The Nature of Physical Science and the Objectives of the Scientist." Philosophy. 27 (1952) 125-37.

648. Fleck, Ludwick. Genesis and Development of a Scientific Fact. Thaddeus J. Trenn and Robert K. Merton, eds. Fred Bradley and Thaddeus J. Trenn, trs. Foreward by T. S. Kuhn. Chicago: University of Chicago Press, 1979. xxviii + 203 pp.

   a. Review essay by Barbara G. Rosenkrantz. Isis. 72 (1981) 96-99.

649. Fourastié, Jean. Les conditions de l'esprit scientifique. Paris: Gallimard, 1966. 256 pp.

650. Fox, Russell, S. Banigan, et al. The Science of Science: Methods of Interpreting Physical Phenomena. New York: Walker, 1964. 243 pp.

651. Frank, Philipp. Modern Science and Its Philosophy. Cambridge: Harvard University Press, 1941. Reprint: New York: Collier Books, 1961. 316 pp.

652. _____. "Metaphysical Interpretations of Science." BJPS 1 (1950) 60-74. 77-91.

   a. "Note on Philipp Frank's Interpretation of Science," by Owen Potter. 2 (1951) 58-60.

653. _____. "The Logical and Sociological Aspects of Science." Proceedings of the American Academy of Arts and Sciences. 80 (1951) 16-30.

654. _____. Philosophy of Science: The Link Between Science and Philosophy. Englewood Cliffs, NJ: Prentice-Hall, 1957. 394 pp.

655. Freistadt, Hans. "Dialectical Materialism: A Friendly Interpretation." PS 23 (1956) 97-110.

656. Frey, Gerhard. "Zum Naturwissenschaftlichen Systembegriff." PN 1 (1952) 480-92.

657. _____. Erkenntnis der Wirklichkeit: Philosophische Folgerungen der modernen Naturwissenschaften. Stuttgart: W. Kohlhammer, 1965. 179 pp.

658. Friedlander, Michael W. The Conduct of Science. Englewood Cliffs, NJ: Prentice-Hall, 1972. xiv + 153 pp.

659. Fries, Horace S. "Science, Causation and Value." PS 14 (1947) 179-80.

660. Funke, Gerhard. "Ontologischer oder prognostizistischer Sinn der Naturerkenntnis?" PN 3 (1956) 403-46.

661. Gallie, W. B. "What Makes a Subject Scientific?" BJPS 8 (1957) 118-39.

     a. "What Makes a Subject Scientific?" by S. Pollock. BJPS 9 (1958) 130-32.

     b. "Gallie and the Scientific Tradition," by D. Harrah. BJPS 10 (1959) 234-39.

     c. "Reply to David Harrah's Discussion Note," by W.B. Gallie. 239-40.

662. Gauthier, Yvon. "Theory of Science and Metatheoretics. Towards a Language of Structures in Science." Science et Esprit. 23 (1971) 379-87.

663. Gérard, Robert. Les chemins divers de la connaissance. Paris: Presses Universitaires de France, 1945.

664. Gerhard, William A. "Natural Science and the Imagination." Thomist. 16 (1953) 190-216.

665. Geymonat, Ludovico. Filosofia e filosofa della scienza. Milan: Feltrinelli, 1975. 198 pp.

666. Ghită, Simion. "Les dimensions philosophiques de la révolution scientifique moderne." Noesis. 8 (1981) 171-82.

667. Gilbert, E. J. Language de la science. Paris: Biologica, 1945. 320 pp.

668. Goldman, Stanford. "The Mechanics of Individuality in Nature." FP I: 1 (1970-71) 395-408; II: 3 (1973) 203-28.

669. Gonseth, F. "Connaitre par la science." Dialectica. 8 (1954) 183-98. 9 (1955) 123-36.

670. Goodman, Nelson. The Structure of Appearance. Cambridge: Harvard University Press, 1951. 392 pp. Second edition: Indianapolis: The Bobbs-Merrill Co., 1966. Third edition: BSPS 53. SL 107. Dordrecht: Reidel, 1977.

     a. Review essay by W.V. Quine. J. Phil. 48 (1951) 556-63.

     b. Review essay by G. H. Müller. Dialectica. 7 (1953) 70-77.

c. Review essay, "Reflections on Nelson Goodman's The Structure of Appearance," by Carl G. Hempel. Phil. Review. 62 (1953) 108-16.

d. Review essay by M. E. Dummett. Mind. 64 (1955) 101-09.

e. Review essay by W. Stegmüller. Philosophische Rundschau. 5 (1957) 280-92.

671. _____. Ways of Worldmaking. Indianapolis: Hackett Press, 1978.

a. "Reflections on Goodman's Ways of Worldmaking," by Hilary Putnam. The Journal of Philosophy. 76 (1979) 603-18.

b. "Comments on Goodman's Ways of Worldmaking," by Carl G. Hempel. Synthese. 45 (1980) 193-99.

c. "The Wonderful Worlds of Goodman," by Israel Scheffler. 201-09.

d. "On Starmaking," by Nelson Goodman. 211-15.

e. Review essay by Paul Ricoeur. Philosophy and Literature. 4 (1980) 107-20.

672. Grassi, Ernesto, and T. von Uexkühl. Die Einheit unseres Wirklichkeitsbildes und die Grenzen der Einzelwissenschaften. Bern: Francke, 1951. 196 pp.

673. _____. Von Ursprung and Grenzen der Geisteswissenschaften und Naturwissenschaften. Bern: Verlag A. Francke AG, 1950. 252 pp.

674. Greenwood, David C. The Nature of Science. London: Vision Press, 1960. xii + 95 pp.

675. Grene, Marjorie, ed. Knowing and Being: Essays by Michael Polanyi. Chicago: University of Chicago Press, 1969. 246 pp.

676. Gruner, Rolf. Theory and Power: On the Character of Modern Sciences. Amsterdam: B. R. Grüner, 1977. 238 pp.

677. Gryasnov, B. "On the Logical Analysis of the Concept 'Object of Scientific Investigation'." Organon. 5 (1968) 49-55.

678. Guzzo, Augusto. La scienza. Torino: Edizioni di 'Filosofia', 1955. cxlii + 528 pp.

679. Häberlin, P. "Die philosophische Frage und die Naturphilosophie." PN 5 (1958-59) 221-28.

680. Habermas, Jürgen. Knowledge and Human Interests. Jeremy Shapiro, ed. Boston: Beacon Press, 1971. 356 pp.

a. "Critical Notice" by Sara Ruddick. Canadian Journal of Philosophy. 2 (1973) 545-69.

b. Review essay by Robert E. Innis. IPQ 13 (1973) 555-64.

681. _____. Erkenntnis and Interesse, mit einem neuen Nachwort. Frankfort am main: Suhrkamp Verlag, 1973. 420 pp.

682. Hanson, Norwood Russell. Patterns of Discovery: An Inquiry into the Conceptual Foundations of Science. Cambridge: Cambridge University Press, 1958. ix + 241 pp.

a. Review essay by P. K. Feyerabend. Phil. Review. 69 (1960) 247-52.

b. Review essay, "Thinking and Discovery," by Magnus Pike. Discovery. 21 (1959) 74-76.

683. _____. "Science and the Human Imagination." Philosophical Review. 67 (1958) 565-69.

684. _____. "Science and the Modern Mind." Physics Today. 13 (1960) 68, 70, 72.

685. Harré, R. An Introduction to the Logic of the Sciences. London: Macmillan, 1960. viii + 180 pp.

686. _____. Theories and Things. London: Sheed & Ward, 1961. 114 pp.

a. Review essay, "On What There is in Physics," by Mary B. Hesse. BJPS 13 (1962) 234-44.

687. _____. Matter and Method. London: Macmillan; New York: St. Martin's Press, 1964. 124 pp.

688. _____. The Anticipation of Nature. London: Hutchinson, 1965. 112 pp.

689. _____. The Method of Science. London: Wykeham, 1970. xii + 123 pp.

690. _____. The Principles of Scientific Thinking. Chicago: University of Chicago Press, 1970. 324 pp.

a. Review essay: "Discussion: R. Harré's The Principles of Scientific Thinking," by Edward H. Madden. Southern Journal of Philosophy. 10 (1972) 23-32.

b. Review essay, "Back to Aristotle?" by David Miller. BJPS 23 (1972) 69-78.

c. "Harre and Nonlogical Necessity," by Barry Cohen and Edward H. Madden. BJPS 24 (1973) 176-82.

d. "Parmenidean Particulars and Vanishing Elements," by Edward H. Madden and Mendel Sachs. SHPS 3 (1972) 151-66.

691. _____. The Philosophies of Science: An Introduction Survey. London: Oxford University Press, 1972. 191 pp.

692. Harris, Errol E. Nature, Mind and Modern Science. London: Allen & Unwin, 1954. xvi + 455 pp.

a. "Nature, Mind, and Modern Science," by Brand Blanshard. Phil. Quart. 5 (1955) 166-74.

b. "The Mind-Dependence of Objects," by E. E. Harris. 6 (1956) 223-35.

693. Hartmann, Max. Die philosophischen Grundlagen der Naturwissenschaften. Erkenntnistheorie und Methodologie. Jena: Gustav Fischer, 1948. xii + 238 pp.

694. _____. "The Philosophical Foundations of the Natural Sciences," in Science and Freedom. Boston: Beacon Press, 1955. 30-34.

695. Hartmann, Nicolai. Philosophie der Natur. Berlin: Walter de Gruyter, 1950. xx + 709 pp.

a. "Professor Hartmann's Philosophy of Nature," by Paul K. Feyerabend. Ratio. 5 (1963) 91-106.

696. Hartung, Frank E. "Science as an Institution." PS 18 (1951) 35-54.

697. Hawkins, David. The Language of Nature: An Essay in the Philosophy of Science. San Francisco: W. H. Freeman and Co., 1964. 372 pp.

698. Hayek, F. A. The Counter Revolution of Science. Glencoe, Ill.: The Free Press, 1952. 255 pp.

699. Heelan, Patrick A. "The Search for Perfect Science in the West." Thought 43 (1968) 165-86.

700. _____. "Hermeneutics of Experimental Science in the Context of the Life-World." Philosophia Mathematica. 9 (1972) 101-44. Also in ZAW 5 (1974) 123-24.

a. "Commentary on Patrick Heelan's 'Hermeneutics of Experimental Science in the Context of the Life-World'," by Theodore Kisiel. ZAW 5 (1974) 124-35.

b. "Comments on Professor Kisiel's Commentary," by P. A. Heelan. 135-37.

701. Heisenberg, Werner. Schritte über Grenzen; gesammelte Reden und Aufsätze. München: R. Piper, 1971. 313 pp. English tr.:

Across the Frontiers. New York: Harper & Row, 1974. xxii + 229 pp.

702. _____. Die Bedeutung des Schönen in der exakten Naturwissenschaft. Stuttgart: Belser-Presse, 1971. 79 pp.

703. Heitler, Walter. Naturphilosophische Streifzüge. Vortrage und Aufsätze. Braunchweig: Vieweg, 1970. 128 pp.

704. Helm, E. "Granzheit als naturphilosophisches Phänomen. Versuch einer kritischen Analyse als Grundlegung einer dynamischen Ganzheitslehre." PN 12 (1970) 297-344.

705. Hempel, Carl G. "Science Unlimited?" AJAPS 4, No. 3 (1973) 31-46.

706. Hesse, Mary B. Science and the Human Imagination. London: The S.C.M. Press, 1954. 171 pp.

707. Hildebrand, Joel H. Science in the Making. New York: Columbia University Press, 1962. 116 pp.

708. Hodgson, Peter. "Presuppositions and Limits of Science." BSPS 59. SL 136. (1979) 133-41.

709. Holton, Gerald. Thematic Origins of Scientific Thought: Kepler to Einstein. Cambridge, MA: Harvard University Press, 1973. 495 pp.

710. Holzkamp, Klaus. Wissenschaft als Handlung: Versuch einer neuen Grundlegung der Wissenschaftslehre. Berlin: de Gruyter, 1968. xi + 397 pp.

711. Hook, Sidney. "Science and Dialectical Materialism," in Science and Freedom. Boston: Beacon Press, 1955. 182-95.

712. Hooker, C. A. "The Metaphysics of Science." International Logic Review. 5 (1974) 111-46.

713. Horigan, James E. Chance or Design? New York: Philosophical Library, 1981.

714. Hörz, Herbert, et al. Philosophische Probleme der Physik. Berlin: Deutscher Verlag der Wissenschaften, 1978. Revised English tr.: Philosphical Problems in Physical Science. Minneapolis: Marxist Education Press, 1980. 190 pp.

715. Hörz, Herbert, Rolf Löther, and Siegfried Wollgast, eds. Philosophie und Naturwissenschaften: Wörterbuch zu den philosophischen Fragen der Naturwissenschaften. Berlin: Dietz, 1978. xxvi + 1044 pp.

716. Hübner, Kurt. Kritik der wissenschaftlichen Vernunft. Freiburg, München: Verlag Alber, 1978. 442 pp.

a. Review essay by F.-W. Korff. Allgemeine Zeitschrift für Philosophie. 4 No. 2 (1979) 65-71.

717. Husserl, Edmund. Die Krisis der europäischen Wissenschaften und die transzendentle Phänomenologie: Eine Einleitung in die phänomeno-logische Philosophie. The Hague: Martinus Nijhoff, 1954. English translation by David Carr: The Crisis of European Sciences and Transcendental Phenomenology: An Introduction to Phenomonological Philosophy. Evanston: Northwestern University Press, 1970. 405 pp.

718. Hutten, Ernest H. The Language of Modern Physics: An Intro-duction to the Philosophy of Science. London: George Allen and Unwin; New York: Macmillan, 1956. 278 pp.

719. _____. The Origins of Science. London: George Allen and Unwin, 1962. 241 pp.

720. _____. "Uncertainty and Information." Scientia. 99 (1964) 199-206. French tr.: Supplement, 120-28.

721. _____. "Creative Science." Scientia. 101 (1966) 3-9. French tr.: Supplement, 1-7.

722. Huxley, Julian. Religion Without Revelation. New York: Harper & Row, 1957. Reprint: New York: New American Library. 222 pp.

723. Itzkoff, Seymour W. Scientific Knowledge and the Concept of Man. Notre Dame, IN: University of Notre Dame Press, 1971. xi + 286 pp.

724. Jaki, Stanley L. The Road of Science and the Ways to God. Chicago: University of Chicago Press, 1978. 478 pp.

725. _____. The Origin of Science and the Science of Its Origin. South Bend, IN: Regnery-Gateway, 1979. viii + 160 pp.

726. Janich, Peter, Friedrich Kambartel, and Jürgen Mittelstrass. Wissenschaftstheorie als Wissenschaftskritik. Frankfurt am Main: Aspekte Verlag, 1974. 168 pp.

727. Janssen, Paul. Grundlagen der wissenschaftlichen Welterken-ntnis. Frankfurt am Main: Klostermann, 1977. vii + 251 pp.

728. Jöhr, Walter Adolf. Gespräche über Wissenschaftstheorie. Tübingen: Mohr, 1973. vii + 109 pp.

729. Joseph, Geoffrey. "Conventionalism and Physical Holism." J. Phil. 74 (1977) 439-62.

730. Juhos, Béla. "Die Methode der fiktiven Prädikate." Archiv für Philosophie. 9 (1959) 140-56; 314-47; 10 (1960) 114-61; 228-89.

731. _____. "Die zwei logischen Ordnungsformen der naturwis-senschaftlichen Beschreibung." Studium Generale. 18 (1965) 581-601.

732. Jung, C. G., and Pauli, W. The Interpretation of Nature and
Psyche. London: Routledge and Kegan Paul; New York: Pantheon
Books, 1955. 247 pp.

733. Kambartel, Friedrich, and Jürgen Mittelstrass, eds. Zum norma-
tiven Fundament der Wissenschaft. Frankfurt am Main: Athenäum
Verlag, 1973. 330 pp.

734. Kanitscheider, B. "Zur Frage der Naturbeschreibung." PN 10
(1967-68) 419-41.

735. Kantor, Jacob R. The Logic of Modern Science. Bloomington,
IN: Principia Press, 1953. xvi + 359 pp.

736. Kapitza, Peter L. Experiment, Theory, Practice. BSPS 46.
Dordrecht: Reidel, 1980.

737. Kattsoff, Louis O. Physical Science and Physical Reality.
The Hague: Martinus Nijhoff, 1957. 311 pp.

738. Kaufman, Arnold S. "The Aims of Scientific Activity." Monist.
52 (1968) 374-89.

739. Kedrow, B. M. "Zu einigen philosophischen Fragen der modernen
Naturwissenschaft." DZP 12 (1964) 279-89.

740. Kemeny, John G. A Philosopher Looks at Science. Princeton,
NJ: Van Nostrand, 1959. 273 pp.

741. Khatchadourian, Haig. "Some Metaphysical Presuppositions of
Science." PS 22 (1955) 194-204.

742. Kneller, George F. Science as a Human Endeavor. New York:
Columbia University Press, 1978. 333 pp.

743. Knorr, Karin D. The Manufacture of Knowledge: An Essay on
the Constructivist and Contextual Nature of Science. New York:
Pergamon Press, 1981. xiv + 189 pp.

744. Kockelmans, Joseph J. Phenomenology and Physical Science: An
Introduction to the Philosophy of Physical Science. Duquesne Studies,
Philosophical Series, 21, Pittsburgh: Duquesne University Press,
1966. 208 pp.

745. _____. The World in Science and Philosophy. Milwaukee:
Bruce, 1969. xx + 184 pp.

746. Köhler, Gustav. Die Wissenschaft und das Unwahrscheinliche.
Darmstadt: Bläschke, 1974. 133 pp.

747. Kojève, Alexandre. "L'Origine chretienne de la science moderne:
l'aventure de l'esprit," in Mélanges Alexandre Koyré. 2 Vols.
Paris: Hermann, 1964. II, 295-306.

a. "Alexander Kojève on the Origin of Modern Science: Sociological Modelling Gone Awry," by Steven Louis Goldman. SHPS 6
(1975) 113-24.

748. Kopnin, P. V. "Contemporary Science and the Marxist-Leninist
Theory of Knowledge." Soviet Studies in Philosophy. 10 (1971-72)
218-30. [English reprint from Voprosy filosofii. (1971, No. 3)].

749. Korch, H. Die wissenschaftliche Hypothese. Berlin: VEB
Deutscher Verlag, 1972. 431 pp.

750. Korn, Jakob. "Die geisteswissenschaftliche Wendung in der
exakten Naturwissenschaften." PN 2 (1954) 383-417.

751. Körner, Stephan. Experience and Theory: An Essay in the
Philosophy of Science. London: Routledge & Kegan Paul; New York:
Humanities Press, 1966. xii + 250 pp.

752. Kraft, V. Erkenntnislehre. Wien: Springer-Verlag, 1960.
viii + 379 pp.

753. Krampf, Wilhelm. "Die Möglichkeit einer operativen Begründung
der exakten Wissenschaften." ZPF 25 (1971) 485-99.

754. Kroy, Moshe. "Applications of Epistemic Logic to the Philosophy
of Science." Logique et analyse. 13 (1970) 413-37.

755. Kutschera, Franz von. Wissenschaftstheorie I & II. München:
Wilhelm Fink Verlag, 1972. 570 pp.

756. Kyburg, Jr., Henry E. Philosophy of Science: A Formal Approach.
New York: Macmillan, 1968. 332 pp.

757. Lachman, Sheldon J. The Foundations of Science. Detroit:
Hamilton Press, 1956. 130 pp.

758. Laer, P. Henry van. Philosophy of Science: Part One: Science
in General. Duquesne Studies: Philosophical Series, 6. Pittsburgh:
Duquesne University Press, 1956. xvii + 164 pp.

759. _____. Philosophico-Scientific Problems. Duquesne Studies:
Philosophical Series, 3. Pittsburgh: Duquesne University Press,
1953. x + 168 pp.

760. Laszlo, Ervin. Introduction to Systems Philosophy: Toward a
New Paradigm of Contemporary Thought. London: Gordon and Breach,
1972. 328 pp.

761. _____. "The Case for Systems Philosophy." Metaphilosophy
3 (1972) 123-41.

a. Comments by Ludwig von Bertalanffy, 142-45; by Lee Thayer,
146-50; by Stephen C. Pepper, 151-53; by Ralph Wendell Burhoe,
154-55.

762. _____. The Systems View of the World: The Natural Philosophy of the New Developments in the Sciences. Oxford: Blackwell,
1975. 131 pp.

763. Lauer, Quentin. "The Marxist Conception of Science." BSPS
14, SL 60 (1974) 377-96.

764. Lawden, D. F. "Modelling Physical Reality." Philosophical
Journal. 5 (1968) 87-104.

765. Lay, Rupert. Grundzüge einer komplexen Wissenschaftstheorie.
2 Vols. Frankfurt am Main: J. Knecht, 1971. 354, 631 pp.

766. Le Cog, John P. "The Limitations of Science." Personalist.
35 (1954) 251-66.

767. Leinfellner, Werner. Einführung in die Erkenntnis-und Wissenschaftstheorie. Mannheim: Bibliographisches Institut Ag., 1965.
207 pp.

768. Lindholm, Lynn M. "Demarcating Science from Confusion."
Scientia. 116 (1981) 49-66. Italian tr.: 67-82.

769. Lorenzen, Paul. Methodisches Denken. Reihe Theorie 2.
Frankfurt a. M.: Suhrkamp Verlag, 1968. 162 pp.

770. _____. Konstruktive Wissenschaftstheorie. Frankfurt am
Main: Suhrkamp, 1974. 236 pp.

771. Luchins, A. S., and E. H. Luchins. "Two Philosophies of
Science: A Study in Contrasts." Synthese. 15 (1963) 292-316. [A
comparison of the views of R. B. Braithwaite and S. Toulmin.]

772. MacKay, Donald. Science, Chance, and Providence. Oxford
University Press, 1978.

    a. Review essay, "A Religious Form of Scientific Life," by
    Anthony Flew. BJPS 30 (1979) 183-86.

773. MacKinnon, Edward. "Cognitional Analysis and the Philosophy
of Science." Continuum. 2 (1964) 343-68.

774. _____. "Epistemological Problems in the Philosophy of
Science." Review of Metaphysics. 22 (1968) 113-37; 329-58.

775. McLaughlin, P. J. Modern Science and God. New York: Philosophical Library, 1954. 89 pp.

776. McNicholl, O. P., A. J. "The Uneasiness of Science." New
Scholasticism. 34 (1950) 57-68.

777. Magnini, Carlomagno. Introduzione alla critica della scienza.
Bologna: R. Pàtron, 1969. ix + 201 pp.

778. Marcuse, Herbert. "On Science and Phenomenology." BSPS 2
(1965) 279-90.

    a. "Comment on the Paper by H. Marcuse," by Aron Gurwitsch,
    291-306.

779. Martin, Richard M. Primordiality, Science, and Value. Albany:
State University of New York Press, 1980. 336 pp.

780. Maslow, Abraham H. The Psychology of Science. South Bend,
IN: Gateway Editions, 1966. 168 pp.

781. May, Eduard. Kleiner Grundriss der Naturphilosophie. Meisen-
heim-
am-Glan: Westkulturverlag Anton Hain, 1949. 106 pp.

782. _____. "Wissenschaft als Aggregat und System." PN 1
(1951) 348-60, 465-79, 2 (1952) 19-34, 332-49.

783. Maziarz, Edward A. "Sciences and Myths as Symbolic Structures."
Proc. Am. Cath. Phil. Assn. 45 (1971) 58-66.

784. Mehlberg, Henryk. The Reach of Science. Oxford University
Press, 1958. xii + 356 pp.

785. Meslen, Andrew G. van. The Philosophy of Nature. Duquesne
Studies: Philosophical Series, 2. Pittsburgh: Duquesne University
Press, 1953. xii + 253 pp.

    a. Review essay by Ernan McMullin. Philosophical Studies
    (Ireland). 5 (1955) 127-38.

786. _____. "La connaissance scientifique." Dialectica. 15
(1961) 55-73.

787. _____. "The Philosophical Value of Scientific Knowledge."
Archives de Philosophie. 27 (1964) 408-23.

788. Meurers, Joseph. "Philosophie und Naturwissenschaft." PN 1
(1951) 337-47.

789. _____. "Wort und Formel in den exakten Naturwissenschaften."
PN 11 (1969) 151-61.

790. _____. Metaphysik und Naturwissenschaft: Eine philosophische Studie über naturwissenschaftliche Problemkreise der Gegenwart.
Darmstadt: Wissenschaftliche Buchgesellschaft, 1976. viii + 127 pp.

791. Meynell, Hugo. "The 'Transcendental Precepts' and the Philosophy
of Science." Philosophical Inquiry. 1 (1978) 29-38.

792. Miller, Hugh. "The Science of Creation." Proc. & Add. Am.
Phil. Assn. 24 (1950-51) 31-47.

793. Mittelstrass, Jürgen. Neuzeit und Aufklärung: Studien zur

Entstehung der Neuzeitlichen Wissenschaft und Philosophie. Berlin and New York: Walter de Gruyter, 1970. 651 pp.

794. _____. Das praktische Fundament der Wissenschaft und die Aufgabe der Philosophie. Kontanz: Universitätsverlag, 1972. 104 pp.

795. _____. Die Möglichkeit von Wissenschaft. Frankfurt/Main: Suhrkamp, 1974. 268 pp.

796. Mohr, H. Lectures on Structure and Significance of Science. New York/Heidelberg/Berlin: Springer Verlag, 1977. xii + 227 pp.

797. Moles, Abraham. La création scientifique. Geneva: Editions René Kister, 1957. 237 pp.

798. Morin, Edgar. La nature de la nature. Paris: Seuil, 1977. 398 pp.

799. Mosedale, Frederick E. Philosophy and Science. Englewood Cliffs, NJ: Prentice-Hall, 1979. xix + 436 pp.

800. Mostowski, Andrzej, et al. Scientific Thought: Some Underlying Concepts, Methods, and Procedures. Paris, The Hague: Mouton/Unesco, 1972. 252 pp.

801. Mulkay, Michael. Science and the Sociology of Knowledge. London: George Allen & Unwin, 1979. 132 pp.

802. Müller-Markus, Siegfried. Protophysik. Part I. The Hague: Martinus Nijhoff, 1971. 427 pp.

803. Munévar, Gonzalo. Radical Knowledge: A Philosophical Inquiry into the Nature and Limits of Science. Indianapolis, IN: Hackett Publishing Co., 1981. 200 pp.

804. Naess, Arne. The Pluralist and Possibilist Aspect of the Scientific Enterprise. Oslo: Universitetsforlaget; London: Allen & Unwin, 1972. 148 pp.

805. Nagel, Ernest. "Some Reflections on the Use of Language in the Natural Sciences." J. Phil. 42 (1945) 617-29.

806. _____. Sovereign Reason. Glencoe, Ill.: The Free Press, 1954. 315 pp.

807. _____. Logic without Metaphysics, and other Essays in the Philosophy of Science. Glencoe, Ill.: The Free Press, 1956. xviii + 433 pp.

808. _____. "The Nature and Aim of Science," in Philosophy of Science Today. S. Morgenbesser, ed. New York: Basic Books, 1967. 3-13.

809. Nalimov, V. V. Faces of Science. Robert G. Colodny, ed.
Philadelphia: ISI Press, 1981. 298 pp.

810. Nash, Leonard K. The Nature of the Natural Sciences. Boston:
Little, Brown & Co., 1963. 406 pp.

811. Nowak, L. "The Model of Empirical Sciences in the Concepts of
the Creators of Marxism." SL 87 (1977) 499-539. [First published
in Studia Filozoficzne. 2 (75) (1972).]

812. _____. "Essence-Idealization-Praxis. An Attempt at a
Certain Interpretation of the Marxist Concept of Science." Poznan
Studies. 2, No. 3 (1976) 1-28.

813. _____. The Structure of Idealization: Towards a Systematic Interpretation of the Marxian Idea of Science. SL 139. Dordrecht: Reidel, 1979. xii + 271 pp.

814. Nyasani, Joseph M. "The Philosophical Problematic in Interpreting Nature." Epistemologia. 4 (1981) 473-88.

815. Oeser, Erhard. Wissenschaft und Information. Vol. 1: Wissenschaftstheorie und empirische Wissenschaftsforschung. Vol 2:
Erkenntnis als Informationsprozess. Vol. 3: Struktur und Dynamik
erfahrungswissenschaftlicher Systeme. Wien and München: Oldenbourg,
1976. 158 + 144 + 166 pp.

816. _____. Wissenschaftstheorie als Rekonstruktion der Wissenschaftsgeschichte. Fallstudien zu einer Theorie der Wissenschaftsentwicklung. Band I: Metrisierung, Hypothesenbildung, Theoriendynamik. Wien: R. Oldenbourg, 1979. 198 pp.

817. Ohmori, Shozo. "Natural Sciences are a Natural History."
AJAPS 3, No. 2 (1967) 5-18.

818. Oppenheim, Paul. "Dimensions of Knowldge." RIP 11 (1957)
151-91.

      a. "A Comment on Dr. Paul Oppenheim's 'Dimensions of Know-
      ledge'," by Charles Morris. 192-93.

      b. "An Empirical Study Related to 'Dimensions of Knowledge',"
      by Frederick R. Kling. 194-205.

      c. "A Note on 'Fort et Étroit' and 'Ample et Faible'," by
      Sylvain Bromberger. 206-10.

      d. "A Natural Order of Scientific Disciplines," by Paul
      Oppenheim. 13 (1959) 354-60.

819. Overington, Michael A. "The Scientific Community as Audience:
Toward a Rhetorical Analysis of Science." Philosophy and Rhetoric.
10 (1977) 143-64.

820. Palcos, A. "Hacia una definición integral de la ciencia."
Scientia. 94 (1959) 75-78. French tr.: Supplement, 47-50.

821. Pap, Arthur. An Introduction to the Philosophy of Science.
Glencoe, Ill.: Free Press, 1962. xiv + 444 pp.

822. Piaget, Jean. "L'analyse psychogénétique et l'épistémologie
des sciences exactes." Synthese. 7 (1948-49) 32-49.

823. _____. Introduction à l'épistémologie genetique. 3
Vols. Paris: Presses Universitaries de France, 1950.

   a. Review essay by F. Gonseth. Dialectica. 4 (1950) 5-20.

824. _____. Le structuralisme. Paris: Presses Universitaires
de France, 1968. Tr. by C. Maschler as Structuralism. New York:
Harper & Row, 1970. 153 pp.

825. _____. "L'Epistémologie génétique. Paris: Presses Universitaires de France, 1970. Tr. by Wolfe Mays: The Principles of
Genetic Epistemology. New York: Basic Books, 1972. 98 pp.

826. Pietschmann, H. "Die drei Grenzen physikalischer Erkenntnis."
PN 17 (1978-79) 90-98.

827. Pinto, Alvaro Vieira. Ciencia e existencia: Problemas filo-
sóficos da pesquisa cientifica. Rio de Janeiro: Paz e Terra, 1969.
537 pp.

828. Polanyi, Michael. Science, Faith and Society. London:
Oxford University Press, 1946. 80 pp.

829. _____. "Pure and Applied Science and Their Appropriate
Forms of Organization," in Science and Freedom. Boston: Beacon
Press, 1955. 36-46. Reprinted in Dialectica. 10 (1956) 231-42.

830. _____. "Beauty, Elegance, and Reality in Science," in
Observation and Interpretation. S. Körner, ed. New York: Academic
Press; London, Butterworths, 1957. 102-06.

831. _____. Personal Knowledge: Towards a Post-Critical
Philosophy. Chicago: University of Chicago Press; London: Routledge
& Kegan Paul, 1958. Revised edition, 1962. Reprint: New York:
Harper & Row, 1964. 416 pp.

   a. Review essay, "Science and Persons," by J. H. Woodger.
   BJPS 11 (1960) 65-71.

   b. Review essay by Greville Norburn. Philosophy. 35 (1960)
   344-49.

832. _____. "Knowing and Being." Mind. 70 (1961) 458-70.
Reprinted in Knowing and Being. M. Grene, ed. Chicago: University
of Chicago Press, 1969. 123-37.

833. _____. "The Unaccountable Element in Science." Philosophy.
37 (1962) 1-14. Reprinted in Knowing and Being. M. Grene, ed.
Chicago: University of Chicago Press, 1969. 105-20, and in Philosophy
Today. 6 (1962) 171-82.

834. _____. "Tacit Knowing: Its Bearing on Some Problems of
Philosophy." Reviews of Modern Physics. 34 (1962) 601-16. Reprinted
in Knowing and Being. M. Grene, ed. Chicago: University of Chicago
Press, 1969. 159-80; and in Philosophy Today. 6 (1962) 239-62.

835. _____. The Tacit Dimension. Garden City, NY: Doubleday,
1966. 108 pp.

   a. Review essay by P. M. C. Davies. Philosophical Studies.
   (Ireland) 17 (1968) 222-34.

836. _____. "The Logic of Tacit Inference." Philosophy. 41
(1966) 1-18. Reprinted in Knowing and Being. M. Grene, ed. Chicago:
University of Chicago Press, 1969. 138-58.

837. Pollock, John L. Knowledge and Justification. Princeton:
Princeton University Press, 1975. 348 pp.

838. Prigogine, Ilya, and Isabelle Stenger. "La nouvelle alliance."
Scientia. 112 (1977) 287-304, 617-30. Italian tr.: 305-18, 631-41.
English tr.: 319-32, 643-53.

839. Provençal, Yvon. "La conscience de l'observateur: de la
physique théorique a la logique mathématique." Dialogue. 16 (1977)
228-44.

840. Pugliese, Orlando. "Die 'Funktionen-und Strukturontologie'
als 'Hintergrund der modernen Wissenschaft'." ZPF 25 (1971) 202-25.

841. Rantala, Veikko. "On the Logical Basis of the Structuralist
Philosophy of Science." Erkenntnis. 15 (1980) 268-86.

842. Ravetz, Jerome R. Scientific Knowledge and Its Social Problems.
New York: Oxford University Press, 1971.

843. Redondi, Pietro. Epistemologia e storia delle scienza: Le
svolte teoriche da Duhem a Bachelard. Milan: Feltrinelli, 1978.
255 pp.

844. Reichenbach, Hans. The Rise of Scientific Philosophy. Berkeley:
University of California Press, 1951. 341 pp.

   a. Review essay, "Certainty and Certitude," by Walter Cerf.
   PPR 13 (1953) 515-24.

   b. Review essay by Kurt Hübner. Philosophische Rundschau. 3
   (1955) 239-43.

845. Rescher, Nicholas. "On First Principles and Their Legitimation."
Allgemeine Zeitschrift für Philosophie. 1, No. 2 (1976) 1-16.

846. Richardson, William J. "Heidegger's Critique of Science."
New Scholasticism 42 (1968) 511-36.

847. Richter, Maurice N. Science as a Cultural Process. Cambridge,
MA: Schenkman, 1972. v + 130 pp.

848. Robert, J.-D. "Philosophie des sciences, intelligibilité de
l'univers matérial et preuve de Dieu. Archives de Philosophie. 23
(1960) 327-87.

849. _____. "Essai de specification des savoirs de type
positif et experimental." Archives de Philosophie. 27 (1964) 5-48,
206-37; 28 (1965) 424-38; 29 (1966) 109-33, 268-80, 397-429.

850. Rosenblueth, Arturo. Mind and Brain: A Philosophy of Science.
Cambridge, MA: MIT Press, 1970. 128 pp.

    a. Review essay by Abraham S. Luchins. "A Biologist's Philo-
    sophy of Science." SHPS 2 (1971) 287-94.

851. Rosenkrantz, Roger D. Inference, Method and Decision: Towards
a Bayesian Philosophy of Science. SL 115 Dordrecht: Reidel, 1977.
xii + 262 pp.

    a. Review essay by Stepen Spielman. J. Phil. 78 (1981)
    356-67.

852. Ross, Stephen David. The Scientific Process. New York:
Humanities Press, 1971. 156 pp.

853. Rossi, Paolo. Immagini della scienza. Rome: Riuniti, 1977.
326 pp.

854. Russell, Bertrand. Human Knowledge: Its Scope and Limits.
New York: Simon and Schuster; 1948. xvi + 524 pp.

    a. Review essay by L. J. Russell. Philosophy. 24 (1949)
    253-60.

    b. Review essay, "Russell and Human Knowledge," by Edward A.
    Maziarz. New Scholasticism. 23 (1949) 318-25.

855. _____. The Scientific Outlook. New York: W. W. Norton,
1931, 1959. 277 pp.

856. Russell, S. J., John. Science and Metaphysics. London:
Sheed and Ward, 1958. 35 pp.

857. Sachkov, Iu. V. "The Evolution of the Style of Thought in
Science." Soviet Studies in Philosophy 7, No. 3 (1968-69) 30-40.
[English reprint from Voprosy filosofii. (1968, No. 4)].

858. Sachsse, Hans. Naturerkenntnis und Wirklichkeit. Braunschweig:
Verlag Friedrich Vieweg & Sohn, 1967. 232 pp.

859. Sakai, H. "Eine neue Moglichkeit der Wissenschaftstheorie."
PN 15 (1974-75) 66-78.

860. Samuel, Viscount. "Man's Ideas About the Universe." Philosophy. 28 (1953) 195-206.

861. _____. In Search of Reality. New York: Philosophical
Library, 1957. 229 pp.

862. Sarton, George. The Life of Science. New York: Henry Schuman,
1948. 197 pp.

863. Schapp, Wilhelm. Metaphysik der Naturwissenschaft. Den Haag:
M. Nijhoff, 1965. x + 141 pp.

864. Schlegel, Richard. Inquiry into Science: Its Domain and Its
Limits. New York: Doubleday, 1971. x + 108 pp.

865. Schlick, Moritz. Grundzuege der Naturphilosophie. Vienna:
Gerold, 1948. x + 115 pp.

866. _____. Philosophy of Nature. New York: Philosophical
Library, 1949.

867. Schneider, Friedrick. "Die Problemsituation der Erkenntnistheorie und das naturwissenschaftliche Erkennen." PN 4 (1957)
245-65.

868. Schnell, Walter. Die Erkenntnis der Natur. Stuttgart: W.
Kohlhammer, 1955. 246 pp.

869. Schock, Rolf. "What is a Science?" Notre Dame Journal of
Formal Logic. 6 (1965) 51-53.

870. Schrödinger, E. Science and Humanism. Physics in our Time.
Cambridge: Cambridge University Press, 1951. ix + 68 pp.

871. _____. Mind and Matter. Cambridge: At the University
Press, 1958. vii + 104 pp.

872. _____. Science, Theory and Man. New York: Dover Publications, 1961. xxiv + 223 pp.

873. Seeger, Raymond J. "On Understanding Physical Phenomena."
Proc. Am. Cath. Phil. Assn. 38 (1964) 47-73.

874. Seiffert, Helmut. Einführung in die Wissenschaftstheorie.
München: Beck, 1970.

875. Selvaggi, F. "Rassegna di filosofia delle scienze."
Gregorianum. 39 (1958) 611-27.

876. Serrano, Jorge A. Filosofía de la ciencia. Mexico City:
Centro de Estudio Educativos, 1980. 291 pp.

877. Simon, Herbert A. The Sciences of the Artificial. 2nd Edition. Cambridge, MA: MIT Press, 1981. 192 pp.

878. Simon, Yves. The Great Dialogue of Nature and Space. Gerard J. Dalcourt, ed. Albany, NY: Magi Books, 1970. 204 pp.

879. Simons, Joseph H. A Structure of Science. New York: Philosophical Library, 1960. 269 pp.

880. Simpson, George. "The Scientist-Technician or Moralist?" PS 17 (1950) 95-108.

881. Singer, Jr., Edgar A. Experience and Reflection. C. West Churchman, ed. Philadelphia: University of Pennsylvania Press, 1959.
    a. Review essay, "Singer's Philosophy of Experimentalism," by Y. H. Krikorian. PS 29 (1962) 81-91.

882. Slichter, Charles S. Science in a Tavern: Essays and Diversions on Science in the Making. Madison: University of Wisconsin Press, 1958. 206 pp.

883. Smart, J. J. C. Between Science and Philosophy: An Introduction to the Philosophy of Science. New York: Random House, 1968. 363 pp.

884. Snell, Bruno. "Science and Dogma," in Science and Freedom. Boston: Beacon Press, 1955. 134-40.

885. Sommerville, John. The Ways of Science. New York: Henry Schuman, 1953. 172 pp.

886. Spinner, Helmut F. Begründung, Kritik und Rationalität. Band I. Wissenschaftstheorie. Wissenschaft und Philosophie, 12. Braunschweig: Vieweg, 1977. xi + 307 pp.

887. Spirito, Ugo. Scienza e filosofia. Firenza: Sansoni, 1950. 305 pp.

888. _____. Ideale del dialogo o ideale della scienza? Roma: Edizioni dell'Ateneo, 1966. 396 pp.

889. Standen, Anthony. Science is a Sacred Cow. New York: Dutton, 1950. 221 pp.

890. Starostin, Boris A. Parametry Razvitiia Nauki. (Boundaries of the Development of Science) Moscow: Nauka, 1980. 280 pp.

891. Stegmüller, Wolfgang. Metaphysik, Skepsis, Wissenschaft. Zweite, verbesserte Aufl. Berlin: Springer, 1969. xii + 460 pp.

892. _____. Probleme und Resulte der Wissenschaftstheorie und analytischen Philosophie. Band I: Wissenschaftliche Erklärung und Begründung. Berlin: Springer-Verlag, 1969. 812 pp.

a. Review essay by Joseph J. Kockelmans. PS 38 (1971) 126-32.

b. Review essay by H. Scheichert. ZAW 1 (1970) 142-50.

c. Review essays by J. C. Marek, J. Gotschl, I. Stelzl, and H. G. Knapp. Conceptus. 4 (1970) 74-89.

d. Review essay by E. Stroker. Philosophische Rundschau. 18 (1972) 1-35.

893. _____. Probleme und Resulte der Wissenschaftstheorie und analytischen Philosophie. Band II: Theorie und Erfahrung. Berlin: Springer-Verlag, 1970. 485 pp.

a. Review essay, "Stegmüller on the Relationship between Theory and Experience," by Joseph J. Kockelmans. PS 39 (1972) 397-420.

b. Review essay, "Changing Patterns of Reconstruction," by Paul Feyerabend. BJPS 28 (1977) 351-69.

c. Review essay by E. Ströker. Philosophische Rundschau. 19 (1972) 1-25.

894. _____. Theorie und Erfahrung. Zweiter Halbband: Theorienstrukturen und Theoriendynamik. Berlin: Springer Verlag, 1973. xix + 327 pp. Eng. tr. by W. Wohlhueter. The Structure and Dynamics of Theories. New York/Heidelberg/Berlin: Springer-Verlag, 1976.

a. Review essay by Joseph J. Kockelmans. PS 43 (1976) 293-97.

b. Review essay by Werner Diederich. Philosophische Rundschau. 21 (1974) 209-28.

895. _____. Probleme und Resulte der Wissenschaftstheorie und analytischen Philosophie. Band IV: Personelle und statistische Wahrscheinlichkeit. Berlin: Springer, 1973. 560 pp.

896. _____. The Structuralist View of Theories. New York: Springer Verlag, 1979. 101 pp.

897. _____. Neue Wege der Wissenschaftsphilosophie. Berlin: Springer, 1980. 198 pp.

898. Stern, Alfred. "A Philosopher Looks at Science." Southern Journal of Philosophy. 7 (1969) 127-38.

899. Stock, Wolfgang G. "The Informatics of Science Establishment, Subject Matter and Methods." Ratio. 22 (1980) 145-54.

900. Strombach, Werner. "Ordnungsstrukturen als Voraussetzung logisch-mathematischer Naturerkenntnis." PN 10 (1967-68) 56-82.

901. Supek, I. "Genealogy of Science and Theory of Knowledge."
WOSPS 12 (1977) 173-83.

902. Synge, J. L. Science: Sense and Nonsense. London: Jonathan
Cape Ltd., 1951. 156 pp.

903. Tagliagambe, Silvano. "I linguaggi della scienza." Scientia.
116 (1981) 517-40. English tr.: 541-57.

904. Talbott, George R. Philosophy and Unified Science. Madras:
Ganesh, 1977. 2 Vols. xliv + 1437 pp.

905. Taylor, F. Sherwood. "The Scientific World-Outlook." Philosophy.
22 (1947) 195-207.

906. Thagard, Paul. "Why Astrology is a Pseudoscience." PSA 1978.
I, 223-34.

907. Thom, René. "Formalisme et scientificité." Les Etudes philosophiques. 33 (1978) 171-78.

908. Törnebohm, Håkan. "Scientific Enterprises from a Biological
Point of View." in Biology, History, and Natural Philosophy. A. D.
Breck and W. Yourgrau, eds. New York: Plenum, 1972. 165-69.

909. _____. "An Essay on Knowledge-Formation." ZAW 6 (1975)
37-64.

910. Toulmin, Stephen. The Philosophy of Science, An Introduction.
London: Hutchinson's University Library, 1953. 172 pp. Reprint:
New York: Harper & Row, 1960.

    a. Review essay by Eike v. Savigny. Philosophische Rundschau.
    19 (1972) 198-203.

    b. Review essay by Ernest Nagel. Mind. 63 (1954) 403-12.

911. _____. Foresight and Understanding: An Inquiry into the
Aims of Science. Bloomington: Indiana University Press, 1961. Reprint: New York: Harper & Row, 1963. 115 pp.

912. _____. Human Understanding. Vol. I: The Collective Use
and Evolution of Concepts. Princeton: Princeton University Press,
1972. xii + 520 pp.

    a. Review essay, "Toulmin's Model of an Evolutionary Episte-
    mology," by R. J. Blackwell. Modern Schoolman. 51 (1973)
    62-68.

    b. Review essay, "Is the Progress of Science Evolutionary?"
    by L. Jonathan Cohen. BJPS 24 (1973) 41-61.

    c. Review essay, "Toulmin's Evolutionary Epistemology," by
    Larry Briskman. Phil. Quart. 24 (1974) 160-69.

d. Review essay, "Limitations of an Evolutionist Philosophy of Science," by John Lossee. SHPS 8 (1977) 349-52.

e. Review essay, "A Populational Approach to Scientific Change," by David L. Hull. Science. 182 (1973) 1121-24.

f. Review essay, "Understanding Toulmin," by Imre Lakatos. Minerva. 14 (1976) 126-43.

913. Tummers, J. H. De Evolutie der Wetenschap (The Evolution of Science). Utrecht: Nijmegen, 1946. 58 pp.

914. Turchin, V. F. The Phenomenon of Science. New York: Columbia University Press, 1977. xvii + 348 pp.

915. Turner, Dean. Commitment to Care: An Integrated Philosophy of Science, Education, and Religion. Old Greenwich, CT: Devin-Adair, 1978. xv + 416 pp.

916. Ullmo, Jean. La pensée scientifique moderne. Paris: Flammarion, 1958. 284 pp.

917. Uyemov, A. "Demonstrative and Heuristic Aspects in the Logical Modeling of Science." LMPS - 4 (1971) 407-13.

918. Vandel, A. "Un humanisme scientifique." Dialectica. 14 (1960) 5-20.

919. Waddington, C. H. The Scientific Attitude. West Drayton, Middlesex: Pelican Books, 1948. 175 pp.

920. _____. Behind Appearance. Edinburgh: Edinburgh University Press, 1969. 252 pp.

921. Waelhens, A. de. "Science, phénoménologie, ontologie." RIP 8 (1954) 254-65.

922. Walker, Marshall J. The Nature of Scientific Thought. Englewood Cliffs, NJ: Prentice-Hall, 1963. 184 pp.

923. Walter, Emil J. Erforschte Welt. Die wichtigsten Ergebnisse der naturwissenschaftlichen Forschung. Bern: Francke Verlag, 1953. 363 pp.

924. Wartofsky, Marx W. Conceptual Foundations of Scientific Thought: An Introduction to the Philosophy of Science. New York: Macmillan, 1968. 560 pp.

925. Weinberg, Julius R. Abstraction, Relation, and Induction: Three Essays in the History of Thought. Madison: University of Wisconsin Press, 1965. 156 pp.

926. Weiss, Paul A. L'archipel scientifique: Etudes sur les fondements et les perspectives de la science. J. Rambaud, tr. Paris: Maloine-Doin éditeurs, 1974. 265 pp.

927. Weizsäcker, Carl Friedrich von. Die Geschichte der Natur.
Göttingen: Vandenhoek und Ruprecht, 1948. English tr.: The History
of Nature. Chicago: University of Chicago Press, 1949. 191 pp.

    a. Review essay, "Die Geschichte der Natur", by Hermann Wein.
PN 1 (1950) 151-57. "Zur 'Geschichte der Natur," by Simon
Moser, 435-38. "Ueber die Bemerkungen von Moser und Schrödinger
zu C. F. von Weizsäckers 'Geschichte der Natur'," by Hermann
Wein, 438-46.

928. _____. Die Einheit der Nature. München: Hansen, 1971.
491 pp. Tr. by Francis J. Zucker: The Unity of Nature. New York:
Farrar, Straus, and Giroux, 1980. x + 406 pp.

929. _____. Voraussetzungen das naturwissenschaftlichen
Denkens. Freiburg: Herder, 1972. 141 pp.

930. _____. "The Preconditions of Experience and the Unity of
Physics." SL 133 (1979) 123-58.

    a. "Comment on von Weizsäcker," by Mary Hesse. 159-70.

    b. "Comment on von Weizsäcker," by Peter Mittelstaedt.
171-76.

931. Werkmeister, W. H. The Basis and Structure of Knowledge. New
York: Harper & Bros. 1948. 451 pp.

    a. Review essay: "Coherence Theory Reconsidered: Professor
Werkmeister on Semantics and the Nature of Empircal Laws," by
May Brodbeck. PS 16 (1949) 75-85.

932. _____. "Science, Its Concepts and Laws." J. Phil. 46
(1949) 444-52.

933. _____. A Philosophy of Science. Lincoln: University of
Nebraska Press, 1965. ix + 551 pp.

934. Weyl, Hermann. Philosophy of Mathematics and Natural Science.
(Revised and augmented English edition. Based on a translation by
Olaf Helmer.) Princeton: Princeton University Press, 1949. Reprint:
New York: Atheneum, 1963. x + 311 pp.

935. Whitrow, G. J. "The Epistemological Foundations of Natural
Philosophy." Philosophy. 21 (1946) 5-28.

936. Whyte, L. L. Accent on Form. London: Routledge & Kegan
Paul, 1955. 202 pp.

937. Wisdom, J. O. "Four Contemporary Interpretations of the
Nature of Science." FP 1 (1970-71) 269-84.

938. Witzemann, Edgar J. "The Scope, Objectives and Limitations of
Modern Science as Seen in the Light of its History." PS 14 (1947)
44-55.

939. Wohlgenannt, Rudolf. Was ist Wissenschaft? Braunschweig:
Friedr. Vieweg & Sohn, 1969. xv + 204 pp.

940. Wojciechowski, Jerzy A. "Réflexions sur le mode de savoir des
sciences physiques." Laval Théologique et Philosophique. 20 (1964)
9-34.

941. Wright, Georg Henrik von. Explanation and Understanding.
London: Routledge & Kegan Paul, 1971. 230 pp.

    a. Review essay by Peter Winch. Metaphilosophy. 4 (1973)
    63-75.

    b. Review essay by G. Even-Granboulan. RMM 82 (1977) 108-20.

942. Yildirim, Cemal. "Towards an Understanding of Science." ZAW
1 (1970) 104-18.

943. Yukawa, Hideki. "Intuition and Abstraction in Scientific
Thinking." AJAPS 2, No. 2 (1962) 40-43.

944. Zaragüeta Bengoechea, Juan. Problemática de la filosofía de
las ciencias. Barcelona, 1954. 20 pp.

945. Ziemski, Stefan. "Towards a New Model of Science." ZAW 7
(1976) 340-47.

946. Ziman, John. Public Knowledge. Cambridge: Cambridge University Press, 1968. xii + 154 pp.

947. _____. Reliable Knowledge: An Exploration of the Grounds
for Belief in Science. Cambridge University Press, 1978. ix +
197 pp.

3.2 Collected Papers and Symposia

948. Achinstein, Peter, and S. F. Barker, eds. The Legacy of Logical
Positivism: Studies in the Philosophy of Science. Baltimore:
Johns Hopkins Press, 1969. x + 300 pp.

949. Actes du deuxième Congrès International de l'Union Internationale de Philosophie des Sciences, Zurich, 1954. 5 Vols. La Neu-
veville, Suisse: Editions du Griffon, 1955. 162, 152, 170, 136,
111 pp.

    a. Review essay, "Methodology and Quantum Physics," by Eva
    Cassirer. BJPS 8 (1958) 334-41.

950. Agassi, J., and R. S. Cohen, eds. Scientific Philosophy Today:
Essays in Honor of Mario Bunge. BSPS 67 Dordrecht: Reidel, 1981.
510 pp.

951. Asquith, Peter D., and Henry E. Kyburg, Jr., eds. Current Research in Philosophy of Science. Proceedings of the P.S.A. Critical Research Problems Conference. East Lansing, MI: Philosophy of Science Association, 1979. xiv + 533 pp.

952. Bärmark, Jan, ed. Perspectives in Metascience. Göteborg: Kungl. Vetenskaps-och Vitterhets-Samhället, 1979. 199 pp.

953. Bogdan, Radu J., ed. Patrick Suppes: Profiles, Vol. I. Dordrecht: Reidel, 1979. 264 pp.

954. Boltzmann, Ludwig. Theoretical Physics and Philosophical Problems: Selected Writings. B. McGuinnes, ed. P. Foulkes, tr. Dordrecht: Reidel, 1974. 280 pp. [Volume 5 of the Vienna Circle Collection.]

955. Boring, E. G. History, Psychology, and Science: Selected Papers. New York: John Wiley, 1963.

   a. Review essay, "E. G. Boring's Philosophy of Science," by Edward H. Madden. PS 32 (1965) 194-201.

956. Boston Studies in the Philosophy of Science. Robert S. Cohen and Marx W. Wartofsky, eds. Dordrecht: Reidel, 1963 - .

   At the end of 1981 69 volumes had been announced in the series, which overlaps partially with the Synthese Library series. For a listing of the volumes in the BSPS series, see Appendix I. In this bibliography each relevant volume of the series is listed separately according to subject matter.

957. Bouligand, G., et al. Hommage à Gaston Bachelard. Paris: Presses Universitaires de France, 1957. 216 pp.

958. Breck, Allen D., and Wolfgang Yourgrau, eds. Biology, History, and Natural Philosophy. New York: Plenum Press, 1972. xiv + 355 pp.

959. Broad, C. D. Induction, Probability, and Causation: Selected Papers. SL 15. Dordrecht: Reidel, 1968. 296 pp.

960. Brown, Harcourt, ed. Science and the Creative Spirit. University of Toronto Press, 1958. xxvii + 165 pp.

961. Bunge, Mario, ed. The Critical Approach to Science and Philosophy. New York: Macmillan, 1964. 496 pp.

962. _____. Delaware Seminar in the Foundations of Physics: Studies in the Foundations, Methodology and Philosophy of Science. Vol. I. New York: Springer-Verlag, 1967. 193 pp.

963. _____. Exact Philosophy - Problems, Tools, and Goals. SL 50. Dordrecht: Reidel, 1973. 214 pp.

964. _____. The Methodological Unity of Science. Dordrecht: Reidel, 1973. 264 pp.

965. Butts, R. E. and J. Hintikka, eds. Logic, Foundations of Mathematics, and Computability Theory. WOSPS 9. Dordrecht: Reidel, 1977. x + 406 pp.

    a. Review essay of WOSPS 9-12 by David Miller. Synthese. 43 (1980) 381-410.

966. _____. Foundational Problems in the Special Sciences. WOSPS 10. Dordrecht: Reidel, 1977. x + 427 pp.

967. _____. Basic Problems in Methodology and Linguistics. WOSPS 11. Dordrecht: Reidel, 1977. x + 321 pp.

968. _____. Historical and Philosophical Dimensions of Logic, Methodology and Philosophy of Science. WOSPS 12. Dordrecht: Reidel, 1977. x + 336 pp.

969. Centre International de Synthèse. XXe semaine de synthèse: Notion de structure et structure de la connaissance, 18-27 avril 1956. Paris: Albin Michel, 1957. xxiv + 436 pp.

970. Cohen, Robert S., J. J. Stachel, and Marx W. Wartofsky, eds. For Dirk Struik. Scientific, Historical and Political Essays in Honor of Dirk Struik. BSPS 15. SL 61. Dordrecht: Reidel, 1974. 652 pp.

971. Cohen, R. S., P. K. Feyerabend, and M. W. Wartofsky, eds. Essays in Memory of Imre Lakatos. BSPS 39. SL 99. Dordrecht: Reidel, 1976. 762 pp.

    a. Review essay by Alfred Schramm. Grazer Philosophische Studien. 11 (1980) 167-82.

972. Cohen, R. S. and J. J. Stachel, eds. Selected Papers of Léon Rosenfeld. BSPS 21. SL 100. Dordrecht: Reidel, 1978. 927 pp.

973. Colodny, Robert G., ed. Frontiers of Science and Philosophy. University of Pittsburgh Series in the Philosophy of Science, Vol. I. Pittsburgh: University of Pittsburgh Press, 1962. 288 pp.

974. _____. Beyond the Edge of Certainty: Essays in Contemporary Science and Philosophy. University of Pittsburgh Series in the Philosophy of Science, Vol. II. Englewood Cliffs, NJ: Prentice-Hall, 1965. 287 pp.

    Review essay by I. C. Hinckfuss. AJP 43 (1965) 384-401.

975. _____. Mind and Cosmos: Essays in Contemporary Science and Philosophy. Pittsburgh: University of Pittsburgh Press, 1966. 362 pp.

    Review essay, "Bayesianism and the Rationality of Scientific Inference," by Jon Dorling. BJPS 23 (1972) 181-90.

976. _____. The Nature and Function of Scientific Theories. University of Pittsburgh Series in the Philosophy of Science, Vol. IV. Pittsburgh: University of Pittsburgh Press, 1970. 361 pp.

977. _____. Logic, Laws and Life. Pittsburgh: University of Pittsburgh Press, 1977. 258 pp.

978. Connaissance scientifique et philosophie. Colloque organisé les 16 et 17 mai 1973 par l'Académie Royale de Sciences, des Lettres et des Beaux-Arts de Belgique à l'occasion du deuxième centenaire de sa fondation. Brussels: Palais des Academies, 1975. 382 pp.

979. Dalla Chiara, M. L., ed. Italian Studies in the Philosophy of Science. BSPS 47. Dordrecht: Reidel, 1980.

980. Davidson, D. and J. Hintikka, eds. Words and Objections: Essays on the Work of W. V. Quine. SL 21. Dordrecht: Reidel, 1969. 366 pp. Revised edition: 1975. 373 pp.

    a. Review essay by B. A. Brody. SHPS 2 (1971) 167-75.

    b. "Replies," by W. V. Quine. Synthese. 19 (1968) 264-322. Reprinted in SL 21 (1975) 292-352.

981. Démonstration, vérification, justification. Louvain: Éditions Nauwelaerts, 1968. 356 pp. (Entretiens de l'Institute International de Philosophie, Liège, 4-9 September, 1967).

982. Destouches, Jean Louis, ed. Logic and Foundations of Science. Dordrecht: Reidel, 1968. viii + 140 pp.

983. Deutscher Kongress für Philosophie, 9th, Dusseldorf, 1969: Philosophie und Wissenschaft. Ludwig Landgrebe, hrsg. Meisenheim am Glan: Hain, 1972. 644 pp.

984. Diemer, Alwin, ed. Der Wissenschaftsbegriff: Histor. und systemat. Untersuchungen. Meisenheim am Glan: Hain, 1970. viii + 277 pp.

985. _____. Der Methoden-und Theorienpluralismus in den Wissenschaften. Meisenheim am Glan: Hain, 1971. 325 pp.

986. Enz, Charles P., and Jagdish Mehra, eds. Physical Reality and Mathematical Description. Dordrecht: Reidel, 1974. 552 pp.

987. Études de philosophie des sciences en hommage à F. Gonseth. Neuchâtel: Editions du Griffon, 1950. 175 pp.

988. L'explication dans les sciences. Colloque de l'Académie internationale de philosophie des sciences, avec le concours du Centre international d'épistémologie génétique. (Geneve 25-29 september 1970) Paris: Flammarion, 1973. 233 pp.

989. Feigl, Herbert. Inquiries and Provocations: Selected Writings, 1929-1974. Dordrecht: Reidel, 1981. xii + 453 pp.

990. Feigl, Herbert, and Grover Maxwell, eds. <u>Current Issues in the Philosophy of Science</u>. Proceedings of Section L of the American Association for the Advancement of Science, 1959. New York: Holt, Rinehart and Winston, 1961. 484 pp.

991. Feyerabend, Paul K., and Grover Maxwell, eds. <u>Mind, Matter, and Method: Essays in Philosophy and Science in Honor of Herbert Feigl</u>. Minneapolis: University of Minnesota Press, 1966. v + 524 pp.

   a. Review essay by Robert M. Young, Mary Hesse, and Arthur I. Fine. BJPS 18 (1968) 325-39.

992. Fisher, Alden L., and George B. Murray, eds. <u>Philosophy and Science as Modes of Knowing: Selected Essays</u>. New York: Appleton-Century-Crofts, 1969. 253 pp.

993. Frank, Philipp G., ed. <u>The Validation of Scientific Theories</u>. Boston: Beacon Press, 1954. Reprint: New York: Collier Books, 1961. 220 pp. [Papers presented at 1953 meeting of the American Association for the Advancement of Science.]

994. Friedrich, S. J., L. W., ed. <u>The Nature of Physical Knowledge</u>. Bloomington: Indiana University Press; Milwaukee: Marquette University Press, 1960. 156 pp.

995. Fritz, Jr., Charles A. <u>Bertrand Russell on the Philosophy of Science</u>. Indianapolis: Bobbs-Merrill, 1965. 232 pp.

996. Giedymin, Jerzy, ed. <u>Kazimierz Ajdukiewicz: The Scientific World-Perspective and Other Essays, 1931-1963</u>. SL 108. Dordrecht: Reidel, 1978. 378 pp.

997. Giere, Ronald N., and Richard S. Westfall, eds. <u>Foundations of Scientific Method: The Nineteenth Century</u>. Bloomington and London: Indiana University Press, 1973. ix + 306 pp.

998. Gram, M. S. and E. D. Klemke, eds. <u>The Ontological Turn: Studies in the Philosophy of Gustav Bergmann</u>. Iowa City: University of Iowa Press, 1974. 314 pp.

999. Grandy, Richard E., ed. <u>Theories and Observation in Science</u>. Englewood Cliffs, NJ: Prentice-Hall, 1973. viii + 183 pp.

1000. Gregg, John R. and F. T. C. Harris, eds. <u>Form and Strategy in Science: Studies Dedicated to Joseph Henry Woodger on the Occasion of His Seventieth Birthday</u>. Dordrecht: Reidel, 1964. 476 pp.

1001. Guntau, Martin, and Helge Wendt, eds. <u>Naturforschung und Weltbild. Enie Einführung in philosophische Probleme der modernen Naturwissenschaften</u>. Berlin: Deutscher Verlag der Wissenschaften, 1967. 372 pp.

1002. Gunter, P. A. Y., ed. and tr. <u>Bergson and the Evolution of Physics</u>. Knoxville: University of Tennessee Press, 1969. xi + 348 pp.

a. Review essay by M. Čapek. Process Studies. 2 (1972)
149-59.

1003. Hanson, Norwood Russell. What I Do Not Believe, and Other
Essays. SL 38. Stephen Toulmin and Harry Woolf, eds. Dordrecht:
Reidel, 1971. 390 pp.

1004. _____. Constellations and Conjectures. SL 48. Willard
C. Humphreys, Jr., ed. Dordrecht: Reidel, 1973. 282 pp.

1005. Healey, R., ed. Reduction, Time and Reality: Studies in
the Philosophy of the Natural Sciences. Cambridge University Press,
1981.

1006. Helmholtz, Hermann von. Epistemological Writings. The Paul
Hertz/Moritz Schlick Centenary Edition of 1921 with Notes and Commentary by the Editors. (Newly translated by Malcolm F. Lowe.
Edited, with an Introduction and Bibliography, by Robert S. Cohen
and Yehuda Elkana.) BSPS 37. SL 79. Dordrecht: Reidel, 1977.
204 pp.

1007. Hintikka, Jaakko. Models for Modalities: Selected Essays.
SL 23. Dordrecht: Reidel, 1969. 220 pp.

1008. _____. Rudolf Carnap, Logical Empiricist: Materials
and Perspectives. SL 73. Dordrecht: Reidel, 1975. lxviii +
400 pp.

a. Review essay by Joseph Agassi. Philosophia. 10 (1981)
57-88.

1009. _____. Essays on Wittgenstein in Honour of G. H. von
Wright. Acta Philosophica Fennica, Vol. 28, Nos. 1-3. Amsterdam:
North-Holland, 1976. 516 pp.

1010. Hintkka, Jaakko, David Gruender, and Evandro Agazzi, eds.
Theory Change, Ancient Axiomatics, and Galileo's Methodology. Pro-
ceedings of the 1978 Pisa Conference on the History and Philosophy of
Science. Vol. I. SL 145 Dordrecht: Reidel, 1981.

1011. _____. Probabilistic Thinking, Thermodynamics, and the
Interaction of the History and Philosophy of Science. Proceedings
of the 1978 Pisa Conference on the History and Philosophy of Science.
Vol. II. SL 146. Dordrecht: Reidel, 1981.

1012. Hockney, Donald, William Harper, and Bruce Freed, eds. Contemporary Research in Philosophical Logic and Linguistic Semantics.
WOSPS 4. Dordrecht: Reidel, 1975. 332 pp.

1013. Holton, Gerald. Science and the Modern Mind: A Symposium.
Boston: Beacon Press, 1958. ix + 110 pp.

1014. Holton, Gerald and William Blanpied, eds. Science and Its
Public: The Changing Relationship. BSPS 33. SL 96. Dordrecht:
Reidel, 1976. 289 pp.

1015. Holton, Gerald, and Robert S. Morison, eds. Limits of Scientific Inquiry. New York: W. W. Norton, 1979. 254 pp. First published in 1978 as Vol. 107, No. 2, of Proceedings of the American Academy of Arts and Sciences.

1016. Hook, Sidney, ed. John Dewey: Philosopher of Science and Freedom: A Symposium. New York: Dial Press, 1950. vi + 383 pp.

1017. _____. Dimensions of Mind: A Symposium. New York: New York University Press, 1960. 281 pp.

1018. Hooker, C. A., J. J. Leach, and E. F. McClennen, eds. Foundations and Applications of Decision Theory. WOSPS 13. Dordrecht: Reidel, 1978. Vol. I: Theoretical Foundations, xxiii + 442 pp. Vol. II: Epistemic and Social Applications, xxiii + 206 pp.

1019. Hooker, C. A., ed. Physical Theory as Logico-Operational Structure. WOSPS 7. Dordrecht: Reidel, 1979. 334 pp.

1020. Howson, Colin, ed. Method and Appraisal in the Physical Sciences: The Critical Background to Modern Science, 1800-1905. Cambridge: Cambridge University Press, 1976. vii + 344 pp.

    a. Review essay, "The Halt and the Blind: Philosophy and History of Science," by Thomas Kuhn. BJPS 31 (1980) 181-92.

    b. Review essay by A. F. Chalmers. Erkenntnis. 16 (1981) 167-76.

1021. Hutchings, Jr., Edward, ed. Frontiers in Science. New York: Basic Books, 1958. 362 pp.

1022. Juhos, Béla. Selected Papers on Epistemology and Physics. G. Frey, ed. P. Foulkes, tr. Dordrecht: Reidel, 1976. 350 pp. [Volume 7 of the Vienna Circle Collection.]

1023. Kiala, Eino Sakari. Reality and Experience: Four Philosophical Essays. R. S. Cohen, ed. Dordrecht: Reidel, 1978. 326 pp. [Volume 12 of the Vienna Circle Collection.]

1024. Kiefer, Howard E. and Milton K. Munitz, eds. Mind, Science, and History. Albany: State University of New York Press, 1970. 321 pp.

1025. Klaus, Georg. Beiträge zu philosophischen Problemen der Einzelwissenschaften. Berlin: Akademie-Verlag, 1978. 145 pp.

1026. Klibansky, Raymond, ed. Philosophy in the Mid-Century: A Survey. Florence: La Nuova Italia-editrice, 1958. xi + 336 pp.

1027. Kockelmans, Joseph J., and Theodore J. Kisiel, eds. Phenomonology and the Natural Sciences. Evanston, IL: Northwestern University Press, 1970. xxi + 520 pp.

a. Review essay, "Phenomenology and Scientific Realism," by
Gary Gutting. New Scholasticism. 48 (1974) 253-66.

1028. Kopnin, P. W., and N. W. Popowitsch, eds. Logik der wissenschaftlichen Forschung. Berlin: Akademie-Verlag, 1969. 473 pp.

1029. Körner, S., ed. Observation and Interpretation: A Symposium
of Philosophers and Physicists. New York: Academic Press; London:
Butterworths Scientific Pub., 1957. xiv + 218 pp.

1030. Krajewski, W., ed. Polish Essays in the Philosophy of the
Natural Sciences. BSPS 68 Dordrecht: Reidel, 1982. 515 pp.

1031. Krüger, Lorenz, ed. Erkenntnisprobleme der Naturwissenschaften.
Text zur Einfuhrung in die Philosophie der Wissenschaft. Cologne:
Kiepenheuer und Witsch, 1970. 534 pp.

1032. Lakatos, I., and A. Musgrave, eds. Problems in the Philosophy of Science. Amsterdam: North-Holland, 1968. ix + 448 pp.

1033. Largeault, Jean. Enigmes et controverses. Paris: Aubier
Montaigne, 1980. 192 pp.

1034. Leach, J. J., R. E. Butts, and G. Pearce, eds. Science,
Decision and Value. WOSPS 1. Dordrecht: Reidel, 1973. 213 pp.

1035. Lenk, Hans, ed. Neue Aspekte der Wissenschaftstheorie.
Braunschweig: Vieweg, 1971. 249 pp.

1036. Lerner, Daniel, ed. Parts and Wholes: The Hayden Colloquium
on Scientific Method and Concept. New York: Free Press, 1963.

1037. Logic, Methodology and Philosophy of Science: Proceedings
of the 1960 International Congress. E. Nagel, P. Suppes, and A.
Tarski, eds. Stanford, CA: Stanford University Press, 1962.
661 pp.

a. Review essays by Elliott Mendelson, Richard C. Jeffrey,
and Ernest Adams. J. Phil. 61 (1964) 76-94.

1038. Logic, Methodology and Philosophy of Science: Proceedings
of the 1964 International Congress. Yehoshua Bar-Hillel, ed.
Amsterdam: North-Holland. 1965. viii + 440 pp.

1039. Logic, Methodology and Philosophy of Science III: Proceedings
of the Third International Congress for Logic, Methodology and
Philosophy of Science, Amsterdam 1967. B. van Rootselaar and J. F.
Staal, eds. Amsterdam: North-Holland, 1968. xii + 553 pp.

1040. Logic, Methodology and Philosophy of Science IV: Proceedings
of the Fourth International Congress for Logic, Methodology and
Philosophy of Science, Bucharest, 1971. Patrick Suppes, Leon Henkin,
Athanase Joja, and Gr. C. Moisil, eds. Amsterdam: North-Holland;
New York: American Elsevier, 1973. x + 981 pp.

1041. Logic, Language, and Probability: A Selection of Papers
Contributed to Sections IV, VI, and XI of the Fourth International
Congress for Logic, Methodology, and Philosophy of Science,
Bucharest, September, 1971. Radu J. Bogdan and Ilkka Niiniluoto,
eds. SL 51. Dordrecht: Reidel, 1973. x + 323 pp.

1042. Lorenz, Kuno, ed. Konstruktionen versus Positionen: Beiträge
zur Diskussion um die Konstruktive Wissenschaftstheorie. Band I:
Spezielle Wissenschaftstheorie. Band II: Allgemeine Wissenschaftstheorie. Berlin, New York: Walter de Gruyter, 1979. I: xx +
350 pp.; II: x + 406 pp.

1043. Mach, Ernst. Knowledge and Error: Sketches on the Psychology
of Enquiry. P. Foulkes, tr. Dordrecht: Reidel, 1976. 393 pp.
[Volume 3 of the Vienna Circle Collection.]

1044. Madden, Edward H., ed. The Structure of Scientific Thought:
An Introduction to the Philosophy of Science. Boston: Houghton
Mifflin, 1960. ix + 381 pp.

1045. Manninen, Juha and Raimo Tuomela, eds. Essays on Explanation
and Understanding: Studies in the Foundations of Humanities and
Social Sciences. SL 72. Dordrecht: Reidel, 1976. 440 pp.

    a. "Replies," by Georg Henrik von Wright. 371-413.

1046. Marković, Mihailo, and Gajo Petrović, eds. Praxis: Yugoslav
Essays in the Philosophy and Methodology of the Social Sciences. SL
134. BSPS 36. Dordrecht: Reidel, 1979.

1047. Mauskopf, Seymour H., ed. The Reception of Unconventional
Science. Boulder, Colorado: Westview Press, 1979. 137 pp.

1048. Mellor, D. H., ed. Science, Belief, and Behavior. Cambridge
University Press, 1980. xii + 227.

1049. Menger, Karl. Selected Papers in Logic and Foundations,
Didactics, Economics. Dordrecht: Reidel, 1978. [Volume 10 of the
Vienna Circle Collection.]

1050. La méthode prospective. Colloque de l'Academie internationale
de philosophie des sciences, 10-13 september 1968, Héverlée (Belgique).
Brussels: Office International de Libraire, 1972. 128 pp.

1051. Meyer-Abich, Adolf, ed. Physics: Beiträge zur naturwissenschaftlichen Synthese. Stuttgart: Hippokrates-Verlag, 1949.
206 pp.

1052. Minnesota Studies in the Philosophy of Science. Vol I: The
Foundations of Science and the Concepts of Psychology and Psycho-
analysis. Herbert Feigl and Michael Scriven, eds. Minneapolis:
University of Minnesota Press, 1956. 346 pp.

1053. Minnesota Studies in the Philosophy of Science. Vol II:
Concepts, Theories, and the Mind-Body Problem. Herbert Feigl,

Michael Scriven and Grover Maxwell, eds. Minneapolis: University of Minnesota Press, 1958. 553 pp.

1054. **Minnesota Studies in the Philosophy of Science**. Vol III: Scientific Explanation, Space, and Time. Herbert Feigl and Grover Maxwell, eds. Minneapolis: University of Minnesota Press, 1962. 628 pp.

1055. **Minnesota Studies in the Philosophy of Science**. Vol IV: Analyses of Theories and Methods of Physics and Psychology. Michael Radner and Stephen Winokur, eds. Minneapolis: University of Minnesota Press, 1970. 441 pp.

    a. Review essay, "For and Against Method," by Noretta Koertge. BJPS 23 (1972) 274-90.

    b. Review essay by C. A. Hooker. Canadian Journal of Philosophy. 1 (1972) 393-407, 489-509.

1056. **Minnesota Studies in the Philosophy of Science**. Vol V: Historical and Philosophical Perspectives of Science. Roger H. Stuewer, ed. Minneapolis: University of Minnesota Press, 1970. 384 pp.

    a. Review essay, "History and Philosophy of Science: Intimate Relationship or Marriage of Convenience?" by Ronald N. Giere. BJPS 24 (1973) 282-97.

1057. **Minnesota Studies in the Philosophy of Science**. Vol VI: Induction, Probability, and Confirmation. Grover Maxwell and Robert M. Anderson, Jr., eds. Minneapolis: University of Minnesota Press, 1975. 551 pp.

1058. **Minnesota Studies in the Philosophy of Science**. Vol. VII: Language, Mind, and Knowledge. Keith Gunderson, ed. Minneapolis: University of Minnesota Press, 1975. 424 pp.

1059. **Minnesota Studies in the Philosophy of Science**. Vol. VIII: Foundations of Space-Time Theories. John Earman, Clark Glymour, and John Stachel, eds. Minneapolis: University of Minnesota Press, 1977. 459 pp.

1060. **Minnesota Studies in the Philosophy of Science**. Vol IX: Perception and Cognition: Issues in the Foundations of Psychology. C. Wade Savage, ed. Minneapolis: University of Minnesota Press, 1978. 502 pp.

1061. Morgenbesser, Sidney, ed. Philosophy of Science Today. New York: Basic Books, 1967. xvi + 208 pp.

1062. Morgenbesser, Sidney, Patrick Suppes, and Morton White, eds. Philosophy, Science, and Method: Essays in Honor of Ernest Nagel. New York: St. Martin's Press, 1969. 613 pp.

    a. Review essay by Marshall Spector. Metaphilosophy. 2
(1971) 251-67.

1063. Mulder, Henk L. and Barbara F. B. van de Velde-Schlick, eds.
Moritz Schlick: Philosophical Papers Volume I (1909-1922). Dordrecht:
Reidel, 1978. 370 pp. [Volume 11 of the Vienna Circle Collection.]

1064. Neurath, Otto. Empiricism and Sociology. M. Neurath and R.
S. Cohen, eds. P. Foulkes and M. Neurath, trs. Dordrecht: Reidel,
1973. 473 pp. [Volume 1 of the Vienna Circle Collection.]

1065. Neurath, Otto, Rudolf Carnap, and Charles Morris, eds. International Encyclopedia of Unified Science. Volume I, Nos. 1-10.
Chicago: University of Chicago Press, 1938-1955. 760 pp. [Combined
edition of 1955 published in 2 volumes.]

    a. Review essay by G. Buchdahl. AJP 35 (1957) 60-67.

1066. Nicholas, J. M., ed. Images, Perception and Knowledge.
WOSPS 8. Dordrecht: Reidel, 1977. 309 pp.

1067. Objectivité et realité dans les différentes sciences: Colloque
de L'Académie Internationale de Philosophie des Sciences. Bruxelles:
Publié sous les auspices de la revue Dialectica, Office Internationale
de Librarie, 1966. 242 pp.

1068. Pappas, George S., and Marshall Swain, eds. Essays on Knowledge
and Justification. Ithaca, NY: Cornell University Press, 1978.
380 pp.

1069. Pelč, Jerzy. Semiotics in Poland: 1894-1969. SL 119.
Dordrecht: Reidel, 1977. 504 pp.

1070. Philosophy of Science. The Philosophy of Science Institute
Lectures, St. John's University. Jamaica, NY: St. John's University Press, 1960. 164 pp.

1071. Philosophy of Science: The Delaware Seminar, Vol. I, 1961-1962.
Bernard Baumrin, ed. New York: Interscience Publishers, 1963.
370 pp.

1072. Philosophy of Science: The Delaware Seminar, Vol. II, 1962-1963.
Bernard Baumrin, ed. New York: Interscience Publishers, 1963.
551 pp.

1073. Problèmes de Philosophie des Sciences. (Premier Symposium --
Bruxelles, 1947), Archives de l'Institut International des Sciences
Theorétiques, Série A, Bulletin de l'Académie Internationale de
Phiosophie des Sciences (Actualités Scientifiques et Industrielles.)
Paris: Hermann et c^{te}, 1948. 7 volumes.

1074. Proceedings of the Boston Colloquium for the Philosophy of
Science, 1961-1962. Marx W. Wartofsky, ed. BSPS 1. SL 6. Dordrecht: Reidel, 1963. 212 pp.

1075. Proceedings of the Boston Colloquium for the Philosophy of
Science, 1962-1964. In Honor of Philipp Frank. Robert S. Cohen and
Marx W. Wartofsky, eds. BSPS 2. SL 10. New York: Humanities
Press, 1965. 475 pp.

1076. Proceedings of the Boston Colloquium for the Philosophy of
Science, 1964-1966. In Memory of Norwood Russell Hanson. Robert S.
Cohen and Marx W. Wartofsky, eds. BSPS 3. SL 14. Dordrecht:
Reidel, 1967. 489 pp.

1077. Proceedings of the Boston Colloquium for the Philosophy of
Science, 1966-1968. Robert S. Cohen and Marx W. Wartofsky, eds.
BSPS 4. SL 18. Dordrecht: Reidel, 1969. 537 pp.

1078. Proceedings of the Boston Colloquium for the Philosophy of
Science, 1966-1968. Robert S. Cohen and Marx W. Wartofsky, eds.
BSPS 5. SL 19. Dordrecht: Reidel, 1969. 482 pp.

1079. Proceedings of the Boston Colloquium for the Philosophy of
Science, 1969-1972. Methodological and Historical Essays in the
Natural and Social Sciences. Robert S. Cohen and Marx W Wartofsky,
eds. BSPS 14. SL 60. Dordrecht: Reidel, 1974. 405 pp.

1080. Proceedings of the Tenth International Congress of Philosophy.
(Amsterdam, August 11-18, 1948). E. W. Beth, H. J. Pos, and J. H.
A. Hollak, eds. Amsterdam: North-Holland, 1949.

1081. Proceedings of the XIth International Congress of Philosophy.
(Brussels, August 20-27 1953). 14 Vols. Amsterdam: North-Holland;
Louvain: Editions E. Nauwelaerts, 1953.

1082. Proceedings of the XIIth International Congress of Philosophy.
(Venezia, September 12-18, 1958). 12 Vols. Firenza: Sansoni
Editore, 1958-61.

1083. Proceedings of the XIIIth International Congress of Philosophy.
(Mexico, September 7-14, 1963) 10 Vols. Universidad Nacional
Autónoma de México, 1963.

1084. Proceedings of the XIVth International Congress of Philosophy.
(Vienna, September 2-9, 1968). 6 Vols. Vienna: Herder, 1968.

1085. Proceedings of the XVth World Congress of Philosophy (Varna,
Bulgaria, September 17-22, 1973). 6 Vols. Sofia, 1973.

1086. Przełecki, Marian and Ryszard Wójcicki, eds. Twenty-five
Years of Logical Methodology in Poland. SL 87. Dordrecht: Reidel,
1977. 803 pp.

1087. Przełecki, Marian, Klemens Szaniawski, and Ryszard Wójcicki,
eds. Formal Methods in the Methodology of Empirical Sciences. SL
103. Dordrecht: Reidel, 1976. 455 pp.

1088. PSA 1970: In Memory of Rudolf Carnap. Roger C. Buck and
Robert S. Cohen, eds. BSPS 8. SL 39. Dordrecht: Reidel, 1971.
615 pp.

1089. Proceedings of the 1972 Biennial Meeting, Philosophy of Science Association. Kenneth F. Schaffner and Robert S. Cohen, eds. BSPS 20. SL 64. Dordrecht: Reidel, 1974. 444 pp.

1090. PSA 1974: Proceedings of the 1974 Biennial Meeting of the Philosophy of Science Association. R. S. Cohen, C. A. Hooker, A. C. Michalos, and J. W. van Evra, eds. BSPS 32. SL 101. Dordrecht: Reidel, 1976. 734 pp.

   a. Review essay by W. Balzer. Grazer Philosophiche Studien. 6 (1978) 169-77.

1091. PSA 1976: Proceedings of the 1976 Biennial Meeting of the Philosophy of Science Association. Vol. I: Contributed Papers and Special Sessions. Vol. II: Symposia. F. Suppe and P. D. Asquith, eds. East Lansing, MI: Philosophy of Science Association, 1976, 1977. I: xxiv + 312; II: x + 618 pp.

1092. PSA 1978: Proceedings of the 1978 Biennial Meeting of the Philosophy of Science Association. Vol. I: Contributed Papers. Vol. II: Symposia. P. D. Asquith and Ian Hacking, eds. East Lansing, MI: Philosophy of Science Association, 1978, 1981. I: xxv + 313; II: ix + 578 pp.

1093. PSA 1980: Proceedings of the 1980 Biennial Meeting of the Philosophy of Science Association. Vol. I: Contributed Papers. Vol. II: Symposia. P. D. Asquith and R. N. Giere, eds. East Lansing, MI: Philosophy of Science Association, 1980, 1981. I: xxvi + 370; II: xviii + 678 pp.

1094. Putnam, Hilary. Mathematics, Matter and Method. Philosophical Papers, Vol. I. Mind, Language and Reality. Philosophical Papers, Vol. II. Cambridge: Cambridge University Press, 1975.

   a. Review essay by Robert Farrell. AJP 58 (1980) 109-77.

1095. Rapport, Samuel, and Helen Wright, eds. Science: Method and Meaning. New York: New York University Press, 1963. xii + 258 pp.

1096. Reichenbach, Hans. Modern Philosophy of Science: Selected Essays. Maria Reichenbach, tr. and ed. Foreward by Rudolf Carnap. New York: Humanities Press, 1959. x + 214 pp.

1097. _____. Collected Works. A. Kamlah and M. Reichenbach, eds. Braunschweig, subsequently Wiesbaden: Vieweg, 1977 --. [Nine volumes, in German, are projected and include German translations of writings originally published in English.]

1098. Reichenbach, Maria, and Robert S. Cohen, eds. Hans Reichenbach: Selected Writings, 1909-1953. 2 Vols. Dordrecht: Reidel, 1978. Vol. I: 500 pp. Vol. II: 434 pp. [Volume 4 of the Vienna Circle Collection.]

1099. Rescher, Nicholas, ed. Studies in the Philosophy of Science. American Philosophical Quarterly. Monograph Series, No. 3. Oxford: Basil Blackwell, 1969. 208 pp.

1100. Rescher, Nicholas, et al., eds. Essays in Honors of Carl G.
Hempel. A Tribute on the Occasion of His Sixty-Fifth Birthday. SL
24. Dordrecht: Reidel, 1969. 272 pp.

1101. Rombach, Heinrich, ed. Wissenschaftstheorie. Bd. I: Probleme
und Positionen der Wissenschaftstheorie. Bd. II: Struktur und Methode
der Wissenschaft. Freiburg: Herder, 1974. 183, 190 pp.

1102. Saarinen, Esa, Risto Hilpinen, Ilkka Niiniluoto, and Merrill
Provence Hintikka, eds. Essays in Honor of Jaakka Hintikka on the
Occasion of His Fiftieth Birthday. SL 124. Dordrecht: Reidel,
1978. 378 pp.

1103. Salmon, Wesley C., ed. Hans Reichenbach: Logical Empiricist.
SL 132. Dordrecht: Reidel, 1979. 782 pp.

1104. Schächter, Josef. Prolegomena to a Critical Grammar. P.
Foulkes, tr. Dordrecht: Reidel, 1973. 161 pp. [Volume 2 of the
Vienna Circle Collection.]

1105. Schilpp, Paul Arthur, ed. The Philosophy of Bertrand Russell.
Library of Living Philosophers, Vol. 5. Evanston and Chicago:
Northwestern University Press, 1944. Reprinted in 2 volumes: New
York: Harper & Row, 1963, with addenda and revisions to the biblio-
graphy in fourth edition, 1971. xx + 829 pp.

    a. Review essay by Virgil C. Aldrich. J. Phil. 42 (1945)
    594-607.

    b. Review essay by L. J. Russell. Philosophy. 20 (1945)
    172-82.

1106. _____ . Albert Einstein: Philosopher Scientist. New
York: Tudor, 1949, 1951. Reprint, in two volumes: New York:
Harper & Row, 1959. 781 pp.

    a. Review essay by Leo A. Foley, S. M. New Scholasticism.
    25 (1951) 318-26.

    b. Review essay by F. Kröner. Dialectica. 7 (1953) 61-69.

1107. _____ . The Philosophy of C. D. Broad. New York: Tudor
Publishing Co., 1959. xii + 866 pp.

    a. Review essay, "On Broad's Contribution to Philosophy," by
    J. O. Wisdom. BJPS 13 (1962) 70-75.

    b. Review essay by Robert Brown. AJP 40 (1962) 83-96.

    c. Review essay by K. W. Rankin. Mind. 71 (1962) 117-23.

    d. Review essay by A. C. Ewing. Philosophy. 38 (1963)
    78-82.

e. Review essay, "The Philosophy of C. D. Broad," by R. M.
Yost. Phil. Review. 77 (1968) 474-90.

1108. _____. The Philosophy of Rudolf Carnap. Library of
Living Philosophers, Vol. 11. La Salle, Illinois: Open Court;
London: Cambridge University Press, 1963. xvi + 1088 pp.

a. "The Philosophy of Rudolf Carnap," by N. L. Wilson.
Dialogue. 4 (1965) 102-12.

b. "Rudolf Carnap," by Peter Achinstein. Review of Metaphysics. 19 (1966) I: 517-49; II: 758-79.

c. "Empiricism and Ontology in Rudolf Carnap's Thought," by
Robert H. Kane. IPQ 7 (1967) 138-76.

d. Review essay by J. J. C. Smart. AJP 43 (1965) 84-96.

e. Review essay by Stephan Körner. Mind. 75 (1966) 285-92.

f. Review essay by Henry E. Kyburg, Jr. J. Phil. 65 (1968)
503-15.

1109. Science, philosophie, foi: Colloque de l'Académie Internationale de Philosophie des Sciences, 8-11 septembre 1971, Bienne.
Brussels: Office International de Libraire, 1974. 243 pp.

1110. Seeger. R. J. and Robert S. Cohen, eds. Philosophical Foundations of Science. Proceedings of an AAAS Program, 1969. BSPS 11.
SL 58. Dordrecht: Reidel, 1974. 545 pp.

1111. La sémantique dans les sciences. Colloque de l'Académie Internationale de philosophie des sciences, 30 août-3 september, 1974,
Rixensart. Brussels: Office International de Libraire, 1978.
233 pp.

1112. Shea, William R., ed. Basic Issues in the Philosophy of
Science. New York: Science History Publications, 1976. 215 pp.

1113. Shils, Edward, ed. The Logic of Personal Knowledge: Essays
Presented to Michael Polanyi on His Seventhieth Birthday. Glencoe,
Illinois: Free Press, 1960. 248 pp.

1114. Simposio de lógica y filosofia de la ciencia, Valencia, 1971.
Filosofia y ciencia en al pensamiento español contemporaneo (1960-1970).
Madrid: Editorial Tecnos, 1973. 324 pp.

1115. Smith, Vincent E., ed. The Logic of Science. New York: St.
John's University Press, 1964. 90 pp.

1116. Sosa, Ernest, ed. The Philosophy of Nicholas Rescher. Dordrecht: Reidel, 1979. xi + 236.

1117. Stauffer, Robert C., ed. Science and Civilization. Madison,
Wisconsin: University of Wisconsin Press, 1949. xvi + 212 pp.

1118. Stegmüller, Wolfgang. Collected Papers on Epistemology, Philosophy of Science, and History of Philosophy. 2 Vols. SL 91. Dordrecht: Reidel, 1977. I: xiv + 240; II: xiv + 285 pp.

1119. Stenius, Erik. Critical Essays, in Acta Philosophica Fennica. 25 (1972) 256 pp.

1120. Suppes, Patrick. Studies in the Methodology and Foundations of Science: Selected Papers from 1911 to 1969. SL 22. Dordrecht: Reidel, 1969. 473 pp.

1121. La symétrie comme principe heuristique dans les différentes sciences. Colloque de l'Académie Internationale de Philosophie des Sciences, 1-3 Septembre 1967, Amsterdam. Bruxelles: Office International de Librairie, 1970. 135 pp.

1122. Synthese Library: Studies in Epistemology, Logic, Methodology, and Philosophy of Science. Jaakko Hintikka et al, eds. Dordrecht: Reidel, 1959 -.

    At the end of 1981, 153 volumes had been announced in the series, which overlaps partially with the BSPS series. For a listing of the volumes in the SL series, see Appendix II. In this bibliography each relevant volume in this series is listed separately according to subject matter.

1123. Waismann, Friedrich. Philosophical Papers. B. McGuinness, ed. Dordrecht: Reidel, 1977. 190 pp. [Volume 8 of the Vienna Circle Collection.]

1124. Weingartner, Paul, ed. Deskription, Analytizität und Existenz. Drittes und Viertes Forschungsgespräch des Internationalen Forschungszentrums Salzburg. Salzburg und München: Verlag Anton Pustet, 1966. 414 pp.

1125. _____. Grundfragen der Wissenschaften und ihre Wurzeln in der Metaphysik. Salzburg -- Munchen: Universitats - Verlag Anton Pustet, 1967. 256 pp.

1126. Weingartner, Paul and Gerhard Zecha, eds. Induction, Physics, and Ethics: Proceedings and Discussions of the 1968 Salzburg Colloquium in the Philosophy of Science. SL 31. Dordrecht: Reidel, 1970. 382 pp.

1127. Weisheipl, James A., ed. The Dignity of Science: Studies in the Philosophy of Science Presented to William Humbert Kane, O.P. Baltimore: The Thomist Press, 1961. 526 pp. Reprinted from The Thomist 25, Nos. 2, 3 and 4 (1961). 137-656.

1128. Wiatr, Jerzy J., ed. Polish Essays in the Methodology of the Social Sciences. BSPS 29. SL 131. Dordrecht: Reidel, 1979. 260 pp.

1129. Wiener, P. P., and Noland, A., ed. Roots of Scientific Thought. A Cultural Perspective. New York: Basic Books, 1957. x + 677 pp.

1130. Wigner, E. P. Symmetries and Reflections: Scientific Essays.
Cambridge, MA: MIT Press, 1970.

   a. "Essay Review" by Peter Kirschenmann. SHPS 4 (1973)
   193-207.

1131. Wojciechowski, Jerzy A., ed. Conceptual Basis of the Classification of Knowledge. Proceedings of the Ottawa Conference, 1971.
Les fondements de la classification des savoirs. Actes du Colloque
d'Ottawa 1971. München: Verlag Dokumentation, 1974.

   a. Review essay by Normand Lacharité. Dialgoue. 17 (1978)
   499-512.

1132. Yourgrau, Wolfgang, and Allen D. Breck, eds. Physics, Logic,
and History. New York: Plenum Press, 1970. 336 pp.

3.3 Textbooks

1133. Ackermann, Robert. Philosophy of Science: An Introduction.
New York: Pegasus, 1970. 166 pp.

1134. Brodbeck, May, ed. Readings in the Philosophy of the Social
Sciences. New York: Macmillan, 1968. 789 pp.

1135. Brody, Baruch A., ed. Readings in the Philosophy of Science.
Englewood Cliffs, NJ: Prentice-Hall, 1970. 637 pp.

1136. Brody, Baruch A., and Nicholas Capaldi, eds. Science: Men,
Methods, Goals. A Reader: Methods of Physical Science. New York:
W. A. Benjamin, 1968. 343 pp.

1137. Brown, Harold I. Perception, Theory and Commitment: The New
Philosophy of Science. Chicago: University of Chicago Press, 1977.
203 pp.

1138. Caws, Peter. The Philosophy of Science: A Systematic Account.
Princeton, NJ: D. van Nostrand, 1965. xii + 354 pp.

1139. Danto, Arthur, and Sidney Morgenbesser, eds. Philosophy of
Science. Cleveland and New York: World Publishing Co., 1960.
477 pp.

1140. Durbin, Paul R., ed. Philosophy of Science: An Introduction.
New York: McGraw-Hill, 1968. 271 pp.

1141. Feigl, Herbert, and May Brodbeck, eds. Readings in the Philosophy of Science. New York: Appleton-Century-Crofts, 1953. 811 pp.

1142. Feigl, Herbert and Wilfrid Sellars, eds. Readings in Philosophical Analysis. New York: Appleton-Century-Crofts, 1949.
626 pp.

1143. Gale, George. Theory of Science: An Introduction to the History, Logic, and Philosophy of Science. New York: McGraw-Hill, 1979. xiv + 298 pp.

1144. Giere, Ronald N. Understanding Scientific Reasoning. New York: Holt, Rinehart and Winston, 1979. 371 pp.

1145. Hacking, Ian, ed. Scientific Revolutions. Oxford and New York: Oxford University Press, 1981. 180 pp.

1146. Hegenberg, Leônidas. Explicações Científicas: Introdução à Filosofia da Ciência. São Paulo, Brazil: Editôra Herder, 1969. xv + 308 pp.

1147. Hempel, Carl G. Philosophy of Natural Science. Englewood Cliffs, NJ: Prentice Hall, 1966. 116 pp. French translation by B. Saint-Sernin under title Éléments d'épistémologie. Paris: Armand Colin, 1972. 184 pp.

1148. Kahl, Russell, ed. Studies in Explanation: A Reader in the Philosophy of Science. Englewood Cliffs, NJ: Prentice-Hall, 1963. 363 pp.

1149. Klemke, E. D., Robert Hollinger, and A. David Kline, eds. Introductory Readings in the Philosophy of Science. Buffalo, NY: Prometheus Books, 1980. 393 pp.

1150. Lambert, Karel, and Gordon G. Brittan, Jr. An Introduction to the Philosophy of Science. Englewood Cliffs, NJ: Prentice-Hall, 1970. 113 pp. 2nd Edition: Reseda, CA: Ridgeview, 1979. 164 pp.

1151. Losee, J. P. An Historical Introduction to the Philosophy of Science. Second edition. New York: Oxford University Press, 1980. 250 pp.

1152. Mannoia, Jr., V. James. What is Science? An Introduction to the Structure and Methodology of Science. Washington, D. C.: University Press of America, 1980. 140 pp.

1153. Meyard, Léon. Logique et philosophie des sciences. 4. ed. Paris: E. Belin, 1958. 319 pp.

1154. Michalos, Alex C., ed. Philosophical Problems of Science and Technology. Boston: Allyn and Bacon, 1974. 623 pp.

1155. Nidditch, P. H., ed. The Philosophy of Science. London: Oxford University Press, 1968. 184 pp.

1156. Quine, W. V., and J. S. Ullian. The Web of Belief. New York: Random House, 1970. Second edition, 1978. 147 pp.

1157. Shapere, Dudley, ed. Philosophical Problems of Natural Science. New York: Macmillan; London: Collier-Macmillan, 1965. 124 pp.

1158. Slaatte, Howard A. Modern Science and the Human Condition.
Washington, D. C.: University Press of America, 1981. 230 pp.

1159. Theobald, D. W. An Introduction to the Philosophy of Science.
London: Methuen, 1968. xiii + 145 pp.

1160. Tweney, Ryan D., Michael E. Doherty, and Clifford R. Mynatt,
eds. On Scientific Thinking. Irvington, NY: Columbia University
Press, 1981. 472 pp.

1161. Walcott, Gregory Dexter. Logic and Scientific Method. Ann
Arbor, MI: Edwards Brothers, 1952. xii + 182 pp.

1162. Weingartner, Paul. Wissenschaftstheorie I: Einführung in die
Hauptprobleme; II, 1: Grundlagenprobleme der Logik und Mathematik.
Stuttgart-Bad, Cannstatt: Fromann-Holzboog, 1971, 1976. 246,
364 pp.

1163. Wiener, Philip P., ed. Readings in the Philosophy of Science.
New York: Scribner, 1953. 645 pp.

4.
Aspects of Scientific Method

4.1 Scientific Method—General

1164. Abercrombie, M. L. Johnson. The Anatomy of Judgment. London: Hutchinson, 1960. 156 pp.

1165. Abruzzi, Adam. "Problems of Inference in the Socio-Physical Sciences." J. Phil. 51 (1954) 537-49.

1166. Agassi, Joseph. "The Logic of Scientific Inquiry." Synthese. 26 (1974) 498-514. Reprinted in BSPS 65 (1981) 223-38.

1167. _____. "Between Metaphysics and Methodology." Poznań Studies. 1, No. 4 (1975). 1-8.

1168. _____. "Assurance and Agnosticism." BSPS 32 (1976). 449-57.

1169. Ajdukiewicz, K. "Methodology and Metascience." SL 87 (1977) 1-12. [First published in Życie Nauki (1948)].

1170. _____. Pragmatic Logic. Translated from the Polish by Olgierd Wojtasiewicz. SL 62. Dordrecht: Reidel, 1974. 460 pp.

1171. Akchurin, I. A. "The Place of Mathematics in the System of the Sciences." Soviet Studies in Philosophy. 6, No. 3 (1967-68) 3-13. [English reprint from Voprosy filosofii. (1967, No. 1).]

1172. Anderson, Jr., Robert M. "Paradoxes of Cosmological Self-Reference." MSPS 6 (1975) 530-40.

1173. Aron, Robert. Discours contre la méthode. Paris: Plon, 1974. 317 pp.

1174. Ballard, Edward G. "Concerning Method in Philosophy and Science." Personalist. 34 (1953) 269-77.

1175. Bechert, Karl. "Methoden der Naturerkenntnis. Ein Vortrag."
PN 2 (1952) 224-38.

1176. Beck, Lewis White. "The Distinctive Traits of an Empirical
Method." J. Phil. 44 (1947) 337-44.

1177. _____. "Constructions and Inferred Entities." PS 17
(1950) 74-86.

1178. Bernays, Paul. "Bemerkungen zur Rolle der Methode in den
Wissenschaften." Archives de Philosophie. 34 (1971) 575-80.

1179. Beveridge, W. I. B. The Art of Scientific Investigation.
London: William Heinemann, 1950. xii + 171 pp.

1180. Birjukov, B. V. Kibernetika i metodologija nauki. [Cybernetics
and the Methodology of Science] Moskva: Izd. 'Nauka', 1974.
414 pp.

1181. Black, Max. Problems of Analysis. Ithaca, NY: Cornell
University Press, 1954. 304 pp.

1182. Blanché, Robert, ed. La méthode expérimentale et la philosophie
de la physique. Paris: Librairie Armand Colin, 1969. 384 pp.

1183. Block, N. J. "Fictionalism, Functionalism and Factor Analysis."
BSPS 32 (1976) 127-42.

1184. Bochenski, J. M. Die zeitgenössischen Denkmethoden. Bern:
Franke Verlag. English tr. by Peter Caws under title The Methods
of Contemporary Thought. Dordrecht: Reidel, 1965. Reprint: New
York: Harper & Row, 1968. 134 pp.

1185. Bohm, David. "On the Relationship between Methodology in
Scientific Research and the Content of Scientific Knowledge." BJPS
12 (1961) 103-16.

1186. Bork, Alfred Morton. "Methodology of the Empirical Sciences."
PS 26 (1959) 31-34.

1187. Broglie, Louis de. "Deduction et induction dans la recherche
scientifique." Scientia. 90 (1955) 147-49.

1188. Brown, G. Burniston. Science: Its Method and Its Philosophy.
New York: W. W. Norton, 1950. 189 pp.

1189. Buchler, Justus. The Concept of Method. New York: Columbia
University Press, 1961. 180 pp.

1190. Bunge, Mario. "A General Black Box Theory." PS 30 (1963)
346-58.

1191. Cannavo, Salvator. Nomic Inference: An Introduction to the
Logic of Scientific Inquiry. The Hague: M. Nijhoff, 1974. xii +
331 pp.

1192. Churchman, C. West. _Theory of Experimental Inference_. New York: Macmillan, 1948. 292 pp.

    a. Review essay: "Relativism and Experimental Inference," by Frederick L. Will. PS 18 (1951) 155-89.

1193. _____. _The Design of Inquiring Systems_. New York: Basic Books, 1971. ix + 288 pp.

1194. Churchman, C. West, and Russell L. Ackoff. _Methods of Inquiry: An Introduction to Philosophy and Scientific Method_. Saint Louis: Educational Publishers. 549 pp.

1195. Cohen, Morris R. _Reason and Nature: An Essay on the Meaning of Scientific Method_. 2nd ed. Glencoe, Illinois: The Free Press, 1953. xxiv + 470 pp.

1196. Cohen, Robert S. "Tacit, Social and Hopeful," in _Interpretations of Life and Mind_. Marjorie Grene, ed. New York: Humanities Press, 1971. 137-48.

1197. Corcoran, John. "Gaps between Logical Theory and Mathematical Practice," in _The Methodological Unity of Science_. M. Bunge, ed. Dordrecht: Reidel, 1973. 23-50.

1198. Crescini, Angelo. _Il senso della ricerca scientifica_. Rome: Ed. dell'Ateneo & Bizzarri, 1978. 425 pp.

1199. Curthoys, J., and W. Suchting. "Feyerabend's Discourse against Method: A Marxist Critique." _Inquiry_. 20 (1977) 243-371.

    a. "Marxist Fairytales from Australia," by Paul Feyerabend. 372-97.

    b. "Rising Up from Downunder," by W. Suchtung. 21 (1978) 337-47.

1200. Dantzig, D. van. "General Procedures of Empirical Science." _Synthese_. 5 (1947) 441-55.

1201. Destouches, Jean-Louis. _Méthodologie: Notions géometriques_. Paris: Gauthier-Villars, 1953. xi + 228 pp.

    a. Review essay by G. H. Müller. _Dialectica_. 7 (1953) 255-75.

1202. _____. "Logique et théorie physique." _Synthese_. 16 (1966) 66-73.

1203. _____. "La méthode de la recherche théorique et la construction des théories." _Archives de Philosophie_. 34 (1971) 581-92.

1204. Dürr, Karl. "Logik als Forschungsmethode." _Synthese_. 7 (1948-49) 27-31.

1205. Feibleman, James K. "Mathematics and Its Applications in the
Sciences." PS 23 (1956) 204-15.

1206. _____. "The Logical Structure of the Scientific Method."
Dialectica. 13 (1959) 208-25.

        a. "Quelques remarques sur l'article de M. Feibleman," by F.
        Gonseth, 226-34.

1207. _____. "The Psychology of the Scientist." Synthese.
12 (1960) 79-113.

1208. _____. Scientific Method: The Hypothetico-Experimental
Laboratory Procedure of the Physical Sciences. The Hague: Martinus
Nijhoff, 1972. vi + 246 pp.

1209. Feibleman, James, and Julius Friend. "Normative Organization
and Empirical Fields." PS 12 (1945) 52-56.

1210. Feigl, Herbert. "Scientific Method without Metaphysical
Presuppositions." Phil. Studies. 5 (1954) 17-29.

1211. Feyerabend, Paul K. "On the Improvement of the Sciences and
Arts, and the Possible Identity of the Two." BSPS 3 (1967) 387-415.

        a. "Comments," by Peter Achinstein, 416-24; by Norman Rudich,
        425-32; by Marx W. Wartofsky, 433-39.

1212. _____. "Science, Freedom and the Good Life." Philosophical Forum. 1 (1968) 127-35.

1213. _____. "Against Method: Outline of an Anarchistic
Theory of Knowledge." MSPS 4 (1970) 17-130.

1214. _____. "Von der beschränkten Gültigkeit methodologischer
Regeln." Neue Hefte für Philosophie. 2/3 (1972) 124-71.

1215. _____. Against Method: Outline of an Anarchistic Theory
of Knowledge. Atlantic Highlands, NJ: Humanities Press, 1974.
339 pp.

        a. "Feyerabend's Against Method: The Case for Methodological
        Pluralism," by Paul Tibbetts. Philosophy of the Social Sciences.
        7 (1977) 265-75.

        b. "How Far does Anything Go? Comments on Feyerabend's
        Epistemological Anarchism," by Tomas Kulka. 227-87.

        c. "The Crisis in Methodology: Feyerabend," by J. N. Hattian-
        gadi. 289-302.

        d. Review essay, "Beyond Truth and Falsehood," by Ernest
        Gellner. BJPS 26 (1975) 331-42.

e. "Logic, Literacy, and Professor Gellner," by Paul Feyerabend. 27 (1976) 381-91.

f. Review essay, "Against Bad Method," by Geoffrey Hellman. Metaphilosophy. 10 (1979) 190-202.

g. "Reply to Hellman's Review," by P. K. Feyerabend. 202-06.

h. "Reply to Feyerabend: From Bad to Worse," by Geoffrey Hellman, 206-07.

i. Review essay, "Against Too Much Method," by John Worrall. Erkenntnis. 13 (1978) 279-95.

j. "Life at the LSE?" by Fantomas. 297-304.

k. Review essay by Joseph Agassi. Philosophia. 6 (1976) 165-77. "Comments and Replies," by Paul K. Feyerabend and Joseph Agassi. 177-91.

l. "Against Against Method; or, Consolations for the Rationalist," by Husain Sarkar. Southwestern Journal of Philosophy. 9, No. 1 (1978) 35-44.

m. Review essay, "Feyerabend's Method," by Hugo Meynell. Phil. Quart. 28 (1978) 242-52.

n. Review essay, "Ist die Vernuft am End?" by Klaus Fischer. ZPF 32 (1978) 387-97.

o. Review essay by G. Petersen-Falshöft. Philosophische Rundschau. 27 (1980) 101-17.

1216. _____. Wider den Methodenzwang. Skizze einer anarchistischen Erkenntnistheorie. Frankfurt a. M.: Suhrkamp Verlag, 1976. 443 pp.

a. Review essay by Wulf Rehder. ZAW 9 (1978) 404-13.

1217. _____. "Dialogue on Method." BSPS 59. SL 136. (1979) 63-131.

1218. _____. Realism, Rationalism, and Scientific Method. Cambridge: Cambridge University Press, 1981.

1219. Finetti, B. de. "Expérience et théorie dans l'élaboration d'une doctrine scientifique." RMM 60 (1955) 264-86.

1220. Flach, Werner. Thesen zum Begriff der Wissenschaftstheorie. Bonn: Bouvier, 1979. 106 pp.

1221. Focken, C. M. Dimensional Methods and Their Applications. Foreward by H. Dingle. London: Edward Arnold, 1953. viii + 224 pp.

1222. Fowler, W. S. The Development of Scientific Method. Oxford: Pergamon Press, 1962. xiii + 116 pp.

1223. Freedman, Paul. The Principles of Scientific Research. London: Macdonald & Co., 1949. 222 pp.

1224. Frey, Gerhard. Philosophie und Wissenschaft: Eine Methoden-lehre. Stuttgart: Kohlhammer Verlag, 1970. 144 pp.

1225. _____. "Handlungsbedingungen wissenschaftlicher Forschung." PN 18 (1980-81) 38-49.

1226. Friedlander, M. W. "Some Comments on Velikovsky's Methodology." BSPS 32 (1976) 477-86.

1227. Gellner, Ernest. "An Ethic of Cognition." BSPS 39 (1976) 161-78.

1228. George, Pulivelil M. "Conceptualization: The Central Problem of Science." Organon. 9 (1973) 23-33.

1229. Girill, T. R. "The Logic of Scientific Puzzles." ZAW 4 (1973) 25-40.

1230. Gonseth, F. "De la méthode dans les sciences et en philosophie." RIP 3 (1949) 203-13.

1231. _____. "Recherches methodologiques." Dialectica. 9 (1955) 137-85.

1232. _____. "La méthodologie des sciences peut-elle être élevée au rang de discipline scientifique?" Dialectica. 11 (1957) 9-20.

1233. _____. Le probleme du temps: Essai sur la méthodologie de la recherche. Neuchâtel: Ed. du Griffon, 1964. 382 pp. English tr. by Eva H. Guggenheimer: Time and Method: An Essay on the Methodology of Research. Springfield, IL: Thomas, 1972. xiii + 453 pp.

   a. Review essay by Eric Emery. Dialectica. 23 (1969) 32-44, 115-34.

1234. Good, I. J. "The Function of Speculation in Science." Theoria to Theory. 1 (1966) 28-43.

1235. Gordon, T. J., and M. J. Raffensperger. "A Strategy for Planning Basic Research." PS 36 (1969) 205-18.

1236. Grünfeld, J. "Method and Language." Science et Esprit. 30 (1978) 185-93.

1237. Grzegorczyk, A. "Nieklasyczne rachunki zdań a metodologiczne schematy badania naukowego i definicje pojęć asertywnych." [Non-classical Propositional Calculi in relation to Methodological Patterns of Scientific Investigation] Studia Logica. 20 (1967) 117-32.

1238. Guggenheimer, H. "Some Comments on Gonseth's Principles of a Methodology of Science." Dialectica. 24 (1970) 23-28.

1239. Hampshire, Stuart. "Some Difficulties in Knowing," in Philsophy, Science, and Method. S. Morgenbesser et al. eds. New York: St. Martin's Press, 1969. 26-47.

1240. Hanson, Norwood Russell. Perception and Discovery: An Introduction to Scientific Inquiry. Willard C. Humphreys, ed. San Francisco: Freeman, Cooper & Co., 1969. 435 pp.

1241. Harris, Errol E. Hypothesis and Perception: The Roots of Scientific Method. London: George Allen & Unwin, 1970. 395 pp.

    a. Review essay, "Harris on the Logic of Science," by Harold
    I. Brown. Dialectica. 26 (1972) 227-46.

1242. _____. "Dialectic and Scientific Method." Idealistic Studies. 3 (1973) 1-17.

1243. Hélal, Georges. "L'Herméneutique de la science et son rapport au fondement de la connaissance." Dialogue. 10 (1971) 60-81.

1244. Hempel, Carl G. "Typological Methods in the Natural and the Social Sciences," in Science, Language, and Human Rights. Philadelphia: University of Pennsylvania Press. 1952. Reprinted in C. G. Hempel, Aspects of Scientific Explanation and other Essays in the Philosophy of Science. New York: Free Press, 1965. 155-71.

1245. Hesse, Mary. The Structure of Scientific Inference. London: Macmillan, 1974. viii + 309 pp.

    a. "Essay Review: Epistemology or Psychology," by David
    Bloor. SHPS 5 (1975) 382-95.

    b. Review essay by Jon Dorling. BJPS 26 (1975) 61-71; 27
    (1976) 160-61.

1246. _____. "Models of Method in Natural and Social Science." Methodology and Science. 8 (1975) 163-78.

1247. Heyde, Johannes Erich. "Annahme und Möglichkeit - Vorgedanken zur Wissenschaftstheorie der Hypothese." PN 4 (1957) 223-44.

1248. Hintikka, Jaakko and Patrick Suppes. Information and Inference. SL 28. Dordrecht: Reidel, 1970. 336 pp.

1249. Hintikka, Jaakko and Unto Remes. "Ancient Geometrical Analysis and Modern Logic." BSPS 39 (1976) 253-76.

1250. Holton, Gerald. "Analysis and Synthesis as Methodological Themata." Methodology and Science. 10 (1977) 3-33.

1251. Hooker, C. A. "Methodology and Systematic Philosophy." WOSPS 11 (1977) 3-23.

1252. Howson, Colin. "Methodology in Non-Empirical Disciplines."
BSPS 59. SL 136. (1979) 257-66.

1253. Hübner, K. "Wissenschaftliche und nichtwissenschaftliche
Naturerfahrung." PN 18 (1980-81) 67-86.

1254. Hull, L. W. H. History and Philosophy of Science. New York:
Longmans, Green, 1959. 325 pp.

1255. Ionescu-Pallas, Nicholas, and Liviu Sofonea. "The Manysided-
ness of the Scientific Approach. Historico-epistemological Modelling.
Examples from Modern Physics." Noesis. 5 (1979) 153-68.

1256. Jeffreys, Harold. Scientific Inference. Second edition.
New York: Cambridge University Press, 1957. 236 pp. 3rd edition,
1973. 273 pp.

      a. Review essay, "Probability and Induction," by E. H.
      Hutten. BJPS 9 (1958) 43-51.

1257. Juhos, Bela. Die Erkenntnis und ihre Leistung: Die naturwissenschaftliche Methode. Wien: Springer-Verlag, 1950. vi +
263 pp.

1258. _____. "Die neue Logik als Voraussetzung der wissenschaftlichen Erkenntnis." Studium Generale. 6 (1953) 593-99.

1259. _____. "Die triadische Methode." Studium Generale. 24
(1971) 924-45.

1260. Kaiser, C. Hillis. An Essay on Method. New Brunswick, N.J.:
Rutgers University Press, 1952. vi + 163 pp.

1261. Kattsoff, L. O. "The Role of Hypothesis in Scientific Investigation." Mind. 58 (1949) 222-27.

1262. Kedrov, B. M. "Methodological Problems of Natural Science."
Soviet Studies in Philosophy. 3, No. 2 (1964) 3-14. [English
reprint from Voprosy filosofii. (1964, No. 3)]

1263. Koertge, Noretta. "Rational Reconstructions." BSPS 39
(1976) 359-70.

1264. König, Hans. "Sur la possibilité d'attribuer une signification générale à certaines règles des sciences expérimentales."
Dialectica. 4 (1950) 322-30.

1265. Kotarbiński, T. "Concepts and Problems in General Methodology
and Methodology of the Practical Sciences." SL 87 (1977) 279-89.
[First published in Studia Filozoficzne. 1 (74) (1972).]

1266. Kottinger, Wolfgang. "Das Begreifen von Methode." Wiener
Jahrbuch für Philosophie. 6 (1973) 42-58.

1267. Kraft, Victor. Die Grundformen der wissenschaftlichen Methoden.
2 Neubearbeitete Auflage. Vienna: Verlag der Oesterreichischen
Akademie der Wissenschaften, 1973. 124 pp.

1268. Krajewski, W. "Idealization and Factualization in Science."
Erkenntnis. 11 (1977) 323-39.

1269. Kremmeter, A.-F. "Möglichkeiten wissenschaftlicher Frages-
tellung." PN 11 (1969) 75-84.

1270. Kroger, Joseph. "Polanyi and Lonergan on Scientific Method."
Philosophy Today. 21 (1977) 2-20.

1271. Laitko, H., and R. Bellmann, hrsg. Wege des Erkennens. Berlin: Deutscher Verlag der Wissenschaften, 1969. 294 pp.

1272. Lastrucci, Carlo L. The Scientific Approach: Basic Principles
of the Scientific Method. Cambridge, MA: Schenkman Pub. Co., 1963.
257 pp.

1273. Laudan, L. "The Sources of Modern Methodology." WOSPS 12
(1977) 3-19.

1274. Leclercq, René. Traité de la méthode scientifique. Paris:
Dunod, 1964. 215 pp.

1275. _____. Le raisonnement scientifique et sa mécanisation.
Paris: Dunod, 1969. 84 pp.

1276. Lee, Donald A. "Scientific Method as a Stage Process."
Dialectica. 22 (1968) 28-44.

1277. LeLionnais, Francois. La méthode dans les sciences modernes.
Paris: Éditions Science et Industrie, 1958. 343 pp.

1278. Lenzen, Victor F. "Procedures of Empirical Science," in
International Encyclopedia of Unified Science. Vol. I, No. 5. O.
Neurath, R. Carnap, and C. W. Morris, eds. Chicago: University of
Chicago Press, 1938-55. 279-339.

1279. Levi, Isaac. "Must the Scientist Make Value Judgments?"
J. Phil. 57 (1960) 345-57.

1280. Lins, Mario. The Logico-Systematic Structure of Science.
Rio de Janeiro: Jornal do Commercio-Rodrigues & Cia, 1954. 19 pp.

1281. _____. Perspectives for the Logico-Conceptual Integration of Science. Rio de Janeiro: Jornal do Commercio-Rodrigues &
Cia, 1954. 19 pp.

1282. _____. Functionalization of the New Logico-Conceptual
Forms. Rio de Janeiro: Jornal do Commercio-Rodrigues & Cia, 1954.
16 pp.

1283. _____. Logico-Semantical Forms of Philosophical Inquiry. Rio de Janeiro: Jornal do Commercio-Rodrigues & Cia, 1955. 35 pp. Reprinted from Archivio di Filosofia (1955, No. 3).

1284. _____. Operations of Sociological Inquiry. Rio de Janeiro: Jornal do Commercio-Rodrugues & Cia, 1956. 122 pp.

1285. Lugg, Andrew. "Overdetermined Problems in Science." SHPS 9 (1978) 1-18.

      a. "Discussion: Overdetermined Problems and Anomalies," by Paul van der Vet. 10 (1979) 259-61.

1286. McLaughlin, Andrew. "Method and Factual Agreement in Science." BSPS 8, SL 39 (1971) 459-69.

1287. Magyar, George. "Towards a Methodology of Experimental Science?" Scientia. 116 (1981) 559-76. Italian tr.: 577-89.

1288. Manara, Carlo F. Metodi della scienza dal Rinascimento ad oggi. Milan: Vita e Pensiero, 1975. 134 pp.

1289. Marcus, Solomon. "Steps of Abstraction in the Contemporary Science." Noesis. 2 (1974) 83-87.

1290. Martin, R. M. "On Objective Intensions and the Law of Inverse Variation," in Philosophy, Science, and Method. S. Morgenbesser et al. eds. New York: St. Martin's Press, 1969. 48-73.

1291. Maslow, A. H. "Problem-Centering vs. Means-Centering in Science." PS 13 (1946) 326-31.

1292. May, E. "Das Prinzip des Aktualismus in seiner generellen wissenschaftstheoretischen Bedeutung." Scientia. 90 (1955) 1-4. French tr.: Supplement, 1-4.

1293. Mercier, André. "Unité et méthode dans la science." Synthese. 5 (1946) 288-95.

1294. Mehlberg, Henry. "The Range and Limits of the Scientific Method." J. Phil. 51 (1954) 285-94.

1295. _____. "The Method of Science, its Range and Limits," in Science and Freddom. Boston: Beacon Press, 1955. 124-33.

1296. Meyer, Willi. "Wissenschaftstheorie und Erfahrung: Zur Ueberwindung des methodologischen Dogmatismus." ZAW 3 (1972) 267-84.

1297. Miescher, K. "Elements der Forschung." Dialectica. 4 (1950) 305-21.

1298. Miller, David L. "The Effect of the Concept of Evolution on Scientific Methodology." PS 15 (1948) 52-60.

1299. Misiek, Józef. "The Real and the Unreal in Methodology."
Reports on Philosophy. 2 (1978) 77-80.

1300. Molina, S. J., Antonio Ma. "Einstein's Epistemology of the
Scientific Method." Thomist. 26 (1963) 100-10.

1301. Mouloud, Noël. Les structures, la recherche et la savoir:
Réflexion sur la méthode et la philosophie des sciences exactes.
Paris: Payot, 1968. 307 pp.

1302. _____. "La science et sa logique." RIP 131-132 (1980)
155-78.

1303. Mshvenieradzé, V. V. "Objective Foundations of the Scientific
Method." Diogenes. 63 (1968) 70-88.

1304. Musgrave, A. "Method or Madness?" BSPS 39 (1975) 457-91.

    a. "Musgrave's 'Appraisal and Advice'," by Husain Sarkar.
    PS 45 (1978) 478-83.

1305. Nalimov, V. V. In the Labyrinths of Language: A Mathemati-
cian's Journey. R. G. Colodny, ed. Philadelphia: ISI Press, 1981.
246 pp.

1306. Nielsen, H. A. Methods of Natural Science: An Introduction.
Englewood Cliffs, NJ: Prentice-Hall, 1967. 70 pp.

1307. Onicescu, O. "Fonctions épistémologiques de la science."
Noesis. 7 (1981) 135-40.

1308. Ouemov, A. I. "L'analyse logique du problème des conséquences
négatives et la classification des méthodes de sa résolution." RIP
25 (1971) 528-37.

1309. Percy, Walker. "Culture: The Antinomy of Scientific Method."
New Scholasticism 32 (1958) 443-75.

1310. Petrie, Hugh G. "The Strategy Sense of 'Methodology'." PS
35 (1968) 248-57.

1311. Pfister, Frederico. Il metodo della scienza. Firenze: Sansoni, 1948. 380 pp.

1312. Pilot, Harald. "Skpetischer und kritischer Theorienpluralismus: Zu Paul Feyerabends Epistemologie." Neue Hefte für Philosophie. 6/7 (1974) 67-103.

1313. Polikarov, Azari. Metodologija na nautschnoto poznanie. 2
Vols. Sofia: 1972, 1973. 277, 312 pp.

1314. _____. Probleme der wissenschaftlichen Erkenntnis vom
methodologischen Gesichtspunkt aus. Sofia, 1977. 455 pp. [In
Bulgarian with Enlish summaries.]

1315. Radnitzky, Gérard. "Toward a Theory of Research Which is Neither Logical Reconstruction nor Psychology or Sociology of Science." Quality and Quantity. 6 (1972) 193-238.

1316. _____. "Philosophy of Science in a New Key: A System-Theoretical Approach." Methodology and Science. 6 (1973) 134-79.

1317. _____. "Vom möglichen Nutzen der Forschungstheorie." Neue Hefte für Philosophie. 6/7 (1974) 129-68.

1318. _____. "Towards a System Philosophy of Scientific Research." Philosophy of the Social Sciences. 4 (1974) 369-98.

1319. _____. "Philosophie de la recherche scientifique." Archives de Philosophie. 37 (1974) 5-76.

1320. _____. "Towards a 'System-Philosophy' Approach in Theory of Science." International Studies in Philosophy. 6 (1974) 17-48.

1321. _____. "The Intellectual Environment and Dialogue Partners of the Normative Theory of Science." Organon. 11 (1975) 5-43.

1322. _____. "Panorama critico de las teorías de la normativa de la ciencia." Pensamiento. 32 (1976) 39-83.

1323. _____. "Prinzipielle Problemstellungen der Forschungspolitik." ZAW 7 (1976) 367-403.

1324. _____. "Justifying a Theory versus Giving Good Reasons for Preferring a Theory: On the Big Divide in the Philosophy of Science." BSPS 59. SL 136 (1979) 313-56.

1325. Rashevsky, N. "The Devious Roads of Science." Synthese. 15 (1963) 107-14.

1326. Raven, Chr. P. "Ueber die Idee einer mehrfachen Komplementarität und ihre Anwendung auf die Wissenschaftslehre." Synthese. 7 (1948-49) 321-33.

1327. Reitzer, A. "Methodologische Probleme der Naturwissenschaften. Kritische Analyse der ostwissenschaftlichen und dialektisch-materialistischen Position." PN 16 (1976-77) 85-124.

1328. Rescher, Nicholas. "The Ethical Dimension of Scientific Research," in Beyond the Edge of Certainty: Essays in Contemporary Science and Philosophy. Robert G. Colodny, ed. Englewood Cliffs, NJ: Prentice-Hall, 1965. 261-76.

1329. _____. Plausible Reasoning: An Introduction to the Theory and Practice of Plausiblistic Inference. Assen/Amsterdam: Van Gorcum, 1976. 124 pp.

1330. Ritchie, Arthur David. Studies in the History and Methods of the Sciences. Edinburgh: University Press, 1958. 230 pp.

1331. Rolston, III, Holmes. "Methods in Scientific and Religious Inquiry." Zygon. 16 (1981) 29-63.

1332. Rybicki, P. "Sciences sociales et sciences naturelles. Quelques aspects méthodologiques communs." Organon. 2 (1965) 17-35.

1333. Sachsteder, W. "Structural Variation in Science." Synthese. 15 (1963) 412-23.

1334. Salmon, Wesley C. "The Foundations of Scientific Inference," in Mind and Cosmos. Robert G. Colodny, ed. Pittsburgh: University of Pittsburgh Press, 1966. 135-275.

1335. Sarkar, Husain. "Methodological Appraisals, Advice, and His-toriographical Models." Erkenntnis. 15 (1980) 371-90.

1336. Scheffler, Israel. The Anatomy of Inquiry: Philosophical Studies in the Theory of Science. New York: Alfred A. Knopf, 1963. 332 pp.

1337. Schlesinger, G. "Two Approaches to Mathematical and Physical Systems." PS 26 (1959) 240-50.

1338. _____. Method in the Physical Sciences. New York: Humanities Press, 1963. vii + 140 pp.

1339. Schmidt, Paul F. "Some Merits and Misinterpretations of Scientific Method." Scientific Monthly. 82 (1956) 20-24.

1340. _____. "Ethical Norms in Scientific Method." J. Phil. 56 (1959) 644-52.

1341. Schwartz, J. "The Pernicious Influence of Mathematics in Science." LMPS-1 (1960) 356-60.

1342. Scriven, Michael. "Science, Fact, and Value," in Philosophy of Science Today. S. Morgenbesser, ed. New York: Basic Books, 1967. 175-89.

1343. Searles, Herbert L. Logic and Scientific Method. New York: Ronald Press, 1948. 326 pp. Second edition, 1956. 378 pp.

1344. Segeth, W. "Methodische Regeln." DZP (1967) 792-806.

1345. Seiffert, August. "Aschaffenburg: Irrtum und Methode." PN 4 (1957) 500-30.

1346. Shibata, S. "Die moderne Wissenschaft und marxistische Phil-osophie. Zur Problematik der Methodologie der Wissenschaft." Poznan Studies. 3 (1977) 142-79.

1347. Simard, Emile. "L'hypothèse." Laval Théologique et Philosophique. 3 (1947) 89-120.

1348. _____. La nature et la portée de la méthode scientific.
Quebec: Les Presses Universitaires, Laval, 1956.

1349. Smith, Mapheus. "Hypothesis vs. Problem in Scientific Investigation." PS 12 (1945) 296-301.

1350. Smith, Vincent E. "Abstraction and the Empiriological Method."
Proc. Am. Cath. Phil. Assn. 26 (1952) 35-50.

1351. Stepin, W. S., and A. N. Jelsukov. Methods of Scientific
Cognition. [In Russian] Minsk, 1974. 152 pp.

1352. Subbotin, A. L. "Idealization as a Method of Scientific
Knowledge." SL 25 (1970) 376-93.

1353. Suppes, Patrick. "The Desirability of Formalization in
Science." J. Phil. 65 (1968) 651-64.

1354. _____. "Nagel's Lectures on Dewey's Logic," in Philosophy, Science, and Method. S. Morgenbesser, P. Suppes, and M. White,
eds. New York: St. Martin's Press, 1969. 2-25.

1355. Tavanec, P. V., ed. Problemi Logiki Naučnogo Poznanija.
Moscow: Nauka, 1964. [English translation by T. J. Blakeley:
Problems of the Logic of Scientific Knowledge. SL 25 (1970) 429 pp.]

1356. Tavanets, P. V., and V. S. Shvyrev. "Some Problems in the
Logic of Scientific Knowledge." Soviet Studies in Philosophy. 1,
No. 3 (1962-63) 33-40. [English reprint from Voprosy filosofii.
(1962, No. 10).]

1357. Thomson, Sir George. "Some Thoughts on the Scientific Method."
BSPS 2 (1965) 81-92.

1358. Tondl, Ladislav. Scientific Procedures. BSPS 10. SL 47.
Dordrecht: Reidel, 1973. 268 pp.

1359. Törnebohm, H. "Reflexions on Scientific Research." Scientia.
106 (1971) 225-43.

1360. _____. "Research as an Example of an Innovative System."
Scientia. 107 (1972) 245-54.

1361. _____. "Inquiring Systems." Scientia. 109 (1974)
197-211.

1362. _____. "Inquiring Systems and Paradigms." BSPS 39
(1976) 635-54.

1363. Törnebohm, Håkan, and Gérard Radnitzky. "Forschung als inno-
vatives System: Entwurf einer integrativen Schweise, die Modelle
erstellt zur Beschreibung und Kritik von Forschungsprozessen." ZAW
2 (1971) 239-90.

1364. Trusted, Jennifer. The Logic of Scientific Inference.
London: Macmillan, 1979. x + 145 pp.

1365. Tucker, John. "Against Some Methods." BSPS 39 (1976) 677-90.

1366. Wahl, D. "Zur Theorie der experimentellen Methode." DZP 16
(1968) 107-13.

1367. Weimer, Walter B. Notes on the Methodology of Scientific
Research. Hillsdale, NJ: Erlbaum; New York: Wiley, 1979. xii +
257 pp.

1368. Wilson, Jr., E. Bright. An Introduction to Scientific Research. New York: McGraw-Hill, 1952. 375 pp.

1369. Wisdom, John O. Foundations of Inference in Natural Science.
London: Methuen, 1952. x + 242 pp.

    a. Review essay, "A Critique of Scientific Critiques," by C.
    W. Churchman. Review of Metaphysics. 7 (1953) 89-98.

1370. Withers, R. F. J. "Epistemology and Scientific Strategy."
BJPS 10 (1959) 89-102.

1371. Wójcicki, Ryszard. "Basic Concepts of Formal Methodology of
Empirical Science." Ajatus. 35 (1973) 168-96. Also in SL 87
(1977) 681-708.

1372. _____. Metodologia formalna nauk empirycaych. Warschau:
Verlag der Polnischen Akademie der Wissenschaften, 1974.

1373. _____. Topics in the Formal Methodology of Empirical
Sciences. SL 135. Dordrecht: Reidel, 1979. 240 pp.

1374. _____. "Is There Any Need for Non-Classical Logic in
Science?" Grazer Philosophische Studien. 12/13 (1981) 119-29.

1375. Woodger, J. H. Physics, Psychology and Medicine: A Methodological Essay. London: Cambridge University Press, 1956. x +
146 pp.

1376. _____. "Abstraction in Natural Science." LMPS-1 (1960)
293-302.

1377. Zich, Otakar, Ivan Malek, and Ladislav Tondl. Metodologii
Experimentálních Věd. Praha: Nakladatelství Československi Akademie
Věd, 1959. 342 pp.

1378. Ziemski, S. "Diagnosis: A New Method in the Empirical
Sciences." Scientia. 109 (1974) 767-77.

1379. _____. "The Typology of Scientific Research." ZAW 6
(1975) 276-91.

1380. _____. "Two Types of Scientific Research." ZAW 10 (1979) 338-42.

1381. Zinov'ev, A. A. "On the Application of Modal Logic in the Methodology of Science." Soviet Studies in Philosophy. 3, No. 3 (1964-65) 20-26. [English reprint from Voprosy filosofii. (1964, No. 8).]

1382. _____. Foundations of the Logical Theory of Scientific Knowledge (Complex Logic). (Revised and enlarged English edition) BSPS 9. SL 46. Dordrecht: Reidel, 1973. 301 pp.

     a. Review essay by Soshichi Uchii. Philosopia. 4 (1974) 583-99.

1383. Zinoviev, A. A., and H. Wessel. "Logic and Empirical Sciences." Studia Logica. 35 (1976) 17-44.

4.2 Evidence, Observation, Experiments

1384. Achinstein, Peter. "Concepts of Evidence." Mind. 87 (1978) 22-45. Reprinted in The Philosopher's Annual. 2 (1979) 1-24.

     a. "In Defense of the Classical Notion of Evidence," by Maya Bar-Hillel and Avishai Margalit. 88 (1979) 576-83.

     b. "On Evidence: A Reply to Bar-Hillel and Margalit," by P. Achinstein. 90 (1981) 108-12.

1385. Aeschlimann, Florence. "Le rôle de la convention dans l'espace d'un observateur isolé." Synthese. 13 (1961) 75-85.

1386. Agassi, Joseph. "Positive Evidence as Social Institution." Philosophia. 1 (1971) 143-58.

1387. _____. "When Should We Ignore Evidence in Favour of a Hypothesis?" Ratio. 15 (1973) 183-205.

1388. _____. "Random Versus Unsystematic Observations." Ratio. 15 (1973) 111-13.

1389. Anderson, Jr., Robert M. "The Illusions of Experience." BSPS 32 (1976) 549-61.

1390. Anguera-Argilaga, Teresa. "Observational Typology." Quality and Quantity. 13 (1979) 449-84.

1391. Armstrong, D. M. "Immediate Perception." BSPS 39 (1976) 23-36.

1392. Bacon, John. "The Logical Form of Perception Sentences." Synthese. 41 (1979) 271-308.

1393. Balzer, W. "Holismus und Theorienbeladenheit der Beobachtungssprache (ein Beispiel)." Erkenntnis. 10 (1976) 337-48.

1394. Bechtel, P. William, and Eric Stiffler. "Observationality:
Quine and the Epistemological Nihilists." PSA 1978. I, 93-108.

1395. Bernays, P. "Quelques points de vue concernant le probleme
de l'evidence." Synthese. 5 (1946) 321-26.

1396. Birnbaum, Allan. "Concepts of Statistical Evidence," in
Philosophy, Science, and Method. S. Morgenbesser et al. eds. New
York: St. Martin's Press, 1969. 112-43.

1397. Brain, W. Russell. Mind, Perception and Science. Springfield, IL: Charles C. Thomas, 1952. 90 pp.

1398. _____. The Nature of Experience. Oxford: Oxford University Press, 1959. 73 pp.

1399. _____. "Space and Sense-data." BJPS 11 (1960) 177-191.

    a. "A Reply to Lord Brain," by J. R. Smythies. BJPS 13
    (1963) 161-65.

    b. "A Reply to Professor Smythies," by Lord Brain. 165-67.

1400. Bünning, Erwin. "Das Experiment als Quelle für Natur-und
Geisteserkenntnisse, dargestellt an der Entwicklung der physiologis-
chen Problemstellung." Studium Generale. 1 (1947-48) 10-18.

1401. Butts, Robert E. "Feyerabend and the Pragmatic Theory of Observation." PS 33 (1966) 383-94.

1402. Carleton, Lawrence Richard. "Perceptual Objections." Synthese.
41 (1979) 309-20.

1403. Carloye, Jack C. "Cornman's Definition of Observation Terms."
Phil. Studies 32 (1977) 283-92.

1404. Carrier, L. S. "The Time-Gap Argument." AJP 47 (1969)
263-72.

    a. "Seeing Through a Time Gap," by Charles B. Daniels. 48
    (1970) 354-59.

    b. "Time-Gap Miopia," by L. S. Carrier. 50 (1972) 55-57.

1405. Chisholm, Roderick M. "Reichenbach on Observing and Perceiving." Phil. Studies. 2 (1951) 45-48.

    a. "On Observing and Perceiving," by Hans Reichenbach.
    92-93.

    b. "Reichenbach on Perceiving," by Roderick M. Chisholm. 3
    (1952) 82-83.

1406. Chomsky, Noam. "Perception and Language." BSPS 1 (1963) 199-205.

1407. Churchland, Paul M. "Two Grades of Evidential Bias." PS 42 (1975) 250-59.

1408. Cornman, J. W. "Mental Terms, Theoretical Terms, and Materialism." PS 35 (1968) 45-63.

    a. "A Recent Drawing of the Theory/Observation Distinction," by Peter K. Machamer 38 (1971) 413-14.

    b. "Observing and What it Entails," by James W. Cornman. 415-17.

    c. "A Note on a Definition of 'Observation Term'," by Philip A. Ostien. 42 (1975) 203-07.

1409. _____. Perception, Common Sense, and Science. New Haven, CT: Yale University Press, 1975. xiv + 422 pp.

    a. Review essay by James A. McGilvray. Philosophical Studies (Ireland). 25 (1976) 263-71.

    b. Review essay by J. J. C. Smart. Philosophia. 7 (1977) 163-69.

    c. Review essay, "Scientific Realism and Perception," by Raimo Tuomela. BJPS 29 (1978) 87-104.

1410. _____. "On the Certainty of Reports About What is Given." Nous. 12 (1978) 93-118.

1411. Cowan, Thomas A. "Experience and Experiment." PS 26 (1959) 77-83.

1412. Darwin, C. G. "Observation and Interpretation," in Observation and Interpretation. S. Körner, ed. New York: Academic Press; London: Butterworths, 1957. 209-18.

1413. Domotor, Zoltan. "Qualitative Information and Entropy Structures." SL 28 (1970) 148-94.

1414. Dorolle, M. "La valeur de l'observation." Revue Phil. 135 (1945) 140-56, 222-35.

1415. Dretske, Fred I. "Observational Terms." Phil. Review. 73 (1964) 25-42.

1416. _____. Seeing and Knowing. Chicago: University of Chicago Press, 1969.

    a. Review essay: "Dretske's Conception of Perception and Knowledge," by Gerald Doppelt. PS 40 (1973) 433-46.

1417. Eklund, Harald. "Über Evidenz und Anschaulichkeit." _Theoria_. 23 (1957) 137-51.

1418. Ellson, D. G. "The Scientists' Criterion of True Observation." PS 30 (1963) 41-52.

1419. Falk, Arthur E. "Learning to Report One's Introspections." PS 42 (1975) 223-41.

1420. Fiala, Silvio. "The Experiment and Its Role in the Theory of Knowledge." PS 18 (1951) 253-58.

1421. Foss, Lawrence. "The Myth of the Given." _Review of Metaphysics_. 22 (1968) 36-57.

1422. Fraassen, Bas C. van. "Facts and Tautological Entailments." _J. Phil_. 66 (1969) 477-87.

1423. Franklin, Allan. "What Makes a 'Good' Experiment?" BJPS 32 (1981) 367-74.

1424. Giedymin, J. "On the Theoretical Sense of the So-Called Observational Terms and Sentences." SL 87 (1977) 111-34. [First published in H. Eilstein, ed. _Teoria i doświadczenie_. Warszawa: 1966.]

1425. Goodman, Nelson. "Sense and Certainty." _Phil. Review_. 61 (1952) 160-67.

1426. Grene, Marjorie. "Three Aspects of Perception." BSPS 23, SL 66 (1974) 13-34.

1427. Hager, N. "Zur Rolle des Gedankenexperiments in der physikalischen Erkenntnis." DZP 27 (1979) 233-40.

1428. Hanson, Norwood Russell. "A Note on Statement of Fact." _Analysis_. 13 (1952) 24.

1429. _____. "Facts and Faith." _Mind_. 67 (1958) 272-75.

1430. _____. "On Having the Same Visual Experience." _Mind_. 69 (1960) 340-50.

1431. _____. "Gedankenexperiments." _Science_. 149 (1965) 1048-49.

1432. _____. "Observation and Interpretation," in _Philosophy of Science Today_. S. Morgenbesser, ed. New York: Basic Books, 1967. 89-99.

1433. _____. _Observation and Explanation: A Guide to Philosophy of Science_. New York: Harper & Row, 1971. 84 pp.

1434. Harding, S. G. "Making Sense of Observation Sentences." _Ratio_ 17 (1975) 65-71.

1435. Hartshorne, Charles. "Perception and the 'Concrete Abstractness' of Science." PPR 34 (1974) 465-76.

1436. Hawkins, David. "Taxonomy and Information." BSPS 3 (1967) 41-55.

1437. Herrick, C. Judson. "The Natural History of Experience." PS 12 (1945) 57-71.

1438. Hesse, Mary B. "A Self-Correcting Observation Language." LMPS-3 (1967) 297-310.

1439. _____. "Is There an Independent Observation Language?" in The Nature and Function of Scientific Theories: Essays in Contemporary Science and Philosophy. Robert G. Colodny, ed. Pittsburgh: University of Pittsburgh Press, 1970. 35-78.

1440. Hilpinen, Risto. "On the Information Provided by Observations." SL 28 (1970) 97-122.

1441. Hintikka, J. "The Varieties of Information and Scientific Explanation." LMPS-3 (1967) 311-32.

1442. _____. "On Semantic Information." SL 28 (1970) 3-27.

1443. _____. "Surface Information and Depth Information." SL 28 (1970) 263-97.

1444. Howell, Robert. "Seeing As." Synthese. 23 (1972) 400-22.

1445. Hull, Richard T. "Feyerabend's Attack on Observation Sentences." Synthese. 23 (1972) 374-99.

1446. Jones, G. A. "Feyerabend and Observation Sentences." Philosophical Inquiry. 1 (1979) 215-20.

1447. Jones, William B. "Theory-ladenness and Theory Comparison." PSA 1978. I, 83-92.

1448. Kadish, Mortimer R. "A Note on the Grounds of Evidence." J. Phil. 46 (1949) 229-43.

1449. Kattsoff, Louis O. "Observation and Interpretation in Science." Phil. Review. 56 (1947) 682-89.

1450. Klawiter, A. "Kuhn and Feyerabend on a Scientific Fact and Its Interpretation." Poznan Studies. 1, No. 2 (1975) 81-86.

1451. Kneale, W. C. "What Can We See?" in Observation and Interpretation. S. Körner, ed. New York: Academic Press; London: Butterworths, 1957. 151-59.

1452. Kordig, Carl R. "The Theory-Ladenness of Observation." Review of Metaphysics. 24 (1971) 448-84.

1453. _____. "The Comparability of Scientific Theories." PS 38 (1971) 467-85.

    a. "Kordig and the Theory-Ladenness of Observation," by George Gale and Edward Walter. 40 (1973) 415-32.

    b. "Observational Invariance," by Carl K. Kordig. 558-69.

1454. Krimsky, Sheldon. "The Use and Misuse of Critical Gedanken-experimente." ZAW 4 (1973) 323-34.

1455. Kronfli, N. S. "Integration Theory of Observables." International Journal of Theoretical Physics. 3 (1970) 199-204. Reprinted in WOSPS 5 (1975) 497-502.

1456. Kuhn, Thomas S. "A Function for Thought Experiments," in L'aventure de la science, Mélanges Alexandre Koyré. Paris: Hermann, 1964. 307-34. Reprinted in T. S. Kuhn, The Essential Tension, Chicago: University of Chicago Press, 1977. 240-65.

1457. Lane, N. R., and S. A. Lane. "Paradigms and Perception." Studies in History and Philosophy of Science. 12 (1981) 47-60.

1458. Laymon, Ronald. "Newton's Experimentum Crucis and the Logic of Idealization and Theory Refutation." SHPS 9 (1978) 51-77.

1459. Lehrer, Keith. "Truth, Evidence, and Inference." Am. Phil. Quart. 11 (1974) 79-92.

1460. Lerner, Daniel, ed. Evidence and Inference. Glencoe, Illinois: Free Press, 1959. 164 pp.

1461. Levi, Isaac. "Epistemic Utility and the Evaluation of Experiments." PS 44 (1977) 368-86.

1462. Lewis, C. I. "The Given Element in Empirical Knowledge." Phil. Review. 61 (1952) 168-75.

1463. Linder, Arthur. "Beobachtung, Versuch, und Zufall." Dialectica. 14 (1960) 203-14.

1464. McLaughlin, Andrew. "Rationality and Total Evidence." PS 37 (1970) 271-78.

1465. Mach, Ernst. "On Thought Experiments." W. O. Price and Sheldon Krimsky, trs. Philosophical Forum. 4 (1973) 449-57.

1466. Machamer, Peter K. "Observation." BSPS 8, SL 39 (1971) 187-201.

1467. Margolis, Joseph. "Notes on Feyerabend and Hanson." MSPS. 4 (1970) 193-95.

1468. Mejbaum, W. "The Physical Magnitude and Experience." SL 87 (1977) 375-432. [First published in Studia Filozoficzne. 2 (41) (1965).]

1469. Mellor, D. H. "Experimental Error and Deducibility." PS 32 (1965) 105-22.

1470. Morgenbesser, Sidney. "Perception: Cause and Achievement." BSPS 1 (1963) 206-12.

1471. Mundle, C. W. K. Perception: Facts and Theories. New York: Oxford University Press, 1971. 192 pp.

1472. Myers, C. Mason. "Perceptual Events, States, and Processes." PS 29 (1962) 285-91.

1473. Nyman, Alf. "Das Experiment, Seine Voraussetzungen und Grenzen." ZPF 4 (1949) 80-96.

1474. Omelianovsky, M. E. "On the Principle of Observability in Modern Physics." FP 2 (1972) 223-39.

1475. Ostien, Philip A. "Observationality and the Comparability of Theories." BSPS 32 (1976) 271-89.

1476. Pasch, Alan. "Science, Perception, and Some Dubious Epistemological Motives." Phil. Studies 8 (1957) 55-61.

        a. "Science and Philosophy: A Reply to Mr. Pasch," by Harry G. Frankfurt. 9 (1958) 85-88.

1477. Pinkham, Gordon N. "Some Doubts About Scientific Data." PS 42 (1975) 260-69.

1478. Pitcher, George. A Theory of Perception. Princeton: Princeton University Press, 1971.

        a. Review essay, "Causation and Direct Realism," by M. S. Gram. PS 39 (1972) 388-96.

1479. Polanyi, Michael. "Sense-Giving and Sense-Reading." Philosophy. 42 (1967) 301-25. Reprinted in Knowing and Being. M. Grene, ed. Chicago: University of Chicago Press, 1969. 181-207.

1480. Przełęcki, M. "Wsprawie uzasadniania zdań spostrzezeniowych." [On Validating Observation Statements] Studia Logica. 13 (1962) 213-18.

1481. Puligandla, R. "Paradoxes of Non-Independent Observation Language." Scientia. 108 (1973) 103-09.

1482. Ratliff, Floyd. "On Mach's Contributions to the Analysis of Sensations." BSPS 6 (1970) 23-41.

1483. Reenpää, Y. "Versuch über die Beobachtungsgrundlagen der exakten empirischen Wissenschaften." Dialectica. 10 (1956) 113-47.

1484. Rehder, Wulf. "Versuche zu einer Theorie von Gedankenexperimenten." Grazer Philosophische Studien. 11 (1980) 105-23.

1485. Reichenbach, Hans. "Are Phenomenal Reports Absolutely Certain?"
Phil. Review. 61 (1952) 147-59.

1486. Rosenblatt, Lief, D. "On Categories and Observation."
Synthesis. 1, No. 3 (1973) 8-15.

1487. Rosenkrantz, Roger. "Experimentation as Communication with
Nature." SL 28 (1970) 58-93.

1488. Rosenthal-Schneider, Ilse. "The Interpretation of Scientific
Evidence." Australian Journal of Science. 9 (1947) 161-66.

1489. Rottschaefer, William A. "Believing is Seeing-Sometimes."
New Scholasticism. 49 (1975) 503-09.

1490. _____. "Observation: Theory-Laden, Theory-Neutral or
Theory-Free?" Southern Journal of Philosophy. 14 (1976) 499-510.

1491. Rynin, David. "Evidence." Synthese. 12 (1960) 6-24.

1492. Sambursky, S. "Some Comments on 'Imaginary Experiments'."
BJPS 11 (1960) 62-64.

1493. Schreiber, Alfred. "Auf dem Wege zu einer logischen Analyse
des Evidenzbegriffs." Allgemeine Zeitschrift für Philosophie. 6,
No. 3 (1981) 60-68.

1494. Servien, Pius. "La représentation mathématique des observables." Synthese. 10 (n.d.) 71-75.

1495. Shames, Morris L. "On the Metamethodological Dimension of
the Expectancy Paradox." PS 46 (1979) 382-88.

        a. "On Shames' Experimenter Expectancy Paradox," by D.
        Primeaux. 47 (1980) 634-37.

1496. Shapere, Dudley. "The Influence of Knowledge on the Description of Facts." PSA 1976. II, 281-98.

1497. Shimony, Abner. "Is Observation Theory-Laden? A Problem in
Naturalistic Epistemology." The Philosopher's Annual. 1 (1978)
116-45.

1498. Sleinis, E. E. "Hanson on Observation and Explanation."
Philosophical Papers. 2 (1973) 73-83.

1499. Smart, J. J. C. "Reports of Immediate Experiences." Synthese.
22 (1971) 346-59.

1500. Smith, David Woodruff. "The Case of the Exploding Perception."
Synthese. 41 (1979) 239-69.

1501. Smith, Vincent E. "Toward a Philosophy of Physical Instru-
ments." Thomist. 10 (1947) 307-33.

1502. Suppe, Frederick. "Facts and Empirical Truth." Canadian Journal of Philosophy. 3 (1973) 196-212.

1503. Suppes, Patrick. "Models of Data." LMPS-1 (1960) 252-61.

1504. _____. "The Structures of Theories and the Analysis of Data," in The Structure of Scientific Theories. F. Suppe, ed. Urbana, Illinois: University of Illinois Press, 1974. 266-83.

1505. Szaniawski, K. "Two Concepts of Information." SL 87 (1977) 635-46. [Reprinted from Theory and Decision. 1 (1974).]

1506. Teller, Paul. "Conditionalization and Observation." Synthese. 26 (1973) 218-58.

1507. Teune, Henry. "Concepts of Evidence in Systems Analysis: Testing Macro-System Theories." Quality and Quantity. 15 (1981) 55-70.

1508. Thienemann, August. "Vergleichende Beobachtung und Experiment in der Biologie." Studium Generale. 1 (1947-48) 303-13.

1509. Thompson, John W. "Coombs' Theory of Data." PS 33 (1966) 376-82.

1510. Tibbetts, Paul. "Hanson and Kuhn on Observation Reports and Knowledge Claims." Dialectica. 29 (1975) 145-56.

1511. Tiercy, G. "Réformation et transformation progressives des interprétations scientifiques des faits, et questions restées sans résponse." Scientia. 86 (1951) 189-94.

1512. Townsend, Burke. "Feyerabend's Pragmatic Theory of Observation and the Comparability of Alternative Theories." BSPS 8, SL 39 (1971) 202-11.

1513. Troll, W. "Die Grundlagen des Naturverständnisses." Scientia. 87 (1952) 11-18. French tr.: Supplement, 7-13.

1514. Vallée, Robert. "Un aspect du problème de l'observation." Methodos. 7 (1955) 289-94.

1515. Weiss, Paul. "The Perception of Stars." Review of Metaphysics. 6 (1952) 233-38.

1516. Weizsäcker, Carl Friedrich von. "Das Experiment." Studium Generale. 1 (1947-48) 3-9.

1517. Werth, R. "On the Theory-Dependence of Observations." SHPS 11 (1980) 137-43.

1518. Whitmore, Charles E. "Perception and Experiment." J. Phil. 54 (1957) 401-09.

1519. Wisdom, J. O. "Observations as the Building Blocks of Science in 20th Century Scientific Thought." BSPS 8, SL 39 (1971) 212-22.

1520. Wolff, Étienne. "Possibilités et limitations de l'expérimentation en biologie." Revue Phil. 148 (1958) 71-80.

1521. Yilmaz, Hüseyin. "Perception and Philosophy of Science." BSPS 13 (1974) 1-91.

1522. Ziedins, R. "Conditions of Observation and States of Observers." Phil. Review. 65 (1956) 299-323.

4.3 Scientific Discovery

1523. Achinstein, Peter. "Inference to Scientific Laws." MSPS 5 (1970) 87-104.

   a. "Comment," by Arnold Koslow, 104-07; by Peter A. Bowman, 107-09; "Reply," by Peter Achinstein, 109-11.

1524. _____. "Discovery and Rule-Books." BSPS 56 (1980) 117-32. Also in RIP 131-132 (1980) 109-29.

1525. Agassi, Joseph. "Genius in Science." Philosophy of the Social Sciences. 5 (1975) 145-61.

1526. _____. "Scientists as Sleepwalkers," in The Interaction Between Science and Philosophy. Y. Elkana, ed. New York: Humanities Press, 1975. 391-405. Reprinted in BSPS 65 (1981) 210-22.

1527. _____. "Who Discovered Boyle's Law?" SHPS 8 (1977) 189-250.

1528. _____. "The Rationality of Discovery." BSPS 56 (1980) 185-200.

1529. Alexander, Peter. "On the Logic of Discovery." Ratio 7 (1965) 219-32.

1530. Austin, James H. Chase, Chance, and Creativity: The Lucky Art of Novelty. New York: Columbia University Press, 1978. xvii + 237 pp.

1531. Ballard, Edward G. "The Routine of Discovery." PS 20 (1953) 157-63.

1532. Bantz, David A. "The Structure of Discovery: Evolution of Structural Accounts of Chemical Bonding." BSPS 60 (1980) 291-329.

1533. Bartlett, Steven. "A Metatheoretical Basis for Interpretations of Problem-Solving Behavior." Methodology and Science. 11 (1978) 59-85.

1534. Blackwell, Richard J. "Approaches to the Explanation of Discovery in Science." Proc. Am. Cath. Phil. Assn. 40 (1966) 181-90.

1535. _____. Discovery in the Physical Sciences. Notre Dame,
IN: University of Notre Dame Press, 1969. xii + 240 pp.

1536. _____. "Scientific Discovery and the Laws of Logic."
New Scholasticism. 50 (1976) 333-44.

     a. "Is the Case for a Logic of Discovery Closed?" by Paul
     T. Durbin. 51 (1977) 396-403.

1537. _____. "In Defense of the Context of Discovery." RIP
131-132 (1980) 90-108.

1538. _____. "The Rationality of Scientific Discovery," in
Wissenschaftliche und ausserwissenschaftliche Rationalität. Refer-
ate und Texte des 4. Internationalen Humanistischen Symposiums
1978. Athena: Griechische Humanistische Gesellschaft, 1981.
189-207.

1539. Bouligand, G. "A propos de l'invention dans les champs
théoriques. Observation quotidienne et montée séculaire de l'esprit."
Revue Phil. 155 (1965) 169-82.

1540. Bradie. Michael. "Polanyi on the Meno Paradox." PS 41
(1974) 203.

     a. "Bradie on Polanyi on the Meno Paradox," by Herbert A.
     Simon. 43 (1976) 147-50.

1541. Brannigan, Augustine. The Social Basis of Scientific Discoveries. Cambridge: Cambridge University Press, 1981. xi +
212 pp.

1542. Briskman, Larry. "Creative Product and Creative Process in
Science and Art." Inquiry. 23 (1980) 83-106.

1543. Bromberger, Sylvain. "Science and the Forms of Ignorance,"
in Observation and Theory in Science. E. Nagel et al. eds. Baltimore: The Johns Hopkins Press, 1971. 45-67.

1544. Buchdahl, Gerd. "Descartes's Anticipation of a 'Logic of
Scientific Discovery'," in Scientific Change. A. C. Crombie, ed.
London: Heinemann, 1963. 399-417.

1545. Bunge, Mario. Intuition and Science. Englewood Cliffs, NJ:
Prentice-Hall, 1962. 142 pp.

1546. Burian, Richard. "Why Philosophers Should Not Despair of
Understanding Scientific Discovery." BSPS 56 (1980) 317-36.

1547. Burks, Arthur W. "Peirce's Theory of Abduction." PS 13
(1946) 301-06.

1548. Caloi, P. "L'intuizione nella scienza." Scientia. 95
(1960) 43-47. French tr.: Supplement, 25-29.

1549. Campbell, Donald T. "Unjustified Variation and Selective
Retention in Scientific Discovery," in Studies in the Philosophy of
Biology. F. J. Ayala and T. Dobzhansky, eds. Berkeley: University
of California Press, 1974. 139-61.

1550. Caws, Peter J. "The Structure of Discovery." Science. 166
(1969) 1375-80.

1551. Chauviré, Christiane. "Peirce, Popper et l'abduction. Pour
en finir avec l'idée d'une logique de la découverte." Revue Phil.
171 (1981) 441-59.

1552. Chihara, Charles S. "Mathematical Discovery and Concept
Formation." Phil. Review. 72 (1963) 17-34.

1553. Curd, Martin. "The Logic of Discovery: An Analysis of Three
Approaches." BSPS 56 (1980) 201-20.

1554. Darden, Lindley. "Discoveries and the Emergence of New
Fields in Science." PSA 1978. I, 149-60.

1555. Dijke, Harry van. "Over het 'Discovery-Justification' Onder-
scheid." Algemeen Nederlands Tijdschrift voor Wijsbegeerte. 67
(1975) 245-66.

1556. Ducasse, C. J. "Whewell's Philosophy of Scientific Discovery."
Phil. Review. 60 (1951) 56-69, 213-34.

1557. Duijn, P. van. "A Model for Theory Finding in Science."
Synthese. 13 (1961) 61-67.

1558. Durbin, O. P., Paul R. "A Logic of Scientific Discovery."
Proc. Am. Cath. Phil. Assn. 40 (1966) 191-202.

1559. _____ . Logic and Scientific Inquiry. Milwaukee:
Bruce, 1968. vii + 132 pp.

1560. Falk, Arthur E. "Two Conceptions of a Logic of Discovery."
Proc. Am. Cath. Phil. Assn. 40 (1966) 203-08.

1561. Fang, J. "Per Analogiam vs. Per Definitionem Relative to the
Patterns of Discovery." Philosophia Mathematica. 12 (1975) 5-22.

1562. Farber, Eduard. "Chemical Discoveries by Means of Analogies."
Isis. 41 (1950) 20-26.

1563. Farre, George L. "On the Linguistic Foundations of the
Problem of Scientific Discovery." J. Phil. 65 (1968) 779-94.

1564. Finocchiaro, Maurice. "Scientific Discoveries as Growth of
Understanding: The Case of Newton's Gravitation." BSPS 56 (1980)
235-56.

1565. Frankfurt, Harry G. "Peirce's Notion of Abduction." J. Phil. 55 (1958) 593-97.

1566. Franklin, Allan. "The Discovery and Nondiscovery of Parity Nonconservation." SHPS 10 (1979) 201-57.

1567. Frolov, Y. P. "The Formula of Important Scientific Discoveries." Synthese. 5 (1947) 506-10.

1568. Gabor, Dennis. Innovations: Scientific, Technological, and Social. Oxford: Oxford University Press, 1970. 113 pp.

1569. Garrett, Alfred B. The Flash of Genius. Princeton, NJ: D. van Nostrand, 1963. 249 pp.

1570. Gay, Hannah. "Noble Gas Compounds: A Case Study in Scientific Conservation and Opportunism." SHPS 8 (1977) 61-70.

1571. _____. "The Asymmetric Carbon Atom: (a) A Case Study of Independent Discovery; (b) An Inductivist Model for Scientific Method." SHPS 9 (1978) 207-38.

1572. Giorello, G. "Intuition and Rigor: Some Problems of a 'Logic of Discovery' in Mathematics." BSPS 47 (1981) 113-35.

1573. Goodfield, June. An Imagined World: A Story of Scientific Discovery. New York: Harper and Row, 1981. 240 pp.

1574. Grmek, Mirko D., ed. On Scientific Discovery. BSPS 34. Dordrecht: Reidel, 1981.

1575. _____. "A Plea for Freeing the History of Scientific Discoveries from Myth." BSPS 34 (1981) 9-42.

1576. Gruber, Howard E. Darwin on Man: A Psychological Study of Scientific Creativity. Together with Darwin's Early and Unpublished Notebooks, transcribed and annotated by Paul H. Barrett. Forward by Jean Piaget. New York: E. P. Dutton, 1974. 495 pp.

1577. _____. "The Evolving Systems Approach to Creative Scientific Work: Charles Darwin's Early Thought." BSPS 60 (1980) 113-30.

1578. Gutting, Gary. "A Defense of the Logic of Discovery." Philosophical Forum. 4 (1973) 384-405.

1579. _____. "The Logic of Invention." BSPS 56 (1980) 221-34.

1580. _____. "Science as Discovery." RIP 131-132. (1980) 26-48.

1581. Hanson, Norwood Russell. "The Logic of Discovery." J. Phil. 55 (1958) 1073-89.

a. "Comment on Mr. Hanson's 'The Logic of Discovery'," by Donald Schon. 56 (1959) 500-04.

b. "More on 'The Logic of Discovery'," by N. R. Hanson. 57 (1960) 182-88.

1582. _____. "Is There a Logic of Discovery?", in Current Issues in the Philosophy of Science. H. Feigl and G. Maxwell, eds. New York: Holt, Rinehart and Winston, 1961. 20-35. Also in AJP 38 (1960) 91-106.

a. "Comments" by Paul K. Feyerabend. 35-39.

b. "Rejoinder" by Norwood Russell Hanson. 40-42.

1583. _____. "Discovering the Positron." BJPS 12 (1961-1962) 194-214; 299-313.

a. "Discovering the Positron," by Ernest H. Hutten. 13 (1962) 54-55.

b. "A Reply to Dr. Hutten," by Norwood Russell Hanson. 55

1584. _____. "Retroductive Inference," in Philosophy of Science: The Delaware Seminar, I. New York: Interscience Publishers, 1963. 21-37.

1585. _____. "Notes Towards a Logic of Discovery," in Perspectives on Peirce. R. J. Bernstein, ed. New Haven: Yale University Press, 1965. 42-65.

1586. _____. "The Idea of a Logic of Discovery." Dialogue. 4 (1965) 48-61.

1587. _____. "An Anatomy of Discovery." J. Phil. 64 (1967) 321-52.

1588. Hattiangadi, J. N. "The Structure of Problems." Philosophy of the Social Sciences. I: 8 (1978) 345-65; II: 9 (1979) 49-76.

1589. _____. "The Vanishing Context of Discovery: Newton's Discovery of Gravity." BSPS 56 (1980) 257-66.

1590. Herivel, J. W. "Prerequisites for Creativity in Theoretical Physics." Scientia. 101 (1966) 345-52. French tr.: Supplement, 155-59.

1591. Hesse, Mary. "Logic of Discovery in Maxwell's Electromagnetic Theory," in Foundations of Scientific Method: The Ninettnth Century. R. N. Giere and R. S. Westfall, eds. Bloomington: Indiana University Press, 1973. 86-114.

1592. Hintikka, Jaakko. "On the Logic of an Interrogative Model of Scientific Inquiry." Synthese. 47 (1981) 69-83.

1593. Holton, Gerald. The Scientific Imagination: Case Studies. Cambridge: Cambridge University Press, 1978. xvi + 382 pp.

1594. Ihde, A. J. "The Inevitability of Scientific Discovery." Scientific Monthly. 67 (1948) 427-29.

1595. Irani, Kaikhosrow D. "Conditions of Justification in the History of Scientific Discovery." Organon. 3 (1966) 63-69.

1596. Jaki, Stanley L. "The Metaphysics of Discovery and the Re-discovery of Metaphysics." Proc. Am. Cath. Phil. Assn. 52 (1978) 188-96.

1597. Jason, G. James. "A Concept of Discovery." Journal of Critical Analysis. 7 (1979) 109-18.

1598. Joerges, Bernward. "Wissenschaftliche Kreativität." ZAW 8 (1977) 383-404.

1599. Kedrov, B. M. "Toward the Methodological Analysis of Scien-tific Discoveries." Soviet Studies in Philosophy. 1 (1962) 45-56. [English reprint from Voprosy filosofii. (1960, No. 5).]

1600. _____. "On the Dialectics of Scientific Discovery." Soviet Studies in Philosophy. 6, No. 1 (1967) 16-27. [English re-print from Voprosy filosofii. (1966, No. 12).]

1601. _____. "Sur la psychologie de la création scientifique." Organon. 6 (1969) 49-68.

1602. _____. "Les degrés de la pensée productive (I.P.G.)." RIP 25 (1971) 467-76.

1603. Kekes, John. "The Centrality of Problem-Solving." Inquiry. 22 (1979) 405-21.

1604. Kisiel, Theodore. "Zu einer Hermeneutik naturwissenschaftlicher Entdeckung." ZAW 2 (1971) 195-221.

1605. _____. "Ars inveniendi: A Classical Source for Contem-porary Philosophy of Science." RIP 131-132 (1980) 130-54.

1606. Klee, James B. "A Point of Departure." PS 15 (1948) 61-70.

1607. Kleiner, Scott A. "Problem Solving and Discovery in the Growth of Darwin's Theores of Evolution." Synthese. 47 (1981) 119-62.

1608. Koertge, Noretta. "Analysis as a Method of Discovery During the Scientific Revolution." BSPS 56 (1980) 139-57.

1609. Koertge, Noretta, et al. "Panel Discussion: The Rational Explanation of Historical Discoveries." BSPS 60 (1980) 21-49.

1610. Koestler, Arthur. The Act of Creation. New York: Macmillan, 1964. Reprint: New York: Dell, 1967. 751 pp.

      a. Review essay, "Understanding and the Act of Creation," by Carl R. Hausman. Review of Metaphysics. 20 (1966) 88-112.

1611. Kogan, Zuce. "Methods of Furthering New Ideas." PS 21 (1954) 127-31.

1612. Kordig, Carl R. "Discovery and Justification." PS 45 (1978) 110-17.

1613. Krebs, H. A., and J. H. Shelley, eds. The Creative Process in Science and Medicine. Amsterdam: Excerpta Medica; New York: Elsevier, 1971.

1614. Kuhn, Thomas S. "Energy Conservation as an Example of Simultaneous Discovery," in Critical Problems in the History of Science. M. Clagett, ed. Madison: University of Wisconsin Press, 1959. 321-56. Reprinted in T. S. Kuhn, The Essential Tension. Chicago: University of Chicago Press, 1977. 66-104.

1615. _____. "The Essential Tension: Tradition and Innovation in Scientific Research," in The Third (1959) University of Utah Conference on the Identification of Scientific Talent. C. W. Taylor, ed. Salt Lake City: University of Utah Press, 1959. 162-74. Reprinted in T. S. Kuhn, The Essential Tension. Chicago: University of Chicago Press, 1977. 225-39.

1616. _____. "The Historical Structure of Scientific Discovery." Science. 136 (1962) 760-64. Reprinted in T. S. Kuhn, The Essential Tension. Chicago: University of Chicago Press, 1977. 165-77.

1617. Laudan, Larry. "Why Was the Logic of Discovery Abandoned?" BSPS 56 (1980) 173-84.

1618. Laudan, Rachel. "The Method of Multiple Working Hypotheses and the Development of Plate Tectonic Theory." BSPS 60 (1980) 331-43.

1619. Lesher, J. H. "On the Role of Guesswork in Science." SHPS 9 (1978) 19-33.

1620. Levin, Michael E. "Predicting Discoveries and the Rule-Description Argument." Logique et analyse. 17 (1974) 481-94.

1621. Lockemann, Georg. "The Varieties of Scientific Discovery." Journal of Chemical Education. 36 (1959) 220-24.

1622. MacKinnon, Edward. "The Discovery of a New Quantum Theory." BSPS 60 (1980) 261-72.

1623. Marchi, Peggy. "The Method of Analysis in Mathematics." BSPS 56 (1980) 159-72.

1624. Maxwell, Nicholas. "The Rationality of Scientific Discovery.
Part I: The Traditional Rationality Problem." PS 41 (1974) 123-53.
"Part II: An Aim Oriented Theory of Scientific Discovery." 247-95.

1625. Medawar, P. B. The Art of the Soluble. London: Methuen,
1967. 160 pp.

        a. Review essay by Peter Novak. Philosophische Rundschau.
        21 (1974) 144-47.

1626. Merton, Robert K. "The Ambivalence of Scientists." BSPS 39
(1976) 433-56.

1627. Meyer, Michel. Découverte et justification en science. Kan-
tisme, néo-positivisme et problematologie. Paris: Editions Klinck-
sieck, 1979. 365 pp.

1628. Mikulinskii, Semen R., and M. G. Iaroshevski, eds. Nauchnoe
otkrytie i ego vosprüatie. (Scientific Discovery and its Interpretation). Moscow: Nauka, 1971. 311 pp.

1629. Monk, Robert. "Productive Reasoning and the Structure of
Scientific Research." BSPS 56 (1980) 337-54.

1630. Moran, Bruce T. "Wilhelm IV of Hesse-Kassel: Informal Communication and the Aristocratic Context of Discovery." BSPS 60
(1980) 67-96.

1631. Morgan, Ch. G. "On the Algorithmic Generation of Hypotheses."
Scientia. 108 (1973) 585-98.

1632. Nickels, Thomas. "Scientific Problems and Constraints."
PSA 1978. I, 134-48.

1633. _____. "Scientific Problems: Three Empiricist Models."
PSA 1980. I, 3-19.

1634. _____. Scientific Discovery, Logic, and Rationality.
BSPS 56. Dordrecht: Reidel, 1980. 385 pp.

1635. _____. Scientific Discovery: Case Studies. BSPS 60.
Dordrecht: Reidel, 1980. 379 pp.

1636. _____. "Introductory Essay: Scientific Discovery and
the Future of Philosophy of Science." BSPS 56 (1980) 1-60.

1637. _____. "Can Scientific Constraints Be Violated Rationally?"
BSPS 56 (1980) 285-316.

1638. _____. "What is a Problem That We May Solve It?"
Synthese. 47 (1981) 85-118.

1639. Paszkiewicz, E. "Context of Discovery and Context of Justification -- Opposition or Complementarity?" Poznan Studies. 3
(1977) 256-64.

1640. Pera, Marcello. "Inductive Method and Scientific Discovery." BSPS 34 (1981) 141-66.

1641. Pietschmann, Herbert. "The Rules of Scientific Discovery Demonstrated from Examples of the Physics of Elementary Particles." FP 8 (1978) 905-19.

1642. Polanyi, Michael. "Problem Solving." BJPS 8 (1957) 89-103.

1643. _____. "Genius in Science." Archives de Philosophie. 34 (1971) 593-608. Reprinted in BSPS 14, SL 60 (1974) 57-71.

1644. Polikarov, A. "The Divergent-Convergent Method -- A Heuristic Approach to Problem-Solving." BSPS 14, SL 60 (1974) 211-33.

1645. Pólya, G. "Preliminary Remarks on a Logic of Plausible Inference." Dialectica. 3 (1949) 28-35.

1646. _____. Mathematics and Plausible Reasoning. Vol. I: Induction and Analogy in Mathematics. Vol. II: Patterns of Plausible Inference. Princeton: Princeton University Press, 1954. I: 280 pp. II: 190 pp.

1647. _____. "Methodology or Heuristics, Strategy or Tactics?" Archives de Philosophie. 34 (1971) 623-30.

1648. Prigogine, Ilya et Isabelle Stengers. "Le problème de l'invention et la philosophie des sciences." RIP, 131-132 (1980) 5-25.

1649. Robert, Serge. "Au-delà de l'opposition de la découverte et de la justification." Dialogue. 20 (1981) 269-80.

1650. Rupke, Nicolaas A. "Bathybius Haeckelii and the Pyschology of Scientific Discovery: Theory Instead of Observed Data Controlled the Late 19th Century 'Discovery' of a Primitive Form of Life." SHPS 7 (1976) 53-62.

1651. Ruse, Michael. "Ought Philosophers Consider Scientific Discovery? A Darwinian Case-Study." BSPS 60 (1980) 131-50.

1652. Sarkar, Husain. "Truth, Problem-Solving, and Methodology." Studies in History and Philosophy of Science. 12 (1981) 61-73.

    a. "Remarks on Truth, Problem-Solving, and Methodology," by Maurice A. Finocchiaro. 261-68.

1653. Schaffner, Kenneth. "Logic of Discovery and Justification of Regulatory Genetics." SHPS 4 (1974) 349-85.

1654. _____. "Discovery in the Biomedical Sciences: Logic or Irrational Intuition?" BSPS 60 (1980) 171-205.

    a. "Comment on Schaffner" by Nancy L. Maull. 207-09.

b. "Reply to Maull" by Kenneth F. Schaffner. 211-12.

1655. Schon, Donald A. Displacement of Concepts. London: Tavistock Publications, 1963. 208 ppl. Republished as Invention and the Evolution of Ideas, 1967.

1666. Scott, William T. "The Personal Character of the Discovery of Mechanisms in Cloud Physics." BSPS 60 (1980) 273-90.

1667. Siegel, Harvey. "Justification, Discovery and the Naturalizing of Epistemology." PS 47 (1980) 297-321.

1668. Simon, Herbert A. "Thinking by Computers," in Mind and Cosmos. Robert G. Colodny, ed. Pittsburgh: University of Pittsburgh Press, 1966. 3-21.

1669. _____. "Scientific Discovery and the Psychology of Problem Solving," in Mind and Cosmos. Robert G. Colodny, ed. Pittsburgh: University of Pittsburgh Press, 1966. 22-40.

1670. _____. "Does Scientific Discovery Have a Logic?" PS 40 (1973) 471-80.

1671. _____. Models of Discovery and Other Topics in the Methods of Science. BSPS 54. SL 114. Dordrecht: Reidel, 1977. 456 pp.

1672. Simon, Herbert A., Patrick W. Langley, and Gary L. Bradshaw. "Scientific Discovery as Problem Solving." Synthese. 47 (1981) 1-27.

1673. Smirnov, V. A. "Levels of Knowledge and Stages in the Process of Knowledge." SL 25 (1970) 22-54.

1674. Smith, Cyril Stanley. From Art to Science: Seventy-Two Objects Illustrating the Nature of Discovery. Cambridge, MA: MIT Press, 1980. 118 pp.

1675. Solla Price, Derek de. "The Analytical (Quantitative) Theory of Science and Its Implications for the Nature of Scientific Discovery." BSPS 34 (1981) 179-90.

1676. Somenzi, Vittorio. "Scientific Discovery from the Viewpoint of Evolutionary Epistemology." BSPS 34 (1981) 167-78.

1677. Taton, R. Reason and Chance in Scientific Discovery. London: Hutchinson, 1957. 171 pp.

1678. Tibbetts, Paul. "The Subjective Element in Scientific Discovery: Popper versus 'Traditional Epistemology'." Dialectica. 34 (1980) 155-60.

1679. Wartofsky, Marx. "Metaphysics as Heuristic for Science." BSPS 3 (1967) 123-72.

a. "Comments" by Ruth Anna Putnam. 173-80.

1680. _____. "Scientific Judgement: Creativity and Discovery in Scientific Thought." Dialectics and Humanism. 5, No. 3 (1978) 35-46. Reprinted in BSPS 60 (1980) 1-16.

1681. Wechsler, Judith, ed. Aesthetics in Science. Cambridge, MA: MIT Press, 1978. 180 pp.

1682. Weimer, Walter. "The Psychology of Inference and Expectation: Some Preliminary Remarks." MSPS 6 (1975) 430-86.

1683. Zahar, Elie. "Einstein, Meyerson and the Role of Mathematics in Physical Discovery." BJPS 31 (1980) 1-43.

1684. Zwicky, Fritz. Discovery, Invention, Research Through the Morphological Approach. New York: Macmillan, 1969. ix + 276 pp.

4.4 Models and Analogies

1685. Achinstein, Peter. "Models, Analogies, and Theories." PS 31 (1964) 328-50.

a. "Formal Models and Achinstein's 'Analogies'," by T. R. Girill. 38 (1971) 96-104.

b. "Models and Analogies: A Reply to Girill," by Peter Achinstein. 39 (1972) 235-40.

c. "Analogies and Models Revisited," by T. R. Girill, 241-44.

1686. _____. "Theoretical Models." BJPS 16 (1965) 102-20.

1687. Addison, J., L. Henkin, and A. Tarski, eds. The Theory of Models. Amsterdam: North-Holland Pub. Co., 1965. 494 pp.

1688. Agassi, Joseph. "Analogies as Generalizations." PS 31 (1964) 351-56.

1689. Akchurin, I. A., M. F. Vedenov, and Iu. V. Sachkov. "Methodological Problems of Mathematical Modeling in Natural Science." Soviet Studies in Philosophy. 5, No. 2 (1966) 23-34. [English reprint from Voprosy filosopfii. (1966, No. 14]).

1690. Altschul, Eugen, and Erwin Biser. "The Validity of Unique Mathematical Models in Science." PS 15 (1948) 11-24.

1691. Apostel, Leo. "Towards the Formal Study of Models in the Non-Formal Sciences." Synthese. 12 (1960) 125-61. Reprinted in SL 3 (1961) 1-37.

1692. Auger, Pierre. "Models in Science." Diogenes. 52 (1965) 1-13.

1693. Barr, William F. "A Syntactic and Semantic Analysis of
Idealizations in Science." PS 38 (1971) 258-72.

   a. "Idealizations and Approximations in Physics," by Robert
   John Schwartz. 45 (1978) 595-603.

1694. _____. "A Pragmatic Analysis of Idealizations in Physics."
PS 41 (1974) 48-64.

1695. Beatty, John. "Optimal-Design Models and the Strategy of
Model Building in Evolutionary Biology." PS 47 (1980) 532-61.

1696. Berlinski, David. "Mathematical Models of the World."
Synthese. 31 (1975) 211-28.

1697. Bertalanffy, L. von. "Zur Geschichte theoretischer Modelle
in der Biologie." Studium Generale. 18 (1965) 290-98.

1698. Bianca, Mariano. "Alcune osservazioni sull'uso di modelli
fisici ed iconici e sulle metafore ed analogie nella formulazione di
teorie scientifiche." Proteus. 13 (1974) 123-46.

1699. Black, Max. Models and Metaphors: Studies in Language and
Philosophy. Ithaca, NY: Cornell University Press, 1962. 267 pp.

   a. Review essay by Douglas Odegard. Philosophy. 39 (1964)
   349-56.

1700. Bonner, John Tyler. "Analogies in Biology," in Form and
Strategy in Science. J. R. Gregg and F. T. C. Harris, eds. Dordrecht: Reidel, 1964. 251-55. Reprinted from Synthese. 15 (1963)
275-82.

1701. Braithwaite, R. B. "Models in the Empirical Sciences."
LMPS-1 (1960) 224-31.

1702. Bunge, Mario. "Analogy, Simulation, Representation." RIP 23
(1969) 16-33.

1703. Bushkovitch, A. V. "Models, Theories, and Kant." PS 41
(1974) 86-88.

1704. _____. "The Concept of Model in Scientific Theory."
International Logic Review. 8 (1977) 24-31.

1705. Byerly, Henry. "Model-Structures and Model-Objects." BJPS
20 (1969) 135-44.

1706. Canguilhem, Georges. "The Role of Analogies and Models in
Biological Discovery," in Scientific Change. A. C. Crombie, ed.
London: Heinemann, 1963. 507-20.

1707. Carloye, Jack C. "An Interpretation of Scientific Models
Involving Analogies." PS 38 (1971) 562-69.

1708. _____. "The Role of Analogy in the Explanation of New Phenomena by a Fundamental Scientific Theory." Methodology and Science. 5 (1972) 3-25.

1709. Cuenod, M. "Analogie et notion de rigueur pour l'ingénieur." Dialectica. 17 (1963) 267-83.

1710. Destouches, J.-L. "Sur la notion de modèle en microphysique." Synthese. 12 (1960) 176-81. Reprinted in SL 3 (1961) 52-57.

1711. Deutsch, Karl. "Some Notes on Research on the Role of Models in the Natural and Social Sciences." Synthese. 7 (1948-49) 506-33.

1712. _____. "Mechanism, Organism, and Society: Some Models in Natural and Social Science." PS 18 (1951) 230-52.

1713. Freudenthal, Hans. "Models in Applied Probability." Synthese. 12 (1960) 202-12.

1714. Freudenthal, Hans, ed. The Concept and the Role of the Model in Mathematics and Natural and Social Sciences. Proceedings of a Colloquium held at Utrecht, The Netherlands, January, 1960. SL 3. Dordrecht: Reidel, 1961. 194 pp.

1715. Frey, G. "Symbolische und ikonische Modelle." Synthese. 12 (1960) 213-21. Reprinted in SL 3 (1961) 89-97.

1716. _____. "The Use of the Concepts 'Isomorphic' and 'Homomorphic' in Epistemology and the Theory of Science." Ratio 11 (1969) 1-13.

1717. Fürth, R. "The Role of Models in Theoretical Physics." BSPS 5 (1969) 327-40.

1718. Gonseth, F. "Analogie et modèles mathematiques." Dialectica. 17 (1963) 119-50.

1719. Götlind, Erik. "Two Views About the Function of Models in Empirical Theories." Theoria. 27 (1961) 58-69.

1720. Groenewold, H. J. "The Model in Physics." Synthese. 12 (1960) 222-27. Reprinted in SL 3 (1961) 98-103.

1721. Hager, N., and H. Hörz. "Modelle und Modellmethode in der wissenschaftlichen Erkenntnis." DZP 25 (1977) 164-79.

1722. Harré, R. "Metaphor, Model and Mechanism." Proc. Arist. Soc. 60 N.S. (1959-60) 101-22.

1723. Hesse, Mary. "Operational Definition and Analogy in Physical Theories." BJPS 2 (1952) 281-94.

1724. _____. "Models in Physics." BJPS 4 (1953) 198-214.

1725. _____. "On Defining Analogy." Proc. Arist. Soc. 60
N.S. (1959-60) 79-100.

1726. _____. Models and Analogies in Science. London: Sheed
and Ward, 1963. Expanded edition: Notre Dame, Indiana: University
of Notre Dame Press, 1966. 184 pp.

1727. _____. "The Explanatory Function of Metaphor." LMPS-2
(1964) 249-59.

1728. Hutten, E. H. "The Role of Models in Physics." BJPS 4
(1954) 284-301.

1729. Jammer, M. "Die Entwicklung des Modellsbegriffes in den
physikalischen Wissenschaften." Studium Generale. 18 (1965) 166-74.

1730. Juhos, Béla. "Ueber Analogieschlüsse." Studium Generale. 9
(1956) 126-29.

1731. Kratzer, A. "Das Bild in der Physik." Studium Generale. 9
(1956) 129-36.

1732. Kuipers, A. "Model and Insight." Synthese. 12 (1960)
249-56. Reprinted in SL 3 (1961) 125-32.

1733. Ławniczak, W. "On Different Concepts of a Model." Poznań
Studies. 5 (1979) 185-91.

1734. Leatherdale, W. H. The Role of Analogy, Model and Metaphor
in Science. Amsterdam: North-Holland; New York: American Elsevier,
1974. 276 pp.

1735. Lee, Donald S. "Analogy in Scientific Theory Construction."
Southern Journal of Philosophy. 7 (1979) 107-26.

1736. Leplin, Jarrett. "The Role of Models in Theory Construction."
BSPS 56 (1980) 267-84.

1737. Lewontin, R. C. "Models, Mathematics and Metaphors," in
Form and Strategy in Science. J. R. Gregg and F. T. C. Harris, eds.
Dordrecht: Reidel, 1964. 274-96. Reprinted from Synthese. 15
(1963) 222-44.

1738. Ley, H. "Der Begriff des Modells in der Biologie." DZP 16
(1968) 113-17.

1739. Loeb, J. "Par quel mécanisme l'analogie est-elle une source
de connaissance?" Dialectica. 17 (1963) 151-58.

1740. McMullin, E. "What do Physical Models Tell Us?" LMPS-3
(1967) 385-96.

1741. Mellor, D. H. "Models and Analogies in Science: Duhem
versus Campbell?" Isis. 59 (1968) 282-90.

1742. Mesarovic, E., and M. D. Henning. "Analogy in the Creative Processes and the Objects of Creation in Art and Sciences." Dialectica. 17 (1963) 159-66.

1743. Meyer, Herman. "On the Heuristic Value of Scientific Models." PS 18 (1951) 111-23.

1744. Morgan, Charles G. "Modality, Analogy, and Ideal Experiments according to C. S. Peirce." Synthese. 41 (1979) 65-84.

1745. Nagasaka, G. "Models and Theories in Physics." AJAPS 5, No. 1 (1976) 21-36.

1746. Nalsin, P. "Analogies, homologies et modèles." Dialectica. 17 (1963) 215-39.

1747. North, John D. "Science and Analogy." BSPS 34 (1981) 115-40.

1748. Paul, S. "Wesenzüge der mathematischen Modellierung physikalischer und biologischer Sachverhalte." DZP 25 (1977) 212-18.

1749. Perelman, Ch. "Analogie et métaphore en science, poesie et philosophie." RIP 23 (1969) 3-15.

1750. Peters, H. M. "Modellbeispiele aus der Geschichte der Biologie." Studium Generale. 18 (1965) 298-305.

1751. Pilet, P. E. "L'analogie en biologie." Dialectica. 20 (1966) 42-49.

1752. Pun, L. "Optimum de l'analogie." Dialectica. 17 (1963) 240-59.

1753. Putnam, Hilary. "Models and Reality." Journal of Symbolic Logic. 45 (1980) 464-82.

1754. Redhead, Michael. "Models in Physics." BJPS 31 (1980) 145-63.

1755. Reitzer, Alfons. "Allgemeine Modelltheorie." ZPF 29 (1975) 257-70.

1756. Rosenblueth, Arturo, and Norbert Wiener. "The Role of Models in Science." PS 12 (1945) 316-21.

1757. Ruben, P., and H. Wolter. "Modell, Modellmethode und Wirklichkeit." DZP 17 (1969) 1225-39.

1758. Ruse, Michael. "The Value of Analogical Models in Science." Dialogue. 12 (1973) 246-53.

1759. _____. "The Nature of Scientific Models: Formal v. Material Analogy." Philosophy of the Social Sciences. 3 (1973) 63-80.

1760. Seeliger, Rudolf. "Analogien und Modelle in der Physik."
Studium Generale. 1 (1947-48) 125-37.

[1761-69 inadvertently skipped in numbering the entries.]

1770. Simon, Herbert A. "Comment: The Meaning and Uses of Models."
Synthese. 13 (1961) 173-74.

1771. Spector, Marshall. "Models and Theories." BJPS 16 (1965)
121-42.

1772. Stachowiak, H. "Gedanken zu einer allgemeinen Theorie der
Modelle." Studium Generale. 18 (1965) 436-63.

1773. _____. Allgemeine Modelltheorie. Wien/New York:
Springer, 1973. xv + 494 pp.

1774. _____. "Die Modellbegriff in der Erkenntnistheorie."
ZAW 11 (1980) 53-68.

1775. Stelzl, I. "Beispiel eines probabilistischen Modells, das
deterministisch falsifiziert wurde." Conceptus. 7 (1973) 53-56.

1776. Suppes, Patrick. "A Comparison of the Meaning and Uses of
Models in Mathematics and the Empirical Sciences." Synthese. 12
(1960) 287-301. Reprinted in SL 3 (1961) 163-77.

1777. Swanson, J. W. "On Models." BJPS 17 (1967) 297-312.

    a. "Remarks on Swanson's Theory of Models," by George L.
    Farre. 18 (1967) 140-44.

1778. Tatarkiewicz, K. "Modèles et systèmes déductifs." Dialectica.
16 (1962) 275-98.

1779. Theobald, D. W. "Models and Method." Philosophy. 39 (1964)
260-67.

1780. Thom, René. "D'un modèle de la science à une science des
modèles." Synthese. 31 (1975) 359-74. English tr.: Theoria to
Theory. 10 (1977) 287-302.

1781. Tuomela, Raimo. "Analogy and Distance." ZAW 11 (1980)
276-91.

1782. Turner, Joseph. "Maxwell on the Method of Physical Analogy."
BJPS 6 (1955) 226-38.

1783. Ubbink, J. B. "Model, Description and Knowledge." Synthese.
12 (1960) 302-19. Reprinted in SL 3 (1961) 178-94.

1784. Uemov, A. I. "The Basic Forms and Rules of Inference by
Analogy." SL 25 (1970) 266-311.

1785. Wartofsky, Marx W. Models: Representation and the Scientific Understanding. BSPS 48. SL 129. Dordrecht: Reidel, 1979. 390 pp.

1786. Weingartner, P. "Analogy among Systems." Dialectica. 33 (1979) 355-78.

1787. Willer, Jörg. "Die methodische Funktion von Modellen." Philosophische Perspektiven. 3 (1971) 250-65.

1788. Wilson, P. R. "On the Argument by Analogy." PS 31 (1964) 34-39.

1789. Wohlfahrt, Th. A. "Analogie als Begriff und Methode der vergleichenden Anatomie." Studium Generale. 9 (1956) 136-42.

1790. Wüstneck, K. D. "Zur philosophischen Verallgemeinerung und Bestimmung des Modellsbegriffs." DZP 11 (1963) 1504-23.

1791. _____. "Einige Gestzmässtigkeiten und Kategorien der wissenschaftlichen Modellmethode." DZP 14 (1966) 1468-76.

1792. Zeigler, Bernard P. "The Hierarchy of System Specifications and the Problem of Structural Inference." PSA 1976. I, 227-39.

1793. Zelbstein, U. "Méthodes analogiques et isomorphisme des chaînes fonctionnelles." Dialectica. 17 (1963) 260-66.

4.5 Simplicity

1794. Ackermann, Robert. "Inductive Simplicity." PS 28 (1961) 152-61.

1795. _____. "Inductive Simplicity in Special Cases." Synthese. 15 (1963) 436-44.

1796. _____. "A Neglected Proposal Concerning Simplicity." PS 30 (1963) 228-35.

1797. Albritton, Jr., Claude G., ed. Uniformity and Simplicity: A Symposium on the Principle of the Uniformity of Nature. Boulder, Co.: Geological Society of America, Inc., 1967. 99 pp.

1798. Barker, Stephen. "On Simplicity in Empirical Hypotheses." PS 28 (1961) 162-71.

1799. _____. "The Role of Simplicity in Explanation," in Current Issues in the Philosophy of Science. H. Feigl and G. Maxwell, eds. New York: Holt, Rinehart and Winston, 1961. 265-74.

    a. "Comments," by Wesley Salmon, 274-76; by Paul K. Feyerabend, 278-80; by Richard Rudner, 381-84; "Rejoinder," by Stephen F. Barker, 276-78, 280-81, 285.

1800. Baženov, Lev B. "La proprietà di semplicità di principio dei costrutti teorici." Scientia. 112 (1977) 655-71. English Tr.: 673-85.

1801. Bouligand, G. "Autour de l'idée de simplicité." Revue Phil. 150 (1960) 37-48.

1802. Bunge, Mario. "The Weight of Simplicity in the Construction and Assaying of Scientific Theories." PS 28 (1961) 120-49.

1803. _____. "The Complexity of Simplicity." J. Phil. 59 (1962) 113-35.

1804. _____. The Myth of Simplicity: Problems of Scientific Philosophy. Englewood Cliffs, NJ: Prentice-Hall, 1963. 239 pp.

1805. Caws, Peter. "Science, Computers, and the Complexity of Nature." PS 30 (1963) 158-64.

1806. Fales, Evan. "Theoretical Simplicity and Defeasibility." PS 45 (1978) 273-88.

1807. Feuer, Lewis S. "The Principle of Simplicity." PS 24 (1957) 109-22.

        a. "The Principle of Simplicity and Verifiability," by G. Schlesinger. 26 (1959) 41-42.

        b. "Rejoinder on the Principle of Simplicity," by Lewis S. Feuer. 43-45.

1808. Friedman, Kenneth S. "Empirical Simplicity as Testability." BJPS 23 (1972) 25-33.

        a. "Friedman's Criterion for Simplicity," by Kenneth L. Manders. 27 (1976) 395-97.

1809. Fries, Horace S. "Logical Simplicity: A Challenge to Philosophy and to Social Inquiry." PS 17 (1950) 207-28.

1810. Goodman, Nelson. "On the Simplicity of Ideas." Journal of Symbolic Logic. 8 (1943) 107-21.

1811. _____. "The Logical Simplicity of Predicates." Journal of Symbolic Logic. 14 (1949) 32-41.

        a. "An Improvement in the Theory of Simplicity." by Nelson Goodman. 228-29.

        b. "New Notes on Simplicity," by Nelson Goodman. 17 (1952) 189-91.

1812. _____. "Axiomatic Measurement of Simplicity." J. Phil. 52 (1955) 709-22.

1813. _____. "Recent Developments in the Theory of Simplicity." PPR 19 (1959) 429-46.

1814. _____. "Condensation versus Simplification." Theoria. 27 (1961) 47-48.

1815. _____. "Safety, Strength, Simplicity." PS 28 (1961) 150-51.

1816. _____. "Science and Simplicity," in Philosophy of Science Today. S. Morgenbesser, ed. New York: Basic Books, 1967. 68-78.

1817. Griffin, N. "Einstein's Simplicism." Scientia. 106 (1971) 1029-54.

1818. Harré, R. "Simplicity as a Criterion of Induction." Philosophy. 34 (1959) 229-34.

1819. Hillman, Donald J. "The Measurement of Simplicity." PS 29 (1962) 225-52.

1820. Joussain, A. "Le principe de simplicité opposé au principe d'identite." Archives de Philosophie. 25 (1962) 288-300.

1821. Kapp, R. O. "Ockam's Razor and the Unification of Physical Science." BJPS 8 (1958) 265-80.

   a. "The Principle of Minimum Assumption," by G. Schelsinger. 11 (1960) 55-59.

   b. "Reply to Criticisms by G. Schlesinger," by Reginald O. Kapp. 59-62.

   c. "A Second Note on the Principle of Minimum Assumption," by G. Schlesinger. 11 (1961) 328-29.

   d. "Reply to Note by G. Schlesinger," by Reginald O. Kapp. 329-31.

1822. Kemeny, John G. "The Use of Simplicity in Induction." Phil. Review. 62 (1953) 391-408.

1823. _____. "Two Measures of Complexity." J. Phil. 52 (1955) 722-33.

1824. Kimbrough, Steven O. "On Simplicity as a Guide to Truth." Kinesis. 9 (1979) 55-72.

1825. Kyburg, Jr., Henry E. "A Modest Proposal Concerning Simplicity." Phil. Review. 70 (1961) 390-95.

   a. "Some Remarks on Kyburg's Modest Proposal," by Robert Ackermann. 71 (1962) 236-40.

1826. Lamouche, André. La Théorie harmonique. Vol. I: Le Principe
de simplicité dans les mathématiques et dans les sciences physiques.
Vol. II: Biologie. Paris: Gauthier-Villars, 1955, 1956. 481 +
575 pp.

1827. Lycan, William. "Occam's Razor." Metaphilosophy. 6 (1975)
223-37.

1828. Maxwell, Nicholas. "Induction, Simplicity and Scientific
Progress." Scientia. 114 (1979) 269-53. Italian Tr.: 655-74.

1829. Nelson, R. J. "Structure of Complex Systems." PSA 1976.
II, 523-42.

1830. Nowak, Leszek. "A Note on Simplicity." Poznan Studies. 2,
No. 3 (1976) 115-19.

1831. Post, H. R. "Simplicity in Scientific Theories." BJPS 11
(1960) 32-41.

1832. Rolston, Howard L. "A Note on Simplicity as a Principle for
Evaluating Rival Scientific Theories." PS 43 (1976) 438-40.

1833. Rosenkrantz, R. D. "Simplicity." WOSPS 6-I (1976) 167-96.

1834. Rudner, Richard S. "An Introduction to Simplcity." PS 28
(1961) 109-19.

1835. Schlesinger, G. "Dynamic Simplicity." Phil. Review. 70
(1961) 485-99.

1836. _____. "The Probability of the Simple Hypothesis."
Am. Phil. Quart. 4 (1967) 152-58.

1837. Scriven, Michael. "The Principle of Inductive Simplicity."
Phil. Studies. 6 (1955) 26-30.

1838. Silvers, Stuart. "Some Comments on Quine's Analysis of
Simplicity." PS 31 (1964) 59-61.

1839. Simon, Herbert A. "How Complex are Complex Systems?" PSA
1976. II, 507-22.

1840. Sober, Elliott. Simplicity. Oxford: Clarendon Press, 1975.
x + 189 pp.

1841. _____. "The Principle of Parsimony." BJPS 32 (1981)
145-56.

1842. Suppes, Patrick. "Nelson Goodman on the Concept of Logical
Simplicity." PS 23 (1956) 153-59.

1843. _____. "Some Remarks About Complexity." PSA 1976.
543-47.

1844. Svenonius, Lars. "Definability and Simplicity." Journal of
Symbolic Logic. 20 (1955) 235-50.

1845. Szymilewicz, I. "The Postulate of Simplicity." Poznan
Studies. 4 (1978) 236-51.

1846. Teller, Edward. The Pursuit of Simplicity. Malibu, CA:
Pepperdine University Press, 1980. 173 pp.

1847. Watkins, J. W. N. "Simplicity and the Problem of Induction."
Philosophical Inquiry. 1 (1979) 183-94.

1848. Williamson, Robert B. "Logical Economy in Einstein's 'On the
Electrodynamics of Moving Bodies'." SHPS 8 (1977) 49-60.

4.6 Meaning of Scientific Terms

1849. Achinstein, Peter. "Theoretical Terms and Partial Interpretation." BJPS 14 (1963) 89-105.

1850. _____. "On the Meaning of Scientific Terms." J. Phil.
61 (1964) 497-509.

        a. "On the 'Meaning' of Scientific Terms," by Paul K. Feyera-
        bend. 62 (1965) 266-74.

1851. _____. "The Problem of Theoretical Terms." Am. Phil.
Quart. 2 (1965) 193-203.

1852. Ambrose, Alice. "Metamorphoses of the Principle of Verifi-
ability," in Current Philosophical Issues: Essays in Honor of Curt
John Ducasse. Frederick C. Dommeyer, ed. Springfield, Illinois:
Charles C. Thomas, 1966. 54-78.

1853. Anderson, Alan Ross. "What do Symbols Symbolize?: Platonism,"
in Philosophy of Science: The Delaware Seminar, I. New York:
Interscience Publishers, 1963. 137-51. Reprinted in Philosophia
Mathematica. 11 (1974) 11-39.

1854. Arthur, Richard. "On Reference as a Component of Meaning in
Science." Philosophica. 18 (1976) 59-76.

1855. Ashby, R. W. "Use and Verification." Proc. Arist. Soc. 56
n.s. (1955-56) 149-66.

1856. Bar-Hillel, Yehoshua, and Rudolf Carnap. "Semantic Information." BJPS 4 (1953) 147-57.

1857. Bartlett, Steven J. "Lower Bounds of Ambiguity and Redundancy."
Poznan Studies. 4 (1978) 37-48.

1858. _____. "Self-Reference, Phenomenology, and Philosophy
of Science." Methodology and Science. 13 (1980) 143-67.

1859. Beauregard, Laurent A. "Reichenbach and Conventionalism."
Synthese. 34 (1977) 265-80. Reprinted in SL 132 (1979) 305-20.

1860. Berenda, C. W. "On Verifiability, Simplicity, and Equivalence." PS 19 (1952) 70-76.

1861. Berg, Jan. "A Note on Reduction Sentences." Theoria. 24
(1958) 1-8.

1862. _____. "Some Problems Concerning Disposition Concepts."
Theoria. 26 (1960) 3-16.

1863. _____. "Ueber ein Argument gegen Reduktionssätze." PN
12 (1970) 101-03.

1864. _____. "On an Argument Against Reduction Sentences."
PS 38 (1971) 118-20.

     a. "Supercalifragilistic Reduction: A Reply to Jan Berg,"
     by Israel Scheffler. 121.

1865. Bergmann, Gustav. "Philosophical and Psychological Pragmatics."
PS 14 (1947) 271-73.

1866. _____. "The Revolt Against Logical Atomism." Phil.
Quart. 7 (1957) 323-39; 8 (1958) 1-13.

1867. Berkowitz, Leonard J. "Achinstein on Empirical Significance:
A Matter of Principle." PS 46 (1979) 459-65.

1868. Berry, George. "Logic with Platonism." Synthese. 19 (1968)
215-49. Reprinted in SL 21 (1975) 243-77.

1869. Beth, E. W. "Analyse sémantique des théories physiques."
Synthese. 7 (1948-49) 206-07.

1870. _____. "Semantics of Physical Theories." Synthese. 12
(1960) 172-75.

1871. _____. "The Relationship Between Formalized Languages
and Natural Languages," in Form and Strategy in Science. J. R.
Gregg and F. T. C. Harris, eds. Dordrecht: Reidel, 1964. 66-81.

1872. Black, Max. "Verificationism and Wittgenstein's Views on
Mathematics." RIP 23 (1969) 284-98.

1873. Bunge, Mario. "A Program for the Semantics of Science."
Journal of Philosophical Logic. 1 (1972) 317-28. Reprinted in SL
50 (1973) 51-62.

1874. _____. Treatise on Basic Philosophy. Vol. I: Semantics
I. Sense and Reference. Dordrecht: Reidel, 1974. 183 pp.

1875. _____. Treatise on Basic Philosophy. Vol. 2: Semantics
II. Interpretation and Truth. Dordrecht: Reidel, 1974. 210 pp.

1876. _____. "Meaning in Science." <u>Poznan Studies</u>. 1, No. 4 (1975) 56-64.

1877. Carnap, Rudolf. <u>Meaning and Necessity: A Study in Semantics and Modal Logic</u>. Chicago: University of Chicago Press, 1947. 258 pp. Second edition, 1956.

    a. Review essay by Ernest Nagel. <u>J. Phil.</u> 45 (1948) 467-72.

    b. Review essay, "Carnap's Semantics," by Max Black. <u>Phil. Review</u>. 58 (1949) 257-64.

1878. _____. "Empiricism, Semantics, and Ontology." RIP 4 (1950) 20-40.

    a. "Decision and Belief in Science: Comments on Rudolf Carnap's Views in 'Empiricism, Semantics, and Ontology'," by Anders Wedberg. <u>Danish Yearbook of Philosophy</u>. 1 (1964) 139-58.

1879. _____. "Meaning Postulates." <u>Phil. Studies</u>. 3 (1952) 65-73.

1880. _____. "Meaning and Synonymy in Natural Languages." <u>Phil. Studies</u>. 6 (1955) 33-47.

    a. "A Note on Carnap's Meaning Analysis," by Roderick M. Chisholm. 87-89.

    b. "On Some Concepts of Pragmatics," by Rudolf Carnap. 89-91.

1881. _____. "The Methodological Character of Theoretical Concepts." MSPS 1 (1956) 38-76.

    a. "A Note on Carnap's Meaning Criterion," by William W. Rozeboom. <u>Phil. Studies</u>. 11 (1960) 33-38.

1882. _____. "Beobachtungssprache und theoretische Sprache." <u>Dialectica</u>. 12 (1958) 236-48. [Tr. by H. Bohnert: "Observation Language and Theoretical Language." SL 73 (1975) 75-86.]

1883. _____. "Theoretische Begriffe der Wissenschaft." ZPF 14 (1960) 209-33, 571-98.

1884. _____. "On the Use of Hilbert's epsilon-operator in Scientific Theories," in <u>Essays on the Foundations of Mathematics</u>. A. Robinson, ed. Jerusalem: Manes Press, 1961.

    a. "On Selection Operators and Semanticists," by William W. Rozeboom. PS 31 (1964) 282-85.

1885. Cartwright, Richard L. "Ontology and the Theory of Meaning." PS 21 (1954) 316-25.

1886. Cohen, Jonathan. "A Relation of Counterfactual Conditionals to Statements of What Makes Sense." Proc. Arist. Soc. 55 n.s. (1954-55) 45-82.

1887. _____. "Is a Criterion of Verifiability Possible?" Midwest Studies in Philosophy. 5 (1980) 347-52.

1888. Chaudhury, Pravas Jivan. "Concepts of Meaning and Understanding in New Physics." Philosophical Quarterly (India). 32 (1959) 173-89.

1889. Child, James. "On the Theoretical Dependence of Correspondence Postulates." PS 38 (1971) 170-77.

1890. Chisholm, Roderick M. "Verification and Perception." RIP 5 (1951) 251-67.

1891. Chomsky, Noam. "Quine's Empirical Assumptions." Synthese. 19 (1968) 53-68. Reprinted in SI 21 (1975) 53-68.

1892. Churchman, C. West. "Concepts Without Primitives." PS 20 (1953) 257-65.

1893. Craig, William. "Replacement of Auxiliary Expressions." Phil. Review. 65 (1956) 38-55.

1894. Czeżowski, Tadeusz. "Definitions in Science." Poznań Studies. 1, No. 4 (1975) 9-17.

1895. Davidson, Donald. "On Saying That." Synthese. 19 (1968) 130-46. Reprinted in SL 21 (1975) 158-74.

1896. Derden, Jr., J. K. "Carnap's Definition of 'Analytic Truth' for Scientific Theories." PS 43 (1976) 506-22.

1897. Dummett, Michael. "The Significance of Quine's Indeterminacy Thesis." Synthese. 27 (1974) 351-98.

    a. "Comment on Michael Dummett," by W. V. Quine. 399.

    b. "Comment on Michael Dummett," by Gilbert Harman. 401-04.

    c. "On Translating Logic," by Charles Parsons. 405-12.

    d. "Reply to W. V. Quine," by Michael Dummett. 413-16.

    e. "Postscript" by Michael Dummett. 523-34.

1898. Elgin, Catherine Z. "Quine's Double Standard: Indeterminacy and Quantifying In." Synthese. 42 (1979) 353-78.

1899. Enç, Berent. "Reference of Theoretical Terms." Noûs. 10 (1976) 261-82.

1900. English, Jane. "Underdetermination: Craig and Ramsey." J. Phil. 70 (1973) 453-62.

1901. _____. "Partial Interpretation and Meaning Change."
J. Phil. 75 (1978) 57-76. Reprinted in The Philosopher's Annual.
2 (1979) 67-87.

1902. Ernest, Paul. "A Critique of Some Formal Theories of Meaning."
BJPS 26 (1975) 319-30.

1903. Essler, Wilhelm Karl. "Ueber die Interpretation von Wissen-
schaftssprachen." Philosophisches Jahrbuch. 77 (1970) 117-30.

1904. _____. "An Inductive Solution to the Problem of Dispositional Predicates." Ratio. 12 (1970) 108-15.

1905. _____. "Die Kreativität der bilateralen Reduktionssätze."
Erkenntnis. 9 (1975) 383-92.

1906. _____. "Some Remarks Concerning Partial Definitions in
Empirical Sciences." Pacific Philosophical Quarterly. 61 1980)
455-62.

1907. Essler, Wilhelm K., and Rainer Trapp. "Some Ways of Opera-
tionally Introducing Dispositional Predicates with regard to Scientific and Ordinary Practice." Synthese. 34 (1977) 371-96.

1908. Inadvertently skipped in numbering the entries.

1909. Eyk, B. J. van. "Ueber die Begrenzung der Anwendbarkeit
klassischer wissenschaftlicher Konzeptionen auf dem Gebiete der
Naturwissenschaften." PN 10 (1967-68) 442-67.

1910. Feigl, Herbert. "Confirmability and Confirmation: Some
Comments on the Empiricist Criterion of Meaning and Related Issues."
RIP 5 (1951) 268-79.

1911. Fetzer, James H. "A World of Dispositions." Synthese. 34
(1977) 397-422.

1912. Feuer, Lewis S. "The Paradox of Verifiability." PPR 12
(1951) 24-41.

1913. Feyerabend, Paul. "Carnaps Theorie der Interpretation Theoretischer Systeme." Theoria. 21 (1955) 55-62.

1914. Fine, Arthur I. "Consistency, Derivability, and Scientific
Change." J. Phil. 64 (1967) 231-40.

    a. "Fine's Criteria of Meaning Change," by Mary Hesse. 65
    (1968) 46-52.

    b. "Fine, Mathematics, and Theory Change," by Michael E.
    Levin. 52-56.

    c. "Meaning Variance and the Comparability of Theories," by
    Jarrett Leplin. BJPS 20 (1969) 69-75.

1915. Fitzgerald, Paul. "Meaning in Science and Mathematics."
BSPS 32 (1976) 235-70.

1916. Fraassen, Bas C. van. "On the Extension of Beth's Semantics
of Physical Theories." PS 37 (1970) 325-39.

1917. _____. "Incomplete Assertion and Belnap Connectives."
WOSPS 4 (1975) 43-70.

1918. _____. "Modality," in Current Research in Philosophy
of Science. P. D. Asquith and H. E. Kyburg, Jr., eds. East Lansing,
MI: Philosophy of Science Assn., 1979. 282-90.

1919. Frank, Philipp. "Non-Scientific Symbols in Science." Synthese.
9 (n.d.) 5-9.

1920. Frey, Gerhard. "Ein Satz betreffend das Verhältnis empirischer
und symbolischer Objektbereiche." PN 4 (1957) 380-98.

1921. Gardner, Michael R. "Apparent Conflicts Between Quine's Indeterminacy Thesis and His Philosophy of Science." BJPS 24 (1973)
381-93.

1922. Geach, P. T. "Quine's Syntactical Insights." Synthese. 19
(1968) 118-29. Reprinted in SL 21 (1975) 146-57.

1923. George, F. H. "Meaning and Behaviour." Synthese. 11 (1959)
245-58.

1924. Giedymin, Jerzy. "The Paradox of Meaning Variance." BJPS 21
(1970) 257-68.

    a. "Consolations for the Irrationalist?" by Jerzy Giedymin.
    22 (1971) 39-48.

1925. Glymour, Clark. "Reichenbach's Entanglements." Synthese.
34 (1977) 219-36. Reprinted in SL 132 (1979) 221-38.

1926. Goguen, J. A. "The Logic of Inexact Concepts." Synthese.
19 (1969) 325-73.

1927. Gorski, D. P. "Ueber abstrakte und idealisierte Objekte,
ueber deren methodologischen und gnoseologischen Status." LMPS-4
(1971) 357-66.

1928. _____. "On the Types of Definition and Their Importance
for Science." SL 25 (1970) 312-75.

1929. Götlind, Erik. "Stimulus Meaning." Theoria. 29 (1963)
93-114.

1930. Grice, H. P. "Vacuous Names." SL 21 (1975) 118-45.

1931. Gross, Edward. "Toward a Rationale for Science." J. Phil.
54 (1957) 829-38.

1932. Gruender, C. D. "Science and Verification." Scientia. 101 (1966) 522-27. French Tr.: Supplement, 231-36.

1933. Grünfeld, J. "Background Knowledge." Science et Esprit. 29 (1977) 329-35.

1934. Grzegorczyk, A. "The Pragmatic Foundations of Semantics." SL 87 (1977) 135-63. [First published in Przeglad Filozoficzny. 44 (1948). Reprinted in Synthese. 8 (1950-51).]

1935. Gullvåg,Ingemund. "Criteria of Meaning and Analysis of Usage." Synthese. 9 (n.d.) 341-61.

1936. Haering, Th. "Das Problem der naturwissenschaftlichen und geisteswissenschaftlichen Begriffsbildung und die Erkennbarkeit der Gegenstände." ZPF 2 (1947) 537-79.

1937. Hanna, Joseph F. "An Explication of 'Explication'." PS 35 (1968) 28-44.

1938. Hanson, N. R. "Picturing Atomic Particles." Scientia. 94 (1959) 149-57. French Tr.: Supplement, 95-103.

1939. _____. "The Stratification of Concepts," in Dimensions of Mind. Sidney Hook, ed. New York: Collier Books, 1960, 233-37.

1940. _____. "A Picture Theory of Theory Meaning," in The Nature and Structure of Scientific Theories: Essays in Contemporary Science and Philosophy. Robert G. Colodny, ed. Pittsburgh: University of Pittsburgh Press, 1970. 233-74. Also in MSPS 4 (1970) 131-41.

1941. Harman, Gilbert. "An Introduction to 'Translation and Meaning,' Chapter Two of Word and Object." Synthese. 19 (1968) 14-26. Reprinted in SL 21 (1975) 14-26.

1942. Harrah, David. "A Model for Applying Information and Utility Functions." PS 30 (1963) 267-73.

1943. Hellman, Geoffrey. "The New Riddle of Radical Translation." PS 41 (1974) 227-46.

1944. Hempel, Carl G. "Problems and Changes in the Empiricist Criterion of Meaning." RIP 4 (1950) 41-63.

1945. _____. "The Concept of Cognitive Significance." Proceedings of the American Academy of Arts and Sciences. 80 (1951) 61-77.

       a. "Comments on Professor Hempel's Paper," by Gustav Bergman. 78-86.

1946. _____. "The Meaning of Theoretical Terms: A Critique of the Standard Empiricist Construal." LMPS-4 (1971) 367-78.

1947. Henle, Paul. "The Problem of Meaning." Proc. and Add. Am.
Phil. Assn. 27 (1953-54) 24-39.

1948. _____. "Meaning and Verifiability," in The Philosophy
of Rudolf Carnap. P. A. Schlipp, ed. La Salle, Illinois: Open
Court, 1963. 165-81.

1949. Hintikka, Jaakko. "Behavioral Criteria of Radical Translation." Synthese. 19 (1968) 69-81. Reprinted in SL 21 (1975)
69-81.

1950. _____. "On Semantic Information," in Physics, Logic,
and History. W. Yourgrau and A. D. Breck, eds. New York: Plenum
Press, 1970. 147-68.

1951. _____. "Carnap's Heritage in Logical Semantics." SL 73
(1975) 217-42. [Reprint of paper originally entitled "Carnap's
Semantics in Retrospect." Synthese. 25 (1973) 372-97.]

1952. Hintikka, Jaakko, and Raimo Tuomela. "Towards a General
Theory of Auxiliary Concepts and Definability in First-Order Theories."
SL 28 (1970) 298-330.

1953. Hochberg, Herbert. "Dispositional Properties." PS 34 (1967)
10-17.

1954. Hoche, Hans-Ulrich. "Kausalgefüge, irreale Bedingungssätze
und das Problem der Definierbarkeit von Dispositionsprädikaten."
ZAW 8 (1977) 257-91.

1955. Hockney, Donald. "The Bifurcation of Scientific Theories and
Indeterminacy of Translation." PS 42 (1975) 411-27.

1956. Holton, G. "Science and the Changing Allegory of Motion."
Scientia. 98 (1963) 191-200. French Tr.: Supplement, 103-12.

1957. Hoy, Ronald C. "The Unverifiability of Unverifiability."
PPR 33 (1973) 393-98.

1958. Kalish, Donald. "Meaning and Truth." University of California
Publications in Philosophy. 25 (1950) 99-117.

1959. Kaminsky, Jack. "Can 'Essence' be a Scientific Term?" PS 24
(1957) 173-79.

1960. Kaminsky, Jack, and Raymond J. Nelson. "Scientific Statements
and Statements About Humanly Created Objects." J. Phil. 55 (1958)
641-48.

1961. Kamlah, Andreas. "Metagesetze und theorieunabhangige Bedeutung physikalischer Begriffe." ZAW 9 (1978) 41-62.

1962. Kaplan, David. "Quantifying In." Synthese. 19 (1968)
178-214. Reprinted in SL 21 (1975) 206-42.

1963. Kasher, Asa. "Pragmatic Representations and Language-Games: Beyond Intensions and Extensions." SL 73 (1975) 271-92.

1964. Kent, William. "Scientific Naming." PS 25 (1958) 185-93.

1965. Kleiner, Scott A. "Recent Theories of Theoretical Meaning." Philosophica. 18 (1976) 35-58.

1966. _____. "Referential Divergence in Scientific Theories." SHPS 8 (1977) 87-109.

1967. Kmita, J. "Meaning and Functional Reason." SL 87 (1977) 171-87. [First published in H. Eilstein, ed. Teoria i doswiad-czenie. Warszawa, 1966. Reprinted from Quality and Quantity. 5 (2) (1971).]

1968. Kneale, W. C. "Verifiability." Arist. Soc. Supp. Vol. 19 (1945) 151-64.

1969. Kokoszyńska, M. "On a Certain Condition of Semantic Theory of Knowledge." SL 87 (1977) 188-99. [First published in Przeglad Filozoficzny. 44 (1948).]

1970. Kordig, Carl R. "Feyerabend and Radical Meaning Variance." Nous. 4 (1970) 399-404.

1971. Kotarbińska, J. "On Ostensive Definitions." SL 87 (1977) 233-60. [First published in Fragmenty Filozoficzne. 2 (1959) Also in Philosophy of Science. 27 (1960) 1-22.]

1972. Lacharité, Normand. "Archéologie du savoir et structures du language scientifique." Dialogue. 9 (1970) 35-53.

1973. Ladrière, Jean. "Language scientifique et language spéculatif." Revue Philosophique de Louvain. 69 (1971) 92-132, 250-82.

1974. _____. "Le Statut épistémique des termes théoriques." International Studies in Philosophy. 7 (1975) 7-40.

1975. Lakoff, George. "Hedges: A Study in Meaning Criteria and the Logic of Fuzzy Concepts." WOSPS 4 (1975) 221-72.

   "Comments: Lakoff's Fuzzy Propositional Logic," by Bas C. van Fraassen. 273-78.

1976. Lane, Gilles. "La position des entités théoriques." Dialogue. 16 (1977) 213-27.

1977. Lehrer, Keith. "Reichenbach on Convention." Synthese. 34 (1977) 237-48. Reprinted in SL 132 (1979) 239-50.

1978. Lennenberg, Eric H. "The Relationship of Language to the Formation of Concepts." BSPS 1 (1963) 48-54.

1979. Lewis, David. "How to Define Theoretical Terms." J. Phil. 67 (1970) 427-46.

1980. Linsky, Bernard. "Putnam on the Meaning of Natural Kind Terms." Canadian Journal of Philosophy. 7 (1977) 819-28.

1981. Lloyd, A. C. "Meaning Without Verifiability." Arist. Soc. Supp. Vol. 42 (1968) 1-6.

1982. Lotz, John. "Natural and Scientific Language." Proceedings of the American Academy of Arts and Sciences. 80 (1951) 87-88.

1983. MacCormac, Earl R. "Meaning Variance and Metaphor." BJPS 22 (1971) 145-59.

1984. MacKinnon, D. M. "Verifiability." Arist. Soc. Supp. Vol. 19 (1945) 101-18.

1985. McMahon, William E. "A Generative Model for Translating from Ordinary Language into Symbolic Notation." Synthese. 35 (1977) 99-116. Reprinted in SL 132 (1979) 637-54.

1986. Madden, Edwad H. "Definition and Reduction." PS 28 (1961) 390-405.

1987. Madden, Edward H., and Murray J. Kiteley. "Postulates and Meaning." PS 29 (1962) 66-78.

    a. "The Formalization of Empirical Significance," by G. Schlesinger. 31 (1964) 65-67.

    b. "Mr. Schlesinger on the M-K Theory," by Edward H. Madden and Murray J. Kiteley. 68-70.

1988. Malherbe, Jean-Francois. "Termes théoriques et référence." Archives de Philosophie. 38 (1975) 201-18.

1989. Marhenke, Paul. "The Criterion of Significance." Proc. & Add. Am. Phil. Assn. 23 (1949-50) 1-21.

1990. Martin, Michael. "Achinstein on Semantic Relevance." Canadian Journal of Philosophy. 3 (1973) 77-88.

1991. Martin, R. M. "On Carnap's Conception of Semantics," in The Philosophy of Rudolf Carnap. P. A. Schilpp, ed. La Salle, Illinois: Open Court, 1963. 351-84.

1992. Maxwell, Grover. "Meaning Postulates in Scientific Theories," in CurrentIssues in the Philosophy of Science. H. Feigl and G. Maxwell, eds. New York: Holt, Rinehart and Winston, 1961. 169-83.

    a. "Comments," by Wilfrid Sellers, 183-92. "Rejoinder," by Grover Maxwell, 192-95.

1993. Mellor, D. H. "In Defense of Dispositions." Phil. Review.
83 (1974) 157-81.

    a. "Mellor on Disposition," by James H. Fetzer. Philosophia.
7 (1978) 651-60.

    b. "Reply to Professor Fetzer," by D. H. Mellor. 661-66.

1994. Meyers, Robert G. "Indeterminacy and Positivism." Synthese.
39 (1978) 317-24.

1995. Miller, David L. "The Nature of Scientific Statements." PS
14 (1947) 219-23.

1996. Morris, Charles W. Signs, Language and Behavior. New York:
Prentice Hall, 1946. 365 pp.

1997. _____. "Science and Discourse." Synthese. 5 (1946)
296-308.

1998. _____. "Foundations of the Theory of Signs," in Inter-
national Encyclopedia of Unified Science. Vol. I, No. 2. O. Neurath,
R. Carnap, and C. W. Morris, eds. Chicago: University of Chicago
Press, 1938-55. 77-137.

1999. Mortimer, H. "Definicja probabilistyczna na przykładzie
definicji genotypu." [Probabilistic Definition as Exemplified by a
Definition of Genotype] Studia Logica. 15 (1964) 103-62.

2000. _____. "Conditions for Acceptance of Probabilistic Postulates." Studia Logica. 26 (1970) 87-97.

2001. Nelson, Everett J. "The Verification Theory of Meaning."
Phil. Review. 63 (1954) 182-92.

    a. "A Comment," by C. I. Lewis. 193-96.

2002. Nielsen, Kai. "Facts, Factual Statements and Theoretical
Terms." Philosophical Studies. (Ireland) 23 (1974) 129-51.

2003. Nola, Robert. "Fixing the Reference of Theoretical Terms."
PS 47 (1980) 505-31.

2004. Palmieri, L. E. "Comments on Verification." Theoria. 22
(1956) 43-48.

2005. Pap, Arthur. "Reduction-Sentences and Open Concepts."
Methodos. 5 (1953) 3-28.

2006. _____. "Belief and Propositions." PS 24 (1957) 123-36.

2007. _____. Semantics and Necessary Truth. New Haven: Yale
University Press, 1958. 456 pp.

a. Review essay, "Analytic Propositions, Definitions, and the A Priori," by Hector Neri Castaneda. Ratio. 2 (1959) 80-101.

2008. _____. "Disposition Concepts and Extensional Logic." MSPS 2 (1958) 196-224.

2009. _____. "Reduction Sentences and Disposition Concepts," in The Philosophy of Rudolph Carnap. P. A. Schilpp, ed. La Salle, Illinois: Open Court, 1963. 559-97.

2010. _____. "Theory of Definition." PS 31 (1964) 49-54.

2011. Papineau, D. "Meaning Variance and the Theory of Reference." Methodology and Science. 10 (1977) 189-219.

2012. _____. "The Vis Viva Controversy: Do Meanings Matter?" SHPS 8 (1977) 111-42.

2013. _____. Theory and Meaning. New York: Oxford University Press, 1980. 225 pp.

2014. Parlavecchia, Paolo. "Linguaggio e teoria scientifica nella concezione di Quine." Scientia. 111 (1976) 417-52. English Tr.: 453-79.

2015. Parsons, Kathryn Pyne. "On Criteria of Meaning Change." BJPS 22 (1971) 131-44.

2016. _____. "A Criterion for Meaning Change." Phil. Studies. 28 (1975) 367-96.

2017. Paul, A. M. "Hanson on the Picturability of Micro-Entities." BJPS 22 (1971) 50-53.

a. "The Picturability of Micro-Entities," by Stephen J. Noren. PS 40 (1973) 234-41.

b. "Are Micro-Entities Picturable?" by T. R. Girill. 43 (1976) 570-74.

c. "Micro-Particles and Picturability: A Reply," by Stephen J. Noren. 45 (1978) 484-87.

2018. Pirie, N. W. "Concepts out of Context: The Pied Pipers of Science." BJPS 2 (1952) 269-80.

a. "Concepts out of Context," by G. S. Carter. BJPS 3 (1952) 86-87.

2019. Polanyi, Michael. "Sense-Giving and Sense-Reading." Philosophy. 42 (1967) 301-25.

2020. Popovich, M. V. "Identity by Sense in Empirical Sciences." WOSPS 11 (1977) 25-30.

2021. Pawɫowski, T. "On the Empirical Meaningfulness of Sentences." SL 87 (1977) 541-49. [First published in Fragmenty Filozoficzne. 3 (1967) Warszawa.]

2022. Presley, C. F. "Arguments about Meaninglessness." BJPS 12 (1961) 225-34.

2023. Przeɫecki, Marian. "Empirical Meaningfulness of Quantitative Statements." Synthese. 26 (1974) 344-55. Reprinted in SL 87 (1977) 577-88.

2024. _____. "On Model Theoretic Approach to Empirical Interpretation of Scientific Theories." Synthese. 26 (1974) 401-06.

2025. _____. "On Identifiability in Extended Domains." WOSPS 11 (1977) 81-89.

2026. _____. "A Model-Theoretic Approach to the Problem of the Interpretation of Empirical Languages." SL 87 (1977) 551-76. [First published in Studia Filozoficzne. 1 (1972).]

2027. Putnam, Hilary. "Craig's Theorem." J. Phil. 62 (1965) 251-60.

   a. "A Rejoinder to Putnam," by Ernest Nagel. 429-32.

2028. _____. "How Not to Talk About Meaning." BSPS 2 (1965) 205-22.

2029. _____. "Explanation and Reference." SL 52 (1973) 199-221.

   a. "Referential Indeterminacy: A Response to Professor Putnam," by Robert Barrett. 222-31.

2030. _____. "Philosophy of Language and Philosophy of Science." BSPS 32 (1976) 603-10.

2031. _____. Meaning and the Moral Sciences. London: Routledge and Kegan Paul, 1978. 145 pp.

2032. Quine, Willard Van Orman. From a Logical Point of View. Cambridge, MA: Harvard University Press, 1953. Second, revised edition, 1961. Reprint: New York: Harper & Row, 1963. 184 pp.

   a. Review essay, "A Logician's Landscape," by P. F. Strawson. Philosophy. 30 (1955) 229-37.

   b. Review essay by J. J. C. Smart. AJP 33 (1955) 45-56.

   c. Review essay by W. Stegmüller. Philosophische Rundschau. 5 (1957) 280-92.

2033. _____. "The Scope and Language of Science." BJPS 8 (1957) 1-17.

2034. _____. Word and Object. New York and London: John
Wiley and MIT Press, 1960. 294 pp.

    a. Review essay by Rulon Wells. Review of Metaphysics. 14
    (1961) 695-703.

    b. Review essay by Irwin C. Lieb. IPQ 2 (1962) 92-109.

2035. _____. The Ways of Paradox and Other Essays. New York:
Random House, 1966. 258 pp.

2036. _____. "Stimulus and Meaning," in The Isenberg Memorial
Lecture Series, 1965-1966. East Lansing, Michigan: Michigan State
University Press, 1969. 39-61. Slightly revised version appeared
under the title "Epistemology Naturalized" as Chapter 3 of Ontological
Relativity and Other Essays. New York: Columbia University Press,
1969. 69-90.

2037. Raggio, A. R. "'Family Resemblance Predicates,' modalités et
réductionnisme." RIP 23 (1969) 339-62.

2038. Randall, John A. "Corism, applied to Specifying Operations
Called Scientific." PS 13 (1946) 215-22.

2039. Rantala, Veikko. Aspects of Definability. Amsterdam:
North-Holland Publishing Co., 1977. 236 pp. Reprinted from Acta
Philosophica Fennica. 29, 2-3 (1977).

2040. Rapoport, Anatol. "Mathematical Biophysics, Cybernetics and
Significs." Synthese. 8 (1950-51) 182-93.

2041. Reichenbach, Hans. "The Verifiability Theory of Meaning."
Proceedings of the American Academy of Arts and Sciences. 80 (1951)
46-60.

2042. Rescher, Nicholas. "A Note on a Species of Definition."
Theoria. 20 (1954) 173-75.

2043. _____. "Semantic Paradoxes and the Propositional Analysis of Indirect Discourse." PS 28 (1961) 437-40.

2044. Richman, Robert J. "Concepts Without Criteria." Theoria.
31 (1965) 65-85.

2045. Rorty, Richard. "Indeterminacy of Translation and of Truth."
Synthese. 23 (1972) 443-62.

2046. Rosenberg, Alexander. "The Virtues of Vagueness in the Languages of Science." Dialogue. 14 (1975) 281-305.

    a. "On Vagueness in the Language of Science," by Stephan
    Körner. 306-08.

    b. "Terms of Experience and Theory: A Rejoinder to Körner,"
    by Alexander Rosenberg. 309-11.

2047. Rozeboom, William W. "Studies in the Empiricist Theory of
Scientific Meaning. Part I: Empirical Realism and Classical Seman-
tics: A Parting of the Ways. Part II: On the Equivalence of
Scientific Theories." PS 27 (1960) 359-73.

2048. _____. "The Factual Content of Theoretical Concepts."
MSPS 3 (1962) 273-357.

2049. _____. "The Crisis in Philosophical Semantics." MSPS 4
(1970) 196-219.

2050. _____. "Dispositions Revisited." PS 40 (1973) 59-74.

2051. Rudner, Richard S. "What Do Symbols Symbolize?: Nominalism,"
in Philosophy of Science: The Delaware Seminar, I. New York:
Interscience Publishers, 1963. 159-76. Reprinted in Philosophia
Mathematica. 11 (1974) 41-78.

2052. Ryle, Gilbert. "The Verification Principle." RIP 5 (1951)
243-50.

2053. Rynin, David. "Meaning and Formation Rules." J. Phil. 46
(1949) 373-86.

2054. Saarinen, Esa. "Dialogue Semantics versus Game-Theoretical
Semantics." PSA 1978. II, 41-59.

2055. Sadovsky, V. N., and V. A. Smirnov. "Definability and Identifiability: Certain Problems and Hypotheses." WOSPS 11 (1977) 63-80.

2056. Sastri, P. S. "Meaning and Truth: Logical Positivism and
Reality." Philosophical Quartery (India). 28 (1955) 169-83.

2057. Schlesinger, G. "The Terms and Sentences of Empirical Science."
Mind. 73 (1964) 394-405.

    a. "Empirical Statements about the Absolute," by Wesley C.
    Salmon. 76 (1967) 430-31.

2058. _____. "Verificationism and Scepticism." AJP 56 (1978)
242-50.

2059. Schnädelbach, Herbert "Dispositionsbegriffe der Erkenntnistheorie: Zum Problem ihrer Sinnbedingungen." ZAW 2 (1971) 89-100.

2060. Schneider, Erna F. "Recent Discussion of Subjunctive Conditionals." Review of Metaphysics. 6 (1953) 623-49.

2061. Schock, Rolf. "On Classifications and Hierarchies." ZAW 10
(1979) 98-106.

2062. Sellars, Wilfrid. "Pure Pragmatics and Epistemology." PS 14
(1947) 181-202.

2063. _____. "Concepts as Involving Laws and Inconceivable Without Them." PS 15 (1948) 287-315.

2064. _____. "Counterfactuals, Dispositions, and the Causal Modalities." MSPS 2 (1958) 225-308.

2065. _____. "Naming and Saying." PS 29 (1962) 7-26.

2066. _____. "Meaning as Functional Classification. (A Perspective on the Relation of Syntax to Semantics.)" Synthese. 27 (1974) 417-38.

    a. "Comment on Wilfrid Sellars," by Daniel Dennett. 439-44.

    b. "Comment on Wilfrid Sellars," by Hilary Putnam. 445-56.

    c. "Reply," by Wilfrid Sellars. 457-66.

2067. Shvyrev, V. S. "The Neopositivist Conception of Empirical Significance, and Logical Analysis of Scientific Knowledge." Soviet Studies in Philosophy. 2 (1963) 10-29.

2068. Simon, H. A. "Identifiability and the Status of Theoretical Terms." WOSPS 11 (1977) 43-61.

2069. Smokler, Howard. "Informational Content: A Problem of Definition." J. Phil. 63 (1966) 201-11.

    a. "A Definition of Informational Content," by S. G. O'Hair. 66 (1969) 132-33.

2070. Stachow, E. W. "On a Game-Theoretic Approach to a Scientific Language." PSA 1978. II, 19-40.

2071. Stenius, Erik. "Beginning with Ordinary Things." Synthese. 19 (1968) 27-52. Reprinted in SL 21 (1975) 27-52.

2072. Stich, Stephen P., John Tinnon, and Lawrence Sklar. "Entailment and the Verificationist Program." Ratio 15 (1973) 84-97.

2073. Stonert, A. "Języki i teorie adekwatne z ontologią języka nauki." [Languages and Theories Adequate to the Ontology of Scientific Language] Studia Logica. 15 (1964) 49-78.

2074. Stopes-Roe, Harry V. "Some Considerations Concerning 'Interpretative Systems'." PS 25 (1958) 143-56.

2075. Stroud, Barry. "Conventionalism and the Indeterminacy of Translation." Synthese. 19 (1968) 82-96. Reprinted in SL 21 (1975) 82-96.

2076. Suppe, Frederick. "On Partial Interpretation." J. Phil. 68 (1971) 57-76.

2077. Suppes, Patrick. "Congruence of Meaning." Proc. & Add. Am.
Phil. Assn. 46 (1972-73) 21-38.

2078. _____. "The Essential but Implicit Role of Modal Concepts in Science." BSPS 20, SL 64 (1974) 305-14.

    a. "Comments on Suppes' Paper," by Aldo Bressan. 315-21.

    b. "Bressan and Suppes on Modality," by Bas C. van Fraassen.
    323-30.

    c. "Replies to van Fraassen's Comments," by Aldo Bressan.
    331-34.

2079. Svenonius, Lars. "Translation and Reduction." SL 50 (1973)
31-50.

2080. Švyrev, V. S. "Problems of the Logical-Methodological Analysis of Relations Between the Theoretical and Empirical Planes of
Scientific Knowledge." SL 25 (1970) 55-90.

2081. Swinburne, R. G. "Vagueness, Inexactness, and Imprecision."
BJPS 19 (1969) 281-99.

2082. Tibbetts, Paul. "Feigl on Raw Feels, the Brain, and Knowledge
Claims: Some Problems regarding Theoretical Concepts." Dialectica.
26 (1972) 247-66.

2083. Tondl, Ladislav. "Problems of Empirical Basis of Science."
Organon. 8 (1971) 5-26.

2084. _____. Problems of Semantics: A Contribution to the
Analysis of the Language of Science. BSPS 66 Dordrecht: Reidel,
1981. 417 pp.

2085. Trapp, Rainer. "Eine Verfeinerung des Reduktionssatzverfahrens
zur Einführung von Dispositionsprädikaten." Erkenntnis. 9 (1975)
355-82.

2086. Tuomela, Raimo. "On Empirical Models of Scientific Theories."
Ajatus. 30 (1968) 169-94.

2087. _____. "Identifiability and Definability of Theoretical
Concepts." Ajatus. 30 (1968) 195-220.

2088. _____. "Model Theory and Empirical Interpretation of
Scientific Theories." Synthese. 25 (1972) 165-75.

2089. _____. Theoretical Concepts. Wien and New York:
Springer, 1973. 254 pp.

2090. _____. "Theoretical Concepts in Neobehavioristic Theories,"
in The Methodological Unity of Science. M. Bunge, ed. Dordrecht:
Reidel, 1973. 123-52.

2091. _____. "Empiricist vs. Realist Semantics and Model Theory." Synthese. 26 (1974) 407-08.

2092. _____. Dispositions. SL 113. Dordrecht: Reidel, 1978. 450 pp.

2093. Vuillemin, J. "Le concept de signification empirique (Stimulus Meaning) chez Quine." RIP 30 (1976) 350-75.

2094. Waismann, F. "Verifiability." Arist. Soc. Supp. Vol. 19 (1945) 119-50.

2095. Walton, Kendall L. "Linguistic Relativity." SL 52 (1973) 1-30.

2096. Warnock, G. J. "Verification and the Use of Language." RIP 5 (1951) 307-22.

2097. White, Morton. "Why Does a Causal Conditional Seem to Assert Possibility When in Fact it Does Not?" Synthese. 26 (1974) 391-95.

2098. Whitrow, G. J. "Operational Analysis and the Nature of Some Physical Concepts." Bulletin of the British Society for the History of Science. 1 (1950) 101-04.

2099. Wilson, Fred. "Implicit Definition Once Again." J. Phil. 62 (1965) 364-74.

2100. _____. "Definition and Discovery." BJPS 18 (1968) 287-303; 19 (1968) 43-56.

2101. _____. "Dispositions: Defined or Reduced?" AJP 47 (1969) 184-204.

2102. Wilson, N. L. "On Semantically Relevant Whatsits: A Semantics for Philosophy of Science." SL 52 (1973) 233-45.

2103. Winnie, John A. "Theoretical Terms and Partial Definitions." PS 32 (1965) 324-28.

2104. _____. "The Implicit Definition of Theoretical Terms." BJPS 18 (1967) 223-29.

2105. Wojcicki, R. "Semantical Criteria of Empirical Meaningfulness." Studia Logica. 19 (1966) 75-110. Reprinted in SL 87 (1977) 647-80.

2106. _____. "Towards a General Semantics of Empirical Theories." WOSPS 11 (1977) 31-39.

2107. Woodger, J. H. "Science Without Properties." BJPS 2 (1951) 193-216.

2108. Yourgrau, Wolfgang, and Chandler Works. "A New, Formalized Version of the Verifiability Principle." Ratio 10 (1968) 54-63.

4.7 Scientific Laws

2109. Agassi, J. "What is a Natural Law?" <u>Studium Generale</u>. 24 (1971) 1051-56.

2110. Alexander, Peter. "Subjunctive Conditionals." <u>Arist. Soc.</u> Supp. Vol. 36 (1962) 185-200.

2111. Ayer, A. J. "What is a Law of Nature?" RIP 10 (1956) 144-65.

2112. Ayers, M. R. "Counterfactuals and Subjunctive Conditionals." <u>Mind</u>. 74 (1965) 347-64.

       a. "M. R. Ayers on the Conditional," by R. N. McLaughlin. 77 (1968) 290-92.

2113. Bacon, John. "Purely Physical Modalities." <u>Theoria</u>. 47 (1981) 134-41.

2114. Bass, Robert E. "Causality, Probability and Organization." PPR 12 (1952) 562-64.

2115. Beardsley, Elizabeth L. "'Non-Accidental' and Counterfactual Sentences." <u>J. Phil</u>. 46 (1949) 573-91.

2116. Beauchamp, Tom L. "Cosmic Epochs and the Scope of Scientific Laws." <u>Process Studies</u>. 2 (1972) 296-300.

2117. Berenda, C. W. "Are Natural Laws Simplified Empirically Useful Tautologies?" <u>Southwestern Journal of Philosophy</u>. 4, No. 2 (1973) 93-100.

2118. Berofsky, Bernard. "The Regularity Theory." <u>Nous</u>. 2 (1968) 315-40.

2119. Bertalanffy, Ludwig von. "Chance or Law," in <u>Beyond Reductionism</u>. A. Koestler and J. R. Smythies, eds. New York: Macmillan, 1970. 56-84.

2120. Biser, Erwin. "Invariance and Timeless Laws." <u>Methodos</u>. 7 (1955) 213-32.

2121. Blum, Alex. "Laws and Instantial Statements." BJPS 21 (1970) 371-78.

2122. Bowie, G. Lee. "The Similarity Approach to Counterfactuals: Some Problems." <u>Nous</u>. 13 (1979) 477-98.

2123. Bressan, A. "On Physical Possibility." BSPS 47 (1981) 197-214.

2124. Brown, Robert, and John Watling. "Counterfactual Conditionals." Mind 61 (1952) 222-33.

2125. Buchdahl, Gerd. "Semantic Sources of the Concept of Law." Synthese. 17 (1967) 54-74. Reprinted in BSPS 3 (1967) 272-92.

2126. Bunge, Mario. "Ley y determinación." Scientia. 96 (1961) 3-7. French Tr.: Supplement, 1-5.

2127. _____. "Causality, Chance, and Law." American Scientist. 49 (1961) 432-48.

2128. _____. "Kinds and Criteria of Scientific Law." PS 28 (1961) 260-81.

2129. Burks, A. W. "Laws of Nature and Reasonableness of Regret." Mind. 55 (1946) 170-72.

2130. Byrne, Peter. "Miracles and the Philosophy of Science." Heythrop Journal. 19 (1978) 162-70.

2131. Inadvertently skipped in numbering the entries.

2132. Cartwright, Nancy. "Causal Laws and Effective Strategies." Nous. 13 (1979) 419-37.

2133. _____. "Do the Laws of Physics State the Facts?" Pacific Philosophical Quarterly. 61 (1980) 75-84.

    a. "Causal Explanation and the Reality of Natural Component Forces," by Lewis G. Creary. 62 (1981) 148-57.

2134. Chandra, Sri. "The Nature of Laws in Physical and Social Sciences." Philosophical Quarterly (India). 29 (1956) 145-54.

2135. Chisholm, R. M. "The Contrary-to-Fact Conditional." Mind. 55 (1946) 289-307.

    a. "The Contrary-to-Fact Conditional," by Frederick L. Will. 56 (1947) 236-49.

2136. _____. "Law Statements and Counterfactual Inference." Analysis. 15 (1955) 97-105.

2137. Clark, Romane. "On What is Naturally Necessary." J. Phil. 62 (1965) 613-25.

    a. "Clark on Natural Necessity," by Roger C. Buck. 625-29.

    b. "'Defeasible' Problems," by Peter Achinstein. 629-33.

2138. Clay, J. "Laws and Principles." Synthese. 5 (1946) 338-48.

2139. Cole, Richard. "Nomos." Southwestern Journal of Philosophy. 10, No. 1 (1979) 7-21.

2140. Day, Patrick. "The Uniformity of Nature." _Am. Phil. Quart._ 12 (1975) 1-16.

2141. Diamond, Cora. "Mr. Goodman on Relevant Conditions and the Counterfactual." _Phil. Studies_. 10 (1959) 42-45.

2142. Diggs, B. J. "Counterfactual Conditionals." _Mind_. 61 (1952) 513-27.

2143. Dretske, Fred I. "Laws of Nature." PS 44 (1977) 248-68.

   a. "Dretske on Laws of Nature," by Ilkka Niiniluoto. 45 (1978) 431-39.

   b. "Reply to Niiniluoto," by Fred I. Dretske. 440-44.

2144. Earman, John. "The Universality of Laws. PS 45 (1978) 173-81.

2145. Fetzer, James H. "The Likeness of Lawlikeness." BSPS 32 (1976) 377-92.

2146. Fetzer, James H., and Donald E. Nute. "Syntax, Semantics, and Ontology: A Probabilistic Causal Calculus." _Synthese_. 40 (1979) 453-96.

2147. Feynman, Richard. _The Character of Physical Law_. Cambridge, MA: MIT Press, 1965. 173 pp.

2148. Finch, Henry Albert. "An Explication of Counterfactuals by Probability Theory." PPR 18 (1958) 368-78.

   a. "A Note on Finch's "An Explication of Counterfactuals by Probability Theory," by Richard C. Jeffrey. 20 (1959) 116.

   b. "Due Care in Explicating Counterfactuals: A Reply to Mr. Jeffrey," by Henry Albert Finch. 117-18.

2149. Fraassen, Bas C. van. "The Only Necessity is Verbal Necessity." _J. Phil._ 74 (1977) 71-86.

2150. Franzwa, Gregg. "Supported Counterfactuals in Non-Causal Contexts." _Southwestern Journal of Philosophy_. 11, No. 1 (1980) 97-103.

2151. Fumerton, R. A. "Subjunctive Conditionals." PS 43 (1976) 523-38.

2152. Gardner, Martin. "Order and Surprise." PS 17 (1959) 109-17.

2153. Goldberg, Bruce. "On the Metalinguistic Interpretation of Counterfactuals." _J. Phil._ 60 (1963) 291-95.

2154. Goodman, Nelson. "The Problem of Counterfactual Conditionals." _J. Phil._ 44 (1947) 113-28.

   a. "Goodman and Relevant Conditions," by R. H. Vincent. Phil. Studies. 12 (1961) 28-29.

2155. Grene, Marjorie. "On the Nature of Natural Necessity." BSPS 23, SL 66 (1974) 228-42.

2156. Grünbaum, Adolf. "Law and Convention in Physical Theory," in Current Issues in the Philosophy of Science. H. Feigl and G. Maxwell, eds. New York: Holt, Rinehart and Winston, 1961. 140-55.

   a. "Comments," by Paul K. Feyerabend. 155-61.

   b. "Rejoinder," by Adolf Grünbaum. 161-68.

2157. Hallie, Philip P. "On So-Called 'Counterfactual Conditionals'." J. Phil. 51 (1954) 273-78.

2158. Harré, R. "Dimensions of Generality." Ratio 4 (1962) 143-56.

2159. _____. "Powers." BJPS 21 (1970) 81-101.

2160. Hesse, Mary. "Subjunctive Conditionals." Arist. Soc. Supp. Vol. 26 (1962) 201-14.

2161. _____. "Miracles and the Laws of Nature," in Miracles. C. F. D. Moule, ed. London: A. R. Mowbray, 1965. 32-43.

2162. Hochberg, Herbert. "Natural Necessity and Laws of Nature." PS 48 (1981) 386-99.

2163. Hung, H.-C. "Nomic Necessity is Cross-Theoretic." BJPS 32 (1981) 219-36.

2164. Inwagen, Peter van. "Laws and Counterfactuals." Nous. 13 (1979) 439-53.

2165. Issman, Samuel. "Les énoncés nomologiques." Annals de l'Institut de Philosophie. (Brussels). (1971) 161-83.

2166. Jackson, Frank. "A Causal Theory of Counterfactuals." AJP 55 (1977) 3-21.

   a. "Jackson on Counterfactuals," by Wayne A. Davis. 58 (1980) 62-65.

2167. Jobe, Evan K. "Some Recent Work on the Problem of Law." PS 34 (1967) 363-81.

2168. _____. "Reichenbach's Theory of Nomological Statements." Synthese. 35 (1977) 231-54. Reprinted in SL 132 (1979) 697-720.

2169. Juhos, Béla. "Wie gewinnen wir Naturgesetze?" ZPF 22 (1968) 534-48.

2170. _____. "Makrophänomene und ihre Zusammensetzung aus Mikrophänomene." PN 12 (1970) 413-20.

2171. Kaminsky, Jack. "Corrigibility and Law." PS 21 (1954) 9-15.

2172. Kamlah, Andreas. "Invarianzgesetze und Zeitmetrik." ZAW 4 (1973) 224-60.

2173. Kanitscheider, Bernulf. "Der semantische Status von Natur-gesetzen." Conceptus. 7 (1973) 27-43.

2174. Kar, Robert. "Ein Konflikt zwischen den Natur-und Denkgeset-zen." PN 4 (1957) 12-57.

2175. King, John L. "Coextensiveness and Lawlikeness." Erkenntnis. 14 (1979) 359-63.

2176. Körner, S. "On Laws of Nature." Mind. 62 (1953) 216-29.

2177. Krimerman, Leonard I. "Laws and Counterfactuals." Phil. Studies. 16 (1965) 40-44.

   a. "Laws and Counterfactuals in Nagel: A Reply to Krimerman," by George Coe. 18 (1967) 24-27.

2178. Kröber, G. "Strukturgesetz und Gesetzessturktur." DZP 15 (1967) 202-16.

2179. Lauter, H. A. "An Examination of Reichenbach on Laws." PS 37 (1970) 131-45.

2180. Lewis, David K. Counterfactuals. Cambridge: Harvard University Press, 1973. x + 150 pp.

   a. Review essay by K. Fine. Mind. 84 (1975) 451-58.

2181. _____. "Counterfactuals and Comparative Possibility." WOSPS 4 (1975) 1-30.

2182. _____. "Counterfactual Dependence and Time's Arrow." Nous. 13 (1979) 455-76.

2183. Ley, H. "Differenz von Gesetz und Regel unter einzelwissen-schaftlichem und methodologischem Aspekt." DZP 22 (1974) 1359-73.

2184. Lins, Mario. The Philosophy of Law: Its Epistemological Problems. Rio de Janeiro: Livraria Freitas Bastos S/A, 1971. 164 pp.

2185. Lloyd, A. C. "The Logical Form of Law Statements." Mind. 64 (1955) 312-18.

2186. Long, Peter. "Natural Laws and So-Called Accidental General Statements." Analysis. 13 (1952-53) 18-23.

2187. Lövestad, Ludvig. "The Structure of Physical Laws." Theoria. 11 (1945) 40-70.

2188. Lowe, E. J. "Sortal Terms and Natural Laws -- An Essay on the Ontological Status of the Laws of Nature." Am. Phil. Quart. 17 (1980) 253-60.

2189. Lyon, Ardon. "The Immutable Laws of Nature." Proc. Arist. Soc. 77 n.s. (1976-77) 107-26.

2190. Marković, Mihailo. "The Concept of Physical Necessity." LMPS-4 (1971) 967-76.

2191. Mejbaum, W. "A Law of Science as an Open Formula." Reports on Philosophy. 1 (1977) 43-49.

2192. Mercier, André. "Lois de la nature et constructions de l'esprit." Synthese. 5 (1946) 203-10.

2193. Milmed, B. K. "Counterfactual Statements and Logical Modality." Mind. 66 (1957) 453-70.

2194. Morton, Adam. "If I were a Dry Well-Made Match." Dialogue. 12 (1973) 322-24.

   a. "Morton on Causal Laws," by Anthony Willing. 13 (1974) 577-78.

   b. "Reply to Willing," by Adam Morton. 579.

2195. Müller, Henning. "Gesetz und Denken in der exakten Natur-wissenschaft." PN 7 (1961-62) 167-79.

2196. Nuchelmans, G. "The Analysis of Counterfactual Conditionals." Synthese. 9 (n.d.) 48-63.

2197. Nute, Donald. "Causes, Laws, and Law Statements." Synthese. 48 (1981) 347-69.

2198. O'Connor, D. J. "The Analysis of Conditional Sentences." Mind. 60 (1951) 351-62.

2199. Olsen, Jens Henrik. "Frequence, Cause, and Modality: Reflections on the Nature of Nomic Connection and the Foundation of Scientific Explanation." Danish Yearbook of Philosophy. 15 (1978) 7-33.

2200. Palmieri, L. E. "Law Statements and the Subject-Predicate Form." Methodos. 7 (1955) 209-11.

2201. Parry, William Tuthill. "Re-examination of the Problem of Counterfactual Conditionals." J. Phil. 54 (1957) 85-93.

   a. "Parry on Counterfactuals," by Nelson Goodman. 442-45.

2202. Patryas, W. "An Analysis of the 'Caeteris Paribus' Clause." Poznan Studies. 1, No. 1 (1975) 59-64.

2203. Pears, David. "Hypotheticals." Analysis. 10 (1949-50) 49-63.

     a. "Natural Laws and Contrary to Fact Conditionals," by William Kneale. 121-25.

2204. Peres, Asher. "The Physicist's Role in Physical Laws." FP 10 (1980) 631-34.

2205. Pietarinen, Juhani. "Lawlikeness, Analogy, and Inductive Logic." Acta Philosophica Fennica. 26 (1972) 1-143.

2206. Popper, K. R. "A Note on Natural Laws and So-Called 'Contrary-to-Fact Conditionals'." Mind. 58 (1949) 62-66.

2207. Presley, C. F. "Laws and Theories in the Physical Sciences." AJP 32 (1954) 79-103.

2208. Proctor, G. L. "Scientific Laws and Scientific Objects in the Tractatus." BJPS 10 (1959) 177-93.

2209. Purtill, Richard L. "Toulmin on Ideals of Natural Order." Synthese. 22 (1971) 431-37.

2210. Putnam, Hilary. "On Properties." SL 24 (1970) 235-54.

2211. Reenpää, Yrjö. "Ueber den Begriff der Gesetzesartigheit." ZPF 26 (1972) 253-65.

2212. Reichenbach, Hans. Nomological Statements and Admissible Operations. Amsterdam: North-Holland, 1954. Reprinted under title: Laws, Modalities, and Counterfactuals, with a Foreward by W. C. Salmon. Berkeley: University of California Press, 1976. xliii + 140 pp.

     a. Review essay, "Reichenbach's Theory of Reasonable Assertion," by Evan Jobe. BJPS 31 (1980) 375-84.

2213. Rescher, Nicholas. "A Factual Analysis of Counterfactual Conditionals." Phil. Studies. 11 (1960) 49-54.

2214. _____. "Lawfulness as Mind-Dependent." SL 24 (1970) 178-97.

2215. _____. "Counterfactual Hypotheses, Laws and Dispositions." Nous. 5 (1971) 157-78.

2216. Richter, G. "Zur positivistischen Konzeption des Gesetzes." DZP 22 (1974) 1374-86.

2217. Rosenthal-Schneider, Ilse. "The Laws of Nature in the Light of Modern Physics and Biology." Australian Journal of Science. 8 (1946) 120-23.

2218. Ruddick, Chester T. "Hume on Scientific Law." PS 16 (1949)
89-93.

2219. Saini, Hugo. "Reflexions sur la notion de Loi en Physique."
Studia Philosophica. 6 (1946) 81-119.

2220. Salmon, Wesley C. "Laws, Modalities, and Counterfactuals."
Synthese. 35 (1977) 191-230. Enlarged version printed in SL 132
(1979) 655-96.

2221. Schleichert, Hubert. "Ueber den Mitteilungsgehalt und die
konventionellen Grundlagen von Naturgesetzen." Archiv für Philosophie.
11 (1962) 179-86. English Tr.: Philosophy Today. 7 (1963) 33-38.

2222. Schmutzer, E. "Symmetrien in den physikalischen Naturgesetzen."
Scientia. 106 (1971) 51-65. English Tr., 66-76.

2223. Schrödinger, Erwin. Was ist ein Naturgesetz? Munchen:
Oldenbourg, 1967. 146 pp.

2224. Schulz, D. J. "On General Criteria for Lawlike Propositions."
Methodology and Science. 9 (1976) 9-14.

2225. Scriven, Michael. "The Key Property of Physical Laws -- In-
accuracy," in Current Issues in the Philosophy of Science. H. Feigl
and G. Maxwell, eds. New York: Holt, Rinehart and Winston, 1961.
91-101.

        a. "Comments," by Henryk Mehlberg. 102-03.

        b. "Rejoinder," by Michael Scriven. 103-04.

2226. Seall, Robert E. "Truth-Valued Fluents and Qualitative
Laws." PS 30 (1963) 36-40.

2227. Senechal, Marjorie, and George Fleck, eds. Patterns of Symmetry. Amherst, MA: University of Massachusetts Press, 1977.
160 pp.

2228. Sharpe, R. A. "Laws, Coincidences, Counterfactuals and
Counter-Identicals." Mind. 80 (1971) 572-82.

2229. Skyrms, Brian. "Physical Laws and the Nature of Philosophical
Reduction." MSPS 6 (1975) 496-529.

2230. _____. "Resiliency, Propensities, and Causal Necessity."
J. Phil. 74 (1977) 704-13.

        a. "Some Remarks on the Concept of Resiliency," by Patrick
        Suppes. 513-14.

2231. _____. "Statistical Laws and Personal Propensities."
PSA 1978. II, 551-62.

2232. _____. Causal Necessity: A Pragmatic Investigation of the Necessity of Laws. New Haven, CT: Yale University Press, 1980. 176 pp.

    a. Review essay by James H. Fetzer. PS 48 (1981) 329-35.

2233. Sosa, Ernest. "Hypothetical Reasoning." J. Phil. 64 (1967) 293-305.

2234. Stegmüller, W. "Begriff des Naturgesetzes." Studium Generale. 19 (1966) 649-57.

2235. Stove, D. C. "Laws and Singular Propositions." AJP 51 (1973) 139-43.

2236. Suchting, W. A. "Regularity and Law." BSPS 14, SL 60 (1974) 73-90.

2237. Sudbury, A. W. "Could There Exist a World Which Obeyed no Scientific Laws?" BJPS 24 (1973) 39-40.

2238. _____. "Scientific Laws That are Neither Deterministic nor Probablistic." BJPS 27 (1976) 307-15.

2239. Temple, Dennis. "Nomic Necessity and Counterfactual Force." Am. Phil. Quart. 15 (1978) 221-27.

2240. Thompson, Manley. "What are Law-Statements About?" J. Phil. 52 (1955) 421-33.

2241. Tooley, Michael. "The Nature of Laws." Canadian Journal of Philosophy. 7 (1977) 667-98.

2242. Tredwell, R. F. "The Problem of Counterfactuals." PS 32 (1965) 310-23.

2243. Ushenko, A. P. "The Principles of Causality." J. Phil. 50 (1953) 85-101.

    a. "Universality, Explanation, and Scientific Law," by Albert Hofstadter. 101-15.

    b. "Some Remarks on Professor Ushenko's Interpretation of Causal Law," by Aldolf Grünbaum. 115-20.

    c. "Remarks on Causation and Compulsion," by Donald C. Williams. 120-24.

2244. _____. "The Counterfactual." J. Phil. 51 (1954) 369-83.

2245. Weinberger, Christa. "Das wissenschaftliche Gesetz und die definitorischen Festsetzungen der wissenschaftlichen Sprache." Conceptus. 7 (1973) 45-51.

2246. Weinberger, Ota. "Der nomische Allsatz." Grazer Philosophische Studien. 4 (1977) 31-41.

2247. Weiner, Joan. "Counterfactual Conundrum." Nous. 13 (1979) 499-509.

2248. Wendt, H. "Bemerkungen zum Strukturbegriff und zum Begriff Strukturgesetz." DZP 14 (1966) 545-61.

2249. Wenzlaff, B. "Symmetrien als allgemeine Strukturgesetze der Natur." DZP 11 (1963) 1217-29.

2250. Weyl, Hermann. Symmetry. New Jersey: Princeton University Press, 1952. 163 pp.

    a. Review essay by J. D. Bernal. BJPS 5 (1955) 335-41.

2251. Wilson, Mark. "Generality and Nomological Form." PS 46 (1979) 161-64.

2252. Woolhouse, R. S. "Counterfactuals, Dispositions, and Capacities." Mind. 82 (1973) 557-65.

2253. Zwart, P. J. "The Way of Science." Methodology and Science. 9 (1976) 15-40.

4.8 Theories

2254. Achinstein, Peter. "Macrotheories and Microtheories." LMPS-4 (1971) 533-66.

2255. Ackermann, Robert J. "Confirmatory Models of Theories." BJPS 16 (1966) 312-26.

2256. Ager, Tryg A., Jerrold L. Aronson, and Robert Weingard. "Are Bridge Laws Really Necessary?" Nous. 8 (1974) 119-34.

    a. "Identities and Reduction: A Reply," by Robert L. Causey. 10 (1976) 333-37.

2257. Alder, Michael D. "On Theories." PS 40 (1973) 212-26.

2258. Alexander, Peter. "Theory-construction and Theory-testing." BJPS 9 (1958) 29-38.

2259. _____. "Speculations and Theories," in Form and Strategy in Science. J. R. Gregg and F. T. C. Harris, eds. Dordrecht: Reidel, 1964. 30-46. Reprinted from Synthese. 15 (1963) 187-203.

2260. Altwegg, M. "Theorie und Erfahrung." Dialectica. 7 (1953) 5-21.

2261. Balzer, Wolfgang, and C. Ulises Moulines. "On Theoreticity."
Synthese. 44 (1980) 467-94.

2262. Balzer, Wolfgang, and Joseph D. Sneed. "Generalized Net
Structures of Empirical Theories." Studia Logica. 36 (1977) 195-211,
37 (1978) 167-94.

2263. Beck, Guido. "Mathematical Formalism and the Physical Picture."
PS 12 (1945) 174-78.

2264. Berent, Paul. "Theoretical Terms in Infinite Theories." PS
40 (1973) 129.

       a. "Ramsey Sentences for Infinite Theories," by Herbert E.
       Hendry. 42 (1975) 28.

2265. Blackwell, Richard J. "A Structuralist Account of Scientific
Theories," IPQ 16 (1976) 263-74.

2266. Bohm, David. "Science as Perception-Communication," in The
Structure of Scientific Theories. F. Suppe, ed. Urbana, Illinois:
University of Illinois Press, 1974. 374-91.

       a. "Professor Bohm's View of the Structure and Development
       of Theories," by Robert L. Causey. 392-401.

       b. "Reply to Professor Causey," by Jeffrey Bub. 402-408.

       c. "Discussion." 409-19.

       d. "Reply to Discussion," by David Bohm. 420-23.

2267. Bohnert, Herbert G. "Communication by Ramsey-Sentence Clause."
PS 34 (1967) 341-47.

2268. Braithwaite, R. B. "The Nature of Theoretical Concepts and
the Role of Models in an Advanced Science." RIP 8 (1954) 34-40.

2269. Bromberger, Sylvain. "A Theory about the Theory of Theory
and about the Theory of Theories," in Philosophy of Science: The Delaware Seminar, II. New York: Interscience Publishers, 1963. 79-105.

2270. _____. "Why-Questions," in Mind and Cosmos. Robert G.
Colodny, ed. Pittsburgh: University of Pittsburgh Press, 1966.
86-111.

2271. Brown, Harold I. "A Functional Analysis of Scientific Theories."
ZAW 10 (1979) 119-40.

2272. Brunner, Karl. "'Assumptions' and the Cognitive Quality of
Theories." Synthese. 20 (1969) 501-25.

2273. Buchdahl, Gerd. "Theory Construction: The Work of Norman
Robert Campbell." Isis. 55 (1964) 151-62.

2274. Bunge, Mario. "What are Physical Theories About?" in Studies in the Philosophy of Science. N. Rescher, ed. Oxford: Basil Blackwell, 1969. 61-99.

2275. _____. "Problems Concerning Intertheory Relations." SL 31 (1970) 285-315.

2276. Byerly, Henry C. "Professor Nagel on the Cognitive Status of Theories." PS 35 (1968) 412-23.

2277. Carloye, Jack C. "Fruitfulness of Scientific Theories." Methodology and Science. 7 (1974) 57-73.

2278. Causey, Robert L. "Theory and Observation," in Current Research in Philosophy of Science. P. D. Asquith and H. E. Kyburg, Jr., eds. East Lansing, MI: Philosophy of Science Assn., 1979. 187-206.

2279. Caws, Peter. "The Functions of Definition in Science." PS 26 (1959) 201-28.

2280. Clay, J. "Le rapport entre l'expérience et la théorie." Dialectica. 6 (1952) 266-69.

2281. Cornman, James W. "Theoretical Phenomenalism." Nous. 7 (1973) 120-38.

2282. Cresswell, M. J. "Physical Theories and Possible Worlds." Logique et analyse. 16 (1973) 495-511.

2283. Dalla Chiara Scabia, Maria Luisa. Modelli sintattici e semantici delle teorie elementari. Milan: Feltrinelli Editore, 1968. 240 pp.

2284. Darden, Lindley, and Nancy Maull. "Interfield Theories." PS 44 (1977) 43-64.

2285. Darmstadter, Howard. "Better Theories." PS 42 (1975) 20-27.

2286. Destouches, J.-L. "Intervention d'une logique de modalité dans une théorie physique." Synthese. 7 (1948-49) 411-17.

2287. Destouches-Fevrier, P. "Logique et théories physiques." Synthese. 7 (1948-49) 400-10.

2288. Duhem, Pierre. La théorie physique: Son objet, sa structure. Paris: Marcel Rivière & Cie, 1914 (second edition). English translation: The Aim and Structure of Physical Theory. Foreward by Louis de Broglie. Philip P. Wiener, tr. Princeton: Princeton University Press, 1954. Reprint: New York: Atheneum, 1974. xxii + 344.

2289. Echarri, J. "Expérience et théorie. Niveaux d'expérience." Dialectica. 6 (1952) 291-304.

2290. Egidi, Rosaria. Il linguaggio delle teorie scientifiche. Napoli: Guida Editori, 1979. 233 pp.

2291. Eilstein, H., ed. Teoria i doświadczenie. Warszawa, 1966.

2292. Ellis, Brian. "A Comparison of Process and Non-Process Theories in the Physical Sciences." BJPS 8 (1957) 45-56.

2293. Enç, Berent. "Spiral Dependence Between Theories and Taxonomy." Inquiry. 19 (1976) 41-71.

2294. Faggiani, D. "Le teorie fisiche ed i criterio di normalità." Scientia. 106 (1971) 245-54. English Tr.: 255-63.

2295. Feigl, Herbert. "The 'Orthodox' View of Theories: Remarks in Defense as well as Critique." MSPS 4 (1970) 3-16.

2296. Feyerabend, P. K. "Explanation, Reduction, and Empiricism." MSPS 3 (1962) 28-97.

2297. _____. "Problems of Empiricism," in Beyond the Edge of Certainty: Essays in Contemporary Science and Philosophy. Robert G. Colodny, Ed. Englewood Cliffs, NJ: Prentice-Hall, 1965. 145-260.

2298. _____. "Problems of Empiricism, Part II" in The Nature and Function of Scientific Theories: Essays in Contemporary Science and Philosophy. Robert G. Colodny, ed. Pittsburgh: University of Pittsburgh Press, 1970. 275-353.

2299. Fine, Arthur, I. "Explaining the Behavior of Entities." Phil. Review. 75 (1966) 496-509.

2300. Gagnebin, S. "Théorie et expérience: Le probleme." Dialectica. 6 (1952) 120-29.

2301. Gonseth, F. "Théorie et expérience." Dialectica. 6 (1952) 143-44.

2302. Gorskii, D. P. "On the Means of Generalizing Scientific Theory." Soviet Studies in Philosophy. 5, No. 4 (1967) 29-38. [English reprint from Voprosy filosofii. (1966, No. 8)]

2303. Griaznov. B. S., B. S. Dynine, and E. P. Nikitine. "L'objet du savoir théorique." RIP 25 (1971) 538-46.

2304. Grize, Jean-Blaise. "Théorie et explication." Studia Philosophica. 21 (1961) 51-67.

2305. Haller, Rudolf. "Theories, Fables, and Parables." Grazer Philosophische Studien. 12/13 (1981) 105-17.

2306. Hamann, J. R. "On the Notion of Optimality in Theories and Theory Construction." Scientia. 100 (1965) 185-89. French Tr.: Supplement, 93-97.

2307. Hanson, Norwood Russell. "Equivalence: The Paradox of
Theoretical Analysis." AJP 41 (1963) 217-32. Revised version
published in Mind, Matter and Method. P. Feyerabend and G. Maxwell,
eds. Minneapolis: University of Minnesota Press, 1966. 413-29.

2308. _____. "Number Theory and Physical Theory: An Analogy."
BSPS 2 (1965) 93-119.

         a. "Comments: On Chronically Unsolved Basic Problems in
         Physical Theories," by Armand Siegel. 121-26.

2309. Harris, John H. "On Comparing Theories." Synthese. 32
(1975) 29-76.

2310. _____. "Strong Scientific Theories." PS 45 (1978)
182-205.

2311. _____. "A Semantical Alternative to the Sneed-Stegmüller-
Kuhn Conception of Scientific Theories." Acta Philosophica Fennica.
30, 2-4 (1978) 184-204.

2312. Heisenberg, Werner. "Der Begriff 'Abgeschlossene Theorie' in
der modernen Naturwissenschaft." Dialectica. 2 (1948) 331-36.

2313. Hempel, Carl G. Fundamentals of Concept Formation in Empirical
Science. International Encyclopedia of Unified Science, Vol. II,
No. 7. Chicago: University of Chicago Press, 1952. 93 pp.

2314. _____. "The Theoretician's Dilemma: A Study in the
Logic of Theory Construction." MSPS 2 (1958) 37-98. Reprinted,
with some changes, in C. G. Hempel, Aspects of Scientific Explanation and Other Essays in the Philosophy of Science. New York: Free
Press, 1965. 173-226.

2315. _____. "On the Structure of Scientific Theories," in
The Isenberg Memorial Lecture Series, 1965-1966. East Lansing,
Michigan: Michigan State University Press, 1969. 11-38.

2316. _____. "On the 'Standard Conception' of Scientific
Theories." MSPS 4 (1970) 142-63.

2317. _____. "A Problem in the Empiricist Construal of Theories."
[In Hebrew] Iyyun. 23 (1972) 69-81.

2318. _____. "Formulation and Formalization of Scientific
Theories: A Summary-Abstract," in The Structure of Scientific Theories.
F. Suppe, ed. Urbana, Illinois: University of Illinois Press,
1974. 244-54.

2319. Hesse, Mary B. "Theories, Dictionaries, and Observation."
BJPS 9 (1958) 12-28.

         a. "A Note on 'Theories, Dictionaries, and Observation'," by
         Mary B. Hesse. 128-29.

2320. Hintikka, Jaakko. "On the Different Ingredients of an Empirical Theory." LMPS-4 (1971) 313-22.

2321. Hoering, W. "On Hypotheses Attached to Theoretical Concepts." Acta Philosophica Fennica. 20, 2-4 (1978) 179-83.

2322. Hooker, C. A. "On Global Theories." PS 42 (1975) 152-79.

2323. Hooker, C. A., ed. Physical Theory as a Logico-Operational Structure. Dordrecht: Reidel, 1979. xvii + 334 pp.

2324. Hübner, R. "Theorie und Empirie." PN 10 (1967-68) 198-210.

2325. Humphreys, Willard C. Anomalies and Scientific Theories. San Francisco: Freeman, Cooper & Co., 1968. 318 pp.

2326. Hutten, Ernest H. "On Semantics and Physics." Proc. Arist. Soc. 49 n.s. (1948-49) 115-32.

2327. Kaiser, Walter. "Operative Gesichtspunkte bei der Diskussion des Weberschen Gesetzes." ZAW 8 (1977) 39-47.

2328. Kamber, F. "The Structure of the Propositional Calculus of a Physical Theory." WOSPS 5 (1975) 221-46.

2329. Kamlah, Andreas. "An Improved Definition of 'Theoretical in a Given Theory'." Erkenntnis. 10 (1976) 349-59.

2330. Kaufman, Steven A. "The Preservation of Epistemic Systematization within the Extended Craigian Program." Synthese. 28 (1974) 215-22.

     a. "Inducibility and Existemic Systematization: Rejoinder to Kaufman," by Ilkka Niiniluoto, 223-32.

2331. Kleiner, Scott A. "Ontological and Terminological Commitment and the Methodological Commensurability of Theories." BSPS 8, SL 39 (1971) 507-18.

2332. _____. "Feyerabend, Galileo and Darwin: How to Make the Best out of What You Have -- or Think You Can Get." SHPS 10 (1979) 285-309.

2333. Koningsveld, H. "Observational and Theoretical Terms." Methodology and Science. 5 (1972) 49-76.

2334. Krüger, Lorenz. "Reduction versus Elimination of Theories." Erkenntnis. 10 (1976) 295-309.

2335. Küttner, Michael. "Theorie unter dem Non-Statement View und der Kuhnsche Wissenschaftler." ZAW 12 (1981) 163-77.

2336. Kuznetsov, I. V. "The Structure of the Scientific Theory and the Structure of the Object." Soviet Studies in Philosophy. 7, No. 2 (1968) 15-26. [English reprint from Voprosy filosofii. (1968, No. 5)]

2337. Kyburg, Jr., Henry E. "How to Make Up a Theory." Phil. Review. 85 (1978) 84-87.

2338. Laszlo, Ervin. "The Rise of General Theories in Contemporary Science." ZAW 4 (1973) 335-44.

2339. Laymon, Ronald. "Feyerabend, Brownian Motion, and the Hiddenness of Refuting Facts." PS 44 (1977) 225-47.

2340. Lehrer, Keith. "Theoretical Terms and Inductive Inference," in Studies in the Philosophy of Science. N. Rescher, ed. Oxford: Basil Blackwell, 1969. 30-41.

2341. Leinfellner, Werner. Struktur und Aufbau wissenschaftlicher Theorien. Wien: Physica-Verlag, Rudolf Liebing K G, 1965. 307 pp.

2342. Ludwig, G. Die Grundstrukturen einer physikalischen Theorie. Berlin, 1978.

    a. Review essay by W. Balzer. Erkenntnis. 15 (1980) 391-408.

2343. _____. "Axiomatische Basis einer physikalischen Theorie und theoretische Begriffe." ZAW 12 (1981) 55-74.

2344. Lukes, Steven. "The Underdetermination of Theory by Data." Arist. Soc. Supp. Vol. 52 (1978) 93-107.

2345. McConnell, J. R. "Reflections on Physical Theories." Philosophical Studies (Ireland) 18 (1969) 7-13.

2346. McMullin, Ernan. "The Fertility of Theory and the Unit for Appraisal in Science." BSPS 39 (1976) 395-432.

2347. Machamer, Peter K. "Feyerabend and Galileo: The Interaction of Theories, and the Reinterpretation of Experience." SHPS 4 (1973) 1-46.

    a. "Machamer on Galileo," by Paul Feyerabend. 5 (1974) 297-304.

2348. Madsen, K. B. "The Languages of Science." Theory and Decision. 1 (1970) 138-54.

2349. Marí, Enrique E. "Semántica y filosofía de la ciencia: una critica a Wolfgang Stegmüller." Dianoia. 26 (1980) 260-75.

2350. Martin, Michael. "The Explication of a Theory." Philosophia. 3 (1973) 179-99.

2351. Martin, R. M. "On Theoretical Constructs and Ramsey Constants." PS 33 (1966) 1-13.

2352. _____. "On Theoretical Entities." Ratio 8 (1966) 158-68.

2353. Matteuzzi, Maurizio L. M. "A Note on the Notion of 'Theory'." Quality and Quantity. 11 (1967) 67-71.

2354. Mehlberg, Henryk. "The Theoretical and Empirical Aspects of Science." LMPS-1 (1960) 275-84.

2355. Mittelstaedt, P. "Verborgene Parameter und beobachtbare Grössen in physikalischen Theorien." PN 10 (1967-68) 468-82.

2356. Moberg, Dale W. "Are There Rival, Incommensurable Theories?" PS 46 (1979) 244-62.

2357. Moulines, C. Ulises. "Approximate Application of Empirical Theories: A General Explication." Erkenntnis. 10 (1976) 201-27.

2358. _____. "Intertheoretic Approximation: The Kepler-Newton Case." Synthese. 45 (1980) 387-412.

2359. Naess, Arne. "Pluralistic Theorizing in Physics and Philosophy." Danish Yearbook of Philosophy. 1 (1964) 101-11.

2360. Nagel, Ernest, Sylvain Bromberger, and Adolf Grünbaum. Observation and Theory in Science. Baltimore: The Johns Hopkins Press, 1971. 134 pp.

2361. Nagel, Ernest. "Theory and Observation," in Observation and Theory in Science. E. Nagel et al. Baltimore: The Johns Hopkins Press, 1971. 15-43.

2362. Newton-Smith, W. "The Underdetermination of Theory by Data." Arist. Soc. Supp. Vol. 52 (1978) 71-91.

2363. Niiniluoto, Ilkka. "Empirically Trivial Theories and Inductive Systematization." SL 51 (1973) 108-14.

2364. Niiniluoto, Ilkka and Raimo Tuomela. Theoretical Concepts and Hypothetico-Inductive Inference. SL 53. Dordrecht: Reidel, 1973. 264 pp.

2365. Nogar, O. P., Raymond J. "Toward a Physical Theory." New Scholasticism. 25 (1951) 397-438.

2366. Nowaczyk, Adam. "Numerical Constructs as Theorems of Empirical Theories." Studia Logica. 35 (1976) 55-70.

2367. O'Neil, William M. Fact and Theory: An Aspect of the Philosophy of Science. Sydney: Sydney University Press, 1969. xiv + 193 pp.

2368. Palter, Robert. "Philosophic Principles and Scientific Theory." PS 23 (1956) 111-35.

2369. Papineau, David. "Ideal Types and Empirical Theories." BJPS 27 (1976) 137-46.

2370. Paris, C. "Expérience et théorie en physique." Dialectica.
6 (1952) 264-65.

2371. Pârva, Ilie. "The Convergence of Models in the Philosophy of
Science." Noesis. 8 (1981) 133-41.

2372. Pauli, W. "Theorie und Experiment." Dialectica. 6 (1952)
141-42.

2373. Pearce, David. "Is There Any Theoretical Justification for a
Non-Statement View of Theories?" Synthese. 46 (1981) 1-39.

2374. _____. "Comments on a Criterion of Theoreticity."
Synthese. 48 (1981) 77-86.

2375. Pietarinen, Juhani. "Quantitative Tools for Evaluating
Scientific Systematizations." SI 28 (1970) 123-47.

2376. Polikarov, Azaria. "Einstein's Conception of Physical Theory."
Epistemologia. 2 (1979) 99-122.

2377. Przełęcki, Marian. "Pojęcia teoretyczne a doświadczenie."
[Theoretical Concepts and Experience] Studia Logica. 11 (1961)
91-138.

2378. _____. The Logic of Empirical Theories. London: Routledge and Kegan Paul, 1969. 108 pp.

2379. _____. "A Set Theoretic versus a Model Theoretic Approach
to the Logical Structure of Physical Theories." Studia Logica. 33
(1974) 91-105.

        a. "Comments," by R. Wójcicki, 105-10; by E. Sharzynski,
        110-12.

2380. Putnam, H. "What Theories are Not." LMPS-1 (1960) 240-51.

2381. Quay, S. J., Paul M. "The Estimative Functions of Physical
Theory." SHPS 6 (1975) 125-57.

2382. Quine, Willard van Orman. "On Simple Theories of a Complex
World," in Form and Strategy in Science. J. R. Gregg and F. T. C.
Harris, eds. Dordrecht: Reidel, 1964. 47-50. Reprinted from
Synthese. 15 (1963) 103-06.

2383. _____. "On Empirically Equivalent Systems of the World."
Erkenntnis. 9 (1975) 313-28.

2384. Radermacher, Hans. "Der Begriff der Theorie in der kantischen
und analytischen Philosophie." ZAW 8 (1977) 63-76.

2385. Radner, Michael. "Possible Theories." Synthese. 41 (1979)
397-415.

2386. Rantala, Veikko. "On the Logical Basis of the Structuralist Philosophy of Science." Erkenntnis. 15 (1980) 269-86.

2387. Rapoport, Anatol. "Various Meanings of 'Theory'." American Political Science Review. 52 (1958) 972-88.

2388. Redhead, M. L. G. "Symmetry in Intertheory Relations." Synthese. 32 (1975) 77-112.

2389. Rogers, Robert. Mathematical Logic and Formalized Theories. New York: American Elsevier, 1971. xi + 235 pp.

2390. Rosenthal, S. B. "The Cognitive Status of Theoretical Terms." Dialectica. 22 (1968) 3-17.

  a. "Remarks," by P. Bernays. 18-19.

  b. "A Comment on Some Comments," by S. B. Rosenthal. 318-20.

2391. Rossel, J. "Théorie et expérience." Dialectica. 6 (1952) 260-63.

2392. Roth, Paul A. "Theories of Nature and the Nature of Theories." Mind. 89 (1980) 431-38.

2393. Rothstein, Jerome. "A Physicist's Thoughts on the Formal Structure and Psychological Motivation of Theory and Observation." RIP 11 (1957) 211-26.

2394. Sacksteder, William. "'Theories' and Usage." J. Phil. 59 (1962) 309-20.

2395. Sampson, Geoffrey. "Theory Choice in a Two-level Science." BJPS 26 (1975) 303-18.

  a. "Sampson's 'Dilemma'," by F. B. D'Agostino. 29 (1978) 183-84.

  b. "A Dilemma Defended," by Geoffrey Sampson. 353-55.

2396. Schäfer, Lothar. Erfahrung und Konvention: Zum Theoriebegriff der empirischen Wissenschaften. Stuttgart: Problemata Frommann-Holzboog, 1974. 230 pp.

2397. _____. "Theorien-dynamische Nachlieferung. Anmerkungen zu Kuhn-Sneed-Stegmüller." ZPF 31 (1977) 19-42.

2398. Schaffner, Kenneth F. "Approaches to Reduction." PS 34 (1967) 137-47.

2399. _____. "Correspondence Rules." PS 36 (1969) 280-90.

2400. _____. "Outlines of a Logic of Comparative Theory Evaluation with Special Attention to Pre- and Post- Relativistic Electrodynamics." MSPS 5 (1970) 311-54.

a. "Comment," by Howard Stein, 354-55; by Arnold Koslow, 356-64; by Peter A. Bowman, 364; "Reply," by K. F. Schaffner, 365-73.

2401. Scheffler, Israel. "Reflections on the Ramsey Method." J. Phil. 65 (1968) 269-74.

a. "In Defense of Ramsey's Elimination Method." by Herbert C. Bohnert. 275-81.

2402. Scheibe, E. "On the Structure of Physical Theories." Acta Philosophica Fennica. 30, 2-4 (1978) 205-24.

2403. Schock, Rolf. "Contributions to Syntax, Semantics, and the Philosophy of Science." Notre Dame Journal of Formal Logic. 5 (1964) 241-89.

2404. Seager, William. "The Principle of Continuity and the Evaluation of Theories." Dialogue. 20 (1981) 485-95.

2405. Sellars, Wilfrid. "Philosophy and the Scientific Image of Man," in Frontiers of Science and Philosophy. Robert G. Colodny, ed. Pittsburgh: University of Pittsburgh Press, 1962. 35-78.

2406. Shapere, Dudley. "Scientific Theories and Their Domains," in The Structure of Scientific Theories. F. Suppe, ed. Urbana, Illinois: University of Illinois Press, 1974. 518-70.

a. "Heuristics and Justification in Scientific Research: Comments on Shapere," by Thomas Nickles. 571-89.

2407. Short, T. L. "Peirce and the Incommensurability of Theories." Monist. 63 (1980) 316-28.

2408. Simon, Herbert A. "The Axiomatization of Physical Theories." PS 37 (1970) 16-26.

2409. _____. "Fit, Finite, and Universal Axiomatization of Theories." PS 46 (1979) 295-301.

2410. Simon, Herbert A., and Guy J. Groen. "Ramsey Eliminability and the Testability of Scientific Theories." BJPS 24 (1973) 367-80.

2411. Smart, J. J. C. "Theory Construction." PPR 11 (1951) 457-73.

2412. Sneed, Joseph D. The Logical Structure of Mathematical Physics. SL 35 Dordrecht: Reidel, 1971. xv + 311 pp.

a. Review essay by C. Ulises Moulines. Erkenntnis. 9 (1975) 423-36.

b. Review essay by Werner Diederich. Philosophische Rundschau. 21 (1974) 209-28.

2413. _____. "Philosophical Problems in the Empirical Science of Science: A Formal Approach." Erkenntnis. 10 (1976) 115-46.

2414. _____. "Theoretization and Invariance Principles." Acta Philosophica Fennica. 30, 24 (1978) 130-78.

2415. _____. "Quantities as Theoretical with respect to Qualities." Epistemologia. 2 (1979) 215-50.

2416. Spector, Marshall. "Theory and Observation." BJPS 17 (1966) 1-20; 89-104.

2417. Strauss, M. "Mathematics as Logical Syntax -- A Method to Formalize the Language of a Physical Theory." Erkenntnis. 7 (1937-38) 147-53. Reprinted in WOSPS 5 (1975) 45-52.

2418. _____. "Intertheory Relations." SL 31 (1970) 220-84.

2419. Suppe, Frederick. "Theories, Their Formulations, and the Operational Imperative." Synthese. 25 (1972) 129-64.

2420. _____. "What's Wrong with the Received View on the Structure of Scientific Theories?" PS 39 (1972) 1-19.

2421. _____. The Structure of Scientific Theories. Urbana, Illinois: University of Illinois Press, 1974. 682 pp. Second Edition: 1977. 818 pp.

    a. Review essay by Steven F. Savitt. Dialogue. 16 (1977) 328-45.

    b. Review essay by Peter Skagestad. Nous. 15 (1981) 234-39.

2422. _____. "Theory Structure," in Current Research in Philosophy of Science. P. D. Asquith and H. E. Kyburg, Jr., eds. East Lansing, MI: Philosophy of Science Assn., 1979. 317-38.

2423. Suppes, Patrick. "What is a Scientific Theory?" in Philosophy of Science Today. S. Morgenbesser, ed. New York: Basic Books, 1967. 55-67.

2424. Swijtink, Zeno G. "Two Suggestions for Ramsey-Reducts of Infinite Theories." PS 43 (1976) 575-77.

2425. Toulmin, Stephen. "The Structure of Scientific Theories," in The Structure of Scientific Theories. F. Suppe, ed. Urbana, Illinois: University of Illinois Press, 1974. 600-14.

2426. Tuomela, Raimo. Auxiliary Concepts within First-Order Scientific Theories. Ann Arbor, MI: University Microfilms, 1969. 150 pp.

2427. Turbayne, Colin Murray. The Myth of Metaphor. New Haven, CT: Yale University Press, 1962. ix + 224 pp.

2428. Turner, Stephen. "Modelling and Evaulating Theories Involving Sequences: Description of a Formal Method." Quality and Quantity. 14 (1980) 511-18.

2429. Wessels, Linda. "Laws and Meaning Postulates (in van Fraassen's View of Theories)." BSPS 32 (1976) 215-34.

2430. Williams, P.M. "On the Logical Relations between Expressions of Different Theories." BJPS 24 (1973) 357-67.

2431. Wilson, Mark. "The Observational Uniqueness of Some Theories." J. Phil. 77 (1980) 208-33.

2432. Wisdom, J.O. "Scientific Theory: Empirical Content, Embedded Ontology, and Weltanschauung." PPR 33 (1972) 62-77.

2433. Wójcicki, Ryszard. "The Factual Content of Empirical Theories." SL 73 (1975) 95-122.

2434. Worrall, J. "Is the Empirical Content of a Theory Dependent on its Rivals?" Acta Philosophica Fennica. 30, 2-4 (1978) 298-310.

4.9 Research Programs

2435. Agassi, Joseph. "The Lakatosian Revolution." BSPS 39 (1976) 9-22.

2436. _____. "The Methodology of Research Projects: A Sketch." ZAW 8 (1977) 30-38. Reprinted in BSPS 65 (1981) 273-82.

2437. Agassi, Joseph, and Charles Sawyer. "Was Lakatos an Elitist?" Ratio 22 (1980) 61-63.

2438. Brown, James R. "Lakatos on Appraisal vs. Advice." Methodology and Science. 14 (1981) 127-31.

2439. Chalmers, A. F. "An Improvement and a Critique of Lakatos' Methodology of Scientific Research Programmes." Methodology and Science. 13 (1980) 2-27.

2440. D'Amour, Gene. "Research Programs, Rationality, and Ethics." BSPS 39 (1976) 87-98.

2441. Derr, Patrick G. "Reflexivity and the Methodology of Scientific Research Programmes." The New Scholasticism. 55 (1981) 500-03.

2442. Dorling, Jon. "Bayesian Personalism, the Methodology of Scientific Research Programmes, and Duhem's Problem." SHPS 10 (1979) 177-87.

     a. "Discussion: A Bayesian Reconstruction of the Methodology of Scientific Research Programs," by M.L.G. Redhead. 11 (1980) 341-47.

2443. Giesen, Bernd, and Michael Schmid. "Rationalität und Erkenntnisfortschritt." ZAW 5 (1974) 256-84.

2444. Kantorovich, Aharon. "An Ideal Model for the Growth of Knowledge in Research Programs." PS 45 (1978) 250-72.

2445. Kockelmans, Joseph J. "Reflections on Lakatos' Methodology of Scientific Research Programs." BSPS 59. SL 136. (1979) 187-203.

2446. Konigsveld, H. "Theorievorming: de methodologie van Imre Lakatos." Algemeen Nederlands Tijdschrift voor Wijsbegeerte. 67 (1975) 96-108.

2447. Kulka, Tomas. "Some Problems Concerning Rational Reconstruction: Comments on Elkana and Lakatos." BJPS 28 (1977) 325-44.

2448. Lakatos, Imre. "Proofs and Refutations." BJPS 14 (1963) Part I: 1-25. Part II: 120-39. Part III: 221-45. Part IV: 15 (1964) 296-42.

2449. _____. "Criticism and the Methodology of Scientific Research Programmes." Proc. Arist. Soc. 69 n.s. (1968-69) 149-86.

2450. _____. "Falsification and the Methodology of Scientific Research Programmes," in Criticism and the Growth of Knowledge. I. Lakatos and A. Musgrave, eds. Cambridge: Cambridge University Press, 1970. 91-195.

2451. _____. "History of Science and Its Rational Reconstructions." BSPS 8, SL 39 (1971) 91-136. Reprinted in Method and Appraisal in the Physical Sciences. C Howson, ed. Cambridge: Cambridge University Press, 1976. 1-39.

   a. "Notes on Lakatos," by Thomas S. Kuhn. 137-46.

   b. "Research Programmes and Induction," by Herbert Feigl. 147-50.

   c. "Can We Use the History of Science to Decide Between Competing Methodologies?" by Richard J. Hall. 151-59.

   d. "Intertheoretic Criticism and the Growth of Science," by Noretta Koertge. 160-73.

   e. "Replies to Critics," by Imre Lakatos. 174-82.

2452. _____. "The Role of Crucial Experiments in Science." SHPS 4 (1974) 309-25.

   a. "Discussion: Some Epistemic Implications of 'Crucial Experiments'," by Philip L. Quinn. 5 (1974) 59-72.

2453. _____. Proofs and Refutations: Essays in the Logic of Mathematical Discovery. J. Worrall and E. Zahar, eds. London: Cambridge University Press, 1976. xii + 174 pp.

a. Review essay, "Lakatos's Philosophy of Mathematics," by Gregory Currie. Synthese. 42 (1979) 335-51.

2454. _____. Philosophical Papers. Vol. I: The Methodology of Scientific Research Programmes. Vol. II: Mathematics, Science and Epistemology. John Worrall and Gregory Curie, eds. Cambridge: Cambridge University Press, 1978. I: vii + 250 pp.; II: x + 285 pp.

a. Review essay by Catherine Z. Elgin and Jonathan E. Adler. Synthese. 43 (1980) 411-20.

b. Review essay, "Imre Lakatos's Philosophy of Science," by Ian Hacking. BJPS 30 (1979) 381-402.

c. Review essay, "Appraising Lakatos," by John F. Fox. AJP 59 (1981) 92-103.

d. Review essay by Alfred Schramm. Philosophische Rundschau. 27 (1980) 84-100.

2455. Lakatos, Imre, and Alan Musgrave, eds. Criticism and the Growth of Knowledge. Cambridge: Cambridge University Press, 1970. 282 pp.

a. Review essay by Joseph Agassi. Inquiry. 14 (1971) 152-64.

b. Review essay by Mark R. Berg. Philosophy Forum. 12 (1972) 347-58.

c. Review essay by Vernon Pratt. Theoria to Theory. 6, No. 1 (1972) 58-79.

d. Review essay, "Science, History and Methodology," by J. J. C. Smart. BJPS 23 (1972) 266-74.

e. Review essay by Maurice A. Finocchiaro. SHPS 3 (1973) 357-72.

f. Review essay by T. R. Girill. Metaphilosophy. 4 (1973) 246-60.

g. "A Comment on Girill's Dualistic View of Scientific Know-ledge as a Resolution of the Kuhn-Popper Debate," by Joel Kassiola. 7 (1976) 149-54.

2456. McMullin, Ernan. "Philosophy of Science and Its Rational Reconstructions." BSPS 58, SL 125 (1978) 221-52.

2457. Millman, Arthur B. "The Plausibility of Research Programs." PSA 1976. I, 140-48.

2458. Musgrave, Alan. "Method or Madness?" BSPS 39 (1976) 457-92.

2459. _____. "Why Did Oxygen Supplant Phlogiston? Research Programmes in the Chemical Revolution," in Method and Appraisal in the Physical Sciences. C. Howson, ed. Cambridge: Cambridge University Press, 1976. 181-209.

2460. _____. "Evidential Support, Falsification, Heuristics, and Anarchism." BSPS 58, SL 125 (1978) 181-201.

2461. Post, Heinz. "Objectivism vs. Sociologism." BSPS 58, SL 125 (1978) 311-18.

2462. Quinn, Philip. "Methodological Appraisal and Heuristic Advice: Problems in the Methodology of Scientific Research Programmes." SHPS 3 (1972) 135-49.

2463. Raub, Werner, and Dirk Koppelberg. "Bewertungen und Empfehlungen in der Methodologie wissenschaftlicher Forschungsprogramme." ZAW 9 (1978) 134-48.

2464. Sarkar, Husain. "Imre Lakatos' Meta-Methodology: An Appraisal." Philosophy of the Social Sciences. 10 (1980) 397-416.

2465. Schramm, Alfred. "Demarkation und rationale Rekonstruktion bei Imre Lakatos." Conceptus. 8, No. 24 (1974) 10-16.

2466. Toulmin, Stephen. "History, Praxis and the 'Third World'. Ambiguities in Lakatos' Theory of Methodology." BSPS 39 (1976) 655-76.

2467. Urbach, Peter. "The Objective Promise of a Research Programme." BSPS 58, SL 125 (1978) 99-113.

2468. Worrall, John. "The Ways in Which the Methodology of Scientific Research Programmes Improves on Popper's Methodology." BSPS 58. SL 125 (1978) 45-70.

2469. _____. "Research Programmes, Empirical Support, and the Duhem Problem: Replies to Criticisms." BSPS 58, SL 125 (1978) 321-38.

2470. Zahar, Elie. "'Crucial' Experiments: A Case Study." BSPS 58, SL 125 (1978) 71-97.

4.10 Explanation and Prediction

2471. Achinstein, Peter. "Explanation," in Studies in the Philosophy of Science. N. Rescher, ed. Oxford: Basil Blackwell, 1969. 9-29.

2472. _____. Law and Explanation: An Essay in the Philosophy of Science. Oxford: Clarendon Press, 1971. 168 pp.

    a. "Discussion Review: Achinstein's Law and Explanation," by James H. Fetzer. PS 42 (1975) 320-33.

2472A. _____. "What is an Explanation?" Am. Phil. Quart. 14 (1977) 1-15.

2473. _____. "Can There be a Model of Explanation?" Theory and Decision. 13 (1981) 201-27.

2474. Ackermann, Robert. "Deductive Scientific Explanation." PS 32 (1965) 155-67.

    a. "A Corrected Model of Explanation," by Robert Ackermann and Alfred Stenner. 33 (1966) 168-71.

    b. "On Two Proposed Models of Explanation," by Charles G. Morgan. 39 (1972) 74-81.

2475. Addis, Laird. "On Defending the Covering-Law 'Model'." BSPS 32 (1976) 361-68.

    a. "Dispositional Explanation and the Covering-Law Model: Response to Laird Addis," by Carl G. Hempel. 369-76.

2476. Agassi, Joseph. "On the Limits of Scientific Explanation: Hempel and Evans-Pritchard." Philosophical Forum. 1 (1968) 171-84.

2477. Alexander, Peter. Sensationalism and Scientific Explanation. New York: Humanities Press, 1963. 149 pp.

2478. Alston, William P. "The Place of the Explanation of Particular Facts in Science." PS 38 (1971) 13-34.

2479. Angel, R. B. "Explanation and Prediction: A Plea for Reason." PS 34 (1967) 276-82.

    a. "Angel's Symmetry Thesis," by Ronald C. Hopson. 38 (1971) 308-09.

2480. Aronson, Jerrold L. "Explanations Without Laws." J. Phil. 66 (1969) 541-57.

2481. Aubert, Vilhelm. "Predictability in Life and in Science." Inquiry. 4 (1961) 131-47.

2482. Audi, Robert. "Inductive-Nomological Explanations and Psychological Laws." Theory and Decision. 13 (1981) 229-49.

2483. Boden, Margaret. "The Paradox of Explanation." Proc. Arist. Soc. 62 n.s. (1961-62) 159-78.

2484. Braithwaite, Richard Bevan. "Teleological Explanation." Proc. Arist. Soc. 47 n.s. (1946-47) i-xx.

2485. _____. Scientific Explanation: A Study of the Function of Theory, Probability and Law in Science. New York and London: Cambridge University Press, 1953. Reprint: New York: Harper and Brothers, 1960. 374 pp.

    a. Review essay, "Braithwaite on Scientific Method," by Abner Shimony. Review of Metaphysics. 7 (1954) 644-60.

    b. Review essay, "R. B. Braithwaite on Science and Ethics," by R. J. Hirst. Phil. Quart. 4 (1954) 351-55.

    c. Review essay, "A Budget of Problems in the Philosophy of Science," by Ernest Nagel. Phil. Review. 66 (1957) 205-25.

2486. Brodbeck, May. "Explanation, Prediction, and 'Imperfect' Knowledge." MSPS 3 (1962) 231-72.

2487. Brody, B.A. "Towards an Aristotelean Theory of Scientific Explanation." PS 39 (1972) 20-31.

    a. "Brody's Defense of Essentialism," by Nathan Stemmer. 40 (1973) 393-96.

    b. "On an Aristotelian Model of Scientific Explanation," by Timothy McCarthy. 44 (1977) 159-66.

2488. Buck, Roger C. "Reflexive Predictions." PS 30 (1963) 359-69.

    a. "Comments on Professor Roger Buck's Paper 'Reflexive Predictions'," by Adolf Grünbaum. 370-72.

    b. "Rejoinder to Grünbaum," by Roger C. Buck. 373-74.

    c. "Reflexive Prediction," by Emile Grunberg and Franco Modigliani. 32 (1965) 173-74.

    d. "Reflexive Predictions," by George D. Romanos. 40 (1973) 97-109.

    e. "More on Reflexive Predictions," by Mary K. Vetterling. 43 (1976) 278-82.

2489. Burch, Robert W. "Functional Explanation and Normalcy." Southwestern Journal of Philosophy. 9, No. 1 (1978) 45-53.

2490. Canfield, John, and Keith Lehrer. "A Note on Prediction and Deduction." PS 28 (1961) 204-08.

    a. "Deduction, Prediction and Completeness Conditions," by Robert W. Beard. 33 (1966) 165-67.

2491. Cartwright, Nancy. "The Truth Doesn't Explain Much." Am. Phil. Quart. 17 (1980) 159-63.

2492. Caws, Peter. "Aspects of Hempel's Philosophy of Science." Review of Metaphysics. 20 (1967) 690-710.

2493. Chaudhury, Pravas Jivan. "Nature of Scientific Explanation." Philosophical Quarterly (India). 25 (1952) 151-64.

2494. Cherry, Christopher. "Explanation and Explanation by Hypothesis." Synthese. 33 (1976) 315-40.

2495. Coffa, José Alberto. "Feyerabend on Explanation and Reduction." J. Phil. 64 (1967) 500-08.

2496. _____. "Deductive Predictions." PS 35 (1968) 279-83.

2497. _____. "Hempel's Ambiguity." Synthese. 28 (1974) 141-64.

   a. "Comments on 'Hempel's Ambiguity' by J. Alberto Coffa," by Wesley C. Salmon. 165-70.

2498. _____. "Probabilities: Reasonable or True?" PS 44 (1977) 186-98.

2499. Cohen, Jonathan. "Teleological Explanation." Proc. Arist. Soc. 51 (1950-51) 255-92.

2500. Cole, Richard. "Possibility Matrices." Theoria. 45 (1979) 8-39.

2501. Collins, Arthur W. "The Use of Statistics in Explanation." BJPS 17 (1966) 127-40.

2502. _____. "Explanation and Causality." Mind. 75 (1966) 482-500.

2503. Cooke, Roger. "A Trivialization of Nagel's Definition of Explanation for Statistical Laws." PS 47 (1980) 644-45.

2504. Cummins, Robert. "Explanation and Subsumption." PSA 1978. I, 163-75.

2505. Cupples, Brian. "Three Types of Explanation." PS 44 (1977) 387-408.

   a. "Four Types of Explanation," by Brian Cupples. 47 (1980) 626-29.

2506. Davis, P. Byard. "Functionalist Analysis and the Theory-Methods Gap: A Proposal." Methodology and Science. 5 (1972) 26-46.

2507. Dietl, Paul. "Paresis and the Alleged Asymmetry Between Explanation and Prediction." BJPS 17 (1967) 313-18.

2508. Dietz, Stephens M. "A Remark on Hempel's Replies to His Critics." PS 37 (1970) 614-17.

2509. Dorling, Jon. "On Explanations in Physics: Sketch of an Alternative to Hempel's Account of the Explanation of Laws." PS 45 (1978) 136-40.

2510. Dougherty, Jude P. "Nagel's Concept of Science." Philosophy Today. 10 (1966) 212-21.

2511. Eberle, Rolf, David Kaplan, and Richard Montague. "Hempel and Oppenheim on Explanation." PS 28 (1961) 418-28.

2512. Economos, John James. "Explanation: What's It All About?" AJP 49 (1971) 139-45.

2513. Ellis, B. "On the Relation of Explanation to Description." Mind. 65 (1956) 498-506.

2514. _____. "Explanation and the Logic of Support." AJP 48 (1970) 177-89.

2515. Engelhardt, Jr., H. Tristram, and Stuart F. Spicker, eds. Evaluation and Explanation in the Biomedical Sciences. Dordrecht: Reidel, 1975. 240 pp.

2516. Essler, Wilhelm K. "A Note on Functional Explanation." Erkenntnis. 13 (1978) 371-76.

2517. Fain, Haskell. "Some Problems of Causal Explanation." Mind. 72 (1963) 519-32.

2518. Fairchild, David. "Revolution and Cause: An Investigation of an Explanatory Concept." New Scholasticism. 50 (1976) 277-92.

2519. Fetzer, James H. "A Single Case Propensity Theory of Explanation." Synthese. 28 (1974) 171-98.

2520. _____. "Statistical Explanations." BSPS 20, SL 64 (1974) 337-47.

2521. _____. "Grünbaum's 'Defense' of the Symmetry Thesis." Phil. Studies. 25 (1974) 173-87.

2522. _____. "Probability and Explanation." Synthese. 48 (1981) 371-08.

2523. Finocchiaro, Maurice A. History of Science as Explanation. Detroit: Wayne State University Press, 1973. 286 pp.

   a. Review essay by Thomas Nickles. Erkenntnis. 14 (1979) 93-102.

2524. _____. "Cause, Explanation, and Understanding in Science: Galileo's Case." Review of Metaphysics. 2 (1975) 117-28.

2525. Forge, John. "The Structure of Physical Explanation." PS 47 (1980) 203-26.

2526. Fraassen, Bas C. van. "The Pragmatics of Explanation." Am. Phil. Quart. 14 (1977) 143-50.

2527. Friedman, Michael. "Explanation and Scientific Understanding." J. Phil. 71 (1974) 5-19.

   a. "Explanation, Conjunction, and Unification," by Philip Kitcher. 73 (1976) 207-12.

2528. Gale, Richard M., and Irving Thalberg. "The Generality of Predictions." J. Phil. 62 (1965) 195-210.

2529. Gallie, W. B. "Explanations in History and the Genetic Sciences." Mind. 64 (1955) 160-80.

   a. "Professor Gallie on Necessary and Sufficient Conditions," by Alan Montefiore. 65 (1956) 534-41.

   b. "In Reply to Mr. Montefiore," by W. B. Gallie. 67 (1958) 92-96.

2530. _____. "The Limits of Prediction," in Observation and Interpretation. S. Körner, ed. New York: Academic Press, London: Butterworths, 1957. 160-64.

2531. Gärdenfors, Peter. "Relevance and Redundancy in Deductive Explanations." PS 43 (1976) 420-31.

2532. _____. "A Pragmatic Approach to Explanations." PS 47 (1980) 404-23.

2533. Gardner, Martin. "A New Prediction Paradox." BJPS 13 (1962) 51.

   a. "A Comment on the New Prediction Paradox," by K. R. Popper. 51.

2534. Gaukroger, Stephen. Explanatory Structures: A Study of Concepts of Explanation in Early Physics and Philosophy. Atlantic Highlands, NJ: Humanities Press, 1977. viii + 262 pp.

2535. Girill, T. R. "Identity, Causality, and the Regressiveness of Micro-Explanations." Dialectica. 28 (1974) 223-38.

2536. _____. "Evaluating Micro-Explanations." Erkenntnis. 10 (1976) 387-405.

2537. _____. "The Problem of Micro-Explanation." PSA 1976. I, 47-55.

2538. _____. "Approximative Explanation." PSA 1978. I, 186-96.

2539. Gluck, Samuel E. "Do Statistical Laws Have Explanatory Efficacy?" PS 22 (1955) 34-38.

2540. Glymour, Clark. "Explanations, Tests, Unity and Necessity." Nous. 14 (1980) 31-50.

   a. "Comments," by Wesley C. Salmon. 51-52.

2541. Goh, S. T. "Some Observations on the Deductive-Nomological Theory." Mind. 79 (1970) 408-14.

2542. Gorovitz, Samuel. "Aspects of the Pragmatics of Explanation." Nous. 3 (1969) 61-72.

2543. _____. "Inscriptionalism and the Objects of Explanation." BJPS 21 (1970) 247-56.

   a. "Explanations, Desires, and Inscriptions," by Israel Scheffler. 22 (1971) 362-69.

2544. Goudge, T. A. "Causal Explanation in Natural History." BJPS 9 (1958) 194-202.

2545. Greeno, James G. "Theoretical Entities in Statistical Explanation." BSPS 8, SL 39 (1971) 3-26.

   a. "Explanation and Relevance," by Wesley C. Salmon. 27-39.

   b. "Remarks on Explanatory Power," by Richard C. Jeffrey. 40-46.

2546. Greenstein, Harold. "The Logic of Functional Explanations." Philosophia. 3 (1973) 247-64.

2547. Gruender, David. "Scientific Explanation and Norms in Science." PSA 1980. I, 329-35.

2548. Grünbaum, Adolf. "Temporally Asymmetric Principles, Parity between Explanation and Prediction, and Mechanism versus Teleology." PS 29 (1962) 146-70. Reprinted in Philosophy of Science: The Delaware Seminar, I. New York: Interscience Publishers, 1963. 57-96.

   a. "Comments on Professor Grünbaum's Remarks at the Wesleyan Meeting," by Michael Scriven PS 29 (1962) 171-74.

   b. "Professor Grünbaum and Teleological Mechanisms," by Brian Skyrms. 31 (1964) 62-64.

2549. Gruner, Rolf. "Teleological and Functional Explanations." Mind. 75 (1966) 516-26.
   a. "Teleological Explanations and the Animal World," by M. Ruse. 82 (1973) 433-36.

"A Belated Reply to Gruner," by David L. Hull. 437-38.

2550. Halfpenny, Peter. "Two-Variable and Three-Variable Functional Explanations." Philosophy of the Social Sciences. 11 (1981) 27-32.

2551. Hanna, Joseph F. "Explanation, Prediction, Description, and Information Theory." Synthese. 20 (1969) 308-34.

2552. _____. "On Transmitted Information as a Measure of Explanatory Power." PS 45 (1978) 531-62.

2553. _____. "An Interpretive Survey of Recent Research on Scientific Explanation," in Current Research in Philosophy of Science. P. D. Asquith and H. E. Kyburg, Jr., eds. East Lansing, MI: Philosophy of Science Assn., 1979. 291-316.

2554. _____. "Single Case Propensities and the Explanation of Particular Events." Synthese. 48 (1981) 409-36.

2555. Hanson, N. R. "On the Symmetry between Explanation and Prediction." Phil. Review. 68 (1959) 349-58.

     a. "Mr. Hanson on the Symmetry of Explanation and Prediction," by Richard G. Henson. PS 30 (1963) 60-61.

2556. Harman, Gilbert. "Knowledge, Inference, and Explanation." Am. Phil. Quart. 5 (1968) 164-73.

2557. Harnatt, J. "Zum Beweis der Nichtexistenz von 'störenden Bedingungen." ZAW 6 (1975) 108-112.

     a. "Reply to Harnatt," by J. Alberto Coffa. 7 (1976) 357-58.

2558. Hausman, David B. "Another Defense of the Deductive Model." Southwestern Journal of Philosophy. 7, No. 1 (1976) 111-17.

2559. Hayek, F. A. "Degrees of Explanation." BJPS 6 (1955) 209-25.

2560. Heath, A. F., ed. Scientific Explanation. Oxford: Clarendon Press, 1981. x + 123 pp.

2561. Hempel, Carl G. "The Function of General Laws in History." J. Phil. 39 (1942) 35-48. Reprinted in slightly modified form in C. G. Hempel, Aspects of Scientific Explanation and Other Essays in the Philosophy of Science. New York: Free Press, 1965. 231-43.

2562. Hempel, Carl G., and Paul Oppenheim. "Studies in the Logic of Explanation." PS 15 (1948) 135-75. Reprinted, with some changes, in C. G. Hempel, Aspects of Scientific Explanation and Other Essays in the Philosophy of Science. New York: Free Press, 1965. 245-90. To which is added "Postscript (1964) to Studies in the Logic of Explanation." 291-95.

     a. "Comments on 'Studies in the Logic of Explanation'," by David L. Miller. PS 15 (1948) 348-49.

b. "Reply to David L. Miller's Comments," by Carl G. Hempel and Paul Oppenheim. 350-52.

2563. Hempel, Carl G. "The Logic of Functional Analysis," in Symposium on Sociological Theory. L. Gross, ed. New York: Harper & Row, 1959. Reprinted with some changes in C. G. Hempel, Aspects of Scientific Explanation and Other Essays in the Philosophy of Science. New York: Free Press, 1965. 297-330. Reprinted in Purpose in Nature. J. V. Canfield, ed. Englewood Cliffs, NJ: Prentice-Hall, 1966. 89-108.

2564. _____ . "Explanation in Science and History," in Frontiers of Science and Philosophy. Robert G. Colodny, ed. Pittsburg: University of Pittsburgh Press, 1962. 7-33.

2565. _____ . "Deductive-Nomological vs. Statistical Explanation." MSPS 3 (1962) 98-169.

2566. _____ . "Explanation and Prediction by Covering Laws," in Philosophy of Science: The Delaware Seminar, I. New York: Interscience Publishers, 1963. 107-33.

2567. _____ . Aspects of Scientific Explanation and Other Essays in the Philosophy of Science. New York: Free Press; London: Collier-Macmillan, 1965. 505 pp.

a. Review essay by Edward J. Nell. History and Theory. 7 (1968) 224-40.

2568. _____ . "Scientific Explanation," in Philosophy of Science Today. S. Morgenbesser, ed. New York: Basic Books, 1967. 79-88.

2569. _____ . "Maximal Specificity and Lawlikeness in Probabilistic Explanation." PS 35 (1968) 116-33.

a. "More on Maximal Specificity," by Henry E. Kyburg, Jr. 37 (1970) 295-300.

b. "A Paradox in Hempel's Criterion of Maximal Specificity," by Roger M. Cooke. 48 (1981) 327-28.

2570. _____ . "Formen und Grenzen des wissenschaftlichen Verstehens." Conceptus. 6 (1972) 5-18.

2571. Hesse, Mary. "A New Look at Scientific Explanation." Review of Metaphysics. 17 (1963) 98-108.

2572. Hirschmann, David. "Function and Explanation." Arist. Soc. Supp. Vol. 47 (1973) 19-38.

2573. Hofstadter, Albert. "Explanation and Necessity." PPR. 11 (1951) 339-47.

2574. Hooker, C. A. "Explanation, Generality and Understanding." AJP 58 (1980) 284-90.

2575. Hopson, Ronald C. "The Objects of Acceptance: Competing Scientific Explanations." BSPS 20, SL 64 (1974) 349-63.

2576. Horovitz, Joseph. "Three Related Temporal Aspects of Scientific Arguments." Logique et analyse. 10 (1967) 262-68.

2577. Hospers, John. "On Explanation." J. Phil. 43 (1946) 337-56.

2578. Humphreys, Paul. "Aleatory Explanations." Synthese. 48 (1981) 225-32.

2579. Humphreys, Willard C. "Statistical Ambiguity and Maximal Specificity." PS 35 (1968) 112-15.

2580. Hung Hin-Chung. "Scientific Explanation or Deceptive Explanation?" Methodology and Science. 11 (1978) 191-204.

2581. Hutten, E. H. "On Explanation in Psychology and in Physics." BJPS 7 (1956) 73-85.

2582. Jeffrey, Richard C. "Statistical Explanation vs. Statistical Inference." SL 24 (1970) 104-13.

2583. Jobe, Evan K. "A Puzzle Concerning D-N Explanation." PS 43 (1976) 542-49.

      a. "Two Flagpoles are More Paradoxical Than One," by Clark Glymour. 45 (1978) 118-19.

2584. Kaplan, David. "Explanation Revisited." PS 28 (1961) 429-36.

2585. Käsbauer, M. "Definitionen der wissenschaftlichen Erklärung." Erkenntnis. 10 (1976) 255-73.

2586. Kemeny, John G., and Paul Oppenheim. "Systematic Power." PS 22 (1955) 27-33.

2587. Kim, Jaegwon. "On the Logical Conditions of Deductive Explanation." PS 30 (1963) 286-91.

      a. "Kim on Deductive Explanation," by Charles G. Morgan. 37 (1970) 434-39.

2588. _____. "Inference, Explanation, and Prediction." J. Phil. 61 (1964) 360-68.

2589. _____. "Events and Their Descriptions: Some Considerations." SL 24 (1970) 198-215.

2590. King, John L. "Statistical Relevance and Explanatory Classification." Phil. Studies. 30 (1976) 313-21.

      a. "Homogeneity Conditions on the Statistical Relevance Model of Explanation," by J.-P. Thomas. 36 (1979) 101-05.

2591. Kitchener, R. F. "On Translating Teleological Explanations." International Logic Review. 7 (1976) 50-56.

2592. Kitcher, Philip. "Explanatory Unification." PS 48 (1981) 507-31.

2593. Kitts, David B. "Retrodiction in Geology." PSA 1978. II, 215-26.

2594. Klein, Barbara V. E. "What Should We Expect of a Theory of Explanation?" PSA 1980. I, 319-28.

2595. Kochanski, Zdzislaw. "Conditions and Limitations of Prediction-Making in Biology." PS 40 (1973) 29-50.

2596. Koertge, Noretta. "An Exploration of Salmon's S-R Model of Explanation." PS 42 (1975) 270-74.

2597. Körner, Stephan, ed. Explanation. Oxford: Blackwell, 1975. 219 pp.

2598. Krah, W. "Prognose und Rüchkopplung." DZP 15 (1967) 785-91.

2599. Krimsky, Sheldon. "On Deductive Non-nomological Explanation." Philosophia. 6 (1976) 303-08.

    a. "Reply to Krimsky on D-N Explanation," by Thomas Nickles. 309-15.

2600. Kröber, G. "Prognose, Hypothese, Gesetz-Logish-methodologische Bemerkungen." DZP 15 (1967) 772-84.

    a. "Prognose und Hypothese," by J. Schermer. 16 (1968) 349-52.

2601. Krüger, Lorenz. "Are Statistical Explanations Possible?" PS 43 (1976) 129-46.

2602. Küttner, Michael. "Ein verbesserter deduktiv-nomologischer Erklärungsbegriff." ZAW 7 (1976) 274-97.

2603. Laymon, Ronald. "Idealization, Explanation, and Confirmation." PSA 1980. I, 336-50.

2604. Leach, James. "Explanation and Value Neutrality." BJPS 19 (1968) 93-108.

2605. Lehman, Hugh. "Statistical Explanation." PS 39 (1972) 500-06.

    a. "Reply to Lehman," by Wesley C. Salmon. 40 (1973) 397-402.

    b. "Causation, Explanation, and Statistical Relevance," by Douglas W. Shrader, Jr. 44 (1977) 136-45.

2606. Lenk, Hans. Erklärung-Prognose-Planung. Skizzen zu Brennpunktproblemen der Wissenschaftstheorie. Freiburg i. Brsg.: Rombach Hochschul, 1972. 119 pp.

2607. Levi, Isaac. "Are Statistical Hypotheses Covering Laws?" Synthese. 20 (1969) 297-307.

2608. Levin, Michael E., and Margarita Rosa Levin. "Flagpoles, Shadows and Deductive Explanation." Phil. Studies. 32 (1977) 293-99.

2609. MacKinnon, Edward. "A Reinterpretation of Harré's Copernican Revolution." PS 42 (1975) 67-79.

       a. "Science as Representation: A Reply to Mr. MacKinnon," by Rom Harré. 44 (1977) 146-58.

2610. McMullin, Ernan. "Structural Explanation." Am. Phil. Quart. 15 (1978) 139-47.

2611. Macklin, Ruth. "Explanation and Action: Recent Issues and Controversies." Synthese. 20 (1969) 388-415.

2612. Madden, Edward M. "Scientific Explanations." Review of Metaphysics. 26 (1973) 723-43.

2613. Manser, A. R. "Function and Explanation." Arist. Soc. Supp. Vol. 47 (1973) 39-52.

2614. Margáin, Hugo. "El papel de las leyes generales en la explicación causal." Diánoia. 24 (1978) 1-17.

2615. Margenau, Henry. "Is the Mathematical Explanation of Physical Data Unique?" LMPS-1 (1960) 348-55.

2616. Martin, Raymond. "Singular Causal Explanations." Theory and Decision. 2 (1972) 221-37.

2617. Massey, Gerald J. "Hempel's Criterion of Maximal Specificity." Phil. Studies. 19 (1968) 43-47.

2618. Matthew, Anthony. "Prediction and Predication." BJPS 22 (1971) 171-82.

2619. Matthews, Robert J. "Explaining and Explanation." Am. Phil. Quart. 18 (1981) 71-77.

2620. Meehl, Paul E. Clinical vs. Statistical Prediction. Minneapolis: University of Minnesota Press, 1954. 149 pp.

2621. Meixner, John. "Homogeneity and Explanatory Depth." PS 46 (1979) 366-81.

2622. Mellor, D. H. "Inexactness and Explanation." PS 33 (1966) 345-59.

2623. _____. "Imprecision and Explanation." PS 34 (1967) 1-9.

2624. _____. "Probable Explanation." AJP 54 (1976) 231-41.

2625. Meltzer, B. "The Third Possibility." Mind. 73 (1964) 430-33.

      a. "Two Forms of the Prediction Paradox," by B. Meltzer and I. J. Good. BJPS 16 (1965) 50-51.

2626. Meyer, Michel. "The Contextualist Approach to Explanation." Logique et analyse. 19 (1976) 479-94.

2627. Miller, David L. "Explanation versus Description." Phil. Review. 56 (1947) 306-12.

2628. _____. "The Behavioral Dimension of Prediction and Meaning." PS 17 (1950) 133-41.

2629. Mischel, Theodore. "Pragmatic Aspects of Explanation." PS 33 (1966) 40-60.

2630. Morgenbesser, Sidney. "The Explanatory-Predictive Approach to Science," in Philosophy of Science: The Delaware Seminar, I. New York: Interscience Publishers, 1963. 41-55.

2631. Nagel, Ernest. The Structure of Science: Problems in the Logic of Scientific Explanation. New York: Harcourt, Brace, and World, 1961. 618 pp.

      a. Review essay by J. J. C. Smart. J. Phil. 59 (1962) 216-23.

      b. Review essay by Adolf Grünbaum. PS 29 (1962) 294-305.

      c. Review essay by Mary Hesse. Mind. 72 (1963) 429-40.

      d. Review essay by Edward H. Madden. PS 30 (1963) 64-70.

      e. Review essay, "Professor Nagel on Abstractive Theories and Experimental Laws," by Leo Simons. PS 31 (1964) 163-67.

      f. Review essay by Michael Scriven. Review of Metaphysics. 17 (1964) 403-24.

      g. Review essay by Paul K. Feyerabend. BJPS 17 (1966) 237-49.

      h. Review essay, "Laws of Science, Theories, Measurement," by Leszek Nowak. PS 39 (1972) 533-48.

2632. Nelson, Alvin F. "The HD Method and Scientific Change." Southwestern Journal of Philosophy. 2, No. 1 (1971) 83-92.

2633. Newman, Fred. "Explanation Sketches." PS 32 (1965) 168-72.

    a. "Newman and Explanation-Sketches," by S. T. Goh. 34 (1967) 273-75.

2634. _____. "Two Analyses of Prediction." Theoria. 32 45-55.

    a. "A Third Analysis of Prediction," by Keith Lehrer. 71-74.

2635. Nickles, Thomas. "Covering Law Explanation." PS 38 (1971) 542-61.

    a. "Nickles on Intensionality and the Covering Law Model," by Danny Steinberg. 40 (1973) 403-07.

    b. "Explanation and Description-Relativity," by Thomas Nickles. 408-14.

2636. Niessen, Manfred. "Zur Differenz alltagswelticher und wissens-chaftlicher Erklärungen." ZAW 8 (1977) 369-74.

2637. Niiniluoto, Ilkka. "Statistical Explanation Reconsidered." Synthese. 48 (1981) 437-72.

2638. Nikitina, A. G. "Logical Conditions for Truth in Scientific Prediction." Soviet Studies in Philosophy. 10 (1971) 176-86. [English reprint from Voprosy filosofii. (1971, No.4)].

2639. Norris, Louis William. "Is Science Explanatory?" Personalist. 26 (1945) 143-52.

2640. Nowak, Leszek. "Idealizational Laws and Explanation." Logique et analyse. 15 (1972) 527-45.

2641. Omer, I. A. "On the D-N Model of Scientific Explanation." PS 37 (1970) 417-33.

    a. "On Omer's Model of Scientific Explanation," by Seppo K. Miettinen. 39 (1972) 249-51.

    b. "Omer on Scientific Explanation," by Charles G. Morgan. 40 (1973) 110-17.

2642. _____. "Minimal Law Explanation." Ratio 22 (1980) 155-66.

2643. Pasquinelli, Alberto. "Scientific Explanation and Causality." Quality and Quantity. 1 (1967) 32-52.

2644. Passmore, J. A. "Prediction and Scientific Law." AJP 24 (1946) 1-33.

2645. Patten, Steven C. "Carl Hempel: Explanations by Reasons."
Canadian Journal of Philosophy. 2 (1973) 503-22.

2646. Peacocke, Christopher A. B. Holistic Explanation. New York:
Oxford University Press, 1979. 216 pp.

2647. Pitt, Joseph C. "Hempel versus Sellars on Explanation."
Dialectica. 34 (1980) 95-120.

2648. _____. "The Role of Inductive Generalizations in Sellars'
Theory of Explanation." Theory and Decision. 13 (1981) 345-56.

2649. Ponce, Margarita, and J. A. Robles. "Notes generales sobre
la explicación." Diánoia. 26 (1980) 105-33.

2650. Post, H. R. "Novel Predictions as a Criterion of Merit."
BSPS 39 (1976) 493-96.

2651. Purtill, Richard. "The Purpose of Science." PS 37 (1970)
301-06.

2652. Railton, Peter. "A Deductive-Nomological Model of Probabilistic
Explanation." PS 45 (1978) 206-26.

2653. _____. "Probability, Explanation, and Information."
Synthese. 48 (1981) 233-56.

2654. Rantala, V. "Prediction and Identifiability." WOSPS 11
(1977) 91-102.

2655. Rapoport, Anatol. "Explanatory Power and Explanatory Appeal
of Theories." Synthese. 24 (1972) 321-42.

2656. Reenpää, Yrjö. "Ueber das Problem der Begründung und Letzbeg-
ründung." ZPF 28 (1974) 516-35.

2657. Rescher, N. "On Predication and Explanation." BJPS 8 (1958)
281-90.

2658. _____. "The Stochastic Revolution and the Nature of
Scientific Explanation." Synthese. 14 (1962) 200-15.

2659. _____. "Fundamental Problems in the Theory of Scientific
Explanation," in Philosophy of Science: The Delaware Seminar, II.
New York: Interscience Publishers, 1963. 41-60.

2660. _____. "Discrete State Systems, Markov Chains, and Problems in the Theory of Scientific Explanation and Prediction." PS
30 (1963) 325-45.

2661. _____. Scientific Explanation. New York: The Free
Press, 1970. 242 pp.

2662. Rescher, Nicholas, and F. Brian Skyrms. "A Methodological
Problem in the Evaluation of Explanation." Nous. 2 (1968) 121-29.

2663. Richfield, Jerome, and Irving M. Copi. "Deciding and Predicting." PS 28 (1961) 47-51.

2664. Riedel, Manfred. "Causal and Historical Explanation." SL 72
(1976) 3-26.

2665. Rogers, Ben. "Probabilistic Causality, Explanation, and
Detection." Synthese. 48 (1981) 201-23.

2666. Rosenkrantz, Roger D. "On Explanation." Synthese. 20
(1969) 335-70.

2667. Ruse, Michael. "Narrative Explanation and the Theory of
Evolution." Canadian Journal of Philosophy. 1 (1971) 59-74.

    a. "Narrative Explanation and Redescription," by Herbert
    Burhenn. 3 (1974) 419-25.

    b. "Narrative Explanation Revisited," by Michael Ruse. 4
    (1975) 529-33.

2667. Ryle, G. "Predicting and Inferring," in Observation and
Interpretation. S. Körner, ed. New York: Academic Press; London:
Butterworths, 1957. 165-70.

2668. Salmon, Wesley. "The Predictive Inference." PS 24 (1957)
180-90.

2669. _____. "Statistical Explanation," in The Nature and
Function of Scientific Theories: Essays in Contemporary Science
and Philosophy. Robert G. Colodny, ed. Pittsburgh: University of
Pittsburgh Press, 1970. 173-231.

2670. _____. "Hempel's Conception of Inductive Inference in
Inductive-Statistical Explanation." PS 44 (1977) 180-85.

2671. _____. "Indeterminism and Epistemic Relativization."
PS 44 (1977) 199-202.

2672. _____. "A Third Dogma of Empiricism." WOSPS 11 (1977)
149-66.

2673. _____. "'Why Ask, Why?'? An Inquiry Concerning Scientific Explanation." Proc. and Add. Am. Phil. Assn. 51 (1978) 683-705.
Reprinted in SL 132 (1979) 403-26.

2674. _____. "Rational Prediction." BJPS 32 (1981) 115-25.

2675. Scheffler, Israel. "Explanation, Prediction, and Abstraction."
BJPS 7 (1957) 293-309.

2676. Scheibe, Erhard. "The Approximative Explanation and the Development of Physics." LMPS-4 (1971) 931-42.

2677. _____. "Gibt es Erklärungen von Theorien?" Allgemeine Zeitschrift fur Philosophie. 1, No. 3 (1976) 26-45.

2678. Scriven, Michael. "Definitions, Explanations, and Theories." MSPS 2 (1958) 99-195.

2679. _____. "Explanations, Predictions, and Laws." MSPS 3 (1962) 170-230.

2680. _____. "The Temporal Asymmetry of Explanations and Predictions," in Philosophy of Science: The Delaware Seminar, I. New York: Interscience Publishers, 1963. 97-105.

2681. _____. "The Limits of Physical Explanation," in Philosophy of Science: The Delaware Seminar, II. New York: Interscience Publishers, 1963. 107-35.

2682. Sellars, Wilfrid. "Theoretical Explanation," in Philosophy of Science: The Delaware Seminar, II. New York: Interscience Publishers, 1963. 61-78.

2683. Skarsgård, Lars. "Some Remarks on the Logic of Explanation." PS 25 (1958) 199-207.

2684. Smart, J. J. C. "Conflicting Views About Explanation." BSPS 2 (1965) 157-69.

      a. "Comments: Scientific Realism or Irenic Instrumentalism," by Wilfrid Sellars, 171-204.

      b. "Comments: How Not to Talk About Meaning," by Hilary Putnam, 205-22.

      c. "Reply to Criticism: Comments on Smart, Sellars and Putnam," by Paul K. Feyerabend, 223-61.

2685. Stachowiak, Herbert. "Ueber kausale, konditionale und strukturelle Erklärungsmodelle." PN 4 (1957) 403-33.

2686. Stegmüller, Wolfgang. "Explanation, Prediction, Scientific Systematization and Non-Explanatory Information." Ratio 8 (1966) 1-24.

2687. _____. "Wissenschaft und Erklärung." ZAW 1 (1970) 252-63.

2688. Suchting, W. A. "Deductive Explanation and Prediction Revisited." PS 34 (1967) 41-52.

2689. Suppe, Frederick. "Hilary Putnam's 'Scientific Explanation': An Editorial Summary-Abstract," in The Structure of Scientific Theories. F. Suppe, ed. Urbana, Illinois: University of Illinois Press, 1974. 424-33.

a. "Putnam on the Corroboration of Theories," by Bas C. van Fraassen. 434-36.

2690. Suppes, Patrick, and Mario Zanotti. "When are Probabilistic Explanations Possible?" Synthese. 48 (1981) 191-99.

2691. Thorpe, D. A. "The Quartercentenary Model of D-N Explanation." PS 41 (1974) 188-95.

2692. Thurnher, R. "'Erklärung' in genetischen und systematischen Zusammenhängen." Dialectica. 33 (1979) 189-200.

2693. Tuomela, Raimo. "Deductive Explanation of Scientific Laws." Journal of Philosophical Logic. 1 (1972) 369-92. Reprinted in SL 50 (1973) 103-26.

a. "Tuomela on Deductive Explanation," by Charles G. Morgan. Journal of Philosophical Logic. 5 (1976) 511-25.

b. "Morgan on Deductive Explanation: A Rejoinder," by Raimo Tuomela. 527-43.

2694. _____. "Causes and Deductive Explanation." BSPS 32 (1976) 325-60.

2695. _____. "Dispositions, Realism, and Explanation." Synthese. 34 (1977) 457-78.

2696. _____. "Explaining Explaining." Erkenntnis. 15 (1980) 211-43.

2697. _____. "Inductive Explanation." Synthese. 48 (1981) 257-94.

2698. Turner, Joseph. "Maxwell on the Logic of Dynamical Explanation." PS 23 (1956) 36-47.

2699. Ullmann, Christian. "Ein Schema für Kausalerklarungen." Erkenntnis. 9 (1975) 131-37.

2700. Watkins, J. W. N. "Ideal Types and Historical Explanation." BJPS 3 (1952) 22-43.

2701. Weingartner, Paul. "Einige Fragen zur erfahrungswissenschaftlichen Vorhersage in der Geschichte der Naturphilosophie." Salzburger Jahrbuch für Philosophie. 12-13 (1968/69) 77-93.

2702. Williams, Mary B. "The Logical Structure of Functional Explanations in Biology." PSA 1976. I, 37-46.

2703. Wilson, Fred. "Explanation in Aristotle, Newton, and Toulmin." PS 36 (1969) 291-310; 400-28.

2704. Wimsatt, William C. "Reductive Explanation: A Functional Account." BSPS 32 (1976) 671-710.

2705. Woodward, James. "Scientific Explanation." BJPS 30 (1979) 41-67.

2706. _____. "Developmental Explanation." Synthese. 44 (1980) 443-66.

2707. Workman, Rollin W. "What Makes an Explanation" PS 31 (1964) 241-54.

2708. Wright, Larry. "Rival Explanations." Mind. 82 (1973) 497-515.

2709. Yolton, John W. "Philosophical and Scientific Explanation." J. Phil. 55 (1958) 133-43.

2710. _____. "Explanation." BJPS 10 (1959) 194-208.

2711. Zaffron, Richard. "Identity, Subsumption, and Scientific Explanation." J. Phil. 68 (1971) 849-60.

4.11 Deduction and Proof in Science

2712. Ajdukiewicz, K. "Axiomatic Systems from the Methodological Point of View." SL 87 (1977) 49-63. [First published in Studia Logica. 9 (1960). Revised tr. by J. Giedymin.]

2713. Alexander, H. Gavin. "General Statements as Rules of Inference?" MSPS 2 (1958) 309-29.

2714. Clark, Joseph T. "Contemporary Science and Deductive Methodology." Proc. Am. Cath. Phil. Assn. 26 (1952) 94-131.

2715. Curry, Haskell B. "The Interpretation of Formalized Implication." Theoria. 25 (1959) 1-26.

2716. Dacey, Raymond. "A Theory of Conclusions." PS 45 (1978) 563-74.

2717. Desmonde, William H. "Gödel, Non-Deterministic Systems, and Hermetic Automata." IPQ 11 (1971) 49-74.

2718. Dubarle, R. P. "Est-il possible d'axiomatiser la physique?" Dialectica. 6 (1952) 226-53.

2719. Dummett, Michael. "The Philosophical Significance of Gödel's Theorum." Ratio. 5 (1963) 140-55.

2720. Goodstein, R. L. "The Significance of Incompleteness Theorems." BJPS 14 (1963) 208-20.

a. "Note on Goodstein's 'The Significance of Incompleteness Theorems'," by Arthur I. Fine. BJPS 15 (1964) 140-41.

b. "Reply to Mr. Fine's Note," by R. L. Goodstein. 141.

2721. Goudge, T. A. "The Genetic Fallacy." Synthese. 13 (1961) 41-48.

2722. Granger, G. G. "Vérification et justification comme auxiliaires de la démonstration." Logique et analyse. 11 (1968) 231-69.

2723. Hanson, Norwood Russell. "A Budget of Cross-Type Inferences, or Invention is the Mother of Necessity." Journal of Philosophy. 58 (1961) 449-70.

2724. _____. "Stability Proofs and Consistency Proofs: A Loose Analogy." PS 31 (1964) 301-18.

2725. _____. "The Genetic Fallacy Revisited." American Philosophical Quarterly. 4 (1967) 101-13.

2726. Hintikka, Jaakko and Unto Remes. The Method of Analysis. Its Geometrical Origin and Its General Significance. BSPS 25. SL 75. Dordrecht: Reidel, 1974. 144 pp.

a. Review essay by Hans Jürgen Engfer. Erkenntnis. 13 (1978) 327-37.

2727. Horstmann, R-P. and L. Krüger, eds. Transcendental Arguments and Science. SL 133. Dordrecht: Reidel, 1979. 314 pp.

2728. Kaila, Eino. "Wenn. . .so. . ." Theoria. 11 (1945) 88-98.

2729. Körner, Stephan. "Deductive Unification and Idealisation." BJPS 14 (1964) 274-84.

2730. _____. "On Deductivism as a Philosophy of Science," in Metaphysics and Explanation. W. H. Capitan and D. D. Merrill, eds. Pittsburgh: University of Pittsburgh Press, nd. 9-19.

2731. Kotarbińska, J. "The Controversy: Deductivism versus Inductivism." SL 87 (1977) 261-78. Republished from Studia Filozoficzne. 1 (22) (1961). Also in LMPS-1 (1960) 265-74.

2732. Kyburg, Jr., Henry E. "The Justification of Deduction." Review of Metaphysics. 12 (1958) 19-25.

2733. Ladrière, Jean. Les Limitations internes des formalismes. Paris: Gauthier-Villars, 1957. xiii + 715 pp.

2734. Nagel, Ernest, and James R. Newman. Gödel's Proof. New York: New York University Press, 1958. 118 pp.

2735. Pollock, John L. Subjunctive Reasoning. Dordrecht: Reidel, 1976. 251 pp.

2736. Sadovskij, V. N. "The Deductive Method as a Problem of the Logic of Science." SL 25 (1970) 160-211.

2737. Salmon, Merrilee H. "Consistency Proofs for Applied Mathematics." _Synthese_. 34 (1977) 301-12. Reprinted in SL 132 (1979) 625-36.

2738. Saunders, L. R. "Rational Deduction in Physics: The Parallelogram of Forces." BJPS 14 (1964) 265-73.

2739. Schlegel, Richard. _Completeness in Science_. New York: Appleton-Century-Crofts, 1967. 280 pp.

    a. Review essay, "No More Discovery in Physics?" by Joseph Agassi. _Synthese_. 18 (1968) 103-08.

2740. Stenius, Erik. "Natural Implication and Material Implication." _Theoria_. 13 (1947) 136-56.

2741. Yourgrau, Wolfgang. "Gödel and Physical Theory." _Mind_. 78 (1969) 77-90.

2742. Zinov'ev, A. A. "Logical and Physical Implication." SL 25 (1970) 91-159.

4.12 Induction

2743. Achinstein, Peter. "From Success to Truth." _Analysis_. 21 (1960-61) 6-9.

    a. "Comment on Mr. Achinstein's Paper," by Gilbert Ryle. 9-11.

    b. "Recipes and Induction: Ryle vs. Achinstein," by Harry V. Stopes-Roe. 115-20.

2744. _____ . "The Circularity of Self-Supporting Inductive Argument." _Analysis_. 22 (1961-62) 138-41.

    a. "Self-Support and Circularity: A Reply to Mr. Achinstein," by Max Black. 23 (1962-63) 43-44.

    b. "Circularity and Induction," by Peter Achinstein. 123-27.

2745. Adler, Jonathan E. "Evaluating Global and Local Theories of Induction." _PSA 1976_. I, 212-23.

2746. Agassi, Joseph. "Positive Evidence in Science and Technology." PS 37 (1970) 261-70.

2747. Alexander, H. G. "Convention, Falsification and Induction." _Arist. Soc._ Supp. Vol. 34 (1960) 131-44.

2748. Ambrose, Alice. "The Problem of Justifying Inductive Inference." J. Phil. 44 (1947) 253-72.

2749. Apostel, Leo. "Logique inductive, modalités épistémiques et
logique de la préférence." RIP 25 (1971) 78-100.

2750. Atkinson, Gary. "Rationality and Induction." Southwestern
Journal of Philosophy. 4, No. 1 (1973) 93-100.

2751. Ayer, A. J. "Induction and the Calculus of Probabilities."
Logique et analyse. 11 (1968) 95-142.

2752. Bar-Hillel, Yehoshua. "A Note on Comparative Inductive
Logic." BJPS 3 (1953) 308-10.

2753. Barker, S. F., and Peter Achinstein. "On the New Riddle of
Induction." Phil. Review. 69 (1960) 511-22.

    a. "More on 'Grue' and Grue," by J. S. Ullian. 70 (1961)
    386-89.

    b. "An Unnoticed Flaw in Barker and Achinstein's Solution to
    Goodman's New Riddle of Induction," by Edward S. Shirley. PS
    48 (1981) 611-17.

2754. Bartley, III, W. W. "Goodman's Paradox: A Simple Minded
Solution." Phil. Studies. 19 (1968) 85-88.

    a. "On a Simple-Minded Solution," by James N. Hullett. PS
    37 (1970) 452-54.

2755. Batens, D. Studies in the Logic of Induction and in the Logic
of Explanation: Containing a New Theory of Meaning Relations.
Brugge: De Tempel, Tempelhoff, 37, 1975. 309 pp.

2756. Baumgaertner, William. "Nature of Induction." Proc. Am.
Cath. Phil. Assn. 25 (1951) 130-36.

2757. Benardete, José A. "Induction and Infinity." Ratio. 13
(1971) 139-49.

2758. Bergmann, Gustav. "Some Comments on Carnap's Logic of Induction." PS 13 (1946) 71-78.

2759. Bertolet, R. J. "On the Merits of Entrenchment." Analysis.
37 (1976-77) 29-31.

2760. Black, Max. "Self-Supporting Inductive Arguments." J. Phil.
55 (1958) 718-25.

2761. _____. "Can Induction be Vindicated?" Phil. Studies.
10 (1959) 5-16.

2762. _____. "The Raison d'Être of Inductive Argument." BJPS
17 (1966) 177-204.

2763. _____. "The Justification of Induction," in Philosophy of Science Today. S. Morgenbesser, ed. New York: Basic Books, 1967. 190-200.

2764. _____. "Some Half-Baked Thoughts About Induction," in Philosophy, Science, and Method. S. Morgenbesser et al. eds. New York: St. Martin's Press, 1969. 144-49.

2765. Blackburn, Simon. "Goodman's Paradox," in Studies in the Philosophy of Science. N. Rescher, ed. Oxford: Basil Blackwell, 1969. 128-42.

2766. _____. Reason and Prediction. Cambridge: Cambridge University Press, 1973. viii + 175 pp.

   a. "Blackburn on Induction," by J. S. Edwards. Mind. 86
      (1977) 114-17.

2767. Blanché, Robert. L'induction scientifique et les lois natur-elles. Paris: PUF, 1975. 170 pp.

2768. Bogdan, Radu, ed. Local Induction. SL 93. Dordrecht: Reidel, 1976. 340 pp.

   a. Review essay by Jonathan E. Adler. PS 44 (1977) 173-77.

2769. Boudot, Maurics. Logique inductive et probabilité. Paris: Armand Colin, 1972. 333 pp.

2770. Brodbeck, May. "The New Rationalism: Dewey's Theory of In-duction." J. Phil. 46 (1949) 780-91.

2771. Brown, Harold I. "Need There be a Problem of Induction?" Canadian Journal of Philosophy. 8 (1978) 521-32.

2772. Buchdahl, G. "Induction and Scientific Method." Mind. 60 (1951) 16-34.

2773. _____. "Inductive Process and Inductive Inference." AJP 34 (1956) 164-81.

2774. _____. "Convention, Falsification and Induction." Arist. Soc. Supp. Vol. 34 (1960) 113-30.

2775. Bunge, Mario. "The Place of Induction in Science." PS 27 (1960) 262-70.

2776. Burks, Arthur W. "The Presupposition Theory of Induction." PS 20 (1953) 177-97.

2777. _____. "On the Presuppositions of Induction." Review of Metaphysics. 8 (1955) 574-611.

2778. _____. "On the Significance of Carnap's System of In-ductive Logic for the Philosophy of Induction," in The Philosophy of Rudolf Carnap. P. A. Schilpp, ed. La Salle, Illinois: Open Court, 1963. 739-59.

2779. Butts, Robert E. "Whewell's Logic of Induction," in Foundations of Scientific Method: The Nineteenth Century. R. N. Giere and R. S. Westfall, eds. Bloomington: Indiana University Press, 1973. 53-85.

2780. Campbell, Keith. "One Form of Scepticism About Induction." Analysis. 23 (1963) 80-83. Reprinted in The Justification of Induction. R. Swinburne, ed. London: Oxford University Press, 1974. 145-48.

2781. Capaldi, Nicholas. "Why There is No Problem of Induction." Journal of Critical Analysis. 3 (1971-72) 9-12.

2782. Cardwell, Charles E. "Gambling for Content." J. Phil. 68 (1971) 860-64.

2783. Cargile, James. "On Goodman's Riddle of Induction." Ratio. 12 (1970) 144-47.

2784. Carnap, Rudolf. "On Inductive Logic." PS 12 (1945) 72-97.

    a. "Carnap's 'On Inductive Logic'," by C. West Churchman. 13 (1946) 339-42.

2785. _____. "The Problem of Relations in Inductive Logic." Phil. Studies. 2 (1951) 75-80.

2786. _____. The Continuum of Inductive Methods. Chicago: University of Chicago Press, 1952. v + 92 pp.

2787. _____. "Inductive Logic and Science." Proceedings of the Academy of Arts and Sciences. 80 (1953) 189-97.

2788. _____. "The Aim of Inductive Logic." LMPS-1 (1960) 303-18.

2789. _____. "Inductive Logic and Inductive Intuition," in The Problem of Inductive Logic. I. Lakatos, ed. Amsterdam: North-Holland, 1968. 258-67.

    a. Discussion by M. Bunge, J. W. Watkins, Y. Bar-Hillel, K. R. Popper, J. Hintikka, and R. Carnap. 268-314.

2790. Carnap, Rudolf, and Richard C. Jeffrey, eds. Studies in Inductive Logic and Probability, Volume I. Berkeley: University of California Press, 1971. 240 pp.

    a. Review essay, "The End of the Road for Inductive Logic?" by Colin Howson. BJPS 26 (1975) 143-49.

2791. Carnap, R., and W. Stegmüller. Induktive Logik und Wahrscheinlichkeit. Wien: Springer-Verlag, 1959. 261 pp.

2792. Caws, Peter. "The Paradox of Induction and the Inductive
Wager." PPR 22 (1962) 512-20.

2793. Chatalian, George. "Induction and the Problem of the External
World." J. Phil. 49 (1952) 601-07.

      a. "The External World and Mr. Chatalian," by Donald C.
      Williams. 50 (1953) 13-18.

2794. Chung-Ying Cheng. "Peirce's Probabilistic Theory of Inductive
Validity." Transactions of the Charles S. Peirce Society. 2 (1966)
86-112.

2795. _____. "Charles Peirce's Arguments for the Non-probabil-
istic Validity of Induction." Transactions of the Charles S. Peirce
Society. 3 (1967) 24-39.

      a. "Some Comments on Cheng, Peirce, and Inductive Validity,"
      by Gordon N. Pinkham. 96-107.

2796. Churchman, C. West. "Statistics, Pragmatics, Induction." PS
15 (1948) 249-68.

      a. "Statistical vs. Pragmatic Inference," by John E. Freund.
      16 (1949) 142-47.

      b. "A Note on Churchman's 'Statistics, Pragmatics, Induction,"
      by Thomas A. Cowan. 148-50.

      c. "Reply to Comments on 'Statistics, Pragmatics, Induction',"
      by C. West Churchman. 151-53.

2797. Churchman, C. West, and Bruce G. Buchanan. "On the Design of
Inductive Systems: Some Philosophical Problems." BJPS 20 (1969)
311-23.

2798. Clendinnen, F. John. "Induction and Objectivity." PS 33
(1966) 215-29.

      a. "A Reply to "Induction and Objectivity'," by Frank Jackson.
      37 (1970) 440-43.

      b. "A Response to Jackson," by F. John Clendinnen. 444-48.

      c. "Reply to a Response," by Frank Jackson. 449-51.

      d. "Clendinnen, Jackson, and Induction," by Gary Jones. 46
      (1979) 466-69.

2799. _____. "Inference, Practice and Theory." Synthese. 34
(1977) 89-132. Reprinted in SL 132 (1979) 85-128.

2800. Coburn, Robert C. "Braithwaite's Inductive Justification of
Induction." PS 28 (1961) 65-71.

2801. Cohen, L. J. The Implications of Induction. London: Methuen, 1970. vii + 248 pp.

    a. "Cohen on Evidential Support," by R. G. Swinburne. Mind. 81 (1972) 244-48.

    b. "A Reply to Swinburne," by L. Jonathan Cohen. 249-50.

2801A. _____. "The Inductive Logic of Progressive Problem-Shifts." RIP 25 (1971) 62-77.

2802. _____. "A Note on Inductive Logic." J. Phil. 70 (1973) 27-40.

2803. _____. The Probable and the Provable. Oxford University Press, 1977. ix + 363 pp.

    a. Review essay, "Support and Surprise: L. J. Cohen's View of Inductive Probability," by Isaac Levi. BJPS 30 (1979) 279-92.

2804. Cohen, L. Jonathan, and Mary Hesse, eds. Applications of Inductive Logic. New York: Oxford University Press, 1980. 432 pp.

2805. Cohen, L. Jonathan, and Avishai Margalit. "The Role of Inductive Reasoning in the Interpretation of Metaphor." Synthese. 21 (1970) 469-87.

2806. Cohen, Yael. "A New View of Grue." ZAW 10 (1979) 244-52.

2807. Copeland, Arthur H. "Statistical Induction and the Foundations of Probability." Theoria. 28 (1962) 27-44; 87-109.

2808. Costantini, Domenico. "The Relevance Quotient." Erkenntnis. 14 (1979) 149-57.

2809. _____. "Inductive Logic and Inductive Statistics." BSPS 47 (1981) 169-83.

2810. Cox, L. Hughes. "Are Scientific Induction and Metaphysical Coherence Really Separate Informal Logics?" Southwestern Journal of Philosophy. 4, No. 1 (1973) 109-18.

2811. Creary, Lewis G. "For the Compleat Logical Empiricist: 'Non-Cognitive' Foundations for Inductive Logic." Am. Phil. Quart. 10 (1973) 123-31.

2812. Creath, Richard. "A Query on Entrenchement." PS 45 (1978) 474-77.

2813. Crow, C. "Some Remarks on Induction." Synthese. 15 (1963) 379-88.

2814. Czerwiński, Z. "Zagadnienie probabilistycznego uzasadnienia indukcji enumeracyjnej." [The Problem of a Probabilistic Justifica-

tion of Enumerative Induction] Studia Logica. 5 (1957) 91-104.
English summary: 106-08. Reprinted in English in SL 87 (1977)
65-80.

2815. _____. "Enumerative Induction and the Theory of Games."
Studia Logica. 10 (1960) 29-36. Republished in SL 87 (1977) 81-91.

2816. _____. "Problematyka indukcji w pracach i działalności
Kazimierz Ajdukiewicz." Studia Logica. 16 (1965) 31-38.

2817. Das, R. "Induction and Non-Instantial Hypothesis (A Review
of Dr. Wisdom's Views)." BJPS 8 (1958) 317-25.

   a. "A Reply to Dr. Das's Criticisms," by J. O. Wisdom.
   325-28.

2818. Dorling, Jon. "Demonstrative Induction: Its Significant
Role in the History of Physics." PS 40 (1973) 360-72.

2819. Dorling, Jon, and David Miller. "Bayesian Personalism,
Falsificationism, and the Problem of Induction." Arist. Soc.
Supplementary Vol. 55 (1981) 109-41.

2820. Edwards, Paul. "Russell's Doubts About Induction." Mind.
68 (1949) 141-63. Reprinted in The Justification of Induction. R.
Swinburne, ed. London: Oxford University Press, 1974. 26-47.

2821. Ellis, Brian. "A Vindication of Scientific Inductive Prac-
tices." Am. Phil. Quart. 2 (1965) 296-304.

2822. Essler, Wilhelm K. Induktive Logik: Grundlagen und Voraussetzungen. Freiburg und Munich: Verlag Karl Alber, 1970.

2823. _____. "Hintikka versus Carnap." Erkenntnis. 9 (1975)
229-33. Reprinted in SL 73 (1975) 365-70.

   a. "Carnap and Essler versus Inductive Generalization," by
   J. Hintikka. Erkenntnis. 9 (1975) 235-44.

2824. Ewing, A. C. "Causality and Induction." 12 (1952) 465-84.

2825. Fain, Haskell. "The Very Thought of Grue." Phil. Review.
76 (1967) 61-73.

2826. Farre, George L. "Remarks on the Relevance of Induction to
the Physical Sciences." Proc. Am. Cath. Phil. Assn. 38 (1964)
178-86.

2827. Feibleman, James K. "On the Theory of Induction." PPR 14
(1954) 332-42.

2828. Feigl, H. "Scientific Method Without Metaphysical Presuppositions." Philosophical Studies. 5 (1954) 17-29.

a. "Concerning Mr. Feigl's 'Vindication' of Induction," by Daniel Kading. PS 27 (1960) 405-07.

b. "On the Vindication of Induction," by Herbert Feigl. 28 (1961) 212-16.

2829. Féraud, L. "Induction amplifiante et inférence statistique." Dialectica. 3 (1949) 127-52.

2830. Feyerabend, Paul K. "A Note on the Problem of Induction." J. Phil. 61 (1964) 349-53.

2831. _____. "A Note on Two 'Problems' of Induction." BJPS 19 (1968) 251-53.

a. "Feyerabend's Solution to the Goodman Paradox," by Lawrence Foster. 20 (1969) 259-60.

b. "The Point of Positive Evidence-Reply to Professor Feyerabend," by T. W. Settle. 352-55.

2832. Fisk, Milton. "Are There Necessary Connections in Nature?" PS 37 (1970) 385-404.

2833. Foster, M. H., and M. L. Martin, eds. Probability, Confirmation, and Simplicity: Readings in the Philosophy of Inductive Logic. New York: Odyssey Press, 1966. 470 pp.

2834. Frey, Gerhard. "Ueber die Gültigkeit genereller Sätze." Synthese. 20 (1969) 104-20.

2835. Friedman, Kenneth. "Son of Grue: Simplicity vs. Entrenchment." Nous. 7 (1973) 366-78.

2836. _____. "Another Shot at the Canons of Induction." Mind. 84 (1975) 177-91.

2837. Fritz, Jr., Charles A. "What is Induction?" J. Phil. 57 (1960) 126-38.

2838. Fumerton, R. A. "Induction and Reasoning to the Best Explanation." PS 47 (1980) 589-600.

2839. Giere, Ronald N. "The Epistemological Roots of Scientific Knowledge." MSPS 6 (1975) 212-61.

2840. Goldmann, Eliezer. "Remarks Concerning Induction." [In Hebrew] Iyyun. 5 (1954) 286-96.

2841. Goodman, Nelson. Fact, Fiction, and Forecast. Cambridge, MA: Harvard University Press, 1955. Second edition: Indianapolis: Bobbs-Merrill, 1965. 128 pp.

a. Review essay, "Fact, Fiction, and Forecast," by Arthur Pap. Review of Metaphysics. 9 (1955) 285-99.

b. Review essay, "Fact, Fiction, and Forecast," by G. P. Henderson. Phil. Quart. 6 (1956) 266-72.

c. Review essay by J. Watling. Mind. 65 (1956) 267-73.

d. Review essay, "Another Look at Fact, Fiction, and Forecast," by John W. Sweigart, Jr. and John P. Stewart. Phil. Studies. 10 (1959) 81-89.

e. Review essay by John C. Cooley. J. Phil. 54 (1957) 293-311.

f. "Reply to an Adverse Ally," by Nelson Goodman. 531-35.

g. "A Somewhat Adverse Reply to Professor Goodman," by J. C. Cooley. 55 (1958) 159-66.

2842. _____. "Forward" and Comments" [on the New Riddle of Induction]. J. Phil. 63 (1966) 281, 328-31.

a. "Emeroses by Other Names," by Donald Davidson. 778-80.

b. "Two Replies," by Nelson Goodman. 64 (1967) 286-87.

2843. Gottlieb, Dale. "Rationality and the Theory of Projection." Noûs. 9 (1975) 319-28.

a. "On Projectability," by Joseph S. Ullian. 329-39.

2844. Graves, John C. "Uniformity and Induction." BJPS 25 (1974) 301-18.

2845. Grégoire, Franz. "L'induction." Revue Philosophique de Louvain. 62 (1964) 108-51.

2846. Griffin, N. "The Problem of Induction." Scientia. 104 (1969) 251-65. French Tr.: Supplement, 138-51.

2847. Grunfeld, J. "Inductive Language." Logique et analyse. 19 (1976) 391-404.

2848. Grunstra, Bernard R. "The Plausibility of the Entrenchment Concept," in Studies in the Philosophy of Science. N. Rescher, ed. Oxford: Basil Blackwell, 1969. 100-27.

2849. Gutting, Gary. "Metaphysics and Induction." Process Studies. 1 (1971) 171-78.

a. "'Metaphysics and Induction': Reply and Rejoinder," by James W. Felt, S. J. and Gary Gutting. 179-82.

2850. Hacking, Ian. "Salmon's Vindication of Induction." J. Phil. 62 (1965) 260-66.

a. "Hacking Salmon on Induction," by Isaac Levi. 481-87.

2851. _____. "Linguistically Invariant Inductive Logic." SL
31 (1970) 33-55. Reprinted from Synthese. 20 (1969) 25-47.

a. "Comments" by Isaac Levi. 56-63. Reprinted from Synthese.
20 (1969) 48-55.

2852. _____. "The Leibniz-Carnap Program for Inductive Logic."
J. Phil. 68 (1971) 597-610.

a. "Possibility, Propensity, and Chance: Some Doubts about
the Hacking Thesis," by Margaret D. Wilson. 610-17.

2853. Hanson, N. R. "Good Inductive Reasons." Phil. Quart. 11
(1961) 123-34.

2854. Hao Wang. "Notes on the Justification of Induction." J. Phil.
44 (1947) 701-10.

2855. _____. "On Scepticism about Induction." PS 17 (1950)
333-35.

a. "Scepticism and the Future," by Frederick L. Will.
336-46.

2856. Harman, Gilbert. "The Inference to the Best Explanation."
Phil. Review. 74 (1965) 88-95.

a. "Enumerative Induction and Best Explanation," by Robert
H. Ennis. J. Phil. 65 (1968) 523-29.

b. "Enumerative Induction as Inference to the Best Explanation," by Gilbert H. Harman. 529-33.

2857. _____. "Induction. A Discussion of the Relevance of
the Theory of Knowledge to the Theory of Induction (with a Digression
to the Effect that Neither Deductive Logic nor the Probability
Calculus has Anything to do with Inference)." SL 26 (1970) 83-99.

2858. Harré, R. "Dissolving the 'Problem' of Induction." Philosophy.
32 (1957) 58-64.

2859. _____. "Non-Cumulative Generalizations." Theoria. 23
(1957) 194-97.

2860. Harris, James F., and Kevin Hoover. "Abduction and the New
Riddle of Induction." Monist. 63 (1980) 329-41.

2861. Harrod, R. F. "Induction and Probability." Philosophy. 26
(1951) 37-52.

a. "Induction and Probability," by G. T. Kneebone. 261-62.

2862. _____. Foundations of Inductive Logic. London: Mac-millan, 1956; New York: Harcourt Brace, 1957. xviii + 290 pp.

    a. Review essay, "The Scandal of Philosophy," by J. Bronowski. BJPS 8 (1958) 329-34.

    b. "On Mr. Roy Harrod's New Argument for Induction," by K. R. Popper. 9 (1958) 221-24.

    c. "New Argument for Induction: Reply to Professor Popper," by Roy Harrod. 10 (1960) 309-12.

    d. "Review essay by L. E. Palmieri. Methodos. 11 (1959) 169-72.

2863. _____. "The General Structure of Inductive Argument." Proc. Arist. Soc. 61 n.s. (1960-61) 41-56.

2864. Hausman, Alan. "Goodman's Perfect Communities." Synthese. 41 (1979) 185-237.

2865. Hay, W. H. "Bertrand Russell on the Justification of Induction." PS 17 (1950) 266-77.

2866. Hempel, Carl G. "Inductive Inconsistencies." Synthese. 12 (1960) 439-69. Reprinted, with slight changes, in C. G. Hempel, Aspects of Scientific Explanation and Other Essays in the Philosophy of Science. New York: Free Press, 1965. 53-79.

2867. _____. "Recent Problems of Induction," in Mind and Cosmos. Robert G. Colodny, ed. Pittsburgh: University of Pittsburgh Press, 1966. 112-34.

2868. _____. "Turns in the Evolution of the Problem of Induction." Synthese. 46 (1981) 389-404.

    a. "Instrumentalism and Scientific Scepticism," by Jim Leach. 405-12.

    b. "Epistemic Utility and Theory Acceptance: Comments on Hempel," by Robert Feleppa. 413-20.

2869. Hertzberg, Lars. "Inductive Soundness, Entrenchment, and 'Luck'." Ajatus. 33 (1971) 40-63.

2870. Hesse, Mary. "Induction and Theory-Structure." Review of Metaphysics. 18 (1964) 109-22.

2871. _____. "Consilience of Inductions," in The Problem of Inductive Logic. I. Lakatos, ed. Amsterdam: North-Holland, 1968. 232-46.

    a. Discussion by L. J. Cohen, J. L. Mackie, W. C. Kneale, and M. B. Hesse. 247-57.

2872. _____. "Ramifications of 'Grue'." BJPS 20 (1969) 13-25.

    a. "Further Ramifications of 'Grue'," by G. M. K. Hunt. 257-59.

2873. _____. "An Inductive Logic of Theories." MSPS 4 (1970) 164-80.

2874. _____. "Probability as the Logic of Science." Proc. Arist. Soc. 72 n.s. (1971-72) 257-72.

2875. Hillman, Donald J. "The Probability of Induction." Phil. Studies. 14 (1963) 51-56.

2876. Hilpinen, Risto. "On Inductive Generalization in Monadic First-order Logic with Identity," in Aspects of Inductive Logic. J. Hintikka and P. Suppes, eds. Amsterdam: North-Holland, 1966. 133-54.

2877. _____. Rules of Acceptance and Inductive Logic. Amsterdam: North-Holland Pub. Co., 1968. 134 pp. Also in Acta Philosophica Fennica. 22 (1968) 134 pp.

    a. "Hilpinen's Rules of Acceptance and Inductive Logic," by Alex C. Michalos. PS 38 (1971) 293-302.

    b. "Rules of Acceptance, Indices of Lawlikeness, and Singular Inductive Inference: Reply to a Critical Discussion," by Risto Hilpinen and Jaakko Hintikka. 303-07.

2878. _____. "Relational Hypotheses and Inductive Inference." Synthese. 23 (1971) 266-86.

2879. _____. "Carnap's New System of Inductive Logic." SL 73 (1975) 333-60. Reprinted from Synthese. 25 (1973) 307-33.

2880. Hintikka, Jaakko. "Towards a Theory of Inductive Generalization." LMPS-2 (1964) 274-88.

2881. _____. "A Two-Dimensional Continuum of Inductive Methods," in Aspects of Inductive Logic. J. Hintikka and P. Suppes, eds. Amsterdam: North-Holland, 1966. 113-32.

2882. _____. "Induction by Enumeration and Induction by Elimination," in The Problem of Inductive Logic. I. Lakatos, ed. Amsterdam: North-Holland, 1968. 191-216.

    a. Discussion by J. R. Lucas, R. Carnap, M. B. Hesse, and J. Hintikka. 217-31.

2883. _____. "Carnap and Essler versus Inductive Generalization." SL 73 (1975) 371-80. Reprinted from Erkenntnis. 9 (1975).

2884. Hintikka, Jaakko, and Patrick Suppes, eds. Aspects of Inductive Logic. Amsterdam: North-Holland Pub. Co., 1966. 320 pp.

a. Review essay by Isaac Levy. BJPS 19 (1968) 73-81.

2885. Hintikka, Jaakko, and Risto Hilpinen. "Knowledge, Acceptance, and Inductive Logic," in Aspects of Inductive Logic. J. Hintikka and P. Suppes, eds. Amsterdam: North-Holland, 1966. 1-20.

2886. Hintikka, Jaakko, and Juhani Pietarinen. "Semantic Information and Inductive Logic," in Aspects of Inductive Logic. J. Hintikka and P. Suppes, eds. Amsterdam: North-Holland, 1966. 96-112.

2887. Hoche, Hans-Ulrich. "Does Goodman's 'Grue' Serve its Purpose?" Ratio 21 (1979) 162-73.

2888. Hoppe, Hansgeorg. "Goodmans Schein Rätsel. Ueber die Widersprüchlichkeit und Erfahrungswidrigkeit des sog. 'New Riddle of Induction'." ZAW 6 (1975) 331-39.

2889. Horton, Mary. "In Defence of Francis Bacon: A Criticism of the Critics of the Inductive Method." SHPS 4 (1973) 241-78.

a. "A Note on Ms. Horton's Defense of Bacon," by Rom Harré. 5 (1974) 305-06.

2890. Hunt, G. M. K. "A Conditional Vindication of the Straight Rule." BJPS 21 (1970) 198-99.

2891. Hutten, Ernest H. "Induction as a Semantic Problem." Analysis. 10 (1949-50) 126-36.

2892. Issman, S. "Le problème de l'induction." RIP 11 (1957) 227-30.

2893. _____. "Les problèmes de la déduction logique et des inférences inductives." RIP 13 (1959) 132-34.

2894. Jackson, Frank. "Grue." J. Phil. 72 (1975) 113-31.

2895. Jeffrey, Richard C. "Goodman's Query." J. Phil. 63 (1966) 281-89.

2896. _____. "Carnap's Inductive Logic." SL 73 (1975) 325-32. Reprinted from Synthese. 25 (1973) 299-306.

2897. _____. Studies in Inductive Logic and Probability, Volume II. Berkeley: University of California Press, 1980. 290 pp.

2898. Jessup, John A. "Peirce's Early Account of Induction." Transactions of the Charles S. Peirce Society. 10 (1974) 224-34.

2899. Johnsen, Bredo C. "Black and the Inductive Justification of Induction." Analysis. 32 (1971-72) 110-12.

2900. _____. "Harman on Induction." Phil. Studies. 36 (1979) 77-83.

2901. _____. "Russell's New Riddle of Induction." Philosophy. 54 (1979) 87-97.

2902. Juhos, Béla von. "Deduktion, Induktion und Wahrscheinlich-keit." Methodos. 6 (1954) 259-78.

2903. _____. "Ueber die empirische Induktion." Studium Generale. 19 (1966) 259-72.

2904. Kahane, Howard. "Nelson Goodman's Entrenchment Theory." PS 32 (1965) 377-83.

   a. "Projecting Unprojectibles," by Robert J. Ackermann. 33 (1966) 70-75.

   b. "Reply to Ackermann," by Howard Kahane. 34 (1967) 184-87.

   c. "Conflict and Decision," by Robert J. Ackermann. 188-93.

2905. Katz, J. J. The Problem of Induction and Its Solution. Chicago: University of Chicago Press, 1962. xiii + 125 pp.

   a. Review essay: "Katz on the Vindication of Induction," by F. John Clendinnen. PS 32 (1965) 370-76.

2906. Kelley, Michael H. "Predicates and Projectibility." Canadian Journal of Philosophy. 1 (1971) 189-206.

2907. Kemeny, John G. "Extension of the Methods of Inductive Logic." Phil. Studies. 3 (1952) 38-42.

2908. _____. "A Contribution to Inductive Logic." PPR 13 (1953) 371-74.

   a. "Remarks to Kemeny's Paper," by Rudolf Carnap. 375-76.

2909. Kielkopf, Charles F. "Deduction and Intuitive Induction." PPR 26 (1966) 379-90.

2910. Kneebone, G. T. "Induction and Probability." Proc. Arist. Soc. 50 n.s. (1949-50) 27-42.

2911. Koehn, Donald R. "C. S. Peirces' 'Illustrations of the Logic of Science' and the Pragmatic Justification of Induction." Transactions of the Charles S. Peirce Society. 9 (1973) 157-74.

2912. Köhler, E. "Observation Established by Artificial Intelligence and the Solution to Goodman's Paradox." Poznań Studies. 4 (1978) 12-26.

2913. Kokoszyńska, M. "O 'dobrej' i 'złej' indukcji." [On 'Good' and 'Bad' Induction] Studia Logica. 5 (1957) 43-65. English summary: 68-70.

2914. Kostiouk, V. N. "Probability and the Problem of Induction."
SL 146 (1981) 33-40.

2915. Kraft, Victor. "Das Problem der Induktion." ZAW 1 (1970)
71-82.

  a. "Ist die Zukunft eine Extrapolation? Bemerkungen zu
  einem Aufsatz von Victor Kraft," by Peter Rohs. 3 (1972)
  81-84.

  b. "Ist die Zukunft eine Extrapolation?" by Victor Kraft.
  358.

2916. Kronthaler, E. "Bemerkungen zum zweidimensionalen Kontinuum
induktiver Methoden von J. Hintikka." Theory and Decision. 1
(1971) 387-92.

  a. "Inductive Generalization and its Problems: A Comment on
  Kronthaler's Comments," by J. Hintikka. 393-98.

2917. _____. "Induktion und Gesetzartigkeit. Zu einer Theorie
von Nelson Goodman." Poznan Studies. 4 (1978) 322-25.

2918. Kuipers, T. A. F. "Inductieve waarschijnlijkheid, de basis
van inductieve logica." Algemeen Nederlands Tijdschrift voor Wijsbegeerte. 64 (1972) 291-96.

2919. _____. "A Generalization of Carnap's Inductive Logic."
SL 73 (1975) 361-64. Reprinted from Synthese. 25 (1973) 334-36.

2920. _____. "On the Generalization of the Continuum of Inductive Methods to Universal Hypotheses." Synthese. 37 (1978) 255-84.

2921. Kutschera, Franz von. "Goodman on Induction." Erkenntnis.
12 (1978) 189-207.

2922. Kyburg, Jr., Henry E. "The Justification of Induction." J.
Phil. 53 (1956) 394-400.

2923. _____. Demonstrative Induction." PPR 21 (1960) 80-92.

2924. _____. "Recent Work in Inductive Logic." Am. Phil. Quart.
1 (1964) 249-87.

2925. _____. "The Rule of Detachment in Inductive Logic," in
The Problem of Inductive Logic. I. Lakatos, ed. Amsterdam: North-
Holland, 1968. 98-119.

  a. Discussion by Y. Bar-Hillel, P. Suppes, K. R. Popper, W.
  C. Salmon, J. Hintikka, R. Carnap, and H. E. Kyburg, Jr.
  120-65.

2926. _____. "An Interpolation Theorem for Inductive Relations."
J. Phil. 75 (1978) 93-98.

a. "High Probability and Inductive Systematization," by
Illka Niiniluoto. 737-39.

2927. Kyburg, Jr., H. E., and Nagel, E., eds. Induction: Some
Current Issues. Middletown, Conn.: Wesleyan U. P., 1963. xxi
+ 220 pp.

2928. Laer, P. H. van. "Wijsgerige aspecten van de wettenschappelijke
inductie." Tijdschrift voor Filosofie. 16 (1954) 55-84.

2929. Lakatos, Imre, ed. The Problem of Inductive Logic. Amsterdam:
North-Holland Publishing Co., 1968. 417 pp. [Vol. 2 of Proceedings
of the International Colloquium in the Philosophy of Science, London,
1965.]

2930. _____. "Changes in the Problem of Inductive Logic," in
The Problem of Inductive Logic. I. Lakatos, ed. Amsterdam: North-
Holland, 1968. 315-417.

2931. Laudan, Laurens L. "Induction and Probability in the Nineteenth Century." LMPS-4 (1971) 429-38.

2932. _____. "William Whewell on the Consilience of Inductions."
Monist. 55 (1971) 368-91.

a. "Whewell's Consilience of Inductions and Predictions," by
Mary Hesse. 520-24.

b. "Reply to Mary Hesse," by Larry Laudan. 525.

2933. Lázaro, José M. "Un tercer tipo de induccion: la explicacion
por hipotesis." Dialogos. 2, No. 3 (1965) 55-71.

2934. Lebedev, S. A. "The Role of Induction in the Functioning of
Contemporary Science." Soviet Studies in Philosophy. 19, No. 4
(1981) 70-88. [English reprint from Voprosy filosofii. (1980, No.
6)]

2935. Lee, Harold N. "An Epistemological Analysis of Induction."
Tulane Studies in Philosophy. 2 (1953) 83-94.

2936. Lehrer, Keith. "Descriptive Completeness and Inductive
Methods." Journal of Symbolic Logic. 28 (1963) 157-60.

2937. _____. "Induction: A Consistent Gamble." Nous. 3
(1969) 285-97.

a. "Truth, Content, and Ties," by Isaac Levi. J. Phil. 68
(1971) 865-76.

2938. _____. "Justification, Explanation, and Induction." SL
26 (1970) 100-33.

2939. _____. "Induction, Reason and Consistency." BJPS 21
(1970) 103-14.

a. "Some Objections to Keith Lehrer's Rule IR," by Diderik Batens. 22 (1971) 357-62.

2940. _____. "Induction and Conceptual Change." Synthese. 23 (1971) 206-25.

2941. _____. "Reasonable Acceptance and Explanatory Coherence: Wilfrid Sellars on Induction." Nous. 7 (1973) 81-103.

2942. _____. "Induction, Rational Acceptance, and Minimally Inconsistent Sets." MSPS 6 (1975) 295-323.

2943. Lenz, John W. "Problems for the Practicalist's Justification of Induction." Phil. Studies. 9 (1958) 4-8. Reprinted in The Justification of Induction. R. Swinburne, ed. London: Oxford University Press, 1974. 98-101.

2944. Levi, Isaac. "Deductive Cogency in Inductive Inference." J. Phil. 62 (1965) 68-77.

2945. _____. Gambling with Truth: An Essay on Induction and the Aims of Science. New York: Knopf; London: Routledge & Kegan Paul, 1967. 246 pp.

a. Review essay by Richard C. Jeffrey. J. Phil. 65 (1968) 313-22.

b. Review essay, "Induction: A Consistent Gamble," by Keith Lehrer. Nous. 3 (1969) 285-97.

2946. _____. "Induction and the Aims of Inquiry," in Philosophy, Science, and Method. S. Morgenbesser et al., ed. New York: St. Martin's Press, 1969. 92-111.

2947. _____. "Inductive Appraisal," in Current Research in Philosophy of Science. P. D. Asquith and H. E. Kyburg, Jr., eds. East Lansing, MI: Philosophy of Science Assn., 1979. 339-51.

2948. Lewis, David. "Immodest Inductive Methods." PS 38 (1971) 54-63.

a. "Lewis on Immodest Inductive Models," by Stephen Spielman. 39 (1972) 375-77.

b. "Spielman and Lewis on Inductive Immodesty," by David Lewis. 41 (1974) 84-85.

c. "Inductive Immodesty and Lawlikeness," by Juhani Pietarinen. 196-98.

2949. Łos, J. "The Foundations of a Methodological Analysis of Mill's Methods." SL 87 (1977) 291-325. [First published in Annales UMCS (1947).]

2950. Lungarzo, Carlos. "Características del Método inductivo en Ciencia.' Dialogos. 11, No. 28 (1975) 25-56.

2951. Łuszczewska-Romahnowa, S. "Indukcja a prawdopodobieństwo." [Induction and Probability] Studia Logica. 5 (1957) 71-87. English summary: 89-90.

2952. McGowan, R. S. "Predictive Policies." Arist. Soc. Supp. Vol. 41 (1967) 57-76.

2953. _____. "Counter-Induction." Danish Yearbook of Philosophy. 5 (1968) 31-46.

2954. Madden, Edward H. "The Riddle of Induction." J. Phil. 55 (1958) 705-18.

2955. _____. "Hume and the Fiery Furnace." PS 38 (1971) 64-78.

2956. Margolis, Joseph. "The Demand for a Justification of Induction." Synthese. 11 (1959) 259-64.

2957. Matson, W. I. "Against Induction and Empiricism." Proc. Arist. Soc. 62 n.s. (1961-62) 143-58.

2958. Maxwell, Grover. "An 'Analytic' Vindication of Induction." Phil. Studies. 12 (1961) 43-45.

2959. Maxwell, Nicholas. "Can There be Necessary Connections Between Successive Events?" BJPS 19 (1968) 1-25. Reprinted in The Justification of Induction. R. Swinburne, ed. London: Oxford University Press, 1974. 149-74.

2960. May, E. "Inducktion und Exhaustion." Methodos. 1 (1949) 137-49. [English Tr.: 150-56].

2961. Mayo, Deborah. "In Defense of the Neyman-Pearson Theory of Confidence Intervals." PS 48 (1981) 269-80.

2962. Medawar, Peter Brian. Induction and Intuition in Scientific Thought. Philadelphia: American Philosophical Society, 1969. 62 pp.

    a. "Is There a Philosophy of Science?: An Essay Review," by C. Truesdell. Centaurus. 17 (1972/73) 142-72.

2963. Merrill, G. H. "Peirce on Probability and Induction." Transactions of the Charles S. Peirce Society. 11 (1975) 90-109.

2964. Miller, David. "The Measure of All Things." MSPS 6 (1975) 350-66.

2965. Moore, Asher. "The Principle of Induction." J. Phil. 49 (1952) 741-47.

    a. "An Analytic Principle of Induction?" by May Brodbeck.
    747-50.

    b. "The Principle of Induction (II): A Rejoinder to Miss
    Brodbeck," by Asher Moore. 750-58.

2966. Moreland, John. "On Projecting Grue." PS 43 (1976) 363-77.

2967. Morris, John M. "Some Problems Concerning Projection." AJP
49 (1971) 38-46.

2968. Mortimer, Halina. "A Rule of Acceptance Based on Logical
Probability." Synthese. 26 (1973) 259-63.

2969. Mukherji, S. R. "The Problem of Induction." Philosophical
Quarterly (India). 37 (1964) 85-91.

2970. Nagel, Ernest. "Carnap's Theory of Induction," in The Philosophy of Rudolf Carnap. P. A. Schilpp, ed. La Salle, Illinois:
Open Court, 1963. 785-825.

2971. Nelson, E. J. "Causal Necessity and Induction." Proc. Arist.
Soc. 64 n.s. (1963-64) 289-300.

2972. _____. "Metaphysical Presuppositions of Induction."
Proc. & Add. Am. Phil. Assn. 40 (1966-67) 19-33.

2973. _____. "Metaphysische Voraussetzungen der Induktion."
PN 13 (1971-72) 458-73.

2974. Nelson, J. O. "Are Inductive Generalizations Quantifiable?"
Analysis. 22 (1961-62) 59-65.

    a. "Unquantified Inductive Generalizations," by J. E. Llewelyn.
    134-37.

2975. _____. "How Inductive Conclusions Can be Certain."
Philosophical Investigations. 3, No. 3 (1980) 20-32.

2976. Nielsen, H. A. "Sampling and the Problem of Induction."
Mind. 68 (1959) 474-81.

2977. Niiniluoto, Ilkka. "Can We Accept Lehrer's Inductive Rule?"
Ajatus. 33 (1971) 254-65.

2978. _____. "Inductive Systematization: Definition and a
Critical Survey." Synthese. 25 (1972) 25-81.

2979. _____. "Inductive Explanation, Propensity, and Action."
SL 72 (1976) 335-68.

2980. _____. "On a K-Dimensional System of Inductive Logic."
PSA 1976. II, 425-47.

a. "Comments on Niiniluoto and Uchii," by Paul Teller. 495-504.

2981. _____. "Degrees of Truthlikeness: From Singular Sentences to Generalisations." BJPS 30 (1979) 371-76.

2982. _____. "Analogy and Inductive Logic." Erkenntnis. 16 (1981) 1-34.

a. "Analogy and Inductive Logic: A Note on Niiniluoto," by Wolfgang Spohn. 35-52.

2983. Nowak, Stefan. "Inductive Inconsistencies and Conditional Laws of Science." Synthese. 23 (1972) 357-73.

2984. _____. "Logical and Empirical Assumptions of Validity of Inductions." Synthese. 37 (1978) 321-50.

2985. Nyman, Alf. "Induction et intuition." Theoria. 19 (1953) 21-41.

2986. Öfsti, Audun. "Some Problems of Counter-Inductive Policy as Opposed to Inductive." Inquiry. 5 (1962) 267-83.

2987. Oliver, W. Donald. "A Re-Examination of the Problem of Induction." J. Phil. 49 (1952) 769-80.

2988. Pakswer, S. "Information, Entropy and Inductive Logic." PS 21 (1954) 254-59.

2989. Pal, Ramachandra. "On a Proof that Pure Induction Approaches Certainty as a Limit." Philosophical Quarterly (India). 28 (1955) 203-06.

2990. Pietarinen, J. Lawlikeness, Analogy, and Inductive Logic. Acta Philosophica Fennica 26. Amsterdam: North-Holland Publishing Co., 1972.

2991. Plamondon, Ann. "Metaphysics and 'Valid Inductions'." Process Studies. 3 (1973) 91-99.

a. "The Feeling for the Future: A Comment on Ann Plamondon's Essay," by James W. Felt, S.J. 100-03.

2992. Plato, Jan von. "On Partial Exchangeability as a Generalization of Symmetry Principles." Erkenntnis. 16 (1981) 53-59.

2993. Pole, Nelson. "'Self-supporting' Inductive Arguments." BSPS 8, SL 39 (1971) 496-503.

2994. Pollock, John L. "Counter-Induction." Inquiry. 5 (1962) 284-94.

2995. _____. "The Logic of Projectibility." PS 39 (1972) 302-14.

2996. _____. "Subjunctive Generalizations." Synthese. 28 (1974) 199-214.

2997. Popper, Karl R. "Conjectural Knowledge: My Solution to the Problem of Induction." RIP 25 (1971) 167-97. Reprinted in Karl R. Popper, Objective Knowledge. Oxford: Clarendon Press, 1972. 1-31.

2998. Post, H. R. "Correspondence, Invariance and Heuristics: In Praise of Conservative Induction." SHPS 2 (1971) 213-55.

2999. Priest, Graham. "Gruesome Simplicity." PS 43 (1976) 432-37.

3000. Quine, W. V. "Natural Kinds." SL 24 (1970) 5-23. Reprinted as Chapter 5 of Quine's Ontological Relativity and Other Essays. New York: Columbia University Press, 1969.

3001. Randall, C. H., and D. J. Foulis. "A Mathematical Setting for Inductive Reasoning." WOSPS 6-III (1976) 169-202.

3002. Rankin, K. W. "Linguistic Analysis and the Justification of Induction." Phil. Quart. 5 (1955) 316-28.

3003. Reichenbach, Hans. "A Conversation Between Bertrand Russell and David Hume." J. Phil. 46 (1949) 545-49.

3004. Rescher, Nicholas. "Non-Deductive Rules of Inference and Problems in the Analysis of Inductive Reasoning." Synthese. 13 (1961) 242-51.

3005. _____. "Pragmatic Justification." Philosophy. 39 (1964) 346-48.

3006. _____. "Peirce and the Economy of Research." PS 43 (1976) 71-98.

a. "Rescher on the Goodman Paradox," by B. L. Bunch. 47 (1980) 119-23.

3007. _____. Induction: An Essay on the Justification of Inductive Reasoning. Pittsburgh: University of Pittsburgh Press, 1980. xii + 225 pp.

3008. Rosenkrantz, Roger. "Inductivism and Probabilism." Synthese. 23 (1971) 167-205.

3009. Rossi, Robert Joseph. "Application of Inductive Logic to the Analysis of Construct Validity." Synthese. 37 (1978) 285-320.

3010. Rozeboom, William W. "Ontological Induction and the Logical Typology of Scientific Variables." PS 28 (1961) 337-77.

3011. Salmon, Wesley C. "The Uniformity of Nature." PPR 14 (1953) 39-48.

3012. _____. "Regular Rules of Induction." Phil. Review. 65 (1956) 385-88.

3013. _____. "Should We Attempt to Justify Induction?" Phil. Studies. 8 (1957) 33-48.

    a. "On Justifying Induction," by P. F. Strawson. 9 (1958) 20-21.

3014. _____. "Vindication of Induction," in Current Issues in the Philosophy of Science. H. Feigl and G. Maxwell, eds. New York: Holt, Rinehart and Winston, 1961. 245-56.

    a. "Comments" by Stephen Barker. 257-60.

    b. "Rejoinder" by Wesley Salmon. 260-62.

    c. "Comments" by Richard Rudner. 262-64.

3015. _____. "On Vindicating Induction." PS 30 (1963) 252-61.

    a. "Salmon's Vindication," by Ian Hacking. 32 (1965) 269-71.

3016. _____. "Inductive Inference," in Philosophy of Science: The Delaware Seminar, II. New York: Interscience Publishers, 1963. 341-70.

3017. _____. "Consistency, Transitivity, and Inductive Support." Ratio 7 (1965) 164-69.

3018. _____. "Carnap's Inductive Logic." J. Phil. 64 (1967) 725-39.

3019. _____. "The Justification of Inductive Rules of Inference," in The Problem of Inductive Logic. I. Lakatos, ed. Amsterdam: North-Holland, 1968. 24-43.

    a. Discussion by I. Hacking, W. C. Kneale, J. W. N. Watkins, Y. Bar-Hillel, D. W. Miller, H. E. Kyburg, and W. C. Salmon. 44-97.

    b. "Remarks on Salmon's Paradox of Primes," by Roberto Torretti. PS 39 (1972) 260-62.

3020. _____. "Partial Entailment as a Basis for Inductive Logic." SL 24 (1970) 47-82.

3021. _____. "Unfinished Business: The Problem of Induction." Phil. Studies. 33 (1978) 1-19.

3022. Salmon, W. C., S. F. Barker, and H. E. Kyburg, Jr. "Symposium on Inductive Evidence." Am. Phil. Quart. 2 (1965) 265-80.

3023. Sarndahl, Carl-Erik. "Some Aspects of Carnap's Theory of Inductive Inference." BJPS 19 (1968) 225-46.

3024. Savage, Leonard J. "Implications of Personal Probability for Induction." J. Phil. 64 (1967) 593-607.

3025. Schagrin, Morton L. "An Analytic Justification of Induction." BJPS 14 (1964) 343-44.

3026. Schlaretzki, W. E. "Scientific Reasoning and the Summum Bonum." PS 27 (1960) 48-57.

3027. Schlesinger, George. "Induction and Parsimony." Am. Phil. Quart. 8 (1971) 179-85.

3027A. _____. "The Justification of Empirical Reasoning." Phil. Quart. 29 (1979) 208-19.

3028. _____. "Strawson on Induction." Philosophia. 10 (1981) 199-208.

3029. Schock, Rolf. "On Induction." Notre Dame Journal of Formal Logic. 6 (1965) 235-40.

3030. Schorsch, Alexander. "Scientific Induction." Proc. Am. Cath. Phil. Assn. 25 (1951) 136-41.

3031. Schreiber, Alfred. "Das Induktionsproblem in Lichte der Approximationstheorie der Wahrheit." ZAW 8 (1977) 77-90.

3032. Schwartz, Robert, Israel Scheffler, and Nelson Goodman. "An Improvement in the Theory of Projectibility." J. Phil. 67 (1970) 605-08.

     a. "A Difficulty on Conflict and Confirmation," by Howard Kahane. 68 (1971) 488-89.

     b. "Confirmation and Conflict," by Robert Schwartz. 483-87.

     c. "On Kahane's Confusions," by Nelson Goodman. 69 (1972) 83-84.

3033. Scozzafava, Romano. "Inferenza bayesiana e 'logica' induttiva." Scientia. 115 (1980) 37-46. English Tr.: 47-53.

3034. Sellars, Wilfrid. "Induction as Vindication." PS 31 (1964) 197-231.

3035. _____. "Are There Non-Deductive Logics?" SL 24 (1970) 83-103.

3036. Sharpe, Robert. "Induction, Abduction, and the Evolution of Science." Transactions of the Charles S. Peirce Society. 6 (1970) 17-33.

3037. Shimony, Abner. "Scientific Inference," in The Nature and Function of Scientific Theories: Essays in Contemporary Science and Philosophy. Robert S. Colodny, ed. Pittsburgh: University of Pittsburgh Press, 1970. 79-172.

3038. Shoemaker, Sydney. "On Projecting the Unprojectible." Phil. Review. 84 (1975) 178-219.

3039. Siemens, Jr., David F. "Deduction, Induction, and Causality." Philosophical Inquiry. 3 (1981) 117-25.

3040. Sikora, Joseph J. "The 'Problem' of Induction." Thomist. 22 (1959) 25-36.

3041. Skyrms, Brian. "On Failing to Vindicate Induction." PS 32 (1965) 253-68.

3042. _____. Choice and Chance: An Introduction to Inductive Logic. Belmont, CA.: Dickenson, 1966. 165 pp.

3043. _____. "A Neglected Logical Lapse in Reichenbach's Pragmatic Justification of Induction." Methodology and Science. 1 (1968) 155-58.

3044. Sloman, Aaron. "Predictive Policies." Arist. Soc. Supp. Vol. 41 (1967) 77-94.

3045. Slote, Michael Anthony. "Some Thoughts on Goodman's Riddle." Analysis. 27 (1966-67) 128-32.

        a. "Differential Properties and Goodman's Riddle," by John O'Connor. 28 (1967-68) 59.

        b. "A General Solution to Goodman's Riddle?" by Michael A. Slote. 29 (1968-69) 55-58.

        c. "Primary and Secondary Tests," by R. G. Swinburne. 203-05.

3046. _____. "Entrenchment and Validity." Analysis. 34 (1973-74) 204-07.

3047. Smart, J. J. C. "Excogitation and Induction." AJP 28 (1950) 191-99.

3048. Smith, Mapheus. "A Note on the Progressive Generalization of Data." PS 14 (1947) 116-22.

3049. Smokler, Howard. "Goodman's Paradox and the Problem of Rules of Acceptance." Am. Phil. Quart. 3 (1966) 71-76.

3050. _____. "Semantical Questions in Carnap's Inductive Logic." BJPS 28 (1977) 129-35.

3051. Sollazzo, Gary. "Barker and Achinstein on Goodman." Phil. Studies. 23 (1972) 91-97.

3052. Spielman, Stephen. "Carnap's Robot and Inductive Logic." Journal of Philosophical Logic. 5 (1976) 407-15.

3053. Spilsbury, R. J. "A Note on Induction." Mind. 58 (1949) 215-17.

3054. Stace, W. T. "Are All Empirical Statements Merely Hypotheses?" J. Phil. 44 (1947) 29-38.

   a. "On the Certainty of Empirical Statements," by Paul Henle. 625-32.

3055. Stemmer, Nathan. "Three Problems in Induction." Synthese. 23 (1971) 287-308.

3056. _____. "A Relative Notion of Natural Generalization." PS 42 (1975) 46-48.

3057. _____. "The Goodman Paradox." ZAW 6 (1975) 340-54.

3058. _____. "The Reliability of Inductive Inferences and our Innate Capacities." ZAW 9 (1978) 93-105.

3059. _____. "Inductive Inferences and the Paradoxes." [In Hebrew] Iyyun. 28 (1978) 49-55.

3060. _____. "A Partial Solution to the Goodman Paradox." Phil. Studies. 34 (1978) 177-85.

   a. "Solving Goodman's Paradox: A Reply to Stemmer," by Kenneth Konyndyk, Jr. 37 (1980) 297-305.

3061. _____. "Projectible Predicates." Synthese. 41 (1979) 375-96.

3062. _____. "Generalization Classes as Alternatives for Similarities and Some Other Concepts." Erkenntnis. 16 (1981) 73-102.

3063. Stenner, Alfred J. "A Note on 'Grue'." Phil. Studies. 18 (1967) 76-78.

3064. Stich, Stephen P., and Richard E. Nisbett. "Justification and the Psychology of Human Reasoning." PS 47 (1980) 188-202.

3065. Stove, D. "Hume, Probability, and Induction." Phil. Review. 74 (1965) 160-77.

3066. Swain, Marshall, ed. Induction, Acceptance, and Rational Belief. SL 26. Dordrecht: Reidel, 1970. 232 pp.

3067. Swinburne, R. G. "Grue." Analysis. 28 (1968) 123-28.

   a. "A Note on Goodman's Paradox," by J. E. J. Altham. BJPS 19 (1968) 257.

3068. _____. "Projectible Predicates." Analysis. 30 (1969-70) 1-11.

3069. _____. The Justification of Induction. London: Oxford University Press, 1974. 179 pp.

3070. Szaniawski, K. "Pragmatyczne uzasadniene zawodnych sposobów wnioskowania." [On the Justification of Inductive Rules of Inference] Studia Logica. 13 (1962) 219-26.

3071. Teensma, E. Solypsism and Induction. Assen: Van Gorcum, 1974. 60 pp.

3072. Teller, Paul. "Goodman's Theory of Projection." BJPS 20 (1969) 219-38.

3073. Tennessen, Herman. "The Petrified Skeleton in the Closet of Philosophy of Science." Methodology and Science. 8 (1975) 125-34.

3074. Thagard, Paul R. "The Best Explanation: Criteria for Theory Choice." J. Phil. 75 (1978) 76-92.

3075. Thomson, Judith Jarvis. "Grue." J. Phil. 63 (1966) 289-310.

    a. "Grue: Some Remarks," by James Hullett and Robert Schwartz. 64 (1967) 259-71.

3076. Todd, William. "Counterfactual Conditionals and the Presuppositions of Induction." PS 31 (1964) 101-10.

3077. Törnebohm, Håkan. "Content of Information." Theoria. 21 (1955) 146-57.

3078. _____. "Two Measures of Evidential Strength," in Aspects of Inductive Logic. J. Hintikka and P. Suppes, eds. Amsterdam: North-Holland, 1966. 81-95.

3079. Tosi, Piero. "Complessità dell'induzione transfinita." Epistemologia. 1 (1978) 113-42.

3080. Tuomela, Raimo. "Inductive Generalization in an Ordered Universe," in Aspects of Inductive Logic. J. Hintikka and P. Suppes, eds. Amsterdam: North-Holland, 1966. 155-74.

3081. Uchii, Soshichi. "Inductive Logic with Causal Modalities: A Probabilistic Approach." PS 39 (1972) 162-78.

3082. _____. "Inductive Logic with Causal Modalities: A Deterministic Approach. Synthese. 26 (1973) 264-303.

3083. _____. "Induction and Causality in a Cellular Space." PSA 1976. II, 448-61.

3084. Urmson, J. O. "Some Questions Concerning Validity." RIP 25 (1953) 217-29. Reprinted in The Justification of Induction. R. Swinburne, ed. London: Oxford University Press, 1972. 74-84.

3085. Vickers, John. "Characteristics of Projectible Predicates."
J. Phil. 64 (1967) 280-86.

3086. Vries, J. de. "Die neue Physik und das Problem der Induktion."
Philosophisches Jahrbuch. 60 (1950) 151-60.

3087. Walk, Kurt. "Simplicity, Entropy, and Inductive Logic," in
Aspects of Inductive Logic. J. Hintikka and P. Suppes, eds. Amsterdam: North-Holland, 1966. 66-80.

3088. Wallace, John R. "Goodman, Logic, Induction." J. Phil. 63
(1966) 310-28.

    a. "Lawlikeness = Truth?" by John R. Wallace. 780-81.

    b. "Goodman, Wallace, and the Equivalence Condition," by
    Marsha Hanen. 64 (1967) 271-80.

3089. Wheatley, Jon. "Entrenchment and Engagement." Analysis. 27
(1966-67) 119-27.

3090. Wilkerson, T. E. "A Gruesome Note." Mind. 82 (1973) 276-77.

3091. Will, Frederick L. "Will the Future be Like the Past?"
Mind. 56 (1947) 332-47.

    a. "Induction and the Future," by Donald Williams. 57
    (1948) 226-29.

3092. _____. "Justification and Induction." Phil. Review.
68 (1959) 359-72.

3093. Williams, Donald. The Ground of Induction. Cambridge:
Harvard University Press, 1947.

    a. Review essay by D. S. Miller. J. Phil. 44 (1947) 673-84.

    b. Review essay by Ernest Nagel. J. Phil. 44 (1947) 685-93.

    c. Review essay, "Donald Williams' Theory of Induction," by
    Frederick L. Will. Phil. Review. 57 (1948) 231-47.

3094. _____. "On the Direct Probability of Inductions."
Mind. 62 (1953) 465-83.

3095. Wilson, Mark. "Maxwell's Condition -- Goodman's Problem."
BJPS 30 (1979) 107-23.

3096. Workman, Rollin W. "The Logical Status of the Principle of
Induction." Synthese. 13 (1961) 68-74.

3097. _____. "Two Extralogical Uses of the Principle of Induction." Phil. Studies. 13 (1962) 27-32.

3098. Wright, G. H. von. "Some Principles of Eliminative Induction." Ajatus. 15 (1949) 315-28.

3099. _____. A Treatise on Induction and Probability. London: Routledge and Kegan Paul, 1951. 310 pp.

3100. _____. The Logical Problem of Induction. 2nd ed. London: Blackwell, 1957. xii + 249 pp.

    a. Review essay, "Probability and Induction," by E. H. Hutten. BJPS 9 (1958) 43-51.

3101. _____. "Broad on Induction and Probability," in The Philosophy of C. D. Broad. P. A. Schilpp, ed. New York: Tudor, 1959. 313-52.

3102. Zabludowski, Andrzej. "Concerning a Fiction about How Facts are Forecast." J. Phil. 71 (1974) 97-112.

    a. "An Improvement on Zabludowski's Critique of Goodman's Theory of Projection," by Ralph Kennedy and Charles Chihara. 72 (1975) 137-41.

    b. "Bad Company: A Reply to Mr. Zabludowski and Others," by Joseph Ullian and Nelson Goodman. 142-45.

    c. "Good or Bad, but Deserved: A Reply to Ullian and Goodman," by Andrzej Zabludowski. 779-84.

    d. "Projectibility Unscathed," by Joseph Ullian and Nelson Goodman. 73 (1976) 527-31.

    e. "The Principle of Wanton Embedding," by Ralph Kennedy and Charles Chihara. 74 (1977) 539-40.

    f. "Quod Periit, Periit," by Andrzej Zabludowski, 541-52.

    g. "The Short of It," by Nelson Goodman and Joseph Ullian. 75 (1978) 263-64.

    h. "Beyond Zabludowskian Competitors: A New Theory of Projectibility," by Ralph Kennedy and Charles Chihara. Phil. Studies. 33 (1978) 229-53.

    i. "Wonton Embedding Revised and Secured," by Joseph S. Ullian. J. Phil. 77 (1980) 487-95.

3103. Zanstra, H. "The Construction of Reality." Proc. Arist. Soc. 45 n.s. (1944-45) 167-84.

3104. Ziemba, Z. "Racjonalna wiara i prawdopodobieństwo a zasadność wnioskowania indukcyjnego." [Rational Belief, Probability and the Justification of Induction] Studia Logica. 12 (1961) 99-124. English Tr.: SL 87 (1977) 709-35.

4.13 Probability

3105. Adams, Ernest, W. "Probability and the Logic of Conditionals," in Aspects of Inductive Logic. J. Hintikka and P. Suppes, eds. Amsterdam: North-Holland, 1966. 265-316.

3106. _____. "Prior Probabilities and Counterfactual Conditionals." WOSPS 6-I (1976) 1-21.

3107. Agassi, Joseph. "Subjectivism: From Infantile Disease to Chronic Illness." Synthese. 30 (1975) 3-14.

 a. "Comments," by S. Spielman, B. Loewer, and I. J. Good. 15-32.

 b. "Replies," by J. Agassi. 33-38.

3108. Allen, Edward H. "Uses of Signed Probability Theory." PS 43 (1976) 53-70.

3109. Anderson, O. "Die Begründung des Gesetzes der Grossen Zahlen und die Umkehrung des Theorems von Bernoulli." Dialectica. 3 (1949) 65-77.

3110. Ayer, A. J. "The Conception of Probability as a Logical Relation," in Observation and Interpretation. S. Körner, ed. New York: Academic Press; London: Butterworths, 1957. 12-17.

3111. _____. "On the Probability of Particular Events." RIP 15 (1961) 366-75.

3112. _____. Probability and Evidence. London: Macmillan, 1972. 144 pp.

3113. Baptist, J. H. "Le raisonnement probabilitaire." Dialectica. 3 (1949) 93-103.

3114. Bar-Hillel, Maya. "The Paradox of Ideal Evidence and the Concept of Relevance." [In Hebrew] Iyyun. 27 (1976-77) 203-15.

3115. Bartlett, M. S. "Probability in Logic, Mathematics and Science." Dialectica. 3 (1949) 104-13.

3116. Belis, Mariana. "On the Causal Structure of Random Processes." SL 51 (1973) 65-77.

3117. Benioff, Paul. "On the Correct Definition of Randomness." PSA 1978. II, 63-78.

3118. Bennett, J. F. "Some Aspects of Probability and Induction." BJPS 7 (1956-1957) 220-230, 316-22.

a. "Some Aspects of Probability and Induction: A Reply to Mr. Bennett," by William Kneale. 8 (1957) 57-63.

3119. Bigelow, John C. "Possible Worlds Foundations for Probability." Journal of Philosophical Logic. 5 (1976) 299-320.

3120. Blair, David G. "On Purely Probabilistic Theories of Scientific Inference." PS 42 (1975) 242-49.

3121. Blom, Siri. "Concerning a Controversy on the Meaning of 'Probability'." Theoria. 21 (1955) 65-98.

3122. Bolker, Ethan D. "A Simultaneous Axiomatization of Utility and Subjective Probability." PS 34 (1967) 333-40.

3123. Bonsack, F. "En quel sens les probabilités sont-elles subjectives?" Dialectica. 19 (1965) 329-44.

3124. Borel, E. "Probabilité et certitude." Dialectica. 3 (1949) 24-27.

3125. Born, Max. "Einstein's Statistical Theories," in Albert Einstein: Philosopher-Scientist. P. A. Schilpp, ed. New York: Tudor, 1951. 161-78.

3126. Braithwaite, R. B. "On Unknown Probabilities," in Observation and Interpretation. S. Körner, ed. New York: Academic Press; London: Butterworths, 1957. 3-11.

3127. _____. "Why is it Reasonable to Base a Betting Rate upon an Estimate of Chance?" LMPS-2 (1964) 263-73.

3128. Brown, G. Spencer. Probability and Scientific Inference. London: Longmans, Green & Co., 1957. ix + 154 pp.

a. Review essay by C. W. K. Mundle. Philosophy. 34 (1959) 150-54.

3129. _____. "Randomness." Arist. Soc. Supp. Vol. 31 (1957) 145-50.

3130. Brown, Peter M. "Conditionalization and Expected Utility." PS 43 (1976) 415-19.

3131. Büchel, Wolfgang. "Statistische Wahrscheinlichkeit und statistische Physik." ZAW 6 (1975) 7-18.

3132. Bunge, Mario. "Azar, probabilidad y ley." Dianoia. 15 (1969) 141-60.

3133. _____. "Possibility and Probability." WOSPS 6-III (1976) 17-34.

3134. Burks, Arthur W. "Reichenbach's Theory of Probability and Induction." The Review of Metaphysics. 4 (1951) 377-93.

3135. _____. Chance, Cause, Reason: An Inquiry into the Nature of Scientific Evidence. Chicago: University of Chicago Press, 1977. xvi + 694 pp.

3136. Buxton, Richard. "The Interpretation and Justification of the Subjective Bayesian Approach to Statistical Inference." BJPS 29 (1978) 25-38.

3137. Campbell, Margaret. "The Nature of Statistics." Methodology and Science. 9 (1976) 51-71. Errata: 196-97.

3138. Cannavo, S. "Extensionality and Randomness in Probability Sequences." PS 33 (1966) 134-46.

3139. Carlsson, Gösta. "Sampling, Probability and Causal Inference." Theoria. 18 (1952) 139-54.

3140. Carnap, Rudolf. "Probability as a Guide in Life." J. Phil. 44 (1947) 141-48.

3141. _____. Logical Foundations of Probability. Chicago: University of Chicago Press, 1950.

   a. Review essay, "Carnap on Probability," by John G. Kemeny. Review of Metaphysics. 5 (1951) 145-56.

   b. Review essay, "Carnap's Theory of Probability," by G. H. von Wright. Phil. Review. 60 (1951) 362-74.

   c. Review essay by Arthur W. Burks. J. Phil. 48 (1951) 524-35.

   d. Review essay, "Professor Carnap's Philosophy of Probability," by Donald C. Williams. PPR 13 (1952) 103-21.

   e. Review essay by Stephen Toulmin. Mind. 62 (1953) 86-99.

   f. Review essay by Kurt Hübner. Philosophische Rundschau. 2 (1954-44) 7-16.

   g. Review essay, "Probability," by D. R. Cousin. Phil. Quart. 4 (1954) 82-84.

3142. _____. "Remarks on Probability." Phil. Studies. 14 (1963) 65-75.

3143. _____. "Notes on Probability and Induction." SL 73 (1975) 293-324. Earlier version published in Synthese. 25 (1973) 269-98.

3144. Chatalian, George. "Probability: Inductive versus Deductive." Phil. Studies. 3 (1952) 49-56.

   a. "Mr. Chatalian on Probability and Deduction," by Donald C. Williams. 4 (1953) 28-29.

b. "Deductive Probability Arguments," by C. J. Ducasse. 29-31.

c. "The End of the Probability Syllogism?" by Nicholas Georgescu-Roegen. 5 (1954) 31-32.

3145. Churchman, C. W. "Probability Theory." PS 12 (1945) 147-73.

3146. _____. "Philosophical Aspect of Statistical Theory." Phil. Review. 55 (1946) 81-87.

3147. _____. "Much Ado About Probability." PS 14 (1947) 176-78.

3148. Coffa, J. Alberto. "Randomness and Knowledge." BSPS 20, SL 64 (1974) 103-15.

3149. Cooper, Neil. "The Concept of Probability." BJPS 16 (1965) 226-38.

3150. Costantini, Domenico. "Per una filosofia realista della probabilità." Epistemologia. 3 (1980) 219-44.

3151. Dale, A. I. "Probability and F-Coherence." PS 43 (1976) 254-65.

3152. _____. "Probability, Vague Statements and Fuzzy Sets." PS 47 (1980) 38-55.

3153. Day, John Patrick. Inductive Probability. London: Routledge and Kegan Paul, 1961. xvi + 336 pp.

a. Review essay by Milton Fisk. Philosophical Studies (Ireland). 11 (1961-62) 207-16.

3154. Dantzig, D. van. "Carnap's Foundation of Probability Theory." Synthese. 8 (1950-51) 459-70.

3155. Del-Negro, Walter. "Die Begründung der Wahrscheinlichkeit und das Anwendungsproblem des Apriorischen." ZPF 3 (1948) 28-35.

3156. Dretske, Fred I. "Reasons, Knowledge and Probability." PS 38 (1971) 216-20.

3157. Düsberg, Klaus Jürgen. "Zur Kritik einer neuen Definition von 'Wahrscheinlichkeit'." ZAW 11 (1980) 103-07.

3158. Edwards, A. F. Likelihood. An Account of the Statistical Concept of Likelihood and Its Application to Scientific Inference. Cambridge: Cambridge University Press, 1972. xiii + 235 pp.

a. Review essay, "Likelihood," by Ian Hacking. BJPS 23 (1972) 132-37.

b. "Two Points in the Theory of Statistical Inference," by
G. A. Barnard. BJPS 23 (1972) 329-31.

3159. Edwards, Jim. "Hidden Variables and the Propensity Theory of
Probability." BJPS 30 (1979) 315-28.

3160. Ellis, B. "The Logic of Subjective Probability." BJPS 24
(1973) 125-52.

3161. Erismann, Theodor. "Wahrscheinlichkeit im Sein und Denken."
Studium Generale. 4 (1951) 88-109. Also in Dialectica, 7 (1953)
331-46, and in ZPF 12 (1958) 321-50.

a. "Ueber Ordnung und Unordnung," by P. Nolfi. Dialectica.
7 (1953) 347-48.

3162. Faber, Roger J. "Re-encountering a Counter-intuitive Probability." PS 43 (1976) 283-89.

3163. Fenstad, Jens Erik. "The Structure of Logical Probabilities."
Synthese. 18 (1968) 1-23.

3164. Féraud, L. "Le raisonnement fondé sur les probabilites."
RMM 53 (1948) 113-38.

3165. Fetzer, James H. "Dispositional Probabilities." BSPS 8, SL
29 (1971) 473-82.

3166. _____. "Reichenbach, References Classes, and Single
Case 'Probabilities'." Synthese. 34 (1977) 185-218. Reprinted in
SL 132 (1979) 187-220.

3167. Finch, Henry A. "Bayesian Rules for Rational Reconstruction
of a System of Hypotheses Weakened by Adverse Observations." Logique
et analyse. 7 (1964) 145-51.

3168. Finch, P. D. "Incomplete Descriptions in the Language of
Probability Theory." WOSPS 6-I (1976) 23-28.

3169. _____. "The Poverty of Statisticism." WOSPS 6-II
(1976) 1-44.

3170. _____. "On the Interference of Probabilities." WOSPS
6-III (1976) 105-10.

3171. _____. "On the Role of Description in Statistical
Enquiry." BJPS 32 (1981) 127-44.

3172. Fine, Terrence L. "A Computational Complexity Viewpoint on
the Stability of Relative Frequency and on Stochastic Independence."
WOSPS 6-I (1976) 29-40.

3173. _____. "An Argument for Comparative Probability."
WOSPS 11 (1977) 105-19.

3174. Finetti, B. de. "Le vrai et le probable." Dialectica. 3 (1949) 78-92.

3175. _____. "La decisione nell'incertezza." Scientia. 98 (1963) 61-68. French Tr.: Supplement, 37-43.

3176. _____. "Initial Probabilities: A Prerequisite for any Valid Induction." SL 31 (1970) 3-17. Reprinted from Synthese. 20 (1969) 2-16.

   a. "Discussion of Bruno de Finetti's Paper," by I. J. Good. 18-25. Reprinted from Synthese. 20 (1969) 17-24.

3177. _____. Teoria delle probabilità. Sintesi introduttiva con appendice critica. Turin: 1970. 770 pp.

   a. Review essay, "The Subjective Theory of Probability," by D. A. Gillies. BJPS 23 (1972) 138-57.

3178. _____. Probability, Induction, and Statistics: The Art of Guessing. London: Wiley, 1972. xxiv + 266 pp.

3179. _____. Theory of Probability: A Critical Introductory Treatment. Vol. I. London: Wiley, 1974. 300 pp.

3180. _____. "La probabilità: guardarsi dalle contraffazioni!" Scientia. 111 (1976) 255-81. English Tr.: 283-303.

3181. _____. "Ambiguità di gergo e di fondo nel campo della probabilità." Scientia. 114 (1979) 707-11. English Tr.: 713-16.

3182. Finkelstein, David. "Classical and Quantum Probability and Set Theory." WOSPS 6-III (1976) 111-16.

3183. Fisher, R. A. Statistical Methods and Scientific Inference. London: Oliver and Boyd, 1956. xiii + 175 pp.

   a. Review essay, "Probability and Induction," by E. H. Hutten. BJPS 9 (1958) 43-51.

3184. Fraassen, Bas C. van. "Probabilities of Conditionals." WOSPS 6-I (1976) 261-300.

   a. "Letter by Stalnaker to van Fraassen." 302-06.

   b. "Letter by van Fraassen to Stalnaker." 307-08.

3185. _____. "Representation of Conditional Probabilities." Journal of Philosophical Logic. 5 (1976) 417-30.

3186. _____. "Relative Frequencies." Synthese. 34 (1977) 133-66. Expanded version printed in SL 132 (1979) 129-68.

3187. _____. "A Problem for Relative Information Minimizers in Probability Kinematics." BJPS 32 (1981) 375-79.

3188. Fraser, D. A. S., and Jock Mackay. "On the Equivalence of Standard Inference Procedures." WOSPS 6-II (1976) 47-58.

3189. Fréchet, M. "Les définitions courantes de la probabilité." Revue Phil. 136 (1946) 129-69.

3190. Freudenberg, Gideon. "Probability and Induction in the Light of Modern Physics." [In Hebrew] Iyyun. 4 (1953) 1-22.

3191. Freudenthal, Hans. "Ist die mathematische Statistik paradox?" Dialectica. 12 (1958) 7-32.

    a. "Pour ouvrir la discussion," by F. Gonseth. 33-36.

3192. _____. "Models in Applied Probability." SL 3 (1961) 78-88.

3193. _____. "Abus philosophiques de la statistique." RMM 67 (1962) 237-46.

3194. _____. "Probabilités objectives ou subjectives?" Logique et analyse. 8 (1965) 265-76.

3195. _____. "Realistic Models in Probability," in The Problem of Inductive Logic. I. Lakatos, ed. Amsterdam: North-Holland, 1968. 1-14.

    a. Discussion by P. Suppes, Y. Bar-Hillel, and H. Freudenthal. 15-23.

3196. Friedman, Kenneth S. "Resolving a Paradox of Inductive Probability." Analysis. 35 (1974-75) 183-85.

3197. Gaifman, Haim. "Subjective Probability, Natural Predicates, and Hempel's Ravens." Erkenntnis. 14 (1979) 105-47.

3198. Galavotti, Maria Carla. "Some Recent Views on Probability and Induction." Quality and Quantity. 8 (1974) 347-76.

3199. Gärdenfors, Peter. "Conditionals and Changes of Belief." Acta Philosophica Fennica. 30, 2-4 (1978) 381-404.

3200. _____. "Forecasts, Decisions and Uncertain Probabilities." Erkenntnis. 14 (1979) 159-81.

3201. Geiringer, Hilda. "On the Foundations of Probability Theory." BSPS 3 (1967) 199-227.

    a. "Comments" by Laszlo Tisza. 228-35.

3202. Giere, Ronald N. "Bayesian Statistics and Biased Procedures." Synthese. 20 (1969) 371-87.

3203. _____. "Objective Single-Case Probabilities and the Foundations of Statistics." LMPS-4 (1971) 467-83.

3203A. _____. "A Laplacean Formal Semantics for Single-Case Propensities." Journal of Philosophical Logic. 5 (1976) 321-53.

3204. _____. "Empirical Probability, Objective Statistical Methods, and Scientific Inquiry." WOSPS 6-II (1976) 63-93.

3205. _____. "Propensity and Necessity." Synthese. 40 (1979) 439-52.

3206. _____. "Foundations of Probability and Statistical Inference," in Current Research in Philosophy of Science. P. D. Asquith and H. E. Kyburg, Jr., eds. East Lansing: MI: Philosophy of Science Assn., 1979. 503-33.

3207. Gillies, D. A. "A Falsifying Rule for Probability Statements." BJPS 22 (1971) 231-61.

    a. "A Comment on Gillies's Falsifying Rule for Probability Statements," by Jacques de Maré. 23 (1972) 335.

    b. "Reply to de Maré," by D. A. Gillies. 335-36.

    c. "On the Infirmities of Gillies's Rule," by Stephen Spielman. 25 (1974) 261-65.

    d. "On Neyman's Paradox and the Theory of Statistical Tests," by M. L. G. Redhead. 265-71.

3208. _____. An Objective Theory of Probability. London: Methuen; New York: Barnes & Noble, 1973. x + 250 pp.

    a. Review essay by Colin Howson. Erkenntnis. 14 (1979) 87-92.

3209. Gillispie, C. C. "Intellectual Factors in the Background of Analysis by Probabilities," in Scientific Change. A. C. Crombie, ed. London: Heinemann, 1963. 431-53.

3210. Gini, C. "Concept et mesure de la probabilité." Dialectica. 3 (1949) 36-54.

3211. Godambe, V. P., and M. E. Thompson. "Philosophy of Survey-Sampling Practise." WOSPS 6-II (1976) 103-22.

3212. Good, I. J. "Subjective Probability as the Measure of a Non-Measurable Set." LMPS-I (1960) 319-29.

3213. _____. The Estimation of Probabilities: An Essay on Modern Bayesian Methods. Cambridge, Ma; MIT Press, 1965. 109 pp.

3214. _____. "On the Principle of Total Evidence." BJPS 17 (1967) 319-21.

3215. _____. "Corroboration, Explanation, Evolving Probability, Simplicity and a Sharpened Razor." BJPS 19 (1968) 123-43.

a. "A Correction Concerning Complexity," by I. J. Good.   25 (1974) 289.

3216. _____. "Random Thoughts about Randomness." BSPS 20, SL 64 (1974) 117-35.

3217. _____. "A Little Learning can be Dangerous." BJPS 25 (1974) 340-42.

3218. _____. "The Bayesian Influence, or How to Sweep Subjectivism under the Carpet." WOSPS 6-II (1976) 125-74.

3219. Gonseth, F. "Sur la méthodologie du calcul des probabilités." Dialectica. 19 (1965) 313-28.

3220. Gonseth, F., P. Bernays, H. Jecklin, and P. Nolfi. "Fondements et applications du calcul des probabilités et de statistique." Dialectica. 7 (1953) 303-30.

3221. Gordesch, Johannes, and Hans Georg Knapp. "Zur logischen Problematik der statistischen Hypothesenprüfung." PN 10 (1967-68) 42-55.

3222. Gottinger, H. W. "Review of Concepts and Theories of Probability." Scientia. 109 (1974) 83-110.

3223. Graves, Spencer. "On the Neyman-Pearson Theory of Testing." BJPS 29 (1978) 1-23.

3224. Gröbner, W. "Ueber die Anwendung des Wahrscheinlichkeitsbegriffes in der Physik." Studium Generale. 4 (1951) 72-77.

3225. Guccione, S., and P. Lo Sardo. "Effective Procedures and Randomness." Scientia. 109 (1974) 489-98.

3226. Hacking, Ian. "On the Foundations of Statistics." BJPS 15 (1964) 1-26.

3227. _____. "Guessing by Frequency." Proc. Arist. Soc. 64 n.s. (1963-64) 55-70.

3228. _____. Logic of Statistical Inference. Cambridge: Cambridge University Press, 1965. 232 pp.

a. Review essay by G. A. Bernard. BJPS 23 (1972) 123-32.

3229. _____. "On Falling Short of Strict Coherence." PS 35 (1968) 284-86.

3230. _____. "Propensities, Statistics and Inductive Logic." LMPS-4 (1971) 485-500.

3231. _____. "Equipossibility Theories of Probability." BJPS 22 (1971) 339-55.

3232. _____ . The Emergence of Probability: A Philosophical
Study of Early Ideas about Probability, Induction and Statistical
Inference. London: Cambridge University Press, 1975.

    a. Review essay, "Ex-Huming Hacking," by Larry Laudan.
Erkenntnis. 13 (1978) 417-35.

    b. Review essay by Fred Wilson. Canadian Journal of Philo-
sophy. 8 (1978) 587-97.

    c. Review essay by Henry E. Kyburg, Jr. Theory and Decision.
9 (1978) 205-17.

    d. Review essay, "The Prehistory of Chance," by Colin Howson.
BJPS 29 (1978) 274-80.

3233. _____ . "Strange Expectations," PS 47 (1980) 562-67.

3234. _____ . "Grounding Probabilities from Below." PSA 1980.
I, 110-16.

3235. Hagstroem, K. G. "Connaissance et stochastique." Dialectica.
3 (1949) 153-72.

3236. Hamann, Jon Ray. "A Note on the Foundations of Generalized
Probability Theory." Methodology and Science. 1 (1968) 130-42.

3237. Hansson, Bengt. "The Appropriateness of the Expected Utility
Model." Erkenntnis. 9 (1975) 175-93.

3238. Hanson, William H. "Names, Random Samples, and Carnap."
MSPS 6 (1975) 367-87.

3239. Harman, Gilbert. "Detachment, Probability, and Maximum Likelihood." Nous. 1 (1967) 401-11.

3240. Harper, W. L. and C. A. Hooker, eds. Foundations of Probability Theory, Statistical Inference and Statistical Theories of
Science. WOSPS 6. Dordrecht: Reidel 1976. Vol. I: Foundations
and Philosophy of Epistemic Applications of Probability Theory, 308
pp. Vol. II: Foundations and Philosophy of Statistical Inference,
455 pp. Vol. III: Foundations and Philosophy of Statistical Theories
in the Physical Sciences, 241 pp.

3241. Hay, William H. "Professor Carnap and Probability." PS 19
(1952) 170-77.

3242. Heilig, Klaus. "Carnap and de Finetti on Bets and the Probability of Singular Events: The Dutch Book Argument Reconsidered."
BJPS 29 (1978) 325-46.

    a. "The Role of 'Dutch Books' and of 'Proper Scoring Rules',"
by Bruno de Finetti. 32 (1981) 55-56.

3243. Hein, Carl. A. "Entropy in Operational Statistics and Quantum Logic." FP 9 (1979) 751-86.

3244. Hellman, Geoffrey. "Randomness and Reality." PSA 1978. II, 79-97.

3245. Hermes, H. "Zum Einfachkeitsprinzip in der Wahrscheinlichkeitsrechnung." Dialectica. 12 (1958) 317-31.

3246. Hesse, Mary. "Bayesian Methods and the Initial Probabilities of Theories." MSPS 6 (1975) 50-105.

3247. Hintikka, Jaakko. "Unknown Probabilities, Bayesianism, and de Finetti Representation Theorem." BSPS 8, SL 39 (1971) 325-41.

3248. Hirsh, G., et al. "Sur un aspect paradoxal de la théorie des probabilités." Dialectica. 8 (1954) 125-44.

3249. _____. "Philosophie ouverte et connaissance probable." Dialectica. 14 (1960) 197-202.

3250. Hjorth, Sune. "The Meanings of Probability Statements." Theoria. 25 (1959) 27-30.

3251. Howson, C. "Must the Logical Probability of Laws be Zero?" BJPS 24 (1973) 153-63.

        a. "On the Probability of Laws being Zero," by John F.
        Price. 27 (1976) 392-95.

3252. _____. "The Rule of Succession, Inductive Logic and Probability Logic." BJPS 26 (1975) 187-98.

3253. _____. "The Development of Logical Probability." BSPS 39 (1976) 277-98.

3254. Howson, Colin, and Graham Oddie. "Miller's So-Called Paradox of Information." BJPS 30 (1979) 253-61.

3255. Hudimoto, Hirosi. "Statistical Approach Involving Bayes' Theorem and the Estimation of the Prior Distribution." AJAPS 4, No. 1 (1971) 35-45.

3256. Humburg, Jürgen. "Die Problematik apriorischer Wahrscheinlichkeiten im System der induktiven Logic von Rudolf Carnap." Archiv für Mathematische Logik und Grundlagenforschung. 14 (1971) 135-47.

3257. Hutten, Ernest H. "Probability-Sentences." Mind. 61 (1952) 39-56.

3258. Issman, Samuel. "La notion de probabilité." Annales de l'Institute de Philosophie. (Brussels) (1969) 139-52.

3259. _____. "Evaluation du degré numérique de probabilité d'un type d'énoncés universels." Annales de l'Institute de Philosophie. (Brussels) (1972) 183-95.

3260. Jackson, Frank, and Robert Pargetter. "Indefinite Probability Statements." Synthese. 26 (1973) 205-17.

3261. Jamison, Dean. "Bayesian Information Usage." SL 28 (1970) 28-57.

3262. Jaynes, E. T. "Confidence Intervals vs. Bayesian Intervals." WOSPS 6-II (1976) 175-213.

      a. "Interfaces between Statistics and Context," by Margaret W. Maxfield. 214-17.

      b. "Jaynes' Reply to Margaret Maxfield." 218-19.

      c. "Comments" by Oscar Kempthorne. 220-28.

      d. "Jaynes' Reply to Kempthorne's Comments." 229-57.

3263. Jecklin, H. "Historisches zur Wahrscheinlichkeitsdefinition." Dialectica 3 (1949) 5-15.

3264. Jeffreys, Harold. "The Present Position in Probability Theory." BJPS 5 (1955) 275-89.

3265. Jeffrey, R. C. "New Foundations for Bayesian Decision Theory." LMPS-2 (1964) 289-300.

3266. _____. The Logic of Decision. New York: McGraw-Hill, 1965. 201 pp.

3267. _____. "The Whole Truth." Synthese. 18 (1968) 24-27.

3268. _____. "Probable Knowledge," in The Problem of Inductive Logic. I. Lakatos, ed. Amsterdam: North-Holland, 1968. 166-80.

      a. Discussion by L. Hurwicz, P. Suppes, and R. C. Jeffrey. 181-90.

3269. _____. "Preference among Preferences." J. Phil. 71 (1974) 377-91.

3270. _____. "Mises Redux." WOSPS 11 (1977) 213-22.

3271. Juhos, Béla von. "Wahrscheinlichkeitsschlüsse als syntaktische Schlussformen." Studium Generale. 6 (1953) 206-14.

3272. _____. "Die 'Wahrscheinlichkeit' als physikalische Beschreibungsform." PN 4 (1957) 297-36.

3273. _____. "Das 'Wahrscheinlichkeitsfeld'." Archiv für Philosophie. 7 (1957) 82-95.

3274. _____. "Ueber die 'absolute' Wahrscheinlichkeit." PN 6 (1960-61) 391-412.

3275. _____. "Drei Begriffe der 'Wahrscheinlichkeit'." Studium Generale. 21 (1968) 1153-73.

3276. _____. "Logical and Empirical Probability." Logique et analyse. 12 (1969) 277-82.

3277. _____. "Geometrie und Wahrscheinlichkeit." ZPF 25 (1971) 500-10.

3278. Kalbfleisch, J. G., and D. A. Sprott. "On Tests of Significance." WOSPS 6-II (1976) 259-70.

3279. Keene, G. B. "Randomness." Arist. Soc. Supp. Vol. 31 (1957) 151-60.

3280. Kemeny, John G. "Fair Bets and Inductive Probabilities." Journal of Symbolic Logic. 20 (1955) 263-73.

3281. _____. "Carnap's Theory of Probability and Induction," in The Philosophy of Rudolf Carnap. P. A. Schilpp, ed. La Salle, Illinois: Open Court, 1963. 711-38.

3282. Kempthorne, Oscar. "Statistics and the Philosophers." WOSPS 6-II (1976) 273-309.

3283. Keuth, Herbert. "Poppers Axiome für eine propensity-Theorie der Wahrscheinlichkeit." ZAW 7 (1976) 99-112.

3284. Keynes, John Maynard. A Treatise on Probability. London: Macmillan, 1921. Reprinted 1929, 1943, 1948, 1952, 1957. Reprint: New York: Harper & Row, 1962, with an Introduction by N. R. Hanson. 466 pp.

3285. King-Farlow, John. "Toulmin's Analysis of Probability." Theoria. 29 (1963) 12-26.

3286. Kirschenmann, Peter. "Concepts of Randomness." SL 50 (1973) 129-48.

3287. _____. "The Problem of the Reference Class." AJAPS 4, No. 5 (1975) 21-29.

3288. Knapp, Hans Georg, and Johannes Gordesch. "Gleichkeitserklärungen in der Statistik." PN 11 (1969) 426-39.

3289. Kneale, William. Probability and Induction. Oxford: Clarendon Press, 1949. viii + 264 pp.

    a. Review essay by C. D. Broad. Mind. 59 (1950) 94-115.

    b. "Mr. Kneale on Probability and Induction," by F. J. Anscombe. Mind. 60 (1951) 299-309.

c. "Probability and Induction," by William Kneale. 310-17.

3290. Kossovsky, N. K. "Some Problems in the Constructive Probability Theory." SL 51 (1973) 83-99.

3291. Kuipers, Theo A. F. "Inductive Probability and the Paradox of Ideal Evidence." Philosophica. 17 (1976) 197-205.

3292. _____. Studies in Inductive Probability and Rational Expectation. SL 123. Dordrecht: Reidel, 1978. 145 pp.

3293. Kurth, Rudolf. "Ueber den Begriff der Wahrscheinlichkeit." PN 5 (1958-49) 413-29.

3294. Kutschera, Franz von. "Zur Problematik der naturwissenschaftlichen Verwendung des subjektiven Wahrscheinlichkeitsbegriffs." Synthese. 20 (1969) 84-103.

3295. _____. "An Open Problem in the Subjectivist Theory of Probability." Ratio. 15 (1973) 247-55.

3296. Kyburg, Jr., H. E. "R. B. Braithwaite on Probability and Induction." BJPS 9 (1958) 203-20.

3297. _____. Probability and the Logic of Rational Belief. Middletown, CT.: Wesleyan University Press, 1961. 234 pp.

3298. _____. "Probability and Rationality.: Phil. Quart. 11 (1961) 193-200.

3299. _____. "Probability and Randomness." Theoria. 29 (1963) 27-55.

3300. _____. "Probability, Rationality, and a Rule of Detachment." LMPS-2 (1964) 301-10.

3301. _____. "Probability and Decision." PS 33 (1966) 250-61.

3302. _____. "Bets and Beliefs." Am. Phil. Quart. 5 (1968) 54-63.

3303. _____. Probability and Inductive Logic. London: Macmillan, 1970. 272 pp.

3304. _____. "Two World Views." Nous. 4 (1970) 337-48.

3305. _____. "On a Certain Form of Philosophical Argument." Am. Phil. Quart. 7 (1970) 229-37.

3306. _____. "Epistemological Probability." Synthese. 23 (1971) 309-26.

3307. _____. The Logical Foundations of Statistical Inference. SL 65. Dordrecht: Reidel, 1974. 421 pp.

a. "Kyburg and Fiducial Inference," by Stephen Leeds. PS 48 (1981) 78-91.

b. "Leeds' Infernal Machine," by Henry E. Kyburg, Jr. 92-94.

c. Review essay by Teddy Seidenfeld. J. Phil. 74 (1977) 47-62.

3308. _____. "Randomness." BSPS 20, SL 64 (1974) 137-49.

3309. _____. "Propensities and Probabilities." BJPS 25 (1974) 358-75.

3310. _____. "The Uses of Probability and the Choice of a Reference Class." MSPS 6 (1975) 262-94.

3311. _____. "Chance." Journal of Philosophical Logic. 5 (1976) 355-93.

3312. _____. "Statistical Knowledge and Statistical Inference." WOSPS 6-II (1976) 315-36.

3313. _____. "Randomness and the Right Reference Class." J. Phil. 74 (1977) 501-21.

a. "Confirmational Conditionalization," by Isaac Levi. 75 (1978) 730-37.

3314. _____. "All Acceptable Generalizations are Analytic." Am. Phil. Quart. 14 (1977) 201-10.

3315. _____. "Subjective Probability: Criticisms, Reflections, and Problems. Journal of Philosophical Logic. 7 (1978) 157-80.

3316. _____. "Tyche and Athena." Synthese. 40 (1979) 415-38.

3317. _____. "Acts and Conditional Probabilities." Theory and Decision. 12 (1980) 149-71.

3318. _____. "Principle Investigation." J. Phil. 78 (1981) 772-78.

3319. Kyburg, Jr., Henry E., and Howard E. Smokler, eds. Studies in Subjective Probability. New York: John Wiley and Sons, 1964. vii + 203 pp.

a. Review essay, "Subjective Probability," by Ian Hacking. BJPS 16 (1966) 334-39.

3320. Largeault, J. "Hasards et probabilités." Dialogue. 17 (1978) 634-58.

3321. _____. "Sur des notions du hasard." Revue Phil. 169 (1979) 33-65.

3322. Leblanc, Hugues. "Two Probability Concepts." J. Phil. 53 (1956) 679-88.

3323. _____. "On Requirements for Conditional Probability Functions." Journal of Symbolic Logic. 25 (1960) 238-42.

3324. _____. "Probabilities as Truth Value Estimates." PS 28 (1961) 414-17.

3325. _____. Statistical and Inductive Probabilities. Englewood Cliffs, NJ: Prentice-Hall, 1962. 148 pp.

3326. _____. "What Price Substitutivity? A Note on Probability Theory." PS 48 (1981) 317-22.

3327. Leblanc, Hugues, and Bas C. van Fraassen. "On Carnap and Popper Probability Functions." Journal of Symbolic Logic. 44 (1979) 369-73.

3328. Lehman, R. Sherman. "On Confirmation and Rational Betting." Journal of Symbolic Logic. 20 (1955) 251-62.

3329. Lehrer, Keith. "Evidence and Conceptual Change." Philosophia. 2 (1972) 273-82.

3330. _____. "The Evaluation of Method: A Hierarchy of Probabilities Among Probabilities." Grazer Philosophische Studien. 12/13 (1981) 131-41.

3331. Levi, Isaac. "Recent Work in Probability and Induction (Review Article)." Synthese. 16 (1966) 234-44.

3332. _____. "Probability Kinematics." BJPS 18 (1967) 197-209.

    a. "The Jones Case," by William L. Harper and Henry E. Kyburg. 19 (1968) 247-51.

    b. "If Jones only Knew More," by Isaac Levi. 20 (1969) 153-59.

3333. _____. "Information and Inference." Synthese. 17 (1967) 369-91.

3334. _____. "Utility and Acceptance of Hypotheses," in Philosophy of Science Today. S. Morgenbesser, ed. New York: Basic Books, 1967. 115-24.

3335. _____. "Probability and Evidence." SL 26 (1970) 134-56.

3336. _____. "Certainty, Probability and the Correction of Evidence." Nous. 5 (1971) 299-312.

3337. _____. "On Indeterminate Probabilities." J. Phil. 71 (1974) 391-418.

3338. _____. "Subjunctives, Dispositions and Chances."
Synthese. 34 (1977) 423-56.

3339. _____. "Direct Inference." J. Phil. 74 (1977) 5-29.

3340. _____. "Abduction and Demands of Information." Acta
Philosophica Fennica. 30, 2-4 (1978) 405-29.

3341. _____. "Coherence, Regularity and Conditional Probability."
Theory and Decision. 9 (1978) 1-15.

3342. _____. "Dissonance and Consistency According to Shackle
and Shafer." PSA 1978. II, 466-77.

3343. _____. The Enterprise of Knowledge. Cambridge, MA:
MIT Press, 1980. xvii + 462 pp.

3344. Lévy, Paul. "Les fondements du calcul des probabilités."
Dialectica. 3 (1949) 55-64. Also in RMM 59 (1954) 164-79.

3345. _____. "Le fondement du calcul des probabilités." RMM
68 (1963) 25-56.

3346. Lewis, David. "Probabilities of Conditionals and Conditional
Probabilities." Phil. Review. 85 (1976) 297-315.

3347. Lindley, D. V. "Bayesian Statistics." WOSPS 6-II (1976)
353-62.

3348. Loewer, B., R. Laddaga, and R. Rosenkrantz. "On the Likelihood Principle and an Alleged Antinomy." PSA 1978. I, 279-86.

3349. Lorenzen, Paul. "Eine konstruktive Deutung des Dualismus in
der Wahrscheinlichkeitstheorie." ZAW 9 (1978) 256-75.

3350. Łos, J. "Semantic Representation of the Probability of
Formulas in Formalized Theories." Studia Logica. 14 (1963) 186-96.
Reprinted in WOSPS 5 (1975) 205-20, and in SL 87 (1977) 327-40.

3351. Lucas, J. R. "The One Concept of Probability." PPR 26
(1965) 180-201.

    a. "The One Systematically Ambiguous Concept of Probability,"
    by William H. Baumer. 28 (1967) 264-68.

3352. _____. The Concept of Probability. Oxford: Oxford
University Press, 1970. xx + 403 pp.

    a. Review essay, "The Plain Man's Guide to Probability," by
    Colin Howson. BJPS 23 (1972) 157-69.

3353. Luce, R. Duncan, and Louis Narens. "Qualitative Independence
in Probability Theory." Theory and Decision. 9 (1978) 225-39.

3354. McShane, Philip. Randomness, Statistics and Emergence. Notre Dame, Indiana: University of Notre Dame Press, 1970. 268 pp.

3355. Mahalanobis, P. C. "The Foundations of Statistics." Dialectica. 8 (1954) 95-111.

3356. March, A. "Wahrscheinlichkeit und Physik." Studium Generale. 4 (1951) 78-80.

3357. Marschak, Jacob, et al. "Personal Probabilities of Probabilities." Theory and Decision. 6 (1975) 121-53.

   a. "Probabilities of Probabilities (A Comment)," by Karl Borch. 155-59.

3358. Martin, Norman M. "The Explicandum of the Classical Concept of Probability." PS 18 (1951) 70-84.

3359. Martin-Löf, Per. "The Literature on von Mises' Kollektivs Revisited." Theoria. 35 (1969) 12-37.

3360. Maxwell, Grover. "Induction and Empiricism: A Bayesian-Frequentist Alternative." MSPS 6 (1975) 106-65.

3361. Mayants, L. S. "On Probability Theory and Probabilistic Physics -- Axiomatics and Methodology." FP 3 (1973) 413-33.

3362. Mayo, Deborah G. "The Philosophical Relevance of Statistics." PSA 1980. I, 97-109.

3363. Mays, W. "Probability Models and Thought and Learning Processes," in Form and Strategy in Science. J. R. Gregg and F. T. C. Harris, eds. Dordrecht: Reidel, 1964. 256-73.

3364. Mehlberg, Josephine J. "Is a Unitary Approach to the Foundations of Probability Possible?" in Current Issues in the Philosophy of Science. H. Feigl and G. Maxwell, eds. New York: Holt, Rinehart and Winston, 1961. 287-301.

3365. _____. "On the Set Theoretical Approach to Probability." Methodology and Science. 1 (1968) 179-89.

3366. Mellor, D. H. "Connectivity, Chance, and Ignorance." BJPS 16 (1965) 209-25. 18 (1967) 235-38.

3367. _____. "Chance." Arist. Soc. Supp. Vol. 43 (1969) 11-36.

3368. _____. The Matter of Chance. Cambridge: Cambridge University Press, 1971. xiii + 190 pp.

   a. Review essay, "Propensities: A Discussion Review," by Wesley C. Salmon. Erkenntnis. 14 (1979) 183-216.

3369.   Menges, Günter.   "On Subjective Probability and Related Problems."   Theory and Decision.   1 (1970) 40-60.

3370.   Metz, A.   "Probabilité et realité."   Archives de Philosophie. 19, No. 4 (1956) 98-114.   Partial English Tr.:   Philosophy Today.   1 (1957) 128-32.

3371.   Miller, David.   "A Paradox of Information."   BJPS 17 (1966) 59-61.

    a.   "A Comment on Miller's New Paradox of Information," by Karl R. Popper.   61-69.

    b.   "A Paradox of Zero Information," by Karl R. Popper. 141-43.

    c.   "Miller's So-Called Paradox of Information," by J. L. Mackie.   144-47.

    d.   "On a So-Called So-Called Paradox: A Reply to Professor J. L. Mackie," by D. Miller.   147-49.

    e.   "Miller's Paradox of Information," by Jeffrey Bub and Michael Radner.   19 (1968) 63-67.

    f.   "The Straight and Narrow Rule of Induction: A Reply to Dr. Bub and Mr. Radner," by David Miller.   145-52.

    g.   "New Mysteries for Old: The Transfiguration of Miller's Paradox," by William W. Rozeboom.   19 (1969) 345-53.

    h.   "A Suggested Resolution of Miller's Paradox," by I. J. Good.   21 (1970) 288-89.

3372.   Miller, Richard W.   "Propensity: Popper or Peirce?"   BJPS 26 (1975) 123-32.

    a.   "Popper versus Peirce on the Probability of Single Cases," by Tom Settle.   28 (1977) 177-80.

3373.   Mises, Richard von.   Probability, Statistics and Truth. Second revised English edition, prepared by Hilda Geiringer.   New York: Macmillan, 1957.   244 pp.

3374.   Moch, F.   "Réflexions sur les probabilités."   Dialectica.   11 (1957) 375-91.

3375.   Nagel, Ernest.   "Principles of the Theory of Probability," in International Encyclopedia of Unified Science.   Vol. I, No. 6.   O. Neurath, R. Carnap, and C. W. Morris, eds.   Chicago: University of Chicago Press, 1938-55.   341-422.

3376.   Nejedlý, Rudolf.   "Der Begriff 'Risiko' in der Wissenschaft." Organon.   9 (1973) 61-81.

3377. Niiniluoto, I. "On the Truthlikeness of Generalizations." WOSPS 11 (1977) 121-47.

3378. Nolfi, P. "Die Wahrscheinlichkeitstheorie im Lichte der dialektischen Philosophie." Dialectica. 3 (1949) 16-23.

3379. Odegard, Douglas. "Ignorance and Equiprobability." Dialogue. 20 (1981) 556-65.

3380. Onicescu, Octav. "Extension of the Theory of Probability." LMPS-4 (1971) 439-49.

3381. O'Toole, Edward J. "A Note on Probability." Philosophical Studies (Ireland). 11 (1961-62) 112-27.

3382. Pap, A. "Neuere Kernfragen der Wahrscheinlichkeitstheorie." Dialectica. 17 (1963) 5-22.

3383. Papineau, David. "Salmon, Statistics, and Backwards Causation." PSA 1978. I, 302-13.

3384. Pauli, W. "Wahrscheinlichkeit und Physik." Dialectica. 8 (1954) 112-17. Discussion: 118-24.

3385. Plato, Jon von. "Reductive Relations in Interpretations of Probability." Synthese. 48 (1981) 61-75.

3386. Popper, K. R. "Two Autonomous Axiom Systems for the Calculus of Probabilities." BJPS 6 (1955) 51-57. "Corrigendum,' 351.

3387. _____. "Probability Magic or Knowledge out of Ignorance." Dialectica. 11 (1957) 354-74.

3388. _____. "The Propensity Interpretation of Probability." BJPS 10 (1959) 25-42.

3389. _____. "On Carnap's Version of Laplace's Rule of Succession." Mind. 71 (1962) 69-73.

    a. "Popper on the Rule of Succession," by Richard C. Jeffrey. 73 (1964) 129.

    b. "On an Alleged Contradiction in Carnap's Theory of Inductive Logic," by Y. Bar-Hillel. 265-67.

    c. "The Mysteries of Udolpho: A Reply to Professor Jeffrey and Bar-Hillel," by K. R. Popper. 76 (1967) 103-10.

3390. _____. "Creative and Non-Creative Definitions in the Calculus of Probability," in Form and Strategy in Science. J. R. Gregg and F. T. C. Harris, eds. Dordrecht: Reidel, 1964. 171-90. Reprinted from Synthese. 15 (1963) 167-86.

3391. Rabinowicz, Wlodzimierz. "Reasonable Beliefs." Theory and Decision. 10 (1979) 61-81.

3392. Rakitov, A. N. "The Statistical Interpretation of Fact and the Role of Statistical Methods in the Structures of Empirical Knowledge." SL 25 (1970) 394-425.

3393. Rescher, Nicholas. "The Concept of Randomness." Theoria. 27 (1961) 1-11.

3394. _____. "On the Probability of Nonrecurring Events," in Current Issues in the Philosophy of Science. H. Feigl and G. Maxwell, eds. New York: Holt, Rinehart and Winston, 1961. 228-37.

    a. "Comments," by William W. Rozeboom. 237-41.

    b. "Rejoinder," by Nicholas Rescher. 241-44.

3395. Reichenbach, Hans. The Theory of Probability: An Enquiry into the Logical and Mathematical Foundations of the Calculus of Probability. Second Edition. E. H. Hutten and M. Reichenbach, trs. Berkeley: University of California Press, 1949. 500 pp.

3396. Richter, H. "Zur Begründung der Wahrscheinlichkeitsrechnung." Dialectica. 8 (1954) 48-77.

3397. Rogers, Ben. "The Probabilities of Theories as Frequencies." Synthese. 34 (1977) 167-84. Reprinted in SL 132 (1979) 169-86.

3398. Rose, Lynn E. "Countering a Counter-Intuitive Probability." PS 39 (1972) 523-24.

    a. "On a Problem in Conditional Probability," by A. I. Dale. 41 (1974) 204-06.

3399. Rosenkrantz, R. D. "The Significance Test Controversy." Synthese. 26 (1973) 304-21.

3400. _____. "Probability Magic Unmasked." PS 40 (1973) 227-33.

3401. _____. "Suppes on Probability, Utility, and Decision Theory," in Patrick Suppes. Radu J. Bogdan, ed. Dordrecht: Reidel, 1979. 111-29.

3402. Russell, L. J. "Probability." Arist. Soc. Supp. Vol. 24 (1950) 63-74.

3403. Ruzavin, G. I. "Probability Logic and its Role in Scientific Research." SL 25 (1970) 212-65.

3404. Ryder, J. M. "Consequences of a Simple Extension of the Dutch Book Argument." BJPS 32 (1981) 164-67.

3405. Rynin, David. "Probability and Meaning." J. Phil. 44 (1947) 589-97.

3406. Sagoroff, Slawtscho. Wissenschaft und Stastik: Das statistische Denken in den empirischen Wissenschaften. Vienna: Verlag
der Oesterreichischen Akademie der Wissenschaften, 1973. 146 pp.

3407. Salmon, Wesley C. "The Frequency Interpretation and Antece-
dent Probabilities." Phil. Studies. 4 (1953) 44-48.

3408. _____. "The Short Run." PS 22 (1955) 214-21.

    a. "Salmon on 'The Short Run'," by William C. Pettijohn. 23
    (1956) 149.

    b. "Reply to Pettijohn," by W. C. Salmon. 150-52.

3409. _____. "What Happens in the Long Run." Phil. Review.
74 (1965) 373-78.

3410. _____. "The Status of Prior Probabilities in Statistical Explanation." PS 32 (1965) 137-46.

    a. "Salmon's Paper," by Henry E. Kyburg, Jr. 147-51.

    b. "Reply to Kyburg," by W. C. Salmon. 152-54.

3411. _____. Statistical Explanation and Statistical Relevance.
With contributions by Richard C. Jeffrey and James G. Greeno.
Pittsburgh: University of Pittsburgh Press, 1971. 117 pp.

3412. Savage, Leonard J. "Difficulties in the Theory of Personal
Probability." PS 34 (1967) 305-10.

    a. "Slightly More Realistic Personal Probability," by Ian
    Hacking. 311-25.

    b. "Amplifying Personal Probability Theory: Comments on L.
    J. Savage's 'Difficulties in the Theory of Personal Probabi-
    lity'," by Abner Shimony. 326-32.

3413. _____. "Probability in Science: A Personalistic Account."
LMPS-4 (1971) 417-28.

3414. Schick, Frederic. "Consistency and Rationality." J. Phil.
60 (1963) 5-19.

    a. "A Further Note on Rationality and Consistency," by Henry
    E. Kyburg, Jr. 463-65.

    b. "Consistency and Rationality: A Comment," by Howard
    Smokler. 62 (1965) 77-80.

3415. Schnorr, C. P. "A Survey of the Theory of Random Sequences."
WOSPS 11 (1977) 193-211.

3416. Schuntermann, Michael F. "Zur Adäquatheit des Hacking --
Stegmüllerschen Stützungsbegriffs." ZAW 8 (1977) 375-78.

3417. _____. "Hacking's Law of Likelihood und das Kartenpara-doxen von Kerridge." ZAW 11 (1980) 354-56.

3418. Schütt, Klaus-P. "A Model to Support the Assessment of Sub-jective Probabilities." Theory and Decision. 12 (1980) 173-83.

3419. Scott, Dana, and Peter Krauss. "Assigning Probabilities to Logical Formulas," in Aspects of Inductive Logic. J. Hintikka and P. Suppes, eds. Amsterdam: North-Holland, 1966. 219-64.

3420. Seidenfeld, Teddy. "Statistical Evidence and Belief Functions." PSA 1978. II, 478-89.

3421. _____. "Direct Inference and Inverse Inference." J. Phil. 75 (1978) 709-30.

    a. "Conditionalization," by Henry E. Kyburg, Jr. 77 (1980) 98-114.

    b. "Seidenfeld's Critique of Kyburgian Statistics," by Stephen Spielman. 791-95.

3422. _____. Philosophical Problems of Statistical Inference. Dordrecht: Reidel, 1979.

3423. _____. "Why I am not an Objective Bayesian: Some Reflections Prompted by Rosenkrantz." Theory and Decision. 11 (1979) 413-40.

    a. "Bayesian Theory Appraisal: A Reply to Seidenfeld," by R. D. Rosenkrantz. 441-51.

3424. _____. "On After-Trial Properties of Best Neyman-Pearson Confidence Intervals." PS 48 (1981) 281-91.

3425. Servien, Pius. Hasard et probabilités. Paris: Presses Universitaires de France. 1949. 137 pp.

3426. Settle, Tom. "Are Some Propensities Probabilities?" SL 51 (1973) 115-20.

3427. _____. "Presuppositions of Propensity Theories of Prob-ability." MSPS 6 (1975) 388-415.

3428. Shafer, Glenn. A Mathematical Theory of Evidence. Princeton: Princeton University Press, 1976. xiii + 297 pp.

    a. Review essay, "On a New Theory of Epistemic Probability," by Peter Williams. BJPS 29 (1978) 375-87.

3429. _____. "A Theory of Statistical Evidence." WOSPS 6-II (1976) 365-434.

3430. _____. "Two Theories of Probability." PSA 1978. 441-65.

3431. _____. "Jeffrey's Rule of Conditioning." PS 48 (1981) 337-62.

3432. _____. "Constructive Probability." Synthese. 48 (1981) 1-60.

3433. Sklar, Lawrence. "Is Probability a Dispositional Property?" J. Phil. 67 (1970) 355-67.

3434. _____. "Unfair to Frequencies." J. Phil. 70 (1973) 41-52.

  a. ". . .But Fair to Chance," by Isaac Levi. 52-55.

3435. _____. "Probability as a Theoretical Concept." Synthese. 40 (1979) 409-14.

3436. Smokler, Howard. "Three Grades of Probabilistic Involvement." Phil. Studies. 32 (1977) 129-42.

3437. _____. "Single-Case Propensities, Modality, and Confirmation." Synthese. 40 (1979) 497-506.

3438. _____. "The Collapse of Modal Distinctions in Probabilistic Contexts." Theoria. 45 (1979) 1-7.

3439. Sneed, Joseph D. "Strategy and the Logic of Decision." Synthese. 16 (1966) 270-83.

3440. Spielman, Stephen. "Assuming, Ascertaining, and Inductive Probability," in Studies in the Philosophy of Science. N. Rescher, ed. Oxford: Basil Blackwell, 1969. 143-61.

3441. _____. "A Refutation of the Neyman-Pearson Theory of Testing." BJPS 24 (1973) 201-22.

3442. _____. "The Logic of Tests of Significance." PS 41 (1974) 211-26.

  a. "The Logic of Tests of Significance," by Roger Carlson. 43 (1976) 116-28.

  b. "Statistical Dogma and the Logic of Significance Testing," by Stephen Spielman. 45 (1978) 120-35.

3443. _____. "Bayesian Inference with Indeterminate Probabilities." PSA 1976. I, 185-96.

3444. _____. "Exchangeability and the Certainty of Objective Randomness." Journal of Philosophical Logic. 5 (1976) 399-406.

3445. Srzednicki, Jan. "Statistical Indeterminism and Scientific Explanation." Synthese. 26 (1973) 197-204.

3446. Stalnaker, Robert C. "Probability and Conditionals." PS 37 (1970) 64-80.

3447. Stegmüller, W. "Bemerkungen zum Wahrscheinlichkeitsproblem." Studium Generale. 6 (1953) 563-93.

3448. _____. "Carnap's Normative Theory of Inductive Probability." LMPS-4 (1971) 501-13.

3449. _____. "Personal Probability, Rational Decision, and Statistical Probability." Methodology and Science. 7 (1974) 1-24.

3450. Strauss, M. "Ist die Limes-Theorie der Wahrscheinlichkeit eine sinnvolle Idealisation?" Synthese. 5 (1946) 90-91.

3451. Strohal, R. "Bemerkungen zur Hypothesewahrscheinlichkeit." Studium Generale. 4 (1951) 80-88.

3452. Suppes, Patrick. "Concept Formation and Bayesian Decisions," in Aspects of Inductive Logic. J. Hintikka and P. Suppes, eds. Amsterdam: North-Holland, 1966. 21-48.

3453. _____. "Probabilistic Inference and the Concept of Total Evidence," in Aspects of Inductive Logic. J. Hintikka and P. Suppes, eds. Amsterdam: North-Holland, 1966. 49-65.

3454. _____. "New Foundations of Objective Probability: Axioms for Propensities." LMPS-4 (1971) 515-29.

3455. _____. "Approximate Probability and Expectation of Gambles." Erkenntnis. 9 (1975) 153-61.

3456. Swartz, Norman. "Absolute Probability in Small Worlds: A New Paradox in Probability Theory." Philosophia. 3 (1973) 167-78.

3457. Swinburne, R. G. "The Probability of Particular Events." PS 38 (1971) 327-43.

3458. Szaniawski, K. "A Method of Deciding between N Statistical Hypotheses." Studia Logica. 12 (1961) 135-44. Reprinted in SL 87 (1977) 615-23.

3459. _____. "Interpretations of the Maximum Likelihood Principle." SL 87 (1977) 625-34. [First published in Rozprawy Logiczne. Warszawa, 1964.]

3460. _____. "Questions and Their Pragmatic Value." SL 51 (1973) 121-23.

3461. Tallet, Jorge. "Probability and Credibility." Dialectica. 30 (1976) 135-44.

3462. Toulmin, S. E. "Probability." Arist. Soc. Supp. Vol. 24 (1950) 27-62.

3463. Trautteur, G. "Prediction, Complexity, and Randomness." SL 51 (1973) 124-28.

3464. Trenholme, Russell. "A Physicalist Analysis of Probability." Nous. 12 (1978) 303-16.

3465. Tversky, Amos. "A Critique of Expected Utility Theory: Descriptive and Normative Considerations." Erkenntnis. 9 (1975) 163-73.

3466. Tversky, A., and D. Kahneman. "Causal Thinking in Judgment under Uncertainty." WOSPS 11 (1977) 167-90.

3467. Uchii, Soshichi. "Higher Order Probabilities and Coherence." PS 40 (1973) 373-81.

3468. Ullian, Joseph S. "Peirce, Gambling, and Insurance." PS 29 (1962) 79-80.

3469. Urmson, J. O. "Two Senses of 'Probable'." Analysis. 8 (1947-48) 9-17.

    a. "Probability: A Rejoinder to Mr. Urmson," by B. Mayo. 30-32.

    b. "More About Probability," by C. H. Whitely. 76-80.

3470. Varadarajan, V. S. "Probability in Physics and a Theorem on Simultaneous Observability." Communications in Pure and Applied Mathematics. 15 (1962) 189-217. Corrected in Ibid. 18 (1965). Reprinted in WOSPS 5 (1975) 171-204.

3471. Vetter, Hermann. Wahrscheinlichkeit und logischer Spielraum. Eine Untersuchung zur induktiven Logik. Tübingen: J. C. B. Mohr, 1967. 115 pp.

3472. _____. "Logical Probability, Mathematical Statistics, and the Problem of Induction." SL 31 (1970) 75-90. Reprinted from Synthese 20 (1969) 56-71.

    a. "Statistics, Induction, and Lawlikeness: Comments on Dr. Vetter's Paper," by Jaakko Hintikka. 91-102. Reprinted from Synthese 20 (1969) 72-83.

3473. Vickers, John M. "Some Remarks on Coherence and Subjective Probability." PS 32 (1965) 32-38.

3474. _____. "Utility and Its Ambiguities." Erkenntnis. 9 (1975) 287-311.

3475. _____. Belief and Probability. SL 104. Dordrecht: Reidel, 1976. 202 pp.

3476. _____. "On the Reality of Chance." PSA 1978. II, 563-78.

3477. Vietoris, L. "Wie kann Wahrscheinlichkeit definiert werden?" Studium Generale. 4 (1951) 69-72.

3478. _____. "Zur Axiomatik der Wahrscheinlichkeitsrechnung." Dialectica. 8 (1954) 37-47.

3479. _____. "Häufigkeit und Wahrscheinlichkeit." Studium Generale. 9 (1956) 85-96.

3480. Vincent, R. H. "The Paradox of Ideal Evidence." Phil. Review. 71 (1962) 497-503.

    a. "Evidence and Ideal Evidence," by William H. Baumer. PPR 24 (1964) 567-72.

3481. Waerden, B. L. van der. "Der Begriff der Wahrscheinlichkeit." Studium Generale 4 (1951) 65-68.

3482. Walley, Peter, and Terrence L. Fine. "Varieties of Modal (Classificatory) and Comparative Probability." Synthese. 41 (1979) 321-74.

3483. Watling, J. "Chance." Arist. Soc. Supp. Vol. 43 (1969) 37-48.

3484. Wegener, Ursula. "Sind Ludwigs Chancengewichtungen propensities im Sinne Poppers?" ZAW 11 (1980) 80-85.

3485. White, Alan R. "The Propensity Theory of Probability." BJPS 23 (1972) 35-43.

    a. "Propensity Theories of Probability Unscathed: A Reply to White," by Tom Settle. BJPS 23 (1972) 331-35.

3486. Will, Frederick L. "Kneale's Theories of Probability and Induction." Phil. Review. 63 (1954) 19-42.

3487. Williams, Donald. "On the Derivation of Probabilities from Frequencies." PPR 5 (1945) 449-84.

    a. "Probability and Non-Demonstrative Inference," by Ernest Nagel. 485-507.

    b. "Reply to Donald C. Williams' Criticism of the Frequency Theory of Probability," by Hans Reichenbach. 508-12.

    c. "Two Concepts of Probability," by Rudolf Carnap. 513-32.

    d. "On The Frequency Theory of Probability," by Henry Margenau. 6 (1945) 11-25.

    e. "Frequencies, Probabilities, and Positivism," by Gustav Bergman. 26-44.

f. "Comments on Donald Williams' Paper," by R. von Mises. 45-46.

g. "Scientific Procedure and Probability," by Felix Kaufman. 47-66.

h. "The Challenging Situation in the Philosophy of Probability," by Donald Williams. 67-86.

i. "Remarks on Induction and Truth," by Rudolf Carnap. 590-602.

j. "On the Nature of Inductive Inference," by Felix Kaufman. 602-09.

k. "Rejoinder to Mr. Kaufman's Reply," by Rudolf Carnap. 609-11.

l. "Comments on Donald Williams' Reply," by Richard von Mises. 611-13.

m. "Is the Laplacian Theory of Probability Untenable?" by Ernest Nagel. 614-18.

n. "The Problem of Probability," by Donald Williams. 619-22.

3488. Williams, P. M. "Bayesian Conditionalisation and the Principle of Minimum Information." BJPS 31 (1980) 131-44.

3489. Wright, G. H. von. "Remarks on the Epistemology of Subjective Probability." LMPS-1 (1960) 330-39.

3490. _____. "Wittgenstein's Views on Probability." RIP 23 (1969) 259-83.

4.14 Verification and Confirmation

3491. Achinstein, Peter. "Confirmation Theory, Order, and Periodicity." PS 30 (1963) 17-35.

3492. _____. "Variety and Analogy in Confirmation Theory." PS 30 (1963) 207-21.

a. "Variety, Analogy, and Periodicity in Inductive Logic," by Rudolf Carnap. 222-27.

3493. Ackermann, Robert. "Sortal Predicates and Confirmation." Phil. Studies. 20 (1969) 1-4.

a. "Sortals and Paradox," by Alex Blum. 22 (1971) 33-34.

3494. Agassi, Joseph. "The Mystery of the Ravens." PS 33 (1966) 395-402.

3495. _____. "Testing as a Bootstrap Operation in Physics." ZAW 4 (1973) 1-24.

3496. _____. "The Problems of Scientific Validation." BSPS 34 (1981) 103-14.

3497. Albrecht-Buehler, G. "Numerical Evaluation of the Validity of Experimental Proofs in Biology." Synthese. 33 (1976) 283-314.

3498. Alexander, H. G. "The Paradoxes of Confirmation." BJPS 9 (1958) 227-33.

    a. "Corroboration Versus Induction," by J. Agassi. 9 (1959) 311-17.

    b. "The Paradoxes of Confirmation -- a Reply to Dr. Agassi," by H. G. Alexander. 10 (1959) 229-34.

    c. "Confirmation Without Background Knowledge," by J. W. N. Watkins. 10 (1960) 318-20.

3499. Annis, David B. "A Contextualist Theory of Epistemic Justification." Am. Phil. Quart. 15 (1978) 213-19.

3500. Ayer, A. J. "Le problème de la confirmation." Logique et analyse. 8 (1965) 238-64. Reprinted in RIP 25 (1971) 3-31.

3501. Baillie, Patricia. "That Confirmation May yet be a Probability." BJPS 20 (1969) 41-51.

    a. "Confirmation as a Probability: Dead but It Won't Lie Down," by T. W. Settle. 21 (1970) 200-01.

    b. "Confirmation and Probability: A Reply to Settle," by Patricia Baillie. 22 (1971) 285-86.

3502. _____. "Confirmation and the Dutch Book Argument." BJPS 24 (1973) 393-97.

3503. Barker, Stephen F. Induction and Hypothesis: A Study of the Logic of Confirmation. Ithaca, NY: Cornell University Press, 1957. xvi + 203 pp.

    a. Review essay, "Reply to Barker's Criticism of Formalism," by Jack Henry. PS 26 (1959) 355-61.

    b. Review essay by R. Harré. Mind. 71 (1962) 412-20.

3504. Batens, Diderik. "The Paradoxes of Confirmation." RIP 25 (1971) 101-18.

3505. Inadvertently skipped in numbering the entries.

3506. Baumer, William. H. "Von Wright's Paradoxes." PS 30 (1963) 165-72.

3507. _____. "Confirmation Without Paradoxes." BJPS 15 (1964) 177-95.

    a. "Baumer on the Confirmation Paradoxes," by Howard Kahane. 18 (1967) 52-56.

    b. "Confirmation Still Without Paradoxes," by William H. Baumer. 19 (198) 57-63.

    c. "Eliminative Confirmation and Paradoxes," by Howard Kahane. 20 (1969) 160-62.

3508. _____. "In Defense of a Principal Theorem." Synthese. 20 (1969) 121-42.

3509. Berent, Paul. "Disconfirmation by Positive Instances." PS 39 (1972) 522.

3510. Black, Max. "Notes on the 'Paradoxes of Confirmation'," in Aspects of Inductive Logic. J. Hintikka and P. Suppes, eds. Amsterdam: North-Holland, 1966. 175-97.

3511. Brody, B. A. "Confirmation and Explanation." J. Phil. 65 (1968) 282-99.

    a. "Confirmation and Explanation," by Michael Martin. Analysis. 32 (1971-72) 167-69.

    b. "Martin on Explanation and Confirmation," by Barry Gower. 33 (1972-73) 107-09.

    c. "More on Confirmation and Explanation," by Baruch Brody. Phil. Studies. 26 (1974) 73-75.

    d. "Explanation and Confirmation Again," by Michael Martin. Analysis. 36 (1975-76) 41-42.

3512. Bunge, Mario. "La vérification des théories scientifiques." Logique et analyse. 11 (1968) 145-79.

3513. Burks, Arthur W. "Justification in Science," in Academic Freedom, Logic, and Religion. Morton White, ed. Philadelphia: University of Pennsylvania Press, 1953. 109-25.

3514. Carnap, Rudolf. "On the Application of Inductive Logic." PPR 8 (1947) 133-47.

    a. "On the Infirmities of Confirmation Theory," by Nelson Goodman. 149-51.

b. "Reply to Nelson Goodman," by Rudolf Carnap.   461-62.

3515. _____. "On the Comparative Concept of Confirmation."
BJPS 3 (1953) 311-18.

3516. Chihara, Charles. "Quine and the Confirmational Paradoxes."
Midwest Studies in Philosophy. 6 (1981) 425-52.

a. "Reply to Chihara," by W. V. Quine.   453-54.

3517. Chisholm, Roderick M. "Evidence as Justification." J. Phil.
58 (1961) 739-48.

3518. Churchman, C. West. "Science and Decision Making." PS 23
(1956) 247-49.

3519. Clay, J. "La vérification d'une theorie." RIP 8 (1954)
41-45.

3520. Cohen, L. Jonathan. "What Has Confirmation to do with Probabilities?" Mind. 75 (1966) 463-81.

a. "An Alleged Condition of Evidential Support," by Alex C.
Michalos.   78 (1969) 440-41.

3521. _____. "A Logic for Evidential Support." BJPS 17
(1966) Part I: 21-43.   Part II: 105-26.

a. "A Note on Consilience," by L. Jonathan Cohen.   19 (1968)
70-71.

3522. _____. "The Paradox of Anomaly." SL 51 (1973) 78-82.

3523. _____. "How Can One Testimony Corroborate Another?"
BSPS 39 (1976) 65-78.

3524. Coffa, J. A. "Two Remarks on Hempel's Logic of Confirmation."
Mind. 79 (1970) 591-96.

3525. Copeland, Sr., Arthur H. "Mathematical Proof and Experimental
Proof." PS 33 (1966) 303-16.

3526. Cornman, James W. "Indirectly Verifiable: Everything or
Nothing." Phil. Studies. 18 (1967) 49-56.

3527. _____. "Foundational versus Nonfoundational Theories of
Empirical Justification." Am. Phil. Quart. 14 (1977) 287-97.

3528. Crawshay-Williams, Rupert. "Equivocal Confirmation." Analysis.
11 (1950-51) 73-79.

3529. Czeżowski, Thadée. "De la vérification dans les sciences em-
piriques. (Analyse logique)." RIP 5 (1951) 347-66. English Tr.:
SL 87 (1977) 93-109.

3530. Darlington, Jared. "On the Confirmation of Laws." PS 26
(1959) 14-24.

    a. "Darlington's 'On the Confirmation of Laws'," by H.
    Linhart. 362.

    b. "Reply to Linhart," by Jared L. Darlington. 363.

    c. "Professor Darlington and the Confirmation of Laws," by
    Hugues Leblanc. 364-66.

    d. "Reply to Leblanc," by Jared L. Darlington. 367-68.

3531. Emmerich, David S., and James G. Greeno. "Some Decision
Factors in Scientific Investigation." PS 33 (1966) 262-70.

3532. Englebretsen, George. "Sommers' Theory and the Paradox of
Confirmation." PS 38 (1971) 438-41.

3533. Erwin, Edward. "The Confirmation Machine." BSPS 8, SL 39
(1971) 306-21.

3534. Esser, P. H. "Is it Possible to Verify Statements by Non-
Verbal Testing Methods?" Synthese. 8 (1950-51) 238-42.

3535. Farrell, Robert J. "Material Implication, Confirmation, and
Counterfactuals." Notre Dame Journal of Formal Logic. 20 (1979)
383-94.

3536. Finch, Henry. "Confirming Power of Observations Metricized
for Decisions Among Hypotheses." PS 27 (1960) 293-307; 391-404.

3537. Foley, Richard. "Inferential Justification and the Infinite
Regress." Am. Phil. Quart. 15 (1978) 311-16.

3538. Foster, Lawrence. "Hempel, Scheffler, and the Ravens."
J. Phil. 68 (1971) 107-14.

    a. "Selective Confirmation and the Ravens: A Reply to
    Foster," by Israel Scheffler and Nelson Goodman. 69 (1972)
    78-83.

3539. Freund, John E. "On the Confirmation of Scientific Theories."
PS 17 (1950) 87-94.

3540. _____. "On the Problem of Confirmation." Methodos. 3
(1951) 33-42.

    a. "Discussion," by Vittorio Somenzi. 42.

3541. Freundlich, Yehudah. "Theory Evaluation and the Bootstrap
Hypothesis." SHPS 11 (1980) 267-77.

3542. Giere, Ronald N. "An Orthodox Statistical Resolution of the
Paradox of Confirmation." PS 37 (1970) 354-62.

3543. Glymour, Clark. "Relevant Evidence." J. Phil. 72 (1975) 403-26.

    a. "An Appraisal of Glymour's Confirmation Theory," by Paul Horwich. 75 (1978) 98-113.

    b. "Glymour on Confirmation," by Aron Edidin. PS 48 (1981) 292-307.

3544. _____. Theory and Evidence. Princeton, NJ: Princeton University Press, 1979. 383 pp.

3545. _____. "Bootstraps and Probabilities." J. Phil. 77 (1980) 691-99.

    a. "The Dispensibility of Bootstrap Conditions," by Paul Horwich. 699-702.

3546. Goddard, L. "The Paradoxes of Confirmation and the Nature of Natural Laws." Phil. Quart. 27 (1977) 97-113.

3547. Good, I. J. "The White Shoe is a Red Herring." BJPS 17 (1967) 322.

    a. "The White Shoe: No Red Herring," by Carl G. Hempel. 18 (1967) 239-40.

    b. "The White Shoe Qua Herring is Pink," by I. J. Good. 19 (1968) 156-57.

3548. Goodman, Nelson. "A Query on Confirmation." J. Phil. 43 (1946) 383-85.

3549. Grandy, Richard E. "Some Comments on Confirmation and Selective Confirmation." Phil. Studies. 18 (1967) 19-24.

3550. Granger, G. G. "Confirmation et information metatheorique." Logique et analyse. 16 (1973) 385-91.

3551. Grant, John. "Confirmation of Empirical Theories by Observation Sets." Philosophia. 8 (1978) 367-80.

3552. Greeno, James G. "Evaluation of Statistical Hypotheses Using Information Transmitted." PS 37 (1970) 279-93.

3553. Grier, Brown. "Prediction, Explanation and Testability as Criteria for Judging Statistical Theories." PS 42 (1975) 373-83.

3554. Griffin, N. "Has Harré Solved Hempel's Paradox?" Mind. 84 (1975) 426-30.

    a. "The Identity of Laws: A Reply to Mr. Griffin," by R. Harré. 85 (1976) 597-600.

3555. Gruner, Rolf. "Historical Facts and the Testing of Hypotheses."
Am. Phil. Quart. 5 (1968) 124-29.

3556. Grzegorczyk, A. "Mathematical and Empirical Verifiability."
SL 87 (1977) 165-69. [First published in Rozprawy Logiczne. Warszawa,
1964.]

3557. Hall, Everett W. "On the Nature of the Predicate 'Verified'."
PS 14 (1947) 123-31.

3558. Hanen, Marsha. "Confirmation and Adequacy Conditions." PS
38 (1971) 361-68.

3559. Helmer, Olaf, and Paul Oppenheim. "A Syntactical Definition
of Probability and of Degree of Confirmation." Journal of Symbolic
Logic. 10 (1945) 25-61.

3560. Hempel, Carl G. "A Purely Syntactical Definition of Confirmation." Journal of Symbolic Logic. 8 (1943) 122-43.

3561. _____. "Studies in the Logic of Confirmation." Mind.
54 (1945) 1-26; 97-121. Reprinted, with some changes, in C. G.
Hempel, Aspects of Scientific Explanation and Other Essays in the
Philosophy of Science. New York: Free Press, 1965. 3-46. To
which is added "Postscript (1964) on Confirmation." 47-51.

    a. "Hempel's Paradoxes of Confirmation," by C. H. Whiteley.
    Mind. 54 (1945) 156-58.

    b. "A Note on the Paradoxes of Confirmation," by C. G.
    Hempel. 55 (1946) 79-82.

3562. Hempel, Carl G., and Paul Oppenheim. "A Definition of 'Degree
of Confirmation'." PS 12 (1945) 98-115.

3563. Hendry, Herbert E., and James E. Roper. "Anything Confirms
Anything?" Synthese. 45 (1980) 217-32.

3564. Hesse, Mary. "Analogy and Confirmation Theory." Dialectica.
17 (1963) 284-95. Also in PS 31 (1964) 319-27.

3565. _____. "Confirmation of Laws," in Philosophy, Science,
and Method. S. Morgenbesser et al., ed. New York: St. Martin's
Press, 1969. 74-91.

3566. _____. "Theories and the Transitivity of Confirmation."
PS 37 (1970) 50-63.

3567. _____. "Bayesianism and Scientific Inference." SHPS 5
(1975) 367-70.

3568. Hintikka, Jaakko. "Inductive Independence and the Paradoxes
of Confirmation." SL 24 (1970) 24-46.

3569. Hooker, C. A. "Goodman, 'Grue' and Hempel." PS 35 (1968) 232-47.

3570. _____. "The Ravens, Hempel and Goodman." AJP 49 (1971) 82-89.

3571. Hooker, C. A., and D. Stove. "Relevance and the Ravens." BJPS 18 (1968) 305-15.

  a. "On 'Ravens and Relevance' and a Likelihood Solution of the Paradox of Confirmation," by L. Gibson. 20 (1968) 75-80.

  b. "Mr. Gibson on Ravens and Relevance," by D. Stove. 21 (1970) 287-88.

3572. Horwich, P. "A Peculiar Consequence of Nicod's Criterion." BJPS 29 (1978) 262-63.

  a. "Nicod's Criterion: Subtler Than You Think," by William W. Rozeboom. PS 47 (1980) 638-43.

3573. Issman, Samuel. "Deux problèmes de la théorie de la confirmation." Annales de l'Institut de Philosophie. (Brussels) (1970) 175-86.

3574. Jackson, Frank, and Robert Pargetter. "Confirmation and the Nomological." Canadian Journal of Philosophy. 10 (1980) 415-28.

3575. Jardine, R. "The Resolution of the Confirmation Paradox." AJP 43 (1965) 359-68.

3576. Jeffrey, Richard C. "Valuation and Acceptance of Scientific Hypotheses." PS 23 (1956) 237-46.

3577. _____. "Carnap's Empiricism." MSPS 6 (1975) 37-49.

3578. Jones, Robert M. "The Non-Reducibility of Koopman's Theorems of Probability in Carnap's System for MC." PS 32 (1965) 368-69.

3570. Kaplan, Mark. "A Bayesian Theory of Rational Acceptance." J. Phil. 78 (1981) 305-31.

3580. Kemeny, John G., and Paul Oppenheim. "Degree of Factual Support." PS 19 (1952) 307-24.

3581. Kimbrough, Steven Orla. "On the Use of Likelihood as a Guide to Truth." PSA 1980. I, 117-28.

3582. Leblanc, Hugues. "Evidence logique et degré de confirmation." Revue Philosophique de Louvain. 52 (1954) 619-25.

3583. _____. "On Logically False Evidence Statements." Journal of Symbolic Logic. 22 (1957) 345-49.

3584. _____. "The Problem of the Confirmation of Laws."
Phil. Studies. 12 (1961) 81-84.

3585. _____. "A New Interpretation of c(h,e)." PPR 21 (1961)
373-76.

3586. _____. "A Revised Version of Goodman's Confirmation
Paradox." Phil. Studies. 14 (1963) 49-51.

3587. _____. "That Positive Instances are No Help." J. Phil.
60 (1963) 453-62.

3588. Lehrer, Keith. "Social Information." Monist. 60 (1977)
473-86.

3589. Lehrer, Keith, Richard Roelofs, and Marshall Swain. "Reason
and Evidence: An Unsolved Problem." Ratio. 9 (1967) 38-48.

3590. Lenz, John W. "Carnap on Defining 'Degree of Confirmation'."
PS 23 (1956) 230-36.

3591. Lenzen, Victor F. "Verification in Science." RIP 5 (1951)
323-46.

3592. Lenzen, Wolfgang. Theorien der Bestätigung wissenschaftlicher
Hypothesen. Stuttgart-Bad Connstatt: Frommann-Holzboog, 1974. 217
pp.

3593. Levi, Isaac. "Decision Theory and Confirmation." J. Phil.
58 (1961) 614-25.

3594. _____. "On the Seriousness of Mistakes." PS 29 (1962)
47-65.

3595. _____. "A Paradox for the Birds." BSPS 39 (1976)
371-78.

3596. _____. "Direct Inference and Confirmational Conditionalization." PS 48 (1981) 532-52.

3597. Lin Chao-Tien. "Solutions to the Paradoxes of Confirmation,
Goodman's Paradox, and Two New Theories of Confirmation." PS 45
(1978) 415-19.

3598. Longino, Helen E. "Evidence and Hypothesis: An Analysis of
Evidential Relations." PS 46 (1979) 35-56.

3599. Mackie, J. L. "The Paradox of Confirmation." BJPS 13 (1963)
265-77.

3600. _____. "The Relevance Criterion of Confirmation." BJPS
20 (1969) 27-40.

        a. "On Irrelevant Criteria of Confirmation," by G. Schlesinger.
        21 (1970) 282-87.

3601. Martin, Michael. "An Emplicative Model of Theory Testing." ZAW 1 (1970) 228-42.

3602. Maxwell, Grover. "Some Current Trends in Philosophy of Science: With Special Attention to Confirmation, Theoretical Entities, and Mind-Body." BSPS 32 (1976) 565-84.

3603. Meehl, Paul E. "Theory Testing in Psychology and Physics: A Methodological Paradox." PS 34 (1967) 103-15.

    a. "Parameter Estimation vs. Hypothesis Testing," by M. I. Charles E. Woodson. 36 (1969) 203-04.

    b. "On Prior Probabilities of Rejecting Statistical Hypotheses," by Herbert Keuth. 40 (1973) 538-46.

    c. "Theory Confirmation in Psychology," by Chris Swoyer and Thomas C. Monson. 42 (1975) 487-502.

3604. Merrill, G. H. "Confirmation and Prediction." PS 46 (1979) 98-117.

    a. "Hypothetico-Deductivism is Hopeless," by Clark Glymour. 47 (1980) 322-25. [Correction: 503]

3605. Meyer, Stuart L. "Urning a Resolution of Hempel's Paradox." PS 44 (1977) 292-96.

3606. Michalos, Alex C. "Two Theorems of Degree of Confirmation." Ratio 7 (1965) 196-98.

3607. _____. "Descriptive Completeness and Linguistic Variance." Dialogue. 6 (1967) 224-28.

3608. _____. "The Impossibility of an Ordinal Measure of Acceptability." Philosophical Forum. 2 (1970) 103-06.

3609. _____. The Popper-Carnap Controversy. The Hague: Martinus Nijhoff, 1971. 124 pp.

    a. "Essay Review," by J. G. McEvoy. SHPS 7 (1976) 63-85.

    b. Review essay by Ilkka Niiniluoto. Synthese. 25 (1973) 417-36.

3610. _____. "Cost-Benefit vs. Expected Utility Acceptance Rules." BSPS 8, SL 39 (1971) 342-74. Also in WOSPS 1 (1973) 163-90.

3611. Mirabelli, Andre. "Belief and Incremental Confirmation of One Hypothesis Relative to Another." PSA 1978. I, 287-301.

3612. Moor, James H. "The Cancellation of Symmetrical Contraries and the Principle of Significant Contradictories." PS 43 (1976) 550-59.

a. "On a Matter of Principle," by G. Schlesinger. 45 (1978) 312-17.

b. "Moor and Schlesinger on Explanation," by Brian Cupples. 46 (1979) 645-50.

3613. Morgenbesser, Sidney. "A Note on Justification." J. Phil. 58 (1961) 748-49.

3614. _____. "Goodman on the Ravens." J. Phil. 59 (1962) 493-95.

3615. Musgrave, Alan E. "Logical versus Historical Theories of Confirmation." 25 (1974) 1-23.

3616. Myro, George. "Aspects of Acceptability." Pacific Philosophical Quarterly. 62 (1981) 107-17.

a. "An Unacceptable Aspect of Acceptability," by Richard E. Grandy. 118-22.

3617. Naess, A. "What Does 'Testability' Mean? An Account of a Procedure Developed by Ludvig Lövestad." Methodos. 9 (1957) 229-37.

3618. Nalimov, V. V. "The Receptivity of Hypotheses." Diogenes. 100 (1977) 179-97.

3619. Namer, Emile. "L'hypothèse et sa vérification à la naissance de la physique mathématique." RIP 25 (1971) 44-61.

3620. Nelson, Alvin F. "Simplicity and the Confirmation Paradoxes." Southwestern Journal of Philosophy. 3, No. 2 (1972) 99-107.

3621. Nelson, John O. "The Confirmation of Hypotheses." Phil. Review. 67 (1958) 95-100.

3622. Nemetz, T. "Information Theoretic Concepts in Testing Hypotheses." RIP 25 (1971) 160-66.

3623. Pastin, Mark. "Counterfactuals in Epistemology." Synthese. 34 (1977) 479-96.

a. "On the Analysis of Warranting," by Fred Feldman. 497-512.

b. "Warranting Reconsidered: Response to Feldman," by Mark Pastin. 37 (1978) 459-64.

c. "Final Comments on the Analysis of Warranting," by Fred Feldman. 465-69.

3624. Patryas, W. "The Sense of Empirical Testing." Poznan Studies. 3 (1977) 180-98.

3625. Paulos, John. "A Model-Theoretic Account of Confirmation." Notre Dame Journal of Formal Logic. 20 (1979) 451-57.

3626. Pollock, John L. "Laying the Raven to Rest: A Discussion of Hempel and the Paradoxes of Confirmation." J. Phil. 70 (1973) 747-55.

    a. "Resurrecting the Ravens," by Y. Freundlich. Synthese. 33 (1976) 341-54.

3627. Popper, K. R. "Degree of Confirmation." BJPS 5 (1954) 143-49.

    a. "A Note on 'Degree of Confirmation," 334.

    b. "Comments on 'Degree of Confirmation' by Professor K. R. Popper," by Yehoshua Bar-Hillel. 6 (1955) 155-57.

    c. "'Content' and 'Degree of Confirmation': A Reply to Dr. Bar-Hillel," by K. R. Popper. 157-63.

    d. "Remarks on Popper's Note on Content and Degree of Confirmation," by Rudolf Carnap. 7 (1956) 243-44.

    e. "Reply to Professor Carnap," by K. R. Popper. 244-45.

    f. "Further Comments on Probability and Confirmation," by Yehoshua Bar-Hillel. 245-48.

    g. "Adequacy and Consistency: A Second Reply to Dr. Bar-Hillel," by K. R. Popper. 249-56.

    h. "A Second Note on Degree of Confirmation," by Karl R. Popper. 350-53.

    i. "A Third Note on Degree of Corroboration or Confirmation," by K. R. Popper. 8 (1958) 294-302.

    j. "On So-called Degrees of Confirmation," by Hugues Leblanc. 10 (1960) 312-15.

    k. "Probabilistic Independence and Corroboration by Empirical Tests," by K. R. Popper. 315-18. [cf. "Errata", BJPS 11 (1960) 88, 149].

    l. "The Paradox of Confirmation," by I. J. Good. BJPS. Part I: 11 (1960) 145-49. Part II: 12 (1961) 63-64.

    m. "The Identity Hypothesis," by Peter Achinstein. 13 (1962) 167-71.

3628. Putnam, Hilary. "A Definition of Degree of Confirmation for Very Rich Languages." PS 23 (1956) 58-62.

3629. _____ . "'Degree of Confirmation' and Inductive Logic," in The Philosophy of Rudolf Carnap. P. A. Schilpp, ed. La Salle, Illinois: Open Court, 1963. 761-83.

3630. _____. "Probability and Confirmation," in Philosophy of Science Today. S. Morgenbesser, ed. New York: Basic Books, 1967. 100-14.

3631. Rescher, Nicholas. "A Theory of Evidence." PS 25 (1958) 83-94.

3632. Resnick, Lawrence. "Confirmation and Hypothesis." PS 26 (1959) 25-30.

3633. Rody, Phillip J. "(C)Instances, the Relevance Criterion, and the Paradoxes of Confirmation." PS 45 (1978) 289-302.

3634. Rogers, Ben. "Material Conditions on Tests of Statistical Hypotheses." BSPS 8, SL 39 (1971) 403-12.

3635. Rosenkrantz, R. D. "Probabilistic Confirmation Theory and the Goodman Paradox." Am. Phil. Quart. 10 (1973) 157-62.

3636. Rozeboom, William W. "New Dimensions of Confirmation." PS 35 (1968) 134-55.

3637. _____. "New Dimensions of Confirmation Theory II: The Structure of Uncertainty." BSPS 8, SL 39 (1971) 342-74.

3638. Sagal, Paul T. "Paradox, Confirmation and Inquiry." Philosophy. 51 (1976) 467-70.

3639. Salmon, Wesley C. "Bayes's Theorem and the History of Science." MSPS 5 (1970) 68-86.

3640. _____. "Confirmation and Relevance." MSPS 6 (1975) 3-36.

3641. Schlanger, Jacques. "Esquisse d'une théorie de la vérifica-tion." Logique et analyse. 23 (1980) 107-44.

3642. Schlesinger, G. "Instantiation and Confirmation." BSPS 2 (1965) 1-18.

        a. "Comments," by Carl G. Hempel, 19-24.

3643. _____. "Natural Kinds." BSPS 3 (1967) 108-22.

3644. _____. "Confirmability and Determinism." Phil. Quart. 18 (1968) 29-39.

3645. _____. Confirmation and Confirmability. Oxford: Clarendon Press, 1974. 109 pp.

3646. _____. "Confirmation and Parsimony." MSPS 6(1975) 324-42.

        a. "Comments on 'Confirmation and Parsimony'," by Paul Teller. 343-46.

b. "Rejoinder to Professor Teller," by George Schlesinger. 347-49.

3647. Schoenberg, Judith. "Confirmation by Observation and the Paradox of the Ravens." BJPS 15 (1964) 200-12.

3648. Schreiber, A. Theorie und Rechtfertigung. Braunschweig: Vieweg, 1975. 204 pp.

3649. Schwartz, Robert. "Paradox and Projection." PS 39 (1972) 245-48.

3650. _____. "Approximate Truth and Confirmation." PS 48 (1981) 606-10.

3651. Sen, Pranab Kumar. "Approaches to the Paradox of Confirmation." Ajatus. 34 (1972) 45-83.

3652. Sharpe, R. A. "Validity and the Paradox of Confirmation." Phil. Quart. 14 (1964) 170-73.

3653. Shimony, Abner. "Coherence and the Axioms of Confirmation." Journal of Symbolic Logic. 20 (1955) 1-28.

3654. Simon, Herbert A. "Prediction and Hindsight as Confirmatory Evidence." PS 22 (1955) 227-30.

3655. _____. "On Judging the Plausibility of Theories." LMPS-3 (1967) 439-59.

3656. Sklar, Lawrence. "Methodological Conservatism." Phil. Review. 84 (1975) 374-400.

3657. Skyrms, Brian. "Nomological Necessity and the Paradoxes of Confirmation." PS 33 (1966) 230-49.

a. "On a Claim by Skyrms Concerning Lawlikeness and Confirmation," by Carl G. Hempel. 35 (1968) 274-78.

3658. Slote, Michael A. "Confirmation and Conservatism." Am. Phil. Quart. 18 (1981) 79-84.

3659. Smokler, Howard. "The Equivalence Condition." Am. Phil. Quart. 4 (1967) 300-07.

3660. _____. "Conflicting Conceptions of Confirmation." J. Phil. 65 (1968) 300-12.

3661. Smullyan, Arthur. "The Concept of Empirical Knowledge." Phil. Review. 65 (1956) 362-70.

3662. _____. "The Method of Consequences." Phil. Review. 72 (1963) 48-56.

3663. Sommers, Fred. "Confirmation and the Natural Subject."
Philosophical Forum. 2 (1970) 245-50.

3664. Stelzl, Ingeborg. "Kann der Grad der Bewährung eine Wahrscheinlichkeit sein? Zur Auseinandersetzung zwischen K. Popper und R.
Carnap." PN 12 (1970) 47-50.

3665. Stemmer, Nathan. "The Objective Confirmation of Hypotheses."
Canadian Journal of Philosophy. 11 (1981) 395-404.

3666. Stove, D. "Hempel and Goodman on the Ravens." AJP 43 (1965)
300-10.

3667. _____. "On Logical Definitions of Confirmation." BJPS
16 (1966) 265-72.

3668. _____. "Hempel's Paradox." Dialogue. 4 (1966) 444-55.

3669. Suppes, Patrick. "A Bayesian Approach to the Paradoxes of
Confirmation," in Aspects of Inductive Logic. J. Hintikka and P.
Suppes, eds. Amsterdam: North-Holland, 1966. 198-207.

3670. _____. "Testing Theories and the Foundations of Statistics." WOSPS 6-II (1976) 437-52.

3671. Swinburne, R. G. "Choosing Between Confirmation Theories."
PS 37 (1970) 602-13.

3672. _____. "Probability, Credibility and Acceptability."
Am. Phil. Quart. 8 (1971) 275-83.

3673. _____. "The Paradoxes of Confirmation -- A Survey."
Am. Phil. Quart. 8 (1971) 318-30.

3674. _____. "Confirmability and Factual Meaningfulness."
Analysis. 33 (1972-73) 71-76.

       a. "Confirmability and Meaningfulness," by Richard I. Sikora.
       34 (1973-74) 142-44.

       b. "Meaningfulness without Confirmability -- A Reply," by R.
       G. Swinburne. 35 (1974-75) 22-27.

       c. "Swinburne on Confirmability," by R. I. Sikora. 195.

3675. _____. An Introduction to Confirmation Theory. London:
Methuen; New York: Barnes and Noble, 1973. vi + 218 pp.

3676. Szaniawski, K. "A Note on Confirmation of Statistical Hypotheses." Studia Logica. 10 (1960) 111-18.

       a. "Degree of Confirmation and Critical Region," by Z.
       Czerwiński. 119-22.

b. "Confirmation, Critical Region, and Empirical Content of Hypotheses," by J. Giedymin. 122-25.

3677. Teller, Paul. "Shimony's A Priori Arguments for Tempered Personalism." MSPS 6 (1975) 166-203.

a. "Vindication: A Reply to Paul Teller," by Abner Shimony, 204-11.

3678. Temple, Dennis. "Grue-Green and Some Mistakes in Confirmation Theory." Dialectica. 28 (1974) 197-210.

3679. Todd, William. "Probability and the Theorem of Confirmation." Mind. 76 (1967) 260-63.

3680. Törnebohm, Håkan. Information and Confirmation. Göteborg: Elanders Boktryckeri Aktiebolag, 1964. 80 pp.

3681. _____. "On the Confirmation of Hypotheses about Regions of Existence." Synthese. 18 (1968) 28-45.

3682. Ullian, Joseph. "Luck, License, and Lingo." J. Phil. 58 (1961) 731-39.

3683. Urbach, Peter. "On the Utility of Repeating the 'Same' Experiment." AJP 59 (1981) 151-62.

3684. Vermeulen, R. "Validity of Hypotheses." Synthese. 9 (n.d.) 385-94.

3685. Vincent, R. H. "A Note on Some Quantitative Theories of Confirmation." Phil. Studies. 12 (1961) 91-92.

3686. _____. "Concerning an Alleged Contradiction." PS 30 (1963) 189-94.

3687. _____. "The Paradoxes of Confirmation." Mind. 73 (1964) 273-79.

3688. _____. "Selective Confirmation and the Ravens." Dialogue. 14 (1975) 3-50.

3689. Walker, Edwin Ruthven. "Verification and Probability." J. Phil. 44 (1947) 97-104.

3690. Walton, Gilbert. "Imagination and Confirmation." Mind. 78 (1969) 580-87.

3691. Watkins, J. W. N. "Between Analytic and Empirical." Philosophy. 32 (1957) 112-31.

a. "Empirical Statements and Falsifiability," by Carl G. Hempel. 33 (1958) 342-48.

b. "A Rejoinder to Professor Hempel's Reply," by J. W. N. Watkins. 349-55.

c. "A Note on Confirmation," by Israel Scheffler. <u>Phil. Studies</u>. 11 (1960) 21-23.

d. "Professor Scheffler's Note," by J. W. N. Watkins. 12 (1961) 16-19.

e. "A Rejoinder on Confirmation," by Israel Scheffler. 19-20.

f. "Popperian Confirmation and the Paradox of the Ravens," by D. Stove. AJP 37 (1959) 149-51.

g. "Mr. Stove's Blunders," by J. W. N. Watkins. 240-41.

h. "On Not being Gulled by Ravens," by W. J. Huggett. 38 (1960) 48-50.

i. "A Reply to Mr. Watkins," by D. Stove. 51-54.

j. "Reply to Mr. Stove's Reply," by J. W. N. Watkins. 54-58.

3692. Watling, John. "Confirmation Discomforted." RIP 17 (1963) 155-70.

3693. Whiteley, C. H. "Confirmation." <u>Proc. Arist. Soc.</u> 74 n.s. (1973-74) 1-14.

3694. Will, Frederick L. "The Justification of Theories." <u>Phil. Review</u>. 64 (1955) 370-88.

3695. _____. "Consequences and Confirmation." <u>Phil. Review</u>. 75 (1966) 34-58.

3696. Williams, Peter M. "Goodman's Paradox and Rules of Acceptance." PS 36 (1969) 311-15.

3697. _____. "The Structure of Acceptance and Its Evidential Basis." BJPS 19 (1969) 325-44.

3698. Wilson, P. R. "On the Confirmation Paradox." BJPS 15 (1964) 196-99.

3699. _____. "A New Approach to the Confirmation Paradox." AJP 42 (1964) 393-401.

a. "Mr. Wilson on the Paradox of Confirmation," by G. Nerlich. 401-05.

3700. Wit, Han F. de. "Logische operationalisatie en confirmatie." <u>Algemeen Nederlands Tijdschrift voor Wijsbegeerte</u>. 64 (1972) 110-39.

3701. _____. "Empirical Testability and Confirmation."
Methodology and Science. 7 (1974) 25-47.

3702. Wright, G. H. von. "The Paradoxes of Confirmation," in
Aspects of Inductive Logic. J. Hintikka and P. Suppes, eds. Amsterdam: North-Holland, 1966. 208-18. Also in Theoria. 31 (1965)
255-74.

3703. _____. "A Note on Confirmation Theory and on the Concept
of Evidence." Scientia. 105 (1970) 595-606.

3704. Wrightsman, Bruce. "The Legitimation of Scientific Belief:
Theory Justification by Copernicus." BSPS 60 (1980) 51-66.

3705. Zweig, Arnulf. "Some Consequences of Professor Feigl's Views
on Justification." Phil. Studies. 9 (1958) 67-69.

     a. "A Note on Justification and Reconstruction," by Herbert
     Feigl. 70-72.

4.15 Falsification, Duhem's Thesis, Adhocness

3706. Ardley, Gavin. "The Principle of Falsification." Philosophical
Studies (Ireland) 9 (1959) 66-72.

3707. Auroux, Sylvain. "Falsification et induction." Dialogue.
20 (1981) 281-307.

3708. Baillie, Patricia. "Falsifiability and Probability." AJP 48
(1970) 99-100.

     a. "Falsifiability and Probability: A Comment," by A. E.
     Musgrave. 50 (1972) 58-60.

     b. "Falsifiability and Probability," by Patricia Baillie.
     61.

3709. Balestra, D. J. "Non-Falsifiability: An Inductivist Perspective." International Logic Review. 10 (1979) 118-25.

     a. "Concerning Dudman's and Balestra's Articles," by Oscar
     Brunner. 11 (1980) 129-35.

3710. Barrett, Robert B. "On the Conclusive Falsification of
Scientific Hypotheses." PS 36 (1969) 363-74.

3711. Binns, Peter. "The Supposed Asymmetry between Falsification
and Verification." Dialectica. 32 (1978) 29-40.

3712. Dretske, Fred I. "Reasons and Falsification." Phil. Quart.
15 (1965) 20-34.

3713. Geiringer, Hilda. "Falsification and Theory of Errors."
Synthese. 5 (1946) 86-89.

3714. Giannoni, Carolo. "Quine, Grünbaum, and the Duhemian Thesis."
Nous. 1 (1967) 283-97.

3715. Giedymin, J. "A Generalization of the Refutability Postulate."
Studia Logica. 10 (1960) 97-108.

3716. Goosens, William K. "Duhem's Thesis, Observationality, and
Justification." PS 42 (1975) 286-98.

3717. Grünbaum, Adolf. "The Falsifiability of the Lorentz-Fitzgerald
Contraction Hypothesis." BJPS 10 (1959) 48-50.

     a. "Testability and 'ad-hocness' of the Contraction Hypothesis,"
     by K. R. Popper. 50.

     b. "Falsifiability of the Lorentz-Fitzgerald Contraction
     Hypothesis," by H. Dingle. 228-29.

     c. "More about Lorentz Transformation Equations," by G. H.
     Keswani. BJPS 11 (1960) 50-55.

     d. "The Falsifiability of the Lorentz-Fitzgerald Contraction
     Hypothesis: A Rejoinder to Professor Dingle," by A. Grünbaum.
     BJPS 11 (1960) 143-45.

     e. "Reply to Professor Grünbaum," by Herbert Dingle. 145.

     f. "More about Lorentz Transformation Equations," by M.
     Born. BJPS 12 (1961) 150-51.

     g. "Reply to Professor Born," by G. H. Keswani. 151-53.

     h. "Professor Dingle on Falsifiability: A Second Rejoinder,"
     by A. Grünbaum. 153-56.

     i. "A Reply to Professor Grünbaum's Rejoinder," by H. Dingle.
     156-57.

3718. _____. "The Duhemian Argument." PS 27 (1960) 75-87.

     a. "In Defense of Duhem," by Francis Seaman. 32 (1965)
     287-94.

3719. _____. "The Falsifiability of Theories: Total or
Partial? A Contemporary Evaluation of the Duhem-Quine Thesis."
Synthese. 14 (1962) 17-34. Reprinted in BSPS 1 (1963) 178-95.

     a. "Comments on A. Grünbaum's Paper," by Robert S. Cohen.
     14 (1962) 193-95.

3720. _____. "Ad hoc Auxiliary Hypotheses and Falsificationism."
BJPS 27 (1976) 329-62.

3721. _____ . "Can We Ascertain the Falsity of a Scientific Hypotheses?" in Observation and Theory in Science. E. Nagel et al. Baltimore: The Johns Hopkins Press, 1971. 69-129. Also in Studium Generale. 22 (1969) 1061-93.

3722. Harding, Sandra G., ed. Can Theories Be Refuted? Essays on the Duhem-Quine Thesis. SL 81. Dordrecht: Reidel, 1976. 318 pp.

   a. Review essay by C. A. Hooker. Metaphilosophy. 9 (1978) 58-68.

3723. Herburt, George K. "The Analytic and the Synthetic: The Duhemian Argument and Some Contemporary Philosophers." PS 26 (1959) 104-13.

3724. Hollinger, Robert. "The Philosophical Significance of the Duhemian Argument." Personalist. 59 (1978) 221-40.

3725. Juhos, Bela. "Die methodologische Symmetrie von Verifikation und Falsification." ZAW 1 (1970) 41-70.

   a. "Falsifizierbarkeit oder Falsification?" by Michael Schmid. 3 (1972) 85-87.

3726. Koertge, Noretta. "Towards a New Theory of Scientific Inquiry." BSPS 58, SL 125 (1978) 253-78.

3727. Kordig, Carl R. "Some Statements are Immune to Revision." New Scholasticism. 55 (1981) 69-76.

3728. Krause, Merton. "Disconfirmative Results and Prior Commitments." PS 31 (1964) 237-40.

3729. Laudan, Laurens. "On the Impossibility of Crucial Falsifying Experiments." PS 32 (1965) 295-99.

3730. Leplin, Jarrett. "Contextual Falsification and Scientific Methodology." PS 39 (1972) 476-90.

3731. _____ . "The Concept of an Ad Hoc Hypothesis." SHPS 5 (1975) 309-45.

3732. Levison, Arnold B. "Professor Scheffler on Falsifiability and Meaning." Phil. Studies. 16 (1965) 76-79.

3733. Mitroff, Ian I. "Systems, Inquiry, and the Meanings of Falsification." PS 40 (1973) 255-76.

3734. Peetz, D. W. "Falsification in Science." Proc. Arist. Soc. 69 n.s. (1968-69) 17-32.

3735. Quinn, Philip L. "The Status of the D-Thesis." PS 36 (1969) 381-99.

3736. _____. "What Duhem Really Meant." BSPS 14 (1969)
33-56.

    a. "Quinn on Duhem: An Emendation," by Nancy Tuana. PS 45
(1978) 456-62.

    b. "Rejoinder to Tuana," by Philip L. Quinn. 463-65.

3737. Rapp, Friedrich. "The Methodological Symmetry between Verification and Falsification." ZAW 6 (1975) 139-44. Spanish Tr.:
Dialogos. 15 (1980) 117-30.

    a. "Zur Symmetrie zwischen Verifikation und Falsifikation,"
by Klaus Eichner. 7 (1976) 119-20.

    b. "A Helpful Argument -- Reply to K. Eichner," by F. Rapp.
121-23.

    c. "Falsification, Rejection, and Modification," by Björn
Wittrock. 8 (1977) 379-82.

    d. "A Note on the Controversy between Rapp and Eichner," by
Klaus Eichner and Werner Habermehl. 9 (1978) 337-38.

    e. "Disentangling Counter-Arguments," by F. Rapp. 339-42.

3738. Schlesinger, George. "It is False That Overnight Everything
Has Doubled in Size." Phil. Studies. 15 (1964) 65-71.

    a. "Is a Universal Nocturnal Expansion Falsifiable or Physi-
cally Vacuous?" by Adolf Grünbaum. 71-79.

3739. Sharpe, R. A. "The Logical Status of Natural Laws." Inquiry.
7 (1964) 414-16.

3740. Skyrms, Brian. "Falsifiability in the Logic of Experimental
Tests." Methodos. 14 (1962) 3-13.

3741. Sklar, Lawrence. "The Falsifiability of Geometric Theories."
J. Phil. 64 (1967) 247-53.

3742. Such, J. Does Experimentum Crucis Exist? [In Polish] Warsaw:
Polish Scientific Publishers, 1975.

3743. Swanson, J. W. "On the D-Thesis." PS 34 (1967) 59-68.

3744. Swinburne, R. G. "Falsifiability of Scientific Theories."
Mind. 73 (1964) 434-36.

    a. "Falsifiability," by Richard Cole. 77 (1968) 133-35.

3745. Tournier, F. "La thèse de Duhem-Quine et l'indétermination
de la traduction." RMM 85 (1980) 503-08.

3746. Vuillemin, Jules. "On Duhem's and Quine's Theses." <u>Grazer Philosophische Studien</u>. 9 (1979) 69-96.

3747. Wedeking, Gary. "Duhem, Quine, and Grünbaum on Falsification." PS 36 (1969) 375-80.

3748. Yoshida, R. M. "Five Duhemian Theses." PS 42 (1975) 29-45.

4.16 Popperianism

For a complete listing of the publications of Karl Popper up to 1974, see T. E. Hansen, comp. "Bibliography of the Writings of Karl Popper," in <u>The Philosophy of Karl Popper</u>. P. A. Schilpp, ed. La Salle, IL: Open Court, 1974. 1201-87.

3749. Ackermann, Robert John. <u>The Philosophy of Karl Popper</u>. Amherst, MA: University of Massachusetts Press, 1976. x + 212 pp.

    a. Review essay, "Ackermann on Popper," by S. Cannavo. <u>International Studies in Philosophy</u>. 11 (1979) 141-45.

3750. Agassi, Joseph. "Empiricism and Inductivism." <u>Phil. Studies</u>. 14 (1963) 85-86.

3751. _____. "Science in Flux: Footnotes to Popper." BSPS 3 (1967) 293-323.

    a. "Comments," by Judith Jarvis Thomson. 324-30.

3752. _____. "The Novelty of Popper's Philosophy of Science." IPQ 8 (1968) 442-63.

3753. _____. "Popper on Learning from Experience," in <u>Studies in the Philosophy of Science</u>. N. Rescher, ed. Oxford: Basil Blackwell, 1969. 162-70.

3754. _____. "Modified Conventionalism is More Comprehensive than Modified Essentialism," in <u>The Philosophy of Karl Popper</u>. P. A. Schilpp, ed. La Salle, Illinois: Open Court, 1974. 693-96.

3755. _____. "Wissenschaft und Metaphysik." <u>Grazer Philosophische Studien</u>. 9 (1979) 97-106.

3756. _____. "To Save Verisimilitude." <u>Mind</u>. 90 (1981) 576-79.

3757. Andersson, Gunnar. "The Problem of Verisimilitude." BSPS 58, SL 125 (1978) 291-310.

3758. Antiseri, Dario. <u>Karl R. Popper: Epistemologia e società aperta</u>. Rome: Armando, 1972. 329 pp.

3759. _____. "Il criterio di falsificabilità alle origini della biologia sperimentale." Proteus. 9 (1972) 51-101.

3760. Archibald, G. C. "Refutation or Comparison?" BJPS 17 (1967) 279-96.

3761. Ayer, A. J. "Truth, Verification and Verisimilitude," in The Philosophy of Karl Popper. P. A. Schilpp, ed. La Salle, Illinois: Open Court, 1974. 684-92.

3762. Ball, Terence. "Popper's Psychologism." Philosophy of the Social Sciences. 11 (1981) 65-67.

3763. Baltas, A., and K. Gavroglu. "A Modification of Popper's Tetradic Schema and the Special Relativity Theory." ZAW 11 (1980) 213-37.

3764. Bar-Hillel, Y. "Popper's Theory of Corroboration," in The Philosophy of Karl Popper. P. A. Schilpp, ed. La Salle, Illinois: Open Court, 1974. 332-48.

3765. Bartley, III, W. W. "A Note on Barker's Discussion of Popper's Theory of Corroboration." Phil. Studies. 12 (1961) 5-10.

3766. _____. "Achilles, the Tortiose, and Explanation in Science and History." BJPS 13 (1962) 15-33.

3767. Bernardini, S. Logica della conoscenza scientifica. Secondo la teoria di Karl R. Popper. Napoli. 1980.

3768. Bernsen, Niels Ole. "Karl Popper's Improvements on Empiricism and the Justification of Knowledge." Danish Yearbook of Philosophy. 11 (1974) 7-23.

3769. Bhattacharya, Nikhil. "Popper's Theory of Rationality in Science." Southern Journal of Philosophy. 16 (1978) 139-54.

3770. Bonnor, W. B. "Instrumentalism and Relativity." BJPS 8 (1958) 291-94.

3771. Boon, Louis. "Repeated Tests and Repeated Testing: How to Corroborate Low Level Hypotheses." ZAW 10 (1979) 1-10.

     a. "Bemerkung zur Popper-Diskussion," by Hans Merkens. 11 (1980) 162-63.

3772. Bouveresse, Renée. Karl Popper, ou Le rationalisme critique. Paris: Vrin, 1978. 191 pp.

3773. Braun, Günther E. "Empirischer Gehalt und Falsifizierbarkeit. Eine semiotische Analyse des Popper-Kriteriums." ZAW 6 (1975) 203-16.

3774. Brown, Harold I. "Objective Knowledge in Science and the Humanities." Diogenes. 97 (1977) 85-104.

3775. Bunge, M., ed. The Critical Approach to Science and Philosophy: Essays in Honor of Karl R. Popper. New York: Free Press, 1964. 480 pp.

3776. Burke, T. E. "The Limits of Relativism." Phil. Quart. 29 (1979) 193-207.

3777. Campbell, Donald T. "Evolutionary Epistemology," in The Philosophy of Karl Popper. P. A. Schilpp, ed. La Salle, Illinois: Open Court, 1974. 413-63.

3778. Carr, Brian. "Popper's Third World." Phil. Quart. 27 (1977) 214-26.

3779. Chalmers, A. F. "On Learning from Our Mistakes." BJPS 24 (1973) 164-73.

3780. Chauvire, Christiane. "Vérifier ou falsifier: De Peirce à Popper." Les Etudes philosophiques. 36 (1981) 257-78.

3781. Churchland, Paul M. "Karl Popper's Philosophy of Science." Canadian Journal of Philosophy. 5 (1975) 145-56.

3782. Cohen, L. Jonathan. "Guessing." Proc. Arist. Soc. 74 n.s. (1973-74) 189-210.

3783. _____. "Some Comments on Third World Epistemology." BJPS 31 (1980) 175-80.

3784. Crittenden, P. J. "Evolutionary Epistemology: A Question of Justification." Philosophical Studies. (Ireland) 25 (1976) 228-43.

3785. Croteau, O. M. I., Jacques. "Une Épistémologie sans sujet connaissant." Science et Esprit. 29 (1977) 1-21, 209-27.

3786. Currie, Gregory. "Popper's Evolutionary Epistemology: A Critique." Synthese. 37 (1978) 413-31.

3787. Davison, R. M. "Aspects of the Soviet Response to Popper." Studies in Soviet Thought. 20 (1979) 105-25.

3788. Detel, Wolfgang. "Zwei Fallstudien zur Prüfung des Falsifikationismus." ZAW 5 (1974) 226-46.

3789. Deutscher, Max. "Popper's Problem of an Empirical Base." AJP 46 (1968) 277-88.

        a. "Deutscher's Problem is not Popper's Problem," by T. W. Settle. 47 (1969) 216-19.

        b. "What is Popper's Problem of an Empirical Basis?" by Max Deutscher. 354-55.

3790. Donagan, Alan. "Popper's Examination of Historicism," in The Philosophy of Karl Popper. P. A. Schilpp, ed. La Salle, Illinois: Open Court, 1974. 905-24.

3791. Doumit, E. "Criticisme poppérien, positivisme logique et discours psychanalytique." RMM 86 (1981) 514-44.

3792. Dussen, W. J. van der. "De filosofie van Popper en de methodenstrijd." Algemeen Nederlands Tijdschrift voor Wijsbegeerte. 63 (1971) 246-69.

3793. Ebert, Theodor. "Ueber eine vermeinliche Entdeckung in der Wissenschaftstheorie." ZAW 5 (1974) 308-16.

3794. Eccles, J. C. "The World of Objective Knowledge," in The Philosophy of Karl Popper. P. A. Schilpp, ed. La Salle, Illinois: Open Court, 1974. 349-70.

3795. Fisk, Milton. "Falsifiability and Corroboration." Philosophical Studies. (Ireland). 9 (1959) 49-65.

3796. Galeazzi, Umberto. "Scienza e ragione strumentale nella scuola di Francoforte e in K. Popper." Proteus. 13 (19740) 97-122.

3797. Gay, Hannah. "Radicals and Types: A Critical Comparison of the Methodologies of Popper and Lakatos and Their Use in the Reconstruction of Some 19th Century Chemistry." SHPS 7 (1976) 1-51.

3798. Godlovitch, S. "Universal, Basic and Instantial Statements in the Logic of Scientific Discovery." PS 20 (1969) 355-56.

3799. Good, Irving John. "Explicativity, Corroboration, and the Relative Odds of Hypotheses." Synthese. 30 (1975) 39-74.

  a. "Comments on I. J. Good," by William L. Harper. 75-78.

  b. "Good's Compromise: Comments on I. J. Good," by John R. Wettersten. 79-82.

  c. "Replies," by I. J. Good. 83-94.

3800. Gray, Bennison. "Popper and the 7th Approximation: The Problem of Taxonomy." Dialectica. 34 (1980) 129-53.

3801. Grove, J. W. "Popper 'Demystified': The Curious Ideas of Bloor (and some others) about World 3." Philosophy of the Social Sciences. 10 (1980) 173-80.

3802. Grünbaum, Adolf. "Is the Method of Bold Conjectures and Attempted Refutations Justifiably the Method of Science?" BJPS 27 (1976) 105-36.

3803. _____. "Can a Theory Answer more Questions than one of its Rivals?" BJPS 27 (1976) 1-23.

3804. _____. "Is Falsifiability the Touchstone of Scientific Rationality? Karl Popper versus Inductivism." BSPS 39 (1976) 213-52.

3805. _____. "Is Psychoanalysis a Pseudo-Science? Karl Popper versus Sigmund Freud." ZPF 31 (1977) 333-53; 32 (1978) 49-69.

3806. _____. "Popper vs. Inductivism." BSPS 58, SL 125 (1978) 117-42.

3807. _____. "Is Freudian Psychoanalytic Theory Pseudo-Scientific by Karl Popper's Criterion of Demarcation?" Am. Phil. Quart. 16 (1979) 131-41.

3808. Grünfeld, J. "Science and Philosophy in Popper." Sciences Ecclesiastiques. 19 (1967) 231-54.

3809. _____. "Popper's Logic." Science et Esprit. 31 (1979) 361-66.

3810. Haack, Susan. "Epistemology with a Knowing Subject." Review of Metaphysics. 33 (1979) 309-35.

3811. Hammerton, M. "Bayesian Statistics and Popper's Epistemology." Mind. 77 (1968) 109-12.

3812. Harré, R. "Counter-induction." Theoria. 29 (1963) 245-64.

3813. Harris, Errol E. "Epicyclic Popperism." BJPS 23 (1972) 55-67.

3814. Harris, John H. "Popper's Definitions of Verisimilitude." BJPS 25 (1974) 160-66.

3815. Harsanyi, John C. "Popper's Improbability Criterion for the Choice of Scientific Hypotheses." Philosophy. 35 (1960) 332-40.

3816. Hooker, C. A. "Formalist Rationality: The Limitations of Popper's Theory of Reason." Metaphilosophy. 12 (1981) 248-66.

3817. Hübner, Kurt. "Some Critical Comments on Current Popperianism on the Basis of a Theory of System Sets." BSPS 58, SL 125 (1978) 279-89.

3818. Jeffrey, Richard C. "Probability and Falsification: Critique of the Popper Program." Synthese. 30 (1975) 95-118.

    a. "Popper and the Non-Bayesian Tradition: Comments on Richard Jeffrey," by Ronald N. Giere. 119-32.

    b. "Comments on Ronald Giere," by I. J. Good. 133-34.

    c. "Comments on Richard Jeffrey," by Terry M. Goode. 135-38.

    d. "Making Sense of Method: Comments on Richard Jeffrey," by David Miller. 139-48.

e. "Replies," by Richard C. Jeffrey.   149-58.

3819. Johansson, Ingvar. A Critique of Karl Popper's Methodology.
Stockholm: Scandinavian University Books, 1975. 210 pp.

3820. _____. "Ceteris paribus Clauses, Closure Clauses and
Falsifiability." ZAW 11 (1980) 16-22.

3821. Jones, Gary E. "Popper and Theory Appraisal." SHPS 9 (1978)
239-49.

3822. Jones, K. E. "Verisimilitude versus Probable Verisimilitude."
BJPS 24 (1973) 174-76.

3823. Juffras, Angelo. "Popper and Dewey on Rationality." Journal
of Critical Analysis. 4 (1972-73) 96-103.

3824. Jürgen, Klaus. "Sind empirische Theorien falsifizierbar?"
ZAW 10 (1979) 11-27.

3825. Kamino, Keiichiro. "On Popper's Notion of Verisimilitude."
AJAPS 6, No. 1 (1981) 1-18.

3826. Kassiola, Joel. "Cognitive Relativism, Popper, and the Logic
of Objectivism." Philosophical Forum. 6 (1975) 366-79.

3827. Keene, G. G. "Confirmation and Corroboration." Mind. 70
(1961) 85-87.

3828. Kelly, Derek A. "Popper's Ontology: An Exposition and Critique." Southern Journal of Philosophy. 13 (1975) 71-82.

3829. Keuth, Herbert. "Objective Knowledge out of Ignorance:
Popper on Body, Mind, and the Third World." Theory and Decision. 5
(1974) 391-412.

3830. _____. "Verisimilitude or the Approach to the Whole
Truth." PS 43 (1976) 311-36.

3831. _____. "Methodologische Regeln des kritischen Rationalismus. Eine Kritik." ZAW 9 (1978) 236-55.

3832. Kirk, G. S. "Popper on Science and the Presocratics." Mind.
69 (1960) 318-39.

a. "Kirk on Heraclitus, and on Fire as the Cause of Balance,"
by Karl R. Popper. 72 (1963) 386-92.

3833. Klemke, E. D. "Karl Popper, Objective Knowledge, and the
Third World." Philosophia. 9 (1979) 45-62.

3834. _____. "Popper's Criticisms of Wittgenstein's Tractatus."
Midwest Studies in Philosophy. 6 (1981) 239-61.

3835. Klowski, Joachim. "Der unaufhebbare Primat der Logik, die
Dialektik des Ganzen und die Grenze der Logik." ZAW 4 (1973) 41-53.

3836. _____. "Lässt sich eine Kernlogik konstituieren? Ein
Versuch dazu vom Standpunkt des pankritischen Rationalismus aus."
ZAW 4 (1973) 303-12.

3837. Kneale, William C. "The Demarcation of Science," in The
Philosophy of Karl Popper. P. A. Schilpp, ed. La Salle, Illinois:
Open Court, 1974. 205-17.

3838. Koertge, Noretta. "The Methodological Status of Popper's
Rationality Principle." Theory and Decision. 10 (1979) 83-95.

3839. _____. "The Problem of Appraising Scientific Theories,"
in Current Research in Philosophy of Science. P. D. Asquith and H.
E. Kyburg, Jr., eds. East Lansing, MI: Philosphy of Science Assn.,
1979. 228-51.

3840. Kraft, Victor. "Popper and the Vienna Circle," in The Philosophy of Karl Popper. P. A. Schilpp, ed. La Salle, Illinois: Open
Court, 1974. 185-204.

3841. Krah, Wolfgang. "Zum Falsifikationprinzip in der Wissenschaftstheorie." ZAW 5 (1974) 304-07.

3842. Ladrière, Jean. "Déterminisme et liberté. Nouvelle position
d'un ancien probleme: le modèle de Popper." Revue Philosophique de
Louvain. 65 (1967) 467-96.

3843. Lakatos, Imre. "Popper on Demarcation and Induction," in
The Philosophy of Karl Popper. P. A. Schilpp, ed. La Salle, Illinois:
Open Court, 1974. 241-73.

3844. _____. "Science and Pseudoscience." Conceptus. 8, No.
24 (1974) 5-9.

3845. Langtry, Bruce. "Popper on Induction and Independence." PS
44 (1977) 326-31.

3846. Ławniczak, W. "Anti-Individualism, Scientific Discovery, and
the Third World." Poznań Studies. 5 (1979) 233-41.

3847. Lee, K. K. "Popper's Falsifiability and Darwin's Natural
Selection." Philosophy. 44 (1969) 291-302.

3848. Lejewski, Czesław. "Popper's Theory of Formal or Deductive
Inference," in The Philosophy of Karl Popper. P. A. Schilpp, ed.
La Salle, Illinois: Open Court, 1974. 632-70.

3849. Levi, I. "Corroboration and Rules of Acceptance." BJPS 13
(1963) 307-13.

     a. "Estimated Utility and Corroboration," by Alex C. Michalos.
     16 (1966) 327-31.

3850. Levison, Arnold. "Popper, Hume, and the Traditional Problem of Induction," in The Philosophy of Karl Popper. P. A. Schilpp, ed. La Salle, Illinois: Open Court, 1974. 322-31.

3851. Lloyd, G. E. R. "Popper versus Kirk: A Controversy in the Interpretation of Greek Science." BJPS 18 (1967) 21-38.

3852. Magee, Bryan. Karl Popper. New York: Viking Press, 1973. 115 pp.

3853. Malherbe, J.-F. La philosophie de Karl Popper et le positivisme logique. Namur: Presses Universitaires de Namur; Paris: Presses Universitaires de France, 1976.

    a. Review essay by Philippe van Parijs. Revue Philosophique de Louvain. 76 (1978) 359-70.

3854. _____. "Karl Popper et Claude Bernard." Dialectica. 35 (1981) 373-88.

3855. Mardiros, Anthony M. "Karl Popper as Social Philosopher." Canadian Journal of Philosophy. 5 (1975) 157-71.

3856. Margenau, Henry. "On Popper's Philosophy of Science," in The Philosophy of Karl Popper. P. A. Schilpp, ed. La Salle, Illinois: Open Court, 1974. 750-59.

3857. Maxwell, Grover. "Corroboration without Demarcation," in The Philosophy of Karl Popper. P. A. Schilpp, ed. La Salle, Illinois: Open Court, 1974. 292-321.

3858. Maxwell, Nicholas. "A Critique of Popper's Views on Scientific Method." PS 39 (1972) 131-52.

3859. Meana, Luis. "Inducción, refutación y demarcación en K. R. Popper. Apuntes críticos." Pensamiento. 34 (1978) 25-45.

3860. Medawar, Peter. "Hypothesis and Imagination," in The Philosophy of Karl Popper. P. A. Schilpp, ed. La Salle, Illinois: Open Court, 1974. 274-91.

3861. Meiland, Jack W. "Cognitive Relativism: Popper and the Argument from Language." Philosophical Forum. 4 (1973) 406-19.

3862. Meyers, Robert G. "In Defense of Popper's Verisimilitude." Phil. Studies. 25 (1974) 213-18.

3863. Miller, David. "Popper's Qualitative Theory of Verisimilitude." BJPS 25 (1974) 166-77.

3864. _____. "On the Comparison of False Theories by Their Bases." BJPS 25 (1974) 178-88.

3865. _____. "The Accuracy of Predictions." Synthese. 30 (1975) 159-92.

a. "Truthlikeness: Comment on David Miller," by Roger D. Rosenkrantz. 193-98.

b. "Verisimilitude: Comment on David Miller," by Joseph Agassi. 199-204.

c. "Comments on David Miller," by I. J. Good. 205-06.

d. "The Accuracy of Predictions: A Reply," by David Miller. 207-20.

e. "Accuracy of Prediction: A Note on David Miller's Problem," by Deryck Horton. BJPS 29 (1978) 179-83.

3866. Mitroff, Ian I., Frederick Betz, and Richard O. Mason. "A Mathematical Model of Churchmanian Inquiring Systems with Special Reference to Popper's Measures for 'The Severity of Tests'." Theory and Decision. 1 (1970) 155-78.

3867. Morgenbesser, Sidney. "Psychologism and Methodological Individualism," in Philosophy of Science Today. S. Morgenbesser, ed. New York: Basic Books, 1967. 160-74.

3868. Mott, Peter L. "Verisimilitude by Means of Short Theorems." Synthese. 38 (1978) 247-74.

3869. Muguerza, Javier. "Tres Fronteras de la Ciencia." Dialogos. 7, No. 18 (1970) 45-87.

3870. Mulkay, Michael. "Putting Philosophy to Work: Karl Popper's Influence on Scientific Practice." Philosophy of the Social Sciences. 11 (1981) 389-407.

3871. Musgrave, Alan E. "Falsification and Its Critics." LMPS-4 (1971) 393-406.

3872. _____. "The Objectivism of Popper's Epistemology," in The Philosophy of Karl Popper. P. A. Schilpp, ed. La Salle, Illinois: Open Court, 1974. 560-96.

3873. _____. "Popper and Diminishing Returns from Repeated Tests." AJP 53 (1975) 248-53.

3874. Mussachia, M. Mark. "Some Comments on Scientific Historical Predictability and Karl Popper's Refutation of Its Possibility." International Studies in Philosophy. 9 (1979) 85-92; 11 (1979) 147-48.

3875. Narskii, I. S. "The Philosophy of the Late Karl Popper." Soviet Studies in Philosophy 18 (1980) 53-77. [English reprint from Filosofskie nauki. (1979, No. 4).]

3876. Nerlich, G. C., and W. A. Suchting. "Popper on Law and Natural Necessity." BJPS 18 (1967) 233-35.

a. "A Revised Definition of Natural Necessity," by Karl R. Popper. 18 (1968) 316-21.

b. "Popper's Revised Definition of Natural Necessity," by W. A. Suchting. 20 (1969) 349-52.

3877. Niiniluoto, Ilkka. "Notes on Popper as Follower of Whewell and Peirce." Ajatus. 37 (1978) 272-327.

3878. Nuzzaci, Francesco. Karl Popper: Un epistemologo fallibilista. Napoli: Glaux, 1975. 262 pp.

3879. Oddie, Graham. "Verisimilitude and Distance in Logical Space." Acta Philosophica Fennica. 30, 2-4 (1978) 227-42.

3880. _____. "Verisimilitude Reviewed." BJPS 32 (1981) 237-65.

3881. Oetjens, Hermann. Sprache, Logik, Wirklichkeit: Der Zusammenhang von Theorie und Erfahrung in K. R. Poppers Logik der Forschung. Stuttgart: Frommann-Holzboog, 1975. 184 pp.

3882. O'Hear, Anthony. "Rationality of Action and Theory-Testing in Popper." Mind. 84 (1975) 273-76.

3883. _____. Karl Popper. London: Routledge & Kegan Paul, 1980. 219 pp.

3884. Olding, A. "A Defence of Evolutionary Laws." BJPS 29 (1978) 131-43.

3885. Pahler, Klaus. "Teststrenge und empirische Bewährung in der Popperianischen Wissenschaftstheorie." ZAW 12 (1981) 98-109.

3886. Pandit, G. L. "Two Concepts of Psychologism." Phil. Studies 22 (1971) 85-91.

3887. Passmore, J. A. "Popper's Account of Scientific Method." Philosophy. 35 (1960) 326-31.

3888. Pera, M. Popper e la scienza su palafitte. Bari. 1981.

3889. Perry, Clifton. "Popper, Winch, and Individualism." IPQ 20 (1980) 59-71.

3890. Popper, Karl R. "Logic Without Assumptions." Proc. Arist. Soc. 47 n.s. (1946-47) 251-92.

3891. _____. "Why are the Calculuses of Logic and Arithmetic Applicable to Reality?" in Aristotelian Society, Supplementary Volume XX: Logic and Reality. London: Harrison and Sons, 1946, 40-60. Reprinted in Conjectures and Refutations. New York: Basic Books; London: Routledge & Kegan Paul, 1962; New York: Harper & Row, 1968. 201-14.

3892. _____. "Naturgesetze und theoretische Systeme," in
Gesetz und Wirklichkeit. Simon Moser, ed. Innsbruck: Tyrolia
Verlag, 1949. 43-60. Expanded English version: "The Bucket and
The Searchlight: Two Theories of Knowledge," in K. R. Popper,
Objective Knowledge. Oxford: Clarendon Press, 1972. 341-61.

3893. _____. "A Note on Berkeley as Precursor of Mach." BJPS
4 (1953) 26-36. Reprinted under title "A Note on Berkeley as Precursor
of Mach and Einstein," in Karl R. Popper, Conjectures and Refutations,
Chapter 6.

3894. _____. "Three Views Concerning Human Knowledge," in
Contemporary British Philosophy: Personal Statements. Third Series.
H. D. Lewis, ed. London: George Allen & Unwin; New York: Macmillan,
1956. 355-88. Reprinted in Conjectures and Refutations. New York:
Basic Books; London: Routledge & Kegan Paul, 1962; New York:
Harper & Row, 1968. 97-119. Reprinted in The Foundations of Knowledge.
C. Landesman, ed. Englewood Cliffs, NJ: Prentice-Hall, 1970.
93-123.

3895. _____. The Poverty of Historicism. London: Routledge
& Kegan Paul, 1957. xiv + 166 pp.

     a. "Critical Study," by Patrick Gardiner. Phil. Quart. 9
     (1959) 172-80.

3896. _____. "Philosophy of Science: A Personal Report," in
British Philosophy in the Mid-Century: A Cambridge Symposium. C.
A. Mace, ed. London: George Allen and Unwin, 1957. 155-91.
Reprinted under the title "Science: Conjectures and Refutations,"
in Conjectures and Refutations, New York: Basic Books; London:
Routledge & Kegan Paul, 1962; New York: Harper & Row, 1968. 33-65.

3897. _____. "The Aim of Science." Ratio. 1 (1957) 24-35.
Revised version in K. R. Popper, Objective Knowledge. Oxford:
Clarendon Press, 1972. 191-205.

3898. _____. "On the Status of Science and of Metaphysics."
Ratio 1-2 (1957-58) 97-115. Reprinted in Conjectures and Refutations.
New York: Basic Books; London: Routledge & Kegan Paul, 1962; New
York: Harper and Row, 1968. 184-200.

3899. _____. "Back to the Pre-Socratics." Proc. Arist. Soc.
59 n.s. (1958-59) 1-24.

3900. _____. The Logic of Scientific Discovery. New York:
Basic Books, 1959. 480 pp. [Translation of Logik der Forschung.
Vienna: 1934.]

     a. Review essay, "The Impact of Logik der Forschung," by H.
     Bondi and C. W. Kilmister. BJPS 10 (1959) 55-57.

     b. Review essay by D. Stove. AJP 38 (1960) 173-87.

c. "The Role of Corroboration in Popper's Methodology," by
Joseph Agassi. 39 (1961) 82-91.

d. "On the Severity of Tests," by P. C. Gibbons. 40 (1962)
79-82.

3901. _____. "Some Comments on Truth and the Growth of Knowledge." LMPS-1 (1960) 285-92.

3902. _____. Conjectures and Refutations: The Growth of
Scientific Knowledge. New York: Basic Books; London: Routledge &
Kegan Paul, 1962. Reprint: New York: Harper & Row, 1968. 417 pp.

a. Review essay by Peter Achinstein. BJPS 19 (1968) 159-68.

3903. _____. "The Demarcation between Science and Metaphysics,"
in The Philosophy of Rudolf Carnap. P. A. Schilpp, ed. La Salle,
Illinois: Open Court, 1963. 183-226. Reprinted in Conjectures and
Refutations. New York: Basic Books; London: Routledge & Kegal
Paul, 1962; New York: Harper & Row, 1968. 253-92.

3904. _____. "Epistemology Without a Knowing Subject."
LMPS-3 (1967) 333-73.

3905. _____. "A Realist View of Logic, Physics, and History,"
in Physics, Logic, and History. W. Yourgrau and A. D. Breck, eds.
New York: Plenum Press, 1970. 1-30. Reprinted in Karl R. Popper,
Objective Knowledge: An Evolutionary Approach. Oxford: Clarendon
Press, 1972. 285-318.

3906. _____. "On the Theory of Objective Mind," in Akten des
XIV. Internationalen Kongresses für Philosophie. Vol. I. Vienna:
Verlag Herder, 1968. 25-53. Reprinted with additions in K. R.
Popper, Objective Knowledge. Oxford: Clarendon Press, 1972.
153-90.

3907. _____. Objective Knowledge: An Evolutionary Approach.
Oxford: Clarendon Press, 1972. 380 pp.

a. Review essay, "Popper's Objective Knowledge," by Michael
Krausz. Dialogue. 13 (1974) 347-51.

b. Review essay by Bernulf Kanitscheider. ZAW 4 (1973)
388-98.

c. Review essay by Joseph Agassi. Philosphia. 4 (1974)
163-201.

d. Review essay by Mendel Sachs. Philosophy of the Social
Sciences. 4 (1974) 399-408.

e. Review essay by Paul Feyerabend. Inquiry. 17 (1974)
475-507.

f. Review essay by [S.] Magala. Erkenntnis. 9 (1975) 245-51.

g. Review essay by Harald Pilot. Philosophische Rundschau. 22 (1976) 42-48.

3908. _____. Unended Quest: An Intellectual Autobiography. London: Fontana/Collins, 1976. 255 pp.

3909. _____. "Induction, Deduction, Objective Truth." Methodology and Science. 9 (1976) 163-73.

3910. _____. "Natural Selection and the Emergence of Mind." Dialectica. 32 (1978) 339-56.

a. "Reply to Popper's Attack on Epiphenomenalism," by Gerhard D. Wassermann. Mind. 88 (1979) 572-75.

3911. _____. Die bieden Grundprobleme der Erkenntnistheorie. Hrg. v. T. E. Hansen. Tübingen: J. C. B. Mohr (Paul Siebeck), 1979. xxxv + 476 pp.

a. Review essay by Werner Flach. Philosophische Rundschau. 28 (1981) 84-100.

3912. _____. "The Present Significance of Two Arguments of Poincaré." Methodology and Science. 14 (1981) 260-64.

3913. Post, H. R. "A Criticism of Popper's Theory of Simplicity." BJPS 12 (1962) 328-31.

3914. Putnam, Hilary. "The 'Corroboration of Theories'," in The Philosophy of Karl Popper. P. A. Schilpp, ed. La Salle, Illinois: Open Court, 1974. 221-40.

3915. Quine, W. V. "On Popper's Negative Methodology," in The Philosophy of Karl Popper. P. A. Schilpp, ed. La Salle, Illinois: Open Court, 1974. 218-20.

3916. Radner, Michael. "Popper and Laplace." MSPS 4 (1970) 417-27.

3917. Radnitzky, Gérard. "Popperian Philosophy of Science as an Antidote Against Relativism." BSPS 39 (1976) 505-46.

3918. _____. "Popperian Image of Science." AJAPS 5, No. 1 (1976) 3-19.

3919. _____. "Tres estilos de pensar en la actual teoría de la ciencia. Sus creadores: Wittgenstein I, Popper y Wittgenstein II." Pensamiento. 35 (1979) 5-35.

3920. _____. "Méthodologie poppérienne et recherche scientifique." Archives de Philosophie. 42 (1979) 3-40, 295-325.

3921. _____. "From Justifying a Theory to Comparing Theories and Selecting Questions." RIP 131-132 (1980) 179-228.

3922. _____. "Progress and Rationality in Research: Science from the Viewpoint of Popperian Methodology." BSPS 34 (1981) 43-102.

3923. Radnitzky, Gerard, and Gunnar Andersson. "Objective Criteria of Scientific Progress? Inductivism, Falsificationism, and Relativism." BSPS 58, SL 125 (1978) 3-19.

3924. Robbins, J. C. "Salmon and Red Herring: Does Corroboration Entail Induction?" Telos. 1 (1968) 27-33.

3925. Robinson, G. S. "Popper's Verisimilitude." Analysis. 31 (1970-71) 194-96.

3926. Rodriques, João Resina. "Science, méthodologie et philosophie chez K. Popper." Revue Philosophique de Louvain. 70 (1972) 240-74.

3927. Rossi, Arcangelo. Popper e la filosofia della scienza. Firenze: Sansoni, 1975. 120 pp.

3928. Rothbart, Daniel. "Popper against Inductivism." Dialectica. 34 (1980) 121-28.

3929. Ruse, Michael. "Karl Popper's Philosophy of Biology." PS 44 (1977) 638-61.

3930. Sachs, Mendel. "Popper and Reality." Synthese. 33 (1976) 355-69.

3931. Sarkar, Husain. "Popper's Principle of Transference: A Conjecture Refuted." Southern Journal of Philosophy. 16 (1978) 363-71.

3932. _____. "Putnam's Schemata." Southwestern Journal of Philosophy. 10, No. 1 (1979) 125-37.

3933. _____. "Popper's Third Requirement for the Growth of Knowledge." Southern Journal of Philosophy. 19 (1981) 489-97.

3934. Schäfer, L. "Ueber die Diskrepanz zwischen Methodologie und Metaphysik bei Popper." Studium Generale. 23 (1970) 856-77.

3935. Schilpp, Paul A., ed. The Philosophy of Karl Popper. 2 Vols. The Library of Living Philosophers, Vol. 14. La Salle, Illinois: Open Court, 1974. 1323 pp.

    a. Review essay by W. W. Bartley, III. Philosophia. 6 (1976) 463-94; 7 (1978) 675-716.

    b. Review essay, "The Popper Phenomenon," by D. H. Mellor. Philosophy. 52 (1977) 195-202.

    c. Review essay, "Popper in Perspective," by John Kekes. Metaphilosophy. 8 (1977) 36-61.

d. Review essay, "Failures in Criticism: Popper and His Commentators," by John Mackie. BJPS 29 (1978) 363-75.

e. Review essay by Klaus Pähler. ZAW 9 (1978) 413-23.

3936. Schlesinger, G. "Popper on Self-Reference," in The Philosophy of Karl Popper. P. A. Schilpp, ed. La Salle, Illinois: Open Court, 1974. 671-83.

3937. Settle, Tom W. "Is Corroboration a Non-demonstrative Form of Inference?" Ratio. 12 (1970) 151-54.

3938. _____. "Induction and Probability Unfused," in The Philosophy of Karl Popper. P. A. Schilpp, ed. La Salle, Illinois: Open Court, 1974. 697-749.

3939. Skagestad, Peter. Making Sense of History: The Philosophies of Popper and Collingwood. Oslo: Universitetsforlaget, 1975. 117 pp.

3940. Stich, Stephen P. "Between Chomskian Rationalism and Popperian Empiricism." BJPS 30 (1979) 329-47.

a. "Popperian Language - Acquisition Undefeated," by Geoffrey Sampson. 31 (1980) 63-67.

b. "Can Popperians Learn to Talk?" by Stephen P. Stich. 32 (1981) 157-64.

3941. Stove, D. C. "Popper on Scientific Statements." Philosophy 53 (1978) 81-88.

a. "Stove on Popper's Scientific Statements," by Michael Rowan and Alan Smithson. 55 (1980) 258-62.

3942. Swinburne, R. G. "Popper's Account of Acceptability." AJP 49 (1971) 167-76.

3943. Thakur, S. C. "Popper on Scientific Method." Philosophical Studies. 19 (1970) 71-82.

3944. Thyssen-Rutten, Nicole. "Le concept de simplicité dans la philosophie des sciences de K. Popper." Logique et analyse. 12 (1969) 179-88.

3945. _____. "A la recherche d'un modèle a propos de Of Clouds and Clocks de K. R. Popper." RIP 24 (1979) 117-23.

3946. Tibbetts, Paul. "Popper's Critique of the Instrumentalist Account of Theories and Theoretical Terms: Some Misunderstandings." Southern Journal of Philosophy. 10 (1972) 57-70.

3947. Tichý, Pavel. "On Poppers' Definitions of Verisimilitude." BJPS 25 (1974) 155-60.

a. "A Note on Verisimilitude," by Karl Popper. 27 (1976) 147-59.

3948. _____. "Verisimilitude Redefined." BJPS 27 (1976) 25-42.

a. "Verisimilitude Redeflated," by David Miller. 363-81.

3949. _____. "Verisimilitude Revisited." Synthese. 38 (1978) 175-96.

3950. Tuomela, Raimo. "Theory-Distance and Verisimilitude." Synthese. 38 (1978) 213-46.

3951. Urbach, Peter. "Is Any of Popper's Arguments Against Historicism Valid." BJPS 29 (1978) 117-30.

3952. Vetter, Hermann. "Inductivism and Falsificationism Reconcilable." Synthese. 23 (1971) 226-33.

3953. _____. "A New Concept of Verisimilitude." Theory and Decision. 8 (1977) 369-75.

3954. Vincent, R. H. "Popper on Qualitative Confirmation and Disconfirmation." AJP 40 (1962) 159-66.

3955. _____. "Corroboration and Probability." Dialogue. 2 (1963) 194-205.

3956. Watkins, John W. N. "When are Statements Empirical?" BJPS 10 (1960) 287-308.

3957. _____. "Corroboration and the Problem of Content- Comparison." BSPS 58, SL 125 (1978) 339-78.

a. "The Gong Show -- Popperian Style," by Paul Feyerabend. 387-92.

b. "Reply to Watkins," by Kurt Hübner. 393-96.

3958. _____. "The Popperian Approach to Scientific Knowledge." BSPS 58, SL 125 (1978) 23-43.

3959. _____. "The Unity of Popper's Thought," in The Philosophy of Karl Popper. P. A. Schilpp, ed. La Salle, Illinois: Open Court, 1974. 371-412.

3960. Wettersten, John R. "Traditional Rationality vs. a Tradition of Criticism: A Criticism of Popper's Theory of the Objectivity of Science." Erkenntnis. 12 (1978) 329-38.

3961. Wiebe, Don. "Comprehensively Critical Rationalism and Commitment." Philosophical Studies (Ireland) 21 (1972) 186-201.

3962. Winch, Peter. "Popper on Scientific Method in the Social Sciences," in The Philosophy of Karl Popper. P. A. Schilpp, ed. La Salle, Illinois: Open Court, 1974. 889-904.

3963. Wisdom, J. O. "The Refutability of 'Irrefutable' Laws." BJPS 13 (1963) 303-06.

3964. _____. "The Nature of 'Normal Science'," in The Philosophy of Karl Popper. P. A. Schilpp, ed. La Salle, Illinois: Open Court, 1974. 820-42.

4.17 Operationalism

3965. Ballard, Edward G. "Operational Definitions and Theory of Measurement." Methodos. 5 (1953) 233-39.

    a. "Comments on the Paper of Edward G. Ballard," by P. W. Bridgman. 240-41.

    b. "Operazionismo e tecnica operativa," by S. Ceccato e V. Somenzi. 242-46. [Translation: 247-49].

3966. Benjamin, A. Cornelius. "Operationalism -- A Critical Evaluation." J. Phil. 47 (1950) 439-44.

    a. "Professor Benjamin on Bridgman -- A Rejoinder," by Joseph Turner, Jr. 774-77.

3967. _____. Operationalism. Springfield, Illinois: Charles C. Thomas, 1955. 154 pp.

3968. Bernstein, Jeremy. "P. W. Bridgman, In Revolt Against Formalism." Synthese. 8 (1950-51) 331-41.

3969. Bradley, John. "On the Operational Interpretation of Classical Chemistry." BJPS 6 (1955) 32-42.

3970. Bridgman, P. W. "Some Implications of Recent Points of View in Physics." RIP 3 (1949) 479-501.

3971. _____. Reflections of a Physicist. New York: Philosophical Library, 1950. xii + 392 pp.

3972. _____. "Philosophical Implications of Physics." Bulletin, The American Academy of Arts and Sciences. 3 (1950) 2-6.

3973. _____. "The Operational Aspect of Meaning." Synthese. 8 (1950-51) 251-59.

3974. _____. "Some Philosophical Aspects of Science." Synthese. 10 (n.d.) 318-26.

3975. _____. "Einstein's Theories and the Operational Point of View," in Albert Einstein: Philosopher-Scientist. P. A. Schilpp, ed. New York: Tudor, 1951. 333-54.

3976. _____. The Nature of Some of Our Physical Concepts. New York: Philosophical Library, 1952. 64 pp. Reprinted from BJPS 1 (1951) 257-72; 2 (1951) 25-44, 142-60.

3977. _____. "Quo Vadis?" Daedalus. 87 (1958) 85-93.

3978. _____. The Way Things Are. Harvard University Press, 1959. x + 333 pp.

3979. Curi, Umberto. Analisi operazionale e operazionalismo. Padova: Cedam-Casa Editrice Dott. Antonio Milani, 1970. 242 pp.

3980. Destouches, J.-L. "Descriptions opérationnelles en physique moderne." Synthese. 10 (n.d.) 59-64.

3981. Dingler, H. "Empiricismus und operationalismus: Die beiden Wissenschaftslehren E-Lehre und O-Lehre in ihrem Verhältnis." Dialectica. 7 (1952) 343-76.

3982. Galanti, E. "Sulla interpretazione neoempiristica dell'operazionismo di Bridgman." Epistemologia. 4 (1981) 457-72.

3983. Gillies, D. A. "Operationalism." Synthese. 25 (1972) 1-24.

3984. Hempel, Carl G. "A Logical Appraisal of Operationalism." Scientific Monthly. 79 (1954) 215-20. Reprinted in slightly modified form in C. G. Hempel, Aspects of Scientific Explanation and Other Essays in the Philosophy of Science. New York: Free Press, 1965. 123-33.

3985. Klüver, Jürgen. Operationalismus: Kritik und Geschichte einer Philosophie der exakten Wissenschaft. Stuttgart-Bad Cannstatt: Freidrich Frommann Verlag, 1971. 220 pp.

3986. Palter, Robert. "Operations and the Occult." PS 23 (1956) 297-314.

3987. Pfannenstill, Bertil. "A Critical Analysis of Operational Definitions." Theoria. 17 (1951) 193-209.

3988. Rothstein, Jerome. "Information and Organization as the Language of the Operational Viewpoint." PS 29 (1962) 406-11.

3989. Schlesinger, G. "P. W. Bridgman's Operational Analysis: The Differential Aspect." BJPS 9 (1959) 299-306.

3990. Somenzi, Vittorio. "An Exemplification of 'Operational Methodology'." Synthese. 9 (n.d.) 15-25.

3991. Székely, D. L. "Begriffsbildung und Operationalismus." Synthese. 10 (n.d.) 65-70.

3992. Wilson, Fred. "Is Operationalism Unjust to Temperature?"
Synthese. 18 (1968) 394-422.

4.18 Measurement

3993. Adams, Ernest. "Elements of a Theory of Inexact Measurement."
PS 32 (1965) 205-28.

3994. _____. "On the Nature and Purpose of Measurement."
Synthese. 16 (1966) 125-69.

3995. _____. "Measurement Theory," in Current Research in
Philosophy of Science. P. D. Asquith and H. E. Kyburg, Jr., eds.
East Lansing, MI: Philosophy of Science Assn., 1979. 207-27.

3996. Agassi, Joseph. "Precision in Theory and in Measurement."
PS 35 (1968) 287-90.

3997. Ballard, Edward G. "The Paradox of Measurement." PS 16
(1949) 134-36.

3998. Benardete, José A. "Continuity and the Theory of Measurement."
J. Phil. 65 (1968) 411-31.

        a. "Measurement and Mathematics," by Henry E. Kyburg, Jr.
        66 (1969) 29-43.

3999. Inadvertently skipped in numbering the entries.

4000. Berka, K. Measurement: Its Concepts, Theories and Problems.
BSPS 72. Dordrecht: Reidel, 1982. 256 pp.

4001. Böhme, Gernot. "Quantifizierung--Metrisierung. Versuch
einer Unterscheidung erkenntnistheoretischer und wissenschaftstheo-
retischer Momente im Prozess der Bildung von quantitativen Begriffen."
ZAW 7 (1976) 209-22.

4002. Bory, C. "Le continu et sa représentation dans les sciences."
RMM 62 (1957) 134-56.

4003. Byerly, Henry C. "Realist Foundations of Measurement." BSPS
20, SL 64 (1974) 375-84.

4004. Byerly, Henry, and Vincent A. Lazara. "Realist Foundations
of Measurement." PS 40 (1973) 10-27.

4005. Cartwright, Helen Morris. "Amounts and Measures of Amounts."
Nous. 9 (1975) 143-64.

4006. Causey, Robert L. "Derived Measurement, Dimensions, and Dimensional Analysis." PS 36 (1969) 252-70.

4007. Christian, C. "Die Bedeutung der Mengentheorie als Grund-lagenwissenschaft." PN 16 (1976-77) 238-72.

4008. Churchman, C. West. "Measurement: A Systems Approach." WOSPS 1 (1973) 70-86.

    a. "Comments," by Isaac Levi, 87-94, and by Ronald Giere, 95-97.

4009. Churchman, C. West, and Philburn Ratoosh, eds. Measurement: Definitions and Theories. New York: John Wiley, 1959.

4010. Colonius, Hans. "On a Weak Extensive Measurement." PS 45 (1978) 303-08.

4011. Cyranski, John F. "Measurement Theory for Physics." FP 9 (1979) 641-71.

4012. Dingle, Herbert. "A Theory of Measurement." BJPS 1 (1950) 5-26.

4013. Domotor, Zoltan. "Species of Measurement Structures." Theoria. 38 (1972) 64-81.

4014. Duistermaat, J. J. "Energy and Entropy as Real Morphisms for Addition and Order." Synthese. 18 (1968) 327-93.

4015. Durand III, Loyal. "On the Theory of Measurement in Quantum Mechanical Systems." PS 27 (1960) 115-33.

4016. Elkana, Yehuda, et al, eds. Toward a Metric of Science: The Advent of Science Indicators. New York: John Wiley, 1978. xiv + 354 pp.

4017. Ellis, Brian. "Some Fundamental Problems of Direct Measure-ment." AJP 38 (1960) 37-47.

4018. _____. "Some Fundamental Problems of Indirect Measure-ment." AJP 39 (1961) 13-29.

4019. _____. "Derived Measurement, Universal Constants, and the Expression of Numerical Laws," in Philosophy of Science: The Delaware Seminar, II. New York: Interscience Publishers, 1963. 371-92.

4020. _____. "On the Nature of Dimensions." PS 31 (1964) 357-80.

4021. _____. Basic Concepts of Measurement. Cambridge: Cambridge University Press, 1966. x + 220 pp.

    a. Review essay by J. E. McGechie. AJP 44 (1966) 353-70.

4022. Falmagne, Jean-Claude. "A Set of Independent Axioms for Positive Hölder Systems." PS 42 (1975) 137-51.

4023. _____. "A Probablistic Theory of Extensive Measurement."
PS 47 (1980) 277-96.

4024. Feyerabend, P. K. "On the Quantum-Theory of Measurement," in
Observation and Interpretation. S. Körner, ed. New York: Academic
Press; London: Butterworths, 1957. 121-30.

4025. Fine, Arthur. "The Two Problems of Quantum Measurement."
LMPS-4 (1971) 567-81.

4026. Garstens, Martin A. "Measurement Theory and Biology," in
Biology, History, and Natural Philosophy. A. D. Breck and W. Yourgrau, eds. New York: Plenum, 1972. 123-33.

4027. Gerard, R. W. "Quantification in Biology." Isis. 52 (1961)
334-52.

4028. Ghose, A. "Observations, Measurements and Mathematics."
International Logic Review. 9 (1978) 61-68.

4029. Gonseth, F. "A propos de la mesure de temps. L'aspect
méthodologique du problème de la précision." RMM 67 (1962) 133-41.

4030. Graves, John C., and James E. Roper. "Measuring Measuring
Rods." PS 32 (1965) 39-56.

4031. Grunstra, Bernard R. "On Distinguishing Types of Measurement."
BSPS 5 (1969) 253-303.

4032. Gudder, Stanley. "Generalized Measure Theory." FP 3 (1973)
399-411.

4033. _____. "A Generalized Measure and Probability Theory
for the Physical Sciences." WOSPS 6-III (1976) 121-39.

4034. Guerlac, Henry. "Quantification in Chemistry." Isis. 52
(1961) 194-214.

4035. Holman, Eric W. "Extensive Measurement Without an Order
Relation." PS 41 (1974) 361-73.

4036. Hoppe, Hans-Hermann. "On How Not to Make Inferences about
Measurement Error." Quality and Quantity. 14 (1980) 503-10.

4037. Jones, R. V. "Some Limits of Measurement." Philosophical
Journal. 1 (1964) 3-19.

4038. Jordan, P. "On the Process of Measurement in Quantum Mechanics."
PS 16 (1949) 269-78.

4039. Kanger, Stig. "Measurement: An Essay in Philosophy of
Science." Theoria. 38 (1972) 1-44.

4040. Katz, Michael. "Łukasiewicz Logic and the Foundatons of
Measurement." Studia Logica. 40 (1981) 209-25.

4041. König, H. "Zur Analyse des Größenbegriffs." Dialectica. 11 (1957) 70-87.

4042. Körner, Stephan. "On Empirical Continuity." Monist. 47 (1962) 1-19.

4043. Krantz, David H. "Extensive Measurement in Semiorders." PS 34 (1967) 348-62.

4044. _____. "Fundamental Measurement of Force and Newton's First and Second Laws of Motion." PS 40 (1973) 481-95.

4045. Kuhn, Thomas S. "The Function of Measurement in Modern Physical Science." Isis. 52 (1961) 161-90. Reprinted in T. S. Kuhn, The Essential Tension. Chicago: University of Chicago Press, 1977. 178-224.

4046. Kyburg, Jr., Henry E. "Direct Measurement." Am. Phil. Quart. 16 (1979) 259-72.

4047. Leverett, Hollis M. "The Mind We Measure and Its Dimensions." PS 15 (1948) 39-46.

4048. Lorenzen, P. "Zur Definition der vier fundamentalen Messgrössen." PN 16 (1976-77) 1-9.

4049. Luce, R. Duncan. "A 'Fundamental' Axiomatization of Multiplicative Power among Three Variables." PS 32 (1965) 301-09.

4050. _____. "Similar Systems and Dimensionally Invariant Laws." PS 38 (1971) 157-68.

4051. _____. "Conjoint Measurement." WOSPS 13-I (1978) 311-36.

4052. _____. "Dimensionally Invariant Numerical Laws Correspond to Meaningful Qualitative Relations." PS 45 (1978) 1-16.

4053. _____. "Suppes' Contributions to the Theory of Measurement," in Patrick Suppes. Radu J. Bogdan, ed. Dordrecht: Reidel, 1979. 93-110.

4054. Luce, R. Duncan, and A. A. J. Marley. "Extensive Measurement When Concatenation is Restricted and Maximal Elements may Exist," in Philosophy, Science, and Method. S. Morgenbesser et al, eds. New York: St. Martin's Press, 1969. 235-49.

4055. McKnight, John L. "The Quantum Theoretical Concept of Measurement." PS 24 (1957) 321-30.

4056. _____. "An Extended Latency Interpretation of Quantum Mechanical Measurement." PS 25 (1958) 209-22.

4057. Margenau, Henry. "Philosophical Problems Concerning the Meaning of Measurement in Physics." PS 25 (1958) 23-34.

4058. _____. "Measurements and Quantum States." PS 30 (1963) 1-16; 138-57.

4059. Marley, A. A. J. "An Alternative 'Fundamental' Axiomatization of Multiplicative Power Relations among Three Variables." PS 35 (1968) 185-86.

4060. _____. "Additive Conjoint Measurement with Respect to a Pair of Orderings." PS 37 (1970) 215-22.

4061. Menger, Karl. "Variables, Constants, Fluents," in Current Issues in the Philosophy of Science. H. Feigl and G. Maxwell, eds. New York: Holt, Rinehart and Winston, 1961. 304-13.

    a. "Comments," by Ernest W. Adams. 313-16.

    b. "Rejoinder," by Karl Menger. 316-18.

4062. Moulyn, Adrian C. "The Functions of Point and Line in Time Measuring Operations." PS 19 (1952) 141-55.

4063. Narens, Louis. "Measurement Without Archimedean Axioms." PS 41 (1974) 374-93.

4064. _____. "A General Theory of Ratio Scalability with Remarks about the Measurement - Theoretic Concept of Meaningfulness." Theory and Decision. 13 (1981) 1-70.

4065. Nelson, Thomas M., and S. Howard Bartley. "Numerosity, Number, Arithmetization, Measurement and Psychology." PS 28 (1961) 178-203.

4066. Osborne, Dale K. "Unified Theory of Derived Measurement." Synthese. 33 (1976) 455-81.

4067. _____. "On Dimensional Invariance." Quality and Quantity. 12 (1978) 75-89.

4068. Park, James L. "Quantum Theoretical Concepts of Measurement." PS 35 (1968) 205-31; 389-411.

4069. Pawson, Ray. "Empiricist Measurement Strategies: A Critique of the Multiple Indicator Approach to Measurement." Quality and Quantity. 14 (1980) 651-78.

4070. Pfanzagl, J. Die axiomatischen Grundlagen einer allegmeinen Theorie des Messens. Würzburg: Physica-Verlag, 1959. 63 pp.

4071. Phipps, Jr., T. E. "The Relativity of Physical Size." Dialectica. 23 (1969) 189-216.

4072. Plochmann, George Kimball. "Is Quantity Prior to Quality?" PS 21 (1954) 62-67.

a. "Comments on Professor Plochmann's 'Is Quantity Prior to Quality?'," by Thomas Storer. 68-73.

4073. Prugovečki, Eduard. "A Postulational Framework for Theories of Simultaneous Measurement of Several Observables." FP 3 (1973) 3-18.

4074. Przełecki, M. "Some Approach to Inexact Measurement." Poznań Studies. 4 (1978) 27-36.

4075. Ramsay, J. O. "Algebraic Representation in the Physical and Behavioral Sciences." Synthese. 33 (1976) 419-53.

4076. Rescher, Nicholas. "A Problem in the Theory of Numerical Estimation." Synthese. 12 (1960) 34-39.

4077. Roberts, Fred S. "On Luce's Theory of Meaningfulness." PS 47 (1980) 424-33.

4078. Roberts, Fred S., and R. Duncan Luce. "Axiomatic Thermodynamics and Extensive Measurement." Synthese. 18 (1968) 311-26.

4079. Robinson, Richard. "Measurement and Statistics: Towards a Clarification of the Theory of 'Permissible Statistics'." PS 32 (1965) 229-43.

4080. Rozeboom, William W. "Scaling Theory and the Nature of Measurement." Synthese. 16 (1966) 170-233.

4081. Rusk, George Y. "General Mensurational Gestaltism." PS 16 (1949) 250-59.

4082. Russo, F. "Mesure et objectivité." Archives de Philosophie. 21 (1958) 92-105.

4083. Sachs, Mendel. "On the Elementarity of Measurement in General Relativity: Toward a General Theory." BSPS 3 (1967) 56-80.

4084. Schleichert, Hubert. "Zur Erkenntnislogik des Messens." Archiv fur Philosophie. 12 (1964) 304-27.

4085. Scott, Dana, and Patrick Suppes. "Foundational Aspects of Theories of Measurement." Journal of Symbolic Logic. 23 (1958) 113-28.

4086. Smart, J. J. C. "Measurement." AJP 37 (1959) 1-13.

4087. Suppes, Patrick. "Finite Equal-interval Measurement Structures." Theoria. 38 (1972) 45-63.

4088. Suppes, Patrick, and Mario Zanotti. "Necessary and Sufficient Conditions for Existence of a Unique Measure Strictly Agreeing with a Qualitative Probability Ordering." Journal of Philosophical Logic. 5 (1976) 431-38.

4089. Süssmann, G. "An Analysis of Measurement," in Observation and Interpretation. S. Körner, ed. New York: Academic Press; London: Butterworths, 1957. 131-36.

4090. Thompson, J. W. "Polarity in the Social Sciences and in Physics." PS 35 (1968) 190-94.

4091. Titiev, Robert J. "Multidimensional Measurement and Universal Axiomatizability." Theoria. 38 (1972) 82-88.

4092. Tondl, Ladislav. "Prerequisites for Quantification in the Empirical Sciences." Theory and Decision. 2 (1972) 238-61.

4093. Tuchańska, B. "Factor versus Magnitude." Poznań Studies. 2, No. 3 (1976) 29-39.

4094. _____. "An Idealizational View on Measurement and Indicator-Based Reasoning." Poznań Studies. 3 (1977) 213-34.

4095. Vuillemin, Jules. "Mesure, vérification, langage." Logique et analyse. 11 (1968) 183-227.

4096. Wallace, O. P., William A. "The Measurement and Definition of Sensible Qualities." New Scholasticism. 39 (1965) 1-25.

4097. Wilks, S. S. "Some Aspects of Quantification in Science." Isis. 52 (1961) 135-42.

5.
Philosophical Issues Concerning Science

5.1 Empiricism and the A Priori

4098. Agassi, Joseph. "Sensationalism." Mind. 75 (1966) 1-24.

    a. "On the Use of Historical Examples in Agassi's 'Sensationalism', by T. A. Beckman. SHPS 1 (1970-71) 293-309.

    b. "Agassi's Alleged Arbitrariness," by Joseph Agassi. 2 (1971) 157-65.

4099. Ajdukiewicz, Kazimierz. "Logic and Experience." Synthese. 8 (1950-51) 289-99.

4100. Alston, William P. "Are Positivists Metaphysicians." Phil. Review. 63 (1954) 43-57.

4101. Anton, Peter. "The Status of the Empirical Principle." AJAPS 2, No. 4 (1964) 21-26.

4102. Aster, Ernst von. "Physikalistischer und psychologistischer Positivismus." Theoria. 16 (1950) 1-20.

4103. Ayer, A. J., ed. Logical Positivism. Glencoe, Illinois: Free Press, 1959. 455 pp.

4104. Barzin, Marcel. "L'empirisme logique." RIP 4 (1950) 84-94.

4105. Berenda, Carlton W. "A Five-Fold Scepticism in Logical Empiricism." PS 17 (1950) 123-32.

4106. Bergmann, Gustav. "Sense Data, Linguistic Conventions, and Existence." PS 14 (1947) 152-63.

4107. _____. "Two Cornerstones of Empiricism." Synthese. 8 (1950-51) 435-52.

4108. _____ . The Metaphysics of Logical Positivism. New
York: Longmans, Green and Co., 1954. 341 pp.

4109. Bohnen, Alfred. "On the Critique of Modern Empiricism: Observational Language, Observational Facts and Theories." Ratio 11
(1969) 38-57.

4110. Cerf, Walter. "Logical Positivism and Existentialism." PS
18 (1951) 327-38.

4111. Cohen, L. Jonathan. "How Empirical is Contemporary Logical
Empiricism?" Philosophia. 5 (1975) 299-317. Reprinted in BSPS 43
(1976) 359-76.

4112. Cohen, Robert S. "Dialectical Materialism and Carnap's
Logical Empiricism," in The Philosophy of Rudolf Carnap. P. A.
Schilpp, ed. La Salle, Illinois: Open Court, 1963. 99-158.

4113. Cowan, T. A., and C. W. Churchman. "On the Meaningfulness of
Questions." PS 13 (1946) 20-24.

4114. Crahay, F. "Logique formelle, empiricisme logique et vérité."
Dialectica. 8 (1954) 347-56.

4115. Creary, Lewis G. "Empiricism and Rationality." Synthese.
23 (1971) 234-65.

4116. Crockett, Campbell. "The Short and Puzzling Life of Logical
Positivism." Modern Schoolman. 31 (1954) 85-92.

4117. Darden, Lindley. "The Heritage from Logical Positivism: A
Reassessment." PSA 1976. II, 242-58.

4118. Demos, Raphael. "Doubts about Empiricism." PS 14 (1947)
203-18.

4119. Dettering, Richard W. "Linguistic Superfluity in Science."
PS 26 (1959) 347-54.

4120. Dingler, Hugo. "Probleme des Positivismus." ZPF 5 (1950)
485-513, 6 (1951) 235-57.

        a. "Dinglers 'methodische Philosophie' und der Neopositivis-
mus," by Victor Kraft. 8 (1954) 259-66.

4121. Dinneen, S.J., John A. "The Course of Logical Positivism."
Modern Schoolman. 34 (1956) 1-21.

4122. Eberle, Rolf A. "A Construction of Quality Classes Improved
upon the Aufbau." SL 73 (1975) 55-74.

4123. Feibleman, James K. "The Metaphysics of Logical Positivism."
Review of Metaphysics. 5 (1951) 55-82.

4124. _____. "Philosophical Empiricism from the Scientific Standpoint." Dialectica. 16 (1962) 5-14.

4125. Feigl, Herbert. "Philosophical Tangents of Science," in Current Issues in the Philosophy of Science. H. Feigl and G. Maxwell, eds. New York: Holt, Rinehart and Winston, 1961. 1-17.

4126. _____. The Power of Positivistic Thinking." Proc. & Add. Am. Phil. Assn. 36 (1962-63) 21-41.

4127. _____. "Empiricism at Bay?: Revisions and a New Defense." BSPS 14, SL 60 (1974) 1-20.

4128. Feyerabend, P. K. "How to be a Good Empiricist -- A Plea for Tolerance in Matters Epistemological," in Philosophy of Science: The Delaware Seminar, II. New York: Interscience Publishers, 1963. 3-39.

4129. _____. "Science Without Experience." J. Phil. 66 (1969) 791-95.

4130. Fiala, F. "Essai sur les notions d'ouverture et de fermeture." Dialectica. 1 (1947) 147-58.

4131. Frank, Philipp. "Logical Empiricism I: The Problem of Physical Reality." Synthese. 7 (1948-49) 458-64.

4132. _____. "Introduction to the Philosophy of Physical Science, on the Basis of Logical Empiricism." Synthese. 8 (1950-51) 28-45.

4133. _____. "Einstein, Mach, and Logical Positivism," in Albert Einstein: Philosopher-Scientist. P. A. Schilpp, ed. New York: Tudor, 1951. 269-86.

4134. _____. "The Pragmatic Components in Carnap's 'Elimination of Metaphysics'," in The Philosophy of Rudolf Carnap. P. A. Schilpp, ed. La Salle, Illinois: Open Court, 1963. 159-64.

4135. Franquiz, José A. "Logical Empiricism as a Philosophy of Science." Philosophical Forum. 3 (1945) 2-5.

4136. Galtung, Johan. "Empiricism, Criticism, Constructivism." Synthese. 24 (1972) 343-72.

4137. Goodman, Nelson. "The Significance of Der logische Aufbau der Welt," in The Philosophy of Rudolf Carnap. P. A. Schilpp, ed. La Salle, Illinois: Open Court, 1963. 545-58.

4138. Gonseth, F. "L'ouverture à l'expérience et les a priori." Dialectica. 9 (1955) 5-22.

4139. Hempel, Carl G. "Rudolf Carnap, Logical Empiricist." SL 73 (1975) 1-14. Reprinted from Synthese. 25 (1973) 256-68.

4140. Hennemann, Gerhard. "Zur Ueberwindung des Positivismus in der Physik." PN 5 (1958-59) 503-13.

4141. Hubbeling, H. G. Language, Logic and Criterion: A Defence of Non-Positivistic Logical Empiricism. Amsterdam: Born N.V., 1971. 128 pp.

4142. Joad, C. E. M. A Critique of Logical Positivism. Chicago: University of Chicago Press, 1950. 154 pp.

4143. Johnson, Oliver A. "Denial of the Synthetic A Priori." Philosophy. 35 (1960) 255-64.

     a. "Reply," by E. Gavin Reeve. 36 (1961) 371.

     b. "Observing and Disconfirming Propositions: A Reply," by Oliver A. Johnson. 37 (1962) 163-64.

4144. Jörgensen, Jörgen. Den logiske Empirismes Udvikling. Copenhagen, 1948.

4145. _____. The Development of Logical Empiricism. International Encyclopedia of Unified Science, II, 9. Chicago: University of Chicago Press, 1951. 100 pp.

4146. _____. "On Empiric and A Priori Knowledge." Danish Yearbook of Philosophy. 6 (1969) 72-88.

4147. Juhos, Bela. "Die erkenntnisanalytische Methode." ZPF 6 (1951) 42-53.

4148. Kazemier, Brugt H. "Remarks on Logical Positivism." Synthese. 5 (1946) 327-31.

4149. Kegley, Charles W. "Reflections on Philipp Frank's Philosophy of Science." PS 26 (1959) 35-40.

     a. "Frank's Philosophy of Science Revisited," by F. James Rutherford. 27 (1960) 183-86.

4150. Kraft, Victor. "Logik und Erfahrung." Theoria. 12 (1946) 205-10.

4151. _____. "Konstruktiver Empirismus." ZAW 4 (1973) 313-22.

4152. Kretzmann, Norman. "Empiricism and the Theory of Meaning." Phil. Quart. 6 (1956) 236-44.

4153. Kutschera, F. von. "Induction and the Empiricist Model of Knowledge." LMPS-4 (1971) 345-56.

4154. Lambros, Charles H. "Schlick's Doctrine of the A priori in Allegmeine Erkenntnislehre." Dialectica. 28 (1974) 103-28.

4155. Leinfellner, W. "Die Konzeption der Analytizitat in wissen-schaftlichen Theorien." PN 8 (1964) 397-417.

4156. McMullin, Ernan. "Empiricism at Sea." BSPS 14, SL 60 (1974) 21-32.

4157. Miller, David L. "Recent Speculations in the Positivistic Movement." Review of Metaphysics. 12 (1959) 462-74.

4158. Morick, Harold, ed. Challenges to Empiricism. Belmont, CA: Wadsworth, 1972. 329 pp.

4159. Morris, Charles W. "Scientific Empiricism," in International Encyclopedia of Unified Science. Vol. I, No. 1. O. Neurath, R. Carnap, and C. W. Morris, eds. Chicago: University of Chicago Press, 1938-55. 63-75.

4160. _____. "Pragmatism and Logical Empiricism," in The Philosophy of Rudolf Carnap. P. A. Schilpp, ed. La Salle, Illinois: Open Court, 1963. 87-98.

4161. Naess, Arne. Wie fördert man heute die empirische Bewegung? Eine Auseinandersetzung mit dem Empirismus von Otto Neurath und Rudolf Carnap. Oslo: Universitetsforlaget, 1956. 48 pp.

4162. Nagel, Ernest. "Naturalism Reconsidered." Proc. & Add. Am. Phil. Assn. 28 (1954-55) 5-17.

4163. Pagano, F. "A priori e a posteriori nella scienza fisica." Scientia. 80 (1946) 43-47. French Tr.: Supplement, 21-25.

4164. Palmieri, L. E. "Pragmatism and the Ideal Language." PS 27 (1960) 271-78.

4165. Pap, Arthur. The A Priori in Physical Theory. New York: King's Crown Press; London: G. Cumberlege, 1946. x + 102 pp.

4166. _____. Elements of Analytic Philosophy. New York: Macmillan, 1949. 514 pp.

4167. _____. Analytische Erkenntnistheorie: kritische Ubersicht über die neueste Entwicklung in USA und England. Vienna: Springer-Verlag, 1955. vi + 242 pp.

    a. Review essay by W. Stegmüller. Philosophische Rundschau. 6 (1958) 35-54.

4168. _____. "Logical Empiricism and Rationalism." Dialectica. 10 (1956) 148-66.

4169. _____. "Nominalism, Empiricism, and Universals." Phil. Quart. 9 (1959) 330-40; 10 (1960) 44-60.

4170. Pasch, Alan. "Empiricism: One 'Dogma' or Two?" J. Phil. 53 (1956) 302-10.

4171. _____. Experience and the Analytic. Chicago: University of Chicago Press, 1958. 275 pp.

4172. Piaget, Jean, and Bärbel Inhelder. "The Gaps in Empiricism," in Beyond Reductionism. A. Koestler and J. R. Symthies, eds. New York: Macmillan, 1970. 118-60.

4173. Priest, G. G. "Two Dogmas of Quineanism." Phil. Quart. 29 (1979) 289-301.

    a. "A Plea for Model Theory," by Chris Mortensen. 31 (1981) 152-57.

4174. Putnam, Ruth Anna. "On Empirical Knowledge." BSPS 4 (1969) 392-410.

    a. "Comment," by John Compton. 411-18.

4175. Quine, W. V. "Two Dogmas of Empiricism." Phil. Review. 60 (1951) 20-43.

4176. Reichenbach, Hans. "Rationalism and Empiricism: An Inquiry into the Roots of Philosophical Error." Proc. & Add. Am. Phil. Assn. 21 (1947-48) 330-46. Reprinted in Phil. Review. 57 (1948) 330-46.

4177. Richman, Robert J. "Why are Synthetic A Priori Judgments Necessary?" Theoria. 30 (1964) 5-20.

4178. Rose, Mary Carman. "The Epistemological Structure of Empiricism." Proc. Am. Cath. Phil. Assn. 41 (1967) 196-204.

4179. Russell, Bertrand. "Logical Positivism." RIP 4 (1950) 3-19.

4180. Rynin, David. "Vindication of L*g*c*l P*s*t*v*sm." Proc. & Add. Am. Phil. Assn. 30 (1956-57) 45-67.

4181. Schächter, Josef. "Ueber das Verstehen." Synthese. 8 (1950-51) 367-84.

4182. Scheffler, Israel. "Prospects of a Modest Empiricism." Review of Metaphysics. 10 (1957) 383-400; 602-25.

4183. Schlick, M. "Positivism and Realism." Synthese. 7 (1948-49) 478-505.

    a. "Remarks on M. Schlick's Essay 'Positivism and Realism'," by D. Rynin. 466-77.

4184. _____. Philosophy of Nature. A. von Zeppelin, tr. New York: Philosophical Library, 1949. 136 pp.

4185. Schultzer, Bent. "Empiricism as a Logical Problem." Theoria. 15 (1949) 298-314.

4186. Searles, Herbert L. "Form and Content in Empirical Science." PS 18 (1951) 223-29.

4187. Sellars, Wilfrid. "Epistemology and the New Way of Words." J. Phil. 44 (1974) 645-60.

4188. _____. "Is There a Synthetic A Priori?" PS 20 (1953) 121-38.

4189. Singer, Jr., E. A. "Dialectic of the Schools." PS 21 (1954) 175-92; 297-315.

4190. Wedberg, Anders. "How Carnap Built the World in 1928." SL 73 (1975) 15-54. [Considerably revised version of paper originally in Synthese. 25 (1973) 337-71.]

5.2 Causality

4191. Agazzi E. "Time and Causality." BSPS 47 (1981) 299-321.

4192. Aronson, Jerrold. "The Legacy of Hume's Analysis of Causation." SHPS 2 (1971) 135-56.

4193. Austin, James. "Systematic Causation." SHPS 9 (1978) 83-97

4194. Avishai, Y., and H. Ekstein. "Causal Independence." FP 2 (1972) 257-70.

4195. Ballard, Edward G. "On the Nature and Use of Dialectic." PS 22 (1955) 205-13.

4196. Barr, H. J. "The Epistemology of Causality From the Point of View of Evolutionary Biology." PS 31 (1964) 286-88.

4197. Beckermann, Ansgar. "Einige Bemerkungen zur statistischen Kausalitätstheorie von P. Suppes." ZAW (1975) 292-310.

4198. Bergman, H. "The Controversy Concerning the Law of Causality in Contemporary Physics." BSPS 13 (1974) 395-462.

4199. Berteval, W. "La science moderne et la causalité." Revue Phil. 138 (1948) 180-90.

4200. Bhattacharya, B. Causality in Science and Philosophy. A Historical and Critical Survey. Calcutta: Sanskrit Pustak Bhandar, 1969. 201 pp.

4201. Birnbaum, Ian. "Statistical Inference in Causal Analysis: Some Foundations." Quality and Quantity. 13 (1979) 203-13.

4202. Brand, Myles, ed. The Nature of Causation. Urbana: University of Illinois Press, 1976. 387 pp.

4203. _____. "Causality," in Current Research in Philosophy of Science. P. D. Asquith and H. E. Kyburg, Jr., eds. East Lansing, MI: Philosophy of Science Assn., 1979. 252-81.

4204. Brand, Myles, and Marshall Swain. "On the Analysis of Causation." Synthese. 21 (1970) 222-27.

   a. "Brand and Swain on Causation," by John A. Barker. 26 (1974) 396-400.

4205. Bratoev, Georgy. "Bemerkungen über die Explikation der kausalen Begriffe." ZAW 9 (1978) 207-24.

4206. Bretzel, Philip von. "Concerning a Probabilistic Theory of Causation Adequate for the Causal Theory of Time." Synthese. 35 (1977) 173-90. Reprinted in SL 132 (1979) 385-402.

4207. Brier, Bob. Precognition and the Philosophy of Science: An Essay on Backward Causation. New York: Humanities Press, 1974. xii + 105 pp.

4208. Bunge, Mario. Causality: The Place of the Causal Principle in Modern Science. Cambridge: Harvard University Press, 1959. xx + 380 pp.

   a. Review essay: "Mario Bunge on Causality," by Richard Schlegel. PS 28 (1961) 72-82.

   b. "Causality: A Rejoinder," by Mario Bunge. PS 29 (1962) 306-17.

   c. Review essay, "Metascience," by P. K. Feyerabend. Phil. Review. 70 (1961) 396-405.

4209. Burks, Arthur W. "Dispositional Statements." PS 22 (1955) 175-93.

4210. Byerly, Henry. "Substantial Causes and Nomic Determination." PS 46 (1979) 57-81.

4211. Cartwright, Nancy. "The Reality of Causes in a World of Instrumental Laws." PSA 1980. II, 38-48.

4212. Caws, Peter. "A Quantum Theory of Causality." Synthese. 15 (1963) 317-26.

4213. Chaudhury, Prabas Jiben. "Causality and Statistical Physics." Philosophical Quarterly (India). 21 (1947) 50-55.

4214. Cole, Richard. "Causes and Explanations." Nous. 11 (1977) 347-74.

4215. Domotor, Zoltan. "Probabilistic and Causal Dependence Structures." Theory and Decision. 13 (1981) 275-92.

4216. Dretske, Fred I., and Aaron Snyder. "Causal Irregularity."
PS 39 (1972) 69-71.

     a. "On Causal Irregularity: A Reply to Dretske and Snyder,"
     by Tom L. Beauchamp. 40 (1973) 285-87.

     b. "Causality and Sufficiency: Reply to Beauchamp," by Fred
     I. Dretske and Aaron Snyder. 288-91.

     c. "A Reexamination of Causal Irregularity," by Steven
     Lauwers. 45 (1978) 471-73.

4217. Earman, John. "Causaton: A Matter of Life and Death." J.
Phil. 73 (1976) 5-25.

     a. "Is There Backward Causation in Classical Electrodynamics?"
     by Adolf Grünbaum and Allen I. Janis. 74 (1977) 475-82.

4218. Ehrenberg, W. Dice of the Gods: Causality, Necessity and
Chance. Dundee, Scotland: G. G. Stevenson, 1977. 109 pp.

4219. Fair, David. "Causation and the Flow of Energy." Erkenntnis.
14 (1979) 219-50.

     a. "A Note on Causation and the Flow of Energy," by D.
     Dieks. 16 (1981) 103-08.

4220. Fales, Walter. "Causes and Effects." PS 20 (1953) 67-74.

4221. Fetzer, James H., and Donald E. Nute. "A Probablistic Causal
Calculus: Conflicting Conceptions." Synthese. 44 (1980) 241-46.

4222. inadvertently skipped in numbering entries.

4223. Fisk, Milton. "A Defense of the Principle of Event Causality."
BJPS 18 (1967) 89-108.

4224. _____. "Capacities and Natures." BSPS 8, SL 39 (1971)
49-62.

     a. "Capacities and Natures: An Exercise in Ontology," by
     Ernan McMullin. 63-82.

     b. "Fisk on Capacities and Natures," by Bruce Aune. 83-87.

4225. Fitzgerald, Paul. "Tachyons, Backwards Causation, and Freedom."
BSPS 8, SL 39 (1971) 415-36.

4226. _____. "On Retrocausality." Philosophia. 4 (1974)
513-51.

4227. Fleury, N., and J. Leite-Lopes. "Les Tachyons et le principe
de causalité." Scientia. 109 (1974) 373-88. English Tr.: 389-402.

a. "Tachyons and Causality," by E. Recami. 721-27.

4228. Frankel, Henry. "Harré on Causation." PS 43 (1976) 560-69.

4229. Friedman, K. S. "Analysis of Causality in Terms of Determinism." Mind. 89 (1980) 544-64.

4230. Good, I. J. "A Theory of Causality." BJPS 9 (1959) 307-10.

4231. _____. "A Causal Calculus." BJPS 11 (1961) 305-18, 12 (1961) 43-51; "Errata and Corrigenda," 13 (1962) 88.

4232. Gorovitz, Samuel. "Causal Judgments and Causal Explanations." J. Phil. 62 (1965) 695-711.

a. "Explanation in Terms of 'The Cause'," by Robert K. Shope. 64 (1967) 312-20.

4233. Grene, Marjorie. "Causes." Philosophy. 38 (1963) 149-59. Reprinted in BSPS 23, SL 66 (1974) 1-12.

4234. Grünbaum, Adolf. "Is Preacceleration of Particles in Dirac's Electrodynamics a Case of Backward Causation? The Myth of Retrocausation in Classical Electrodynamics." PS 43 (1976) 165-201. [Corrections: 44 (1977) 177]. Also in Epistemologia. 1 (1978) 353-96. German Tr.: Allgemeine Zeitschrift für Philosophie. 4, No. 1 (1979) 1-39.

a. "On Grünbaum and Retrocausation in Classical Electrodynamics," by Charles Nissim-Sabat. PS 46 (1979) 118-35.

b. "Retrocausation and the Formal Assimilation of Classical Electrodynamics to Newtonian Mechanics: A Reply to Nissim-Sabat's 'On Grünbaum and Retrocausation'," by Adolf Grünbaum and Allen I. Janis. 136-60.

c. "A Reply to Grünbaum and Janis," by Charles Nissim-Sabat. 48 (1981) 127-29.

4235. Hanson, N. R. "Causal Chains." Mind. 64 (1955) 289-311.

4236. Harre, R., and E. H. Madden. "Natural Powers and Powerful Natures." Philosophy. 48 (1973) 209-30.

4237. Harre, R., and E. H. Madden. Causal Powers: A Theory of Natural Necessity. Oxford: Blackwell; Totowa, NJ: Rowman & Littlefield, 1975. 191 pp.

a. Review essay by Edward Pols. IPQ 16 (1976) 369-77.

4238. Hartmann, Max. "Die Kausalität in der Biologie." Studium Generale. 1 (1947-48) 350-56.

4239. Hedman, Carl G. "On When There Must be a Time-Difference between Cause and Effect." PS 29 (1972) 507-11.

4240. Hennemann, Gerhard. "Das Problem der Kausalität in der Physik." ZPF 22 (1968) 369-83.

4241. Henry-Hermann, Grete. "Die Kausalität in der Physik." Studium Generale. 1 (1947-48) 375-83.

4242. Hesslow, Germund. "Two Notes on the Probabilistic Approach to Causality." PS 43 (1976) 290-92.

      a. "In Defense of a Probabilistic Theory of Causality," by Deborah A. Rosen. 45 (1978) 604-13.

4243. _____. "Causality and Determination." PS 48 (1981) 591-605.

4244. Hillinger, Claude. "A Generalization of the Principle of Causality Which Makes it Applicable to Evolutionary Systems." Methodology and Science. 1 (1968) 143-47. Also in Synthese. 18 (1968) 68-74.

4245. Hörz, H. "Das Verhaltnis von Kausalität und Gesetz in der Physik." DZP 22 (1974) 954-67.

4246. Humphreys, Paul. "Probabilistic Causality and Multiple Causation." PSA 1980. II, 25-37.

4247. Hutten, E. H. "Causality and the Theory of Elementary Particles." Scientia. 97 (1962) 223-32. French Tr.: Supplement, 115-24.

4248. Kaila, Eino. Terminalkausalität als die Grundlage eines unitarischen Naturbegriffs. Helsinki: Societas Philosophica, 1956. 122 pp. Also in Acta Philosophica Fennica. 10 (1956) 7-122.

4249. Kim, Jaegwon. "Causation, Nomic Subsumption, and the Concept of Event." J. Phil. 70 (1973) 217-36.

4250. _____. "Causes as Explanations: A Critique." Theory and Decision. 13 (1981) 293-309.

4251. Kline, A. David. "Are There Cases of Simultaneous Causations?" PSA 1980. I, 292-301.

4252. Kreyche, Gerald F. "Some Causes of the Elimination of Causality in Contemporary Science." Thomist. 29 (1965) 60-78.

4253. Kron, Aleksandar. "An Analysis of Causality." SL 72 (1976) 159-82.

4254. Kuhn, Thomas S. "Les notions de causalité dans le developpement de la physique." Etudes d'épistémologie génétique. 25 (1971) 7-18. English version in T. S. Kuhn, The Essential Tension. Chicago: University of Chicago Press, 1977. 21-30.

4255. Lenzen, Victor F. Causality in Natural Science. Springfield,
Illinois: Charles S. Thomas, 1953. vi + 121 pp.

4256. Levin, Michael E. "The Extensionality of Causation and
Causal-Explanatory Contexts." PS 43 (1976) 266-77.

    a. "On the Alleged Extensionality of 'Causal Explanatory
    Contexts'," by Cindy Stern. 45 (1978) 614-25.

    b. "Lavoisier's Slow Burn," by Michael E. Levin and Margarita
    R. Levin. 626-29.

4257. Lyon, Ardon. "Causality." BJPS 18 (1967) 1-20.

4258. McWilliams, S.J., James A. "Cause in Science and Philosophy."
Modern Schoolman. 25 (1947) 11-18.

4259. Mackie, J. L. "The Direction of Causation." Phil. Review.
75 (1966) 441-66.

4260. _____. The Cement of the Universe: A Study of Causation.
Oxford: Oxford University Press, 1974. 329 pp.

    a. Review essay, "Testing the Cement: An Examination of
    Mackie on Causation," by J. A. Foster. Inquiry. 18 (1975)
    487-98.

    b. Review essay by Bernard Berofsky. J. Phil. 74 (1977)
    103-18.

    c. Review essay by Anselm Müller. Grazer Philosophische
    Studien. 3 (1977) 155-84.

    d. Review essay by Tom L. Beauchamp and Alexander Rosenberg.
    Canadian Journal of Philosophy. 7 (1977) 371-404.

4261. _____. "Dispositions, Grounds, and Causes." Synthese.
34 (1977) 361-70.

4262. Madden, Edward H. "Causality and the Notion of Necessity."
BSPS 4 (1969) 450-62.

4263. Majone, G. "Functional Relations in Causal Analysis."
Quality and Quantity. 1 (1967) 153-65.

4264. Marcus, Hugo. "Kausalität." PN 2 (1954) 418-34.

4265. Marquardt, Hans. "Kausalität in der Biologie unter dem
Aspekt der Botanik." Studium Generale. 1 (1947-48) 357-66.

4266. Maruyama, Magoroh. "Morphogenesis and Morphostasis." Methodos.
12 (1960) 251-96.

4267. Maxwell, Nicholas. "Can There by Necessary Connections between Successive Events?" BJPS 19 (1968) 1-25.

4268. Mercier, André. "Note sur l'interpretation de la causalité en physique moderne." Synthese. 11 (1959) 153-59.

4269. Mertz, Donald W. "On Galileo's Method of Causal Proportionality." SHPS 11 (1980) 229-42.

4270. Metz, A. "Causalité scientifique et cause première." Archives de Philosophie. 24 (1961) 517-41.

4271. Meurers, Joseph. "Das Problem der Kausalität im Bereich der grossen Massen und Räume." PN 4 (1957) 209-22.

4272. Morrison, Paul G. "On Partial Identity of Cause and Effect." BJPS 11 (1960) 42-49.

4273. Mittasch, Paul Alwin. "Ueber Kausalität in der Chemie." Studium Generale. 1 (1947-48) 366-75.

4274. _____. "Akausalität." PN 1 (1952) 577-88.

4275. Moyal, J. E. "Causality, Determinism, and Probability." Philosophy. 24 (1949) 310-17.

4276. Nerlich, Graham. "A Problem about Sufficient Conditions." BJPS 22 (1971) 161-70.

4277. Novak, Stefan. "Some Problems of Causal Interpretation of Statistical Relationships." PS 27 (1960) 23-38.

4278. Otte, Richard. "A Critique of Suppes' Theory of Probabilistic Causality." Synthese. 48 (1981) 167-89.

4279. Poncelet, G. "Le principe de causalité: Hume et Heisenberg." International Logic Review. 1 (1970) 167-75.

4280. Puigrefagut, Ramón. "La causalidad en los escritos de Max Planck." Pensamiento. 7 (1951) 321-54.

4281. Rosenberg, Alexander. "Causation and Counterfactuals: Lewis' Treatment Reconsidered." Dialogue. 18 (1979) 209-19.

4282. Ruddick, William. "Causal Connection." Synthese. 18 (1968) 46-67. Reprinted in BSPS 4 (1969) 419-41.

4283. Salmon, Wesley C. "An 'At-At' Theory of Causal Influence." PS 44 (1977) 215-24.

4284. _____. "Causality: Production and Propagation." PSA 1980. II, 49-69.

4285. _____. "Probabilistic Causality." Pacific Philosophical Quarterly. 61 (1980) 50-74.

   a. "Some Comments on Probabilistic Causality," by I. J. Good. 301-04.

b. "Cutting the Causal Chain," by Paul Humphreys. 305-14.

c. "A Further Comment on Probabilistic Causality: Mending the Chain," by I. J. Good. 452-54.

4286. Sayre, Kenneth M. "Statistical Models of Causal Relations." PS 44 (1977) 203-14.

a. "Sayre's Statistical Model of Causal Relations," by Douglas Shrader, Jr. 45 (1978) 630-32.

b. "Masking and Causal Relatedness: An Elucidation," by Kenneth Sayre. 633-37.

4287. Schlick, Moritz. Gesetze, Kausalität und Wahrscheinlichkeit. Vienna: Gerold, 1948. 116 pp.

4288. _____. "Causality in Contemporary Physics." BJPS 12 (1961-1962) 177-93, 281-98.

a. "Schlick's Treatment of Determinism," by H. C. Plaut. BJPS 13 (1963) 315-16.

4289. Scriven, Michael. "Causation as Explanation." Nous. 9 (1975) 3-16.

a. "Scriven on Causation as Explanation," by Ernest Sosa. Theory and Decision. 13 (1981) 357-61.

4290. Simon, Herbert A. "On the Definition of the Causal Relation." J. Phil. 49 (1952) 517-28.

a. "Some Remarks on an Analysis of the Causal Relation," by Nicholas Rescher. 51 (1954) 239-41.

4291. Simon, Herbert A., and Nicholas Rescher. "Cause and Counter-factual." PS 33 (1966) 323-40.

4292. Skotnicky, J. "Die Bedeutung des Kausalitätsprinzips und siene thermodynamische Formulierung." PN 12 (1970) 70-79.

4293. _____. "Ueber Kausalität." PN 16 (1976-77) 161-66.

4294. Sosa, Ernest. "Varieties of Causation." Grazer Philosophische Studien. 11 (1980) 93-103.

4295. Suppes, Patrick. A Probabilistic Theory of Causality. Amsterdam: North-Holland, 1970. 130 pp. Also in Acta Philosophica Fennica. 24 (1970) 1-130.

a. Review essay, in Hebrew, by Daniel Amit. Iyyun. 25 (1974) 83-94.

4296. Vandamme, F. "Synthesis, Realism and Causality." Logique et analyse. 18 (1975) 215-34.

4297. Wallace, William A. Causality and Scientific Explanation:
Vol. I: Mediaeval and Early Classical Science. Vol. II: Classical
and Contemporary Science. Ann Arbor: University of Michigan Press,
1972, 1974. 288. 422 pp. Reprinted by University Press of America,
1981.

4298. Weizsäcker, Carl Friedrich von. "Beitrag zur Diskussion über
Kausalität." Studium Generale. 2 (1949) 126-29.

4299. Wilkie, J. S. "The Problem of the Temporal Relation of Cause
and Effect." BJPS 1 (1950) 211-29.

4300. Wisdom, John O. Causation and the Foundations of Science.
Paris: Hermann, 1946. 54 pp.

4301. Wojciechowski, J. A. "La Notion de cause dans la physique
contemporaine." Dialogue. 1 (1962) 81-92.

4302. Wright, G. H. von. "On the Logic and Epistemology of the
Causal Relation." LMPS-4 (1971) 293-312.

5.3 Mechanism and Teleology

4303. Achinstein, Peter. "Function Statements." PS 44 (1977)
341-67. Reprinted in The Philosopher's Annual. 1 (1978) 1-32.

4304. Ackermann, Robert. "Mechanism and the Philosophy of Biology."
Southern Journal of Philosophy. 6 (1968) 143-51.

4305. _____. "Mechanism, Methodology, and Biological Theory."
Synthese. 20 (1969) 219-29.

4306. Ackoff, Russell L., and Fred E. Emery. On Purposeful Systems.
Chicago and New York: Aldine-Atherton Press, 1972. vii + 288 pp.

4307. Allan, D. Maurice. "Towards a Natural Teleology." J. Phil.
49 (1952) 449-59.

4308. Atlan, Henri. "Life, Physico-Chemistry and Organization."
[In Hebrew] Iyyun. 26 (1975) 207-17.

4309. Ayala, Francisco J. "Teleological Explanations in Evolutionary
Biology." PS 37 (1970) 1-15.

4310. Baublys, Kenneth K. "Comments on Some Recent Analyses of
Functional Statements in Biology." PS 42 (1975) 469-86.

4311. Beckner, Morton. "Function and Teleology." BSPS 27 (1976)
197-212.

4312. Bell, Charles G. "Mechanistic Replacement of Purpose in
Biology." PS 15 (1948) 47-51.

4313. Benton, E. "Vitalism in Nineteenth-Century Scientific Thought: A Typology and Reassessment." SHPS 5 (1974) 17-48.

4314. Birch, Charles. "Chance, Necessity and Purpose," in Studies in the Philosophy of Biology. F. J. Ayala and T. Dobzhansky, eds. Berkeley: University of California Press, 1974. 225-39.

4315. Brandon, Robert N. "Biological Teleology: Questions and Explanations." Studies in History and Philosophy of Science. 12 (1981) 91-105.

4316. Brown, Robert. "Dispositional and Teleological Statements." Phil. Studies. 3 (1952) 73-80.

4317. Canfield, John. "Teleological Explanation in Biology." BJPS 14 (1964) 285-95.

        a. "Teleological Explanation in Biology," by Hugh S. Lehman. BJPS 15 (1965) 327.

        b. "Teleological Explanation in Biology: A Reply," by John Canfield. 327-31.

        c. "Functional Analyses in Biology," by Harry G. Frankfurt and Brian Poole. 17 (1966) 69-72.

        d. "Canfield's Functional Translation Schema," by Lowell Nissen. 21 (1970) 193-95.

4318. _____ . Purpose in Nature. Englewood Cliffs, NJ: Prentice-Hall, 1966. 111 pp.

4319. Carlo, William E. "Mechanism and Vitalism: A Reappraisal." Pacific Philosophy Forum. 6/3 (1968) 57-68.

4320. Chihara, Charles S. "On Alleged Refutations of Mechanism Using Gödel's Incompleteness Results." J. Phil. 69 (1972) 507-27.

4321. Clark, Desmond M. "Teleology and Mechanism: M. Grene's Absurdity Argument." PS 46 (1979) 321-25.

        a. "Comment on Desmond Clarke 'Teleology and Mechanism: M. Grene's Absurdity Argument'," by Marjorie Grene. 326-27.

4322. Collins, Arthur W. "Teleological Reasoning." J. Phil. 75 (1978) 540-50.

        a. "Teleology and Mentalism," by Peter Achinstein. 551-53.

4323. Cummins, Robert. "Functional Analysis." J. Phil. 72 (1975) 741-65.

4324. Dessauer, F. Die Teleologie in der Natur. München: Reinhardt, 1949. 72 pp.

4325. Dobbs, H. A. C. "Diathesis, the Self-Winding Watch, and Photosynthesis." BJPS 8 (1957) 140-50.

a. "Mr. Dobbs on 'Diathesis, the Self-Winding Watch, and Photosynthesis'," by R. O. Kapp. 159-60.

b. "Reply to Professor R. O. Kapp," by H. A. C. Dobbs. BJPS 8 (1958) 306-09.

4326. Ducasse, C. J. "Life, Telism, and Mechanism." PPR 20 (1959) 18-24.

4327. Duchesneau, Francois. "Téléologie et détermination positive de l'order biologique." Dialectica. 32 (1978) 135-54.

4328. Enç, Berent. "Function Attributions and Functional Explanations." PS 46 (1979) 343-65.

a. "Enç on Harvey and Consequence Etiologies," by James Lennox. 48 (1981) 323-26.

4329. Engels, Eve-Marie. "Teleologie-eine 'Sache der Formulierung' oder eine 'Formulierung der Sache'? Ueberlegungen zu Ernest Nagels reduktionistischer Strategie und Versuch ihrer Widerlegung." ZAW 9 (1978) 225-35.

4330. Esposito, Joseph L. "Teleological Causation." Philosophical Forum. 12 (1980-81) 116-27.

4331. Falk, Arthur E. "Purpose, Feedback and Evolution." PS 48 (1981) 198-217.

4332. Feuer, Lewis S. "Teleological Principles in Science." Inquiry. 21 (1978) 377-407.

4333. Frolov, I. T. "Organic Determinism and Teleology in Biological Research." WOSPS 10 (1977) 119-29.

4334. Gagnebin, Elie. "Essai sur la finalité de la nature." Dialectica. 4 (1950) 133-43.

4335. Goldman, Alvin I. "The Compatibility of Mechanism and Purpose." Phil. Review. 78 (1969) 468-82.

4336. Goldstein, Leon J. "Recurrent Structures and Teleology." Inquiry. 5 (1962) 1-10. Reprinted in Philosophy Today. 6 (1962) 183-91.

4337. Grene, Marjorie. "Biology and Teleology." Cambridge Review. (Feb., 1964) 269-73. Reprinted in BSPS 23, SL 66 (1974) 172-79.

4338. Hanson, W. H. "Mechanism and Gödel's Theorem." BJPS 22 (1971) 9-16.

4339. Harris, Errol E. "Teleology and Teleological Explanation."
J. Phil. 56 (1959) 5-25.

4340. Hassenstein, Bernhard. "Biologische Teleonomie." Neue Hefte
für Philosophie. 20 (1981) 60-71.

4341. Hawkins, David. "The Thermodynamics of Purpose," in Beyond
the Edge of Certainty: Essays in Contemporary Science and Philosophy.
Robert G. Colodny, ed. Englewood Cliffs, NJ: Prentice-Hall, 1965.
102-117.

4342. Hein, Hilde. "Mechanism and Vitalism as Meta-Theoretical
Commitments." Philosophical Forum. 1 (1968) 185-205.

4343. _____. "Mechanism, Vitalism, and Biopoesis." Pacific
Philosophy Forum. 6/3 (1968) 4-56.

4344. _____. "Molecular Biology vs. Organicism: The Enduring
Dispute between Mechanism and Vitalism." Synthese. 20 (1969)
238-53.

4345. Jellinghaus, Karl Theodor. "Zum Verhältnis von Kausalität
und Finalität im organischen Geschehen." PN 3 (1955) 194-210.

4346. Khatchadourian, Haig. "Proteins and Probability: A Criticism
of M. Pierre Lecomte du Noüy's Argument for Teleology Based on Some
Probability-Estimates." PPR 16 (1955) 223-28.

4347. Kreisel, G. "A Notion of Mechanistic Theory." SL 78 (1976)
3-18. Reprinted from Synthese. 29 (1974) 11-26.

4348. Krikorian, Y. H. "Singer on Mechanism and Teleology." J.
Phil. 54 (1957) 569-76.

4349. Ladrière, Jean. "Le rôle de la notion de finalité dans une
cosmologie philosophique." Revue Philosophique de Louvain. 67
(1969) 143-81.

4350. Lehman, Hugh. "Functional Explanation in Biology." PS 32
(1965) 1-19.

4351. Loeb, Jacques. The Mechanistic Conception of Life. Donald
Fleming, ed. Cambridge: Belknap Press of Harvard University Press,
1964. 216 pp.

4352. Lowenthal, David. "The Case for Teleology." Independent
Journal of Philosophy. 2 (1978) 95-105.

4353. Lucas, J. R. "Minds, Machines and Gödel." Philosophy. 36
(1961) 112-27. Reprinted in K. M. Sayre and F. J. Crosson, eds.
The Modelling of Mind. Notre Dame, IN: Notre Dame University
Press, 1963. 255-71, and in A. R. Anderson, ed. Minds and Machines.
Englewood Cliffs, NJ: Prentice-Hall, 1964. 43-59.

a. "Lucas Against Mechanism," by David Lewis. <u>Philosophy</u>. 44 (1969) 231-33.

b. "Goedel's Theorem and Mechanism," by David Coder. 234-37.

c. "Mechanism: A Rejoinder," by J. R. Lucas. 45 (1970) 149-51.

4354. Macklin, Ruth. "Action, Causality, and Teleology." BJPS 19 (1969) 301-16.

4355. Malcolm, Norman. "The Conceivability of Mechanism." <u>Phil. Review</u>. 77 (1968) 45-72.

a. "Neurophysiological Laws and Purpose Principles," by William L. Rowe. 80 (1971) 502-08.

b. "On the Conceivability of Mechanism," by Michael Martin. PS 38 (1971) 79-86.

4356. Mayr, Ernst. "Teleological and Teleonomic, a New Analysis." BSPS 14, SL 60 (1974) 91-118.

4357. Montefiore, Alan. "Final Causes." <u>Arist. Soc.</u> Supp. Vol. 45 (1971) 171-92.

4358. Morin, Harald. "Ein Problem der Teleologie." <u>Theoria</u>. 11 (1945) 20-39.

4359. Nagel, Ernest. "Mechanistic Explanation and Organismic Biology." PPR 11 (1951) 327-38.

4360. _____. "Goal-directed Processes in Biology." <u>J. Phil.</u> 74 (1977) 261-79.

4361. _____. "Functional Explanations in Biology." <u>J. Phil.</u> 74 (1977) 280-301.

4362. _____. <u>Teleology Revisited and Other Essays in the Philosophy and History of Science</u>. New York: Columbia University Press, 1979. viii + 352 pp.

4363. Nelson, R. J. "On Mechanical Recognition." PS 43 (1976) 24-52.

4364. _____. "Mechanism, Functionalism, and the Identity Theory." <u>J. Phil.</u> 73 (1976) 365-85.

4365. Nissen, Lowell. "Neutral Functional Schemata." PS 38 (1971) 251-57.

4366. _____. "Nagel's Self-Regulation Analysis of Teleology." <u>Philosophical Forum</u>. 12 (1980-81) 128-38.

4367. Noble, Denis. "Charles Taylor on Teleological Explanation."
Analysis. 27 (1966-67) 96-103.

    a. "Teleological Explanation - A Reply to Denis Noble," by
    Charles Taylor. 141-43.

    b. "The Conceptualist View of Teleology," by Denis Noble.
    28 (1967-68) 62-63.

4368. Piccone, Paul. "Functionalism, Teleology, and Objectivity."
Monist. 52 (1968) 408-23.

4369. Plamondon, Ann. "The Contemporary Reconciliation of Mechanism
and Organicism." Dialectica. 29 (1975) 213-22.

4370. Poli, E. "Considerazioni critiche sul problema del cosiddetto
finalismo biologico." Scientia. 81 (1947) 81-84. French Tr.:
Supplement, 29-32.

4371. Pomerantz, Itzhak. "Teleological and Mechanistic Explanations
in Leibowitz." [In Hebrew] Iyyun. 26 (1975) 218-25.

4372. Ponce, Margarita. "La definición de sistemas teleológicos."
Diánoia. 24 (1978) 168-89.

4373. _____. "Aristóteles y la teleología actual." Diánoia.
25 (1979) 101-25.

4374. Rensch, B. "Drei heterogene Bedeutungen des Begriffs 'Zufall'."
PN 18 (1980-81) 197-208.

4375. Richards, Robert J. "Substantive and Methodological Teleology
in Aristotle and Some Logical Empiricists." Thomist. 37 (1973)
702-33.

4376. Robinson, Andrew. "Is There 'Purpose' in Modern Biology?"
Proc. Am. Cath. Phil. Assn. 46 (1972) 167-76.

4377. Rosenblueth, Arturo, Norbert Wiener, and Julian Bigelow.
"Behavior, Purpose, and Teleology." PS 10 (1943) 18-24. Reprinted
in Purpose in Nature. J. V. Canfield, ed. Englewood Cliffs, NJ:
Prentice-Hall, 1966. 9-16.

4378. Ruse, Michael E. "Functional Statements in Biology." PS 38
(1971) 87-95.

    a. "A Comment on Ruse's Analysis of Function Statements," by
    Larry Wright. 39 (1972) 512-14.

    b. "A Reply to Wright's Analysis of Functional Statements,"
    by Michael Ruse. 40 (1973) 277-80.

4379. Scheffler, Israel. "Thoughts on Teleology." BJPS 9 (1958)
265-84. Reprinted in Purpose in Nature. J. V. Canfield, ed.
Englewood Cliffs, NJ: Prentice-Hall, 1966. 48-66.

4380. Schwartz, Herbert Thomas. "Finality in the Physical Sciences." Proc. Am. Cath. Phil. Assn. 23 (1949) 80-90.

4381. Shelanski, Vivien B. "Nagel's Translation of Teleological Statements: A Critique." BJPS 24 (1973) 397-401.

4382. Simon, Thomas W. "A Cybernetic Analysis of Goal-Directedness." PSA 1976. I, 56-67.

4383. Singer, Jr., Edgar A. "Mechanism, Vitalism, Naturalism." PS 13 (1946) 81-99.

4384. Siwek, S.J., Paul. "The Mechanical Theory of Life According to Julius Schultz." Proc. Am. Cath. Phil. Assn. 23 (1949) 137-43.

4385. Smart, J. J. C. "Gödel's Theorem, Church's Theorem, and Mechanism." Synthese. 13 (1961) 105-10.

4386. Sprigge, Timothy L. S. "Final Causes." Arist. Soc. Supp. Vol. 45 (1971) 149-70.

4387. Stegmüller, W. "Einige Beiträge zum Problem der Teleologie und der Analyse von Systemen mit zielgerichteter Organization." Synthese. 13 (1961) 5-40.

4388. Stocker, Otto. "Das System der biologischen Wissenschaften und das Problem der Finalität in empirischer und transzendentaler Betrachtung." PN 5 (1958-59) 96-112.

4389. Taylor, Charles. "How is Mechanism Conceivable?" in Interpretations of Life and Mind. Marjorie Grene, ed. New York: Humanities Press, 1971. 38-64.

4390. Taylor, Richard. "Comments on a Mechanistic Conception of Purposefulness." PS 17 (1950) 310-17. Reprinted in Purpose in Nature. John V. Canfield, ed. Englewood Cliffs, NJ: Prentice-Hall, 1966. 17-26.

    a. "Purposeful and Non-Purposeful Behavior," by Arturo Rosenblueth and Norbert Wiener. 318-26.

    b. "Purposeful and Non-Purposeful Behavior: A Rejoinder," by Richard Taylor. 327-32.

    c. "Purposeful and Non-Purposeful Behavior," by A. C. Moulyn. 18 (1951) 154.

4391. Thorpe, W. H. Purpose in a World of Chance: A Biologist's View. Oxford: Oxford University Press, 1978. xii + 124 pp.

4392. Toulmin, Stephen. "Teleology in Contemporary Science and Philosophy." Neue Hefte für Philosophie. 20 (1981) 140-52.

4393. Varela, Francisco, and Maturana Humberto. "Mechanism and Biological Explanation." PS 39 (1972) 378-82.

4394. Valcke, Louis. "L'Anti-finalisme de Jacques Monod." Dialogue.
11 (1972) 546-68.

4395. Wallenmaier, Thomas E. "Towards a Quantitative Concept of
Teleological Systems." Nature and System. 1 (1979) 147-55.

4396. Whitmore, Charles E. "Thoughts on the Problem of Mechanism."
J. Phil. 45 (1948) 489-98.

4397. Wicken, Jeffrey S. "Evolutionary Self-Organization and the
Entropy Principle: Teleology and Mechanism." Nature and System. 3
(1981) 129-41.

4398. Wimsatt, William C. "Teleology and the Logical Structure of
Function Statements." SHPS 3 (1972) 1-80.

        a. "Wimsatt on Function Statements," by Lowell Nissen. 8
        (1977) 341-47.

4399. Wolff, Etienne. "Monstruosité et finalité." Les Etudes
philosophiques. 15 (1960) 323-32.

4400. Woodfield, Andrew. Teleology. Cambridge: Cambridge University
Press, 1976. viii + 232 pp.

        a. Review essay, "The Ins and Outs of Teleology: A Critical
        Examination of Woodfield," by Larry Wright. Inquiry. 21
        (1978) 223-45.

        b. "Critical Notice," by Michael Ruse. Canadian Journal
        of Philosophy. 8 (1978) 191-203.

        c. Review essay by Gerhard D. Wassermann. Philosophia. 10
        (1981) 125-32.

4401. Woolhouse, R. S. "The Temporal Structure of Goal-Directedness."
Phil. Quart. 29 (1979) 56-64.

        a. "Reply to Woolhouse on the Temporal Structure of Goal-
        Directedness," by Andrew Woodfield. 65-73.

4402. Wright, Larry. "The Case Against Teleological Reductionism."
BJPS 19 (1968) 211-23.

4403. _____. "Explanation and Teleology." PS 39 (1972)
204-18.

        a. "Wright and Taylor: Empiricist Teleology," by Arthur J.
        Minton. 42 (1975) 299-306.

        b. "On Teleology and Organisms," by Stephen Utz. 44 (1977)
        313-20.

        c. "Rejoinder to Utz," by Larry Wright, 321-25.

4404. _____. "Teleological Etiologies." Philosophical Forum.
4 (1973) 575-84.

4405. _____. "Mechanisms and Purposive Behavior III." PS 41
(1974) 345-60.

4406. _____. Teleological Explanations. Berkeley: University
of California Press, 1976. ix + 153 pp.

        a. "Critical Notice," by Michael Ruse. Canadian Journal of
        Philosophy. 8 (1978) 191-203.

4407. _____. "Functions." Philosophical Review. 82 (1973)
139-68. Reprinted in BSPS 27 (1976) 213-42.

        a. "Wright on Functions," by Patrick Grim. Analysis. 35
        (1974-75) 62-64.

        b. "Reply to Grim," by Larry Wright. 36 (1975-76) 156-57.

4408. Wuketits, Franz M. "On the Notion of Teleology in Contemporary
Life Sciences." Dialectica. 34 (1980) 277-90.

4409. Yamamoto, Makoto. "A Note on Vitalism versus Mechanism."
AJAPS 4, No. 1 (1971) 46-48.

4410. Yourgrau, W. "General System Theory and the Vitalism-Mechanism
Controversy." Scientia 87 (1952) 307-11. French Tr.: Supplement,
168-72.

5.4 Reduction and Emergence

4411. Ackermann, Robert. "The Fallacy of Conjunctive Analysis."
The Monist. 53 (1969) 478-87. Reprinted in Basic Issues in the
Philosophy of Time. E. Freeman and W. Sellars, eds. La Salle,
Illinois: Open Court, 1971. 154-63.

4412. Agazzi, Evandro. "Systems Theory and the Problem of Reductionism." Erkenntnis. 12 (1978) 339-58.

4413. Ayala, Francisco J. "Biology as an Autonomous Science."
American Scientist. 56 (1968) 207-21. Reprinted in BSPS 27 (1976)
312-29.

4414. Bahm, Archie J. "Organic Unity and Emergence." J. Phil. 44
(1947) 241-44.

4415. _____. "Emergence of Purpose." J. Phil. 44 (1947)
633-36.

4416. Beckner, Morton. "Reduction, Hierarchies and Organicism," in
Studies in the Philosophy of Biology. F. J. Ayala and T. Dobzhansky,
eds. Berkeley: University of California Press, 1974. 163-77.

4417. Berenda, C. W. "On Emergence and Prediction." J. Phil. 50
(1953) 269-74.

4418. Bogaard, Paul A. "The Limitations of Physics as a Chemical
Reducing Agent." PSA 1978. II, 345-56.

4419. Bonevac, Daniel. Reduction in the Abstract Sciences. Foreward by Wilfrid Sellars. Indianapolis, IN: Hackett Publishing Co.,
1982. 184 pp.

4420. Brittan, Jr., Gordon G. "Explanation and Reduction." J.
Phil. 67 (1970) 446-57.

4421. Brodbeck, May. "Methodological Individualisms: Definition
and Reduction." PS 25 (1958) 1-22.

4422. Brody, Baruch. "The Reduction of Teleological Sciences."
Am. Phil. Quart. 12 (1975) 69-76.

4423. Campbell, Donald T. "'Downward Causation' in Hierarchically
Organised Biological Systems," in Studies in the Philosophy of Biology.
F. J. Ayala and T. Dobzhansky, eds. Berkeley: University of Cali-
fornia Press, 1974. 179-86.

4424. Caplan, Arthur. "Babies, Bathwater and Derivational Reduction."
PSA 1978. II, 357-70.

4425. Carlo, William E. "Reductionism and Emergence: Mechanism
and Vitalism Revisited." Proc. Am. Cath. Phil. Assn. 40 (1966)
94-103.

4426. Cartwright, Nancy. "Do Token-Token Identity Theories Show
Why We Don't Need Reductionism?" Phil. Studies. 36 (1979) 85-90.

4427. Causey, Robert L. "Polanyi on Structure and Reduction."
Synthese. 20 (1969) 230-37.

4428. _____. "Attribute-Identities in Microreductions."
J. Phil. 69 (1972) 407-22.

4429. _____. "Uniform Microreductions." Synthese. 25 (1972)
176-218.

4430. Commoner, B., G. Holton, E. Nagel, J. R. Platt, and M. Polanyi.
"Do Life Processes Transcend Physics and Chemistry? (A Symposium)."
Zygon. 3 (1968) 442-72.

4431. Eberle, Rolf A. "Replacing One Theory by Another Under Preservation of a Given Feature." PS 38 (1971) 486-501.

4432. Enç, Berent. "Identity Statements and Microreductions."
J. Phil. 73 (1976) 285-306.

4433. Frankl, Viktor E. "Reductionism and Nihilism," in Beyond
Reductionism. A. Koestler and J. R. Smythies, eds. New York:
Macmillan, 1970. 396-427.

4434. Fuchs-Kittowski, K. "Reduktive Methode und Reduktionismus in
den Biowissenschaften." DZP 29 (1981) 503-16.

4435. Girill, T. R. "Criteria for Part-Whole Relation in Micro-
Reductions." Philosophia. 6 (1976) 69-79.

4436. Glymour, Clark. "On some Patterns of Reduction." PS 37
(1970) 340-53.

4437. Goodfield, June. "Changing Strategies: A Comparison of Re-
ductionist Attitudes in Biological and Medical Research in the Nineteenth and Twentieth Centuries," in Studies in the Philosophy of
Biology. F. J. Ayala and T. Dobzhansky, eds. Berkeley: University
of California Press, 1974. 65-86.

4438. Gottlieb, Dale. "Ontological Reduction." J. Phil. 73 (1976)
57-77.

4439. Grene, Marjorie. "Biology and the Problem of Levels of
Reality." New Scholasticism. 41 (1967) 94-123. Reprinted in BSPS
23, SL 66 (1974) 35-52.

4440.   . "Reducibility: Another Side Issue?" in Interpretations of Life and Mind. Marjorie Grene, ed. New York: Humanities Press, 1971. 14-37. Reprinted in BSPS 23, SL 66 (1974) 53-73.

4441.   . Interpretations of Life and Mind: Essays around
the Problem of Reduction. New York: Humanities Press, 1971. 152
pp.

4442. Grossman, Reinhardt. Ontological Reduction. Bloomington:
Indiana University Press, 1973. 215 pp.

4443. Hellman, Geoffrey Paul, and Frank Wilson Thompson. "Physicalism: Ontology, Determination, and Reduction." J. Phil. 72
(1975) 551-64.

    a. "What is Physicalism?" by John Earman. 565-67.

4444. Hempel, Carl G. "Reduction: Ontological and Linguistic
Facets," in Philosophy, Science, and Method. S. Morgenbesser et al.
eds. New York: St. Martin's Press, 1969. 179-99.

4445. Hooker, C. A. "Towards a General Theory of Reduction." Part
I: "Historical and Scientific Setting." Dialogue. 20 (1981)
38-59. Part II: "Identity in Reduction." 201-36. Part III:
Cross-Categorial Reduction." 496-529.

4446. Horgan, Terence. "Supervenient Bridge Laws." PS 45 (1978) 227-49.

4447. _____. "Token Physicalism, Supervenience, and the Generality of Physics." Synthese. 49 (1981) 395-413.

4448. Hovard, Richard B. "Theoretical Reduction: The Limits and Alternatives to Reductive Methods in Scientific Explanation." Philosophy of the Social Sciences. 1. (1971) 83-100.

4449. Hull, David L. "Informal Aspects of Theory Reduction." BSPS 32 (1976) 653-70.

4450. Jubien, Michael. "Two Kinds of Reduction." J. Phil. 66 (1969) 533-41.

4451. Karpinskaia, R. S. "Biological Reductionism and World View." Soviet Studies in Philosophy. 19, No. 1 (1980) 49-68. [English reprint from Voprosy filosofii. (1979, No. 11)].

4452. Kauffman, Stuart. "Elsasser, Generalized Complementarity, and Finite Classes: A Critique of His Anti-Reductionism." BSPS 20, SL 64 (1974) 57-65.

4453. Kekes, John. "Physicalism, The Identity Theory, and the Doctrine of Emergence." PS 33 (1966) 360-75.

4454. Kemeny, John G., and Paul Oppenheim. "On Reduction." Phil. Studies. 7 (1956) 6-19.

    a. "On the Kemeny-Oppenheim Treatment of Reduction," by J. W. Swanson. 13 (1962) 94-96.

4455. Kenny, Anthony J. P. "The Homunculus Fallacy," in Interpretations of Life and Mind. Marjorie Grene, ed. New York: Humanities Press, 1971. 65-74.

    a. "Not Every Homunculus Spoils the Argument," by Amélie Rorty. 75-80.

    b. "Reply to Mrs. Rorty," by Anthony J. P. Kenny. 81-83.

4456. Kim, Jaegwon. "Reduction, Correspondence and Identity." Monist. 52 (1968) 424-38.

4457. Koestler, Arthur. "Beyond Atomism and Holism - The Concept of the Holon," in Beyond Reductionism. A. Koestler and J. R. Smythies, eds. New York: Macmillan, 1970. 192-232.

4458. Koestler, Arthur, and J. R. Symthies, eds. Beyond Reductionism: New Perspectives in the Life Sciences. London: Hutchinson, 1969; New York: Macmillan, 1970. x + 438 pp.

4459. Levy, Monique. "Les relations entre chimie et physique et le problème de la réduction." Epistemologia. 2 (1979) 337-70.

4460. _____. "The 'Reduction by Synthesis' of Biology to
Physical Chemistry." PSA 1980. I, 151-59.

4461. Lowry, Ann. "A Note on Emergence." Mind. 83 (1974) 276-77.

4462. McMullin, Ernan. "The Dialectics of Reduction." Idealistic
Studies. 2 (1972) 95-115.

4463. Makai, Maria. "Against Reductionism and Purism: Tertium
Datur." SL 72 (1976) 27-58.

4464. Matson, Floyd W. The Broken Image: Man, Science and Society.
New York: George Braziller, 1964. 355 pp.

4465. Mayr, Dieter. "Investigations of the Concept of Reduction."
Erkenntnis. 10 (1976) 275-94; 16 (1981) 109-29.

4466. Medawar, Peter. "A Geometric Model of Reduction and Emergence,"
in Studies in the Philosophy of Biology. F. J. Ayala and T. Dobzhansky,
eds. Berkeley: University of California Press, 1974. 57-63.

4467. Meehl, P. E. and Wilfrid Sellars. "The Concept of Emergence."
MSPS 1 (1956) 239-52.

4468. Montalenti, G. "From Aristotle to Democritus via Darwin," in
Studies in the Philosophy of Biology. F. J. Ayala and T. Dobzhansky,
eds. Berkeley: University of California Press, 1974. 3-19.

4469. Nagel, Ernest. "Wholes, Sums, and Organic Unities." Phil.
Studies. 3 (1952) 17-32.

4470. Nelson, Alvin F. "Emergentism Reconsidered and an Alternative."
Southern Journal of Philosophy. 7 (1969) 187-92.

4471. _____. "Emergentism Reconsidered and an Alternative.
Southwestern Journal of Philosophy. 1, No. 1 (1970) 125-32.

4472. Nickles, Thomas. "Two Concepts of Intertheoretic Reduction."
J. Phil. 70 (1973) 181-201.

4473. Pap, Arthur. "The Concept of Absolute Emergence." BJPS 2
(1952) 302-11.

4474. Peacocke, A. R. "Reductionism: A Review of the Epistemological Issues and Their Relevance to Biology and the Problem of Consciousness." Zygon. 11 (1976) 307-34.

      a. "Peacocke's Reductionism," by Mary Hesse. 335-36.

4475. Petersen, Arne Friemuth. "Biological Evolution or Anti-Chaos:
On the Problem of Reduction in Biology and Psychology." Danish
Yearbook of Philosophy. 12 (1975) 65-92.

4476. Pluhar, Evelyn B. "Emergence and Reduction." SHPS 9 (1978)
279-89.

4477. Polanyi, Michael. "Life's Irreducible Structure." BSPS 27 (1976) 128-42.

4478. Popper, Karl R. "Scientific Reduction and the Essential Incompleteness of All Science," in Studies in the Philosophy of Biology. F. J. Ayala and T. Dobzhansky, eds. Berkeley: University of California Press, 1974. 259-84.

4479. Pratt, Vernon. "Explaining the Properties of Organisms." SHPS 5 (1974) 1-15.

4480. Prigogine, Ilya. "Unity of Physical Laws and Levels of Description," in Interpretations of Life and Mind. Marjorie Grene, ed. New York: Humanities Press, 1971. 1-13.

4481. Richardson, Robert C. "Functionalism and Reductionism." PS 46 (1979) 533-58.

    a. "How to Reduce a Functional Psychology?" by Patricia Kitcher. 47 (1980) 134-40.

4482. Roll-Hansen, Nils. "On the Reduction of Biology to Physical Science." Synthese. 20 (1969) 277-89.

4483. Ruse, M. E. "Reduction, Replacement, and Molecular Biology." Dialectica. 25 (1971) 39-72.

4484. _____. "Reduction in Genetics." BSPS 32 (1976) 633-52.

4485. Schievella, P. S. "Emergent Evolution and Reductionism." Scientia. 108 (1973) 323-30.

4486. Schlesinger, G. "The Prejudice of Micro-Reduction." BJPS 12 (1961) 215-24.

4487. Shaffner, Kenneth F. "The Watson-Crick Model and Reductionism." BJPS 20 (1969) 325-48. Reprinted in BSPS 27 (1976) 101-27.

4488. _____. "Reductionism in Biology: Prospects and Problems." BSPS 32 (1976) 613-32.

4489. Simons, Leo. "The Reduction of Temperature." J. Phil. 59 (1962) 365-71.

4490. Sklar, Lawrence. "Types of Inter-Theoretic Reduction." BJPS 18 (1967) 109-24.

4491. _____. "Thermodynamics, Statistical Mechanics, and the Complexity of Reductions." BSPS 32 (1976) 15-32.

4492. Spector, Marshall. "Russell's Maxim and Reduction as Replacement." Synthese. 32 (1975) 135-76.

4493. _____. Concepts of Reduction in Physical Science. Philadelphia: Temple University Press, 1978. xii + 114 pp.

a. Review essay, "Reduction as Replacement," by Ron Yoshida. BJPS 32 (1981) 400-10.

4494. Spinner, Helmut F. "Science without Reduction." Inquiry. 16 (1973) 16-94.

4495. Svenonius, Lars. "Translation and Reduction." Journal of Philosophical Logic. 1 (1972) 297-316.

4496. Thorpe, William H. "Reductionism in Biology," in Studies in the Philosophy of Biology. F. J. Ayala and T. Dobzhansky, eds. Berkeley: University of California Press, 1974. 109-38.

4497. Watanabe, Satoshi. "Logic of the Empirical World -- With Reference to Identity Theory and Reductionism." AJAPS 4, No. 4 (1974) 41-58.

4498. Wimsatt, William C. "Reduction and Reductionism," in Current Research in Philosophy of Science. P. D. Asquith and H. E. Kyburg, Jr., eds. East Lansing, MI: Philosophy of Science Assn., 1979. 352-77.

4499. _____. "Reductionistic Research Strategies and Their Biases in the Units of Selection Controversy." BSPS 60 (1980) 213-60.

4500. Yoshida, Ronald M. Reduction in the Physical Sciences. Philosophy in Canada Monograph Series, 4. Halifax, Nova Scotia: Dalhousie University Press, 1977. 90 pp.

a. Review essay, "Reduction in Physics," by W. Krajewski. BJPS 29 (1978) 280-83.

b. Review essay by C. A. Hooker. Dialogue. 18 (1979) 81-99.

5.5 Realism and Instrumentalism

4501. Ackermann, Robert. "Sellars and the Scientific Image." Nous. 7 (1973) 138-51.

4502. Agassi, Joseph. "Duhem's Instrumentalism and Autonomism." Ratio 12 (1970) 148-50.

4503. _____. "The Future of Berkeley's Instrumentalism." International Studies in Philosophy. 7 (1975) 167-78.

4504. Armstrong, D. H. Universals and Scientific Realism. Vol. I: Nominalism and Realism. Vol. II: A Theory of Universals. Cambridge: Cambridge University Press, 1978. 149, 190 pp.

4505. Bar-Hillel, Yehoshua. "The Status of Theoretical Entities."
[In Hebrew] Iyyun. 11 (1960) 146-51.

4506. Bergmann, Gustav. "Remarks on Realism." PS 13 (1946) 261-73.

4507. Berk, Edwin. "Reference and Scientific Realism." Southwestern
Journal of Philosophy. 10, No. 2 (1979) 139-46.

4508. Berstein, Richard J. "Sellars' Vision of Man-in-the Universe."
Review of Metaphysics. 20 (1966) 113-43; 290-316.

4509. Bertalanffy, L. von. "An Essay on the Relativity of Categories."
PS 22 (1955) 243-63.

    a. "Knowledge and Experience: Comment on a Paper by L. von
    Bertalanffy on 'The Relativity of Categories'," by J. P.
    McKinney. 24 (1957) 349-56.

4510. Bhaskar, Roy. "Forms of Realism." Philosophica. 15 (1975)
99-127.

4511. _____. A Realist Theory of Science. Sussex: Harvester
Press, 1978. 284 pp.

4512. Biser, Erwin. "Entity and Aspects (As Pertaining to Physical
Theory)." PS 14 (1947) 105-15.

4513. Blackmore, John. "On the Inverted Use of the Terms 'Realism'
and 'Idealism' among Scientists and Historians of Science." BJPS 30
(1979) 125-34.

4514. Born, Max. "Physical Reality." Phil. Quart. 3 (1953) 139-49.

4515. _____. "Symbol and Reality." Dialectica. 20 (1966)
143-57.

4516. Boyd, Richard N. "Realism, Underdetermination, and a Causal
Theory of Evidence." Nous. 7 (1973) 1-12.

4517. _____. "Scientific Realism and Naturalistic Epistemology."
PSA 1980. II, 613-62.

4518. Bradie, Michael. "Is Scientific Realism a Contingent Thesis?"
BSPS 20, SL 64. (1974) 367-73.

4519. _____. "Models, Metaphors, and Scientific Realism."
Nature and System. 2 (1980) 3-20.

4520. Brodbeck, May. "Structure of Science. Philosophy of Science,
a Separate Discipline, Meets Philosophy Proper on the Question,
'What Exists'?" Science. 134 (1961) 997-99.

4521. Bunge, M. "Physics and Reality." Dialectica. 19 (1965)
195-222. Also in Dialectica. 20 (1966) 174-95.

4522. _____. "Ontología y ciencia." Diánoia. 21 (1975) 50-59.

4523. Burian, Richard M. "Sellarsian Realism and Conceptual Change in Science." SL 133 (1979) 197-225.

    a. "Some Remarks on Realism and Scientific Revolutions: Comment on Burian," by Lorenz Krüger. 227-33.

    b. "Realism and Underdetermination: Comment on Burian," by Charles Parsons. 235-42.

4524. Čapek, Milič. "Simple Location and Fragmentation of Reality." Monist. 48 (1964) 195-218.

4525. Churchland, Paul. Scientific Realism and the Plasticity of Mind. Cambridge: Cambridge University Press, 1979. x + 157 pp.

    a. "Critical Notice," by Bas C. van Fraassen. Canadian Journal of Philosophy. 11 (1981) 555-67.

4526. Clay, J. "The Problem of Reality." Synthese. 6 (1947-48) 305-16.

4527. Clement, William C. "Russell's Structuralist Thesis." Phil. Review. 62 (1953) 266-75.

    a. "Physics and Structures," by Robert Palter. 65 (1956) 371-84.

4528. Cornman, James W. "Sellars, Scientific Realism, and Sensa." Review of Metaphysics. 23 (1970) 417-51.

    a. "Science, Sense Impressions, and Sensa: A Reply to Cornman," by Wilfrid Sellars. 24 (1971) 391-447.

4529. _____. "Craig's Theorem, Ramsey-Sentences, and Scientific Instrumentalism." Synthese. 25 (1972) 82-128.

4530. _____. "Can Eddington's 'Two' Tables be Identical?" AJP 52 (1974) 22-38.

    a. "Cornman on the Colour of Micro-Entities," by Stephen J. Noren. 53 (1975) 65-67.

4531. _____. "Sellars on Scientific Realism and Perceiving." PSA 1976. II, 344-58.

4532. Destouches, J.-L. "Observation, prévision, invention, objectivite, réalité dans les sciences ayant acquis une forme théorique." Dialectica. 20 (1966) 137-42.

4533. Dubarle, D. "Objectivité et réalité dans le cas de la physique probabiliste." Dialectica. 20 (1966) 158-73.

4534. Ducasse, C. J. "Reality, Science and Metaphysics." Synthese. 8 (1950-51) 9-21.

4535. Earman, John. "Fairy Tales vs. An Ongoing Story: Ramsey's Neglected Argument for Scientific Realism." Phil. Studies. 33 (1978) 195-202.

4536. Elgin, C. Z. "Lawlikeness and the End of Science." PS 47 (1980) 56-68.

4537. Faggiani, D. "Realità fisica ed invarianza." Scientia. 104 (1969) 323-28. French Tr.: Supplement, 173-78.

4538. Feibleman, James K. Inside the Great Mirror. The Hague, 1958. 216 pp.

4539. Feigl, Herbert. "Existential Hypotheses: Realistic versus Phenomenalistic Interpretations." PS 17 (1950) 35-62.

   a. "Logical Reconstructionism," by C. West Churchman. 164-66.

   b. "Comments on Realistic versus Phenomenalistic Interpretations," by Philipp Frank. 166-69.

   c. "A Note on Semantic Realism," by Carl G. Hempel. 169-73.

   d. "Science and Semantic Realism," by Ernest Nagel. 174-81.

   e. "On Feigl's 'Existential Hypotheses'," by A. G. Ramsperger. 182-85.

   f. "Logical Reconstruction, Realism and Pure Semiotic," by Herbert Feigl. 186-95.

4540. Feinberg, G. "On What There May Be in the World," in Philosophy, Science, and Method. S. Morgenbesser et al, eds. New York: St. Martin's Press, 1969. 152-64.

4541. Feyerabend, Paul. "Wittgenstein's Philosophical Investigations." Phil. Review. 64 (1955) 449-83.

4542. _____. "An Attempt at a Realistic Interpretation of Experience." Proc. Arist. Soc. 58 n.s. (1957-58) 143-70.

   a. "Notes on P. K. Feyerabend's Criticism of Positivism," by R. Harré. BJPS 10 (1958) 43-48.

4543. Field, Hartry. Science Without Numbers: The Case for Nominalism. Princeton, NJ: Princeton University Press, 1980. 100 pp.

4544. Folse, Henry J. "Belief and the New Scientific Realism." Tulane Studies in Philosophy. 30 (1981) 37-81.

4545. Fraassen, Bas C. van. "Wilfrid Sellars on Scientific Realism."
Dialogue. 14 (1975) 606-16.

    a. "Sellars' Scientific Realism: A Reply to van Fraassen,"
by Mark Thornton. 20 (1981) 79-83.

4546. _____. "To Save the Phenomena." J. Phil. 73 (1976)
623-32.

    a. "Approximate Truth and Natural Necessity," by Richard N.
Boyd. 633-35.

    b. "To Save the Noumena," by Clark Glymour. 635-37.

4547. _____. "On the Radical Incompleteness of the Manifest
Image." PSA 1976. II, 335-43.

4548. _____. "Theory Construction and Experiment: An Empiricist View." PSA 1980. 663-78.

4549. _____. The Scientific Image. Oxford: Clarendon Press,
1980. 238 pp.

    a. Review essay by Yvon Gauthier. Dialogue. 20 (1981)
579-86.

4550. Gaifman, Haim. "Ontology and Conceptual Frameworks."
Erkenntnis. 9 (1975) 329-53; 10 (1976) 21-85.

4551. Gardner, Michael R. "Realism and Instrumentalism in 19th
Century Atomism." PS 46 (1979) 1-34.

4552. Geymonat, Ludovico. Scienza e realismo. Milano: Feltrinelli,
1977. 180 pp.

4553. Gibson, James J. "New Reasons for Realism." Synthese. 17
(1967) 162-72.

4554. Giedymin, Jerzy. "Instrumentalism and Its Critique: A Reappraisal." BSPS 39 (1976) 179-208.

4555. Glymour, Clark. "Theoretical Realism and Theoretical Equivalence." BSPS 8, SL 39 (1971) 275-88.

4556. Goldman, Alan H. "Realism." Southern Journal of Philosophy.
17 (1979) 175-92.

4557. Gonseth, F. "La preuve dans les sciences du réel." RIP 8
(1954) 25-33.

4558. _____. "Stratégie de fondement et stratégie d'engagement."
Dialectica. 22 (1968) 171-86.

4559. Goodman, Nelson. "Words, Works, Worlds," Erkenntnis. 9
(1975) 57-73.

4560. Gorski, D. P. "The Scientific Representation of Reality: Its Difficulties." Diogenes. 60 (1967) 20-34.

4561. Gruender, D. "Constructs and Fictions." Dialectica. 22 (1968) 20-27.

4562. Gutting, Gary. "Philosophy of Science," in The Synoptic Vision: Essays on the Philosophy of Wilfrid Sellars. Notre Dame, Indiana: University of Notre Dame Press, 1977. 73-104.

4563. _____. "Husserl and Scientific Realism." PPR 39 (1978) 42-56.

4564. Hacking, Ian. "Do We See Through a Microscope?" Pacific Philosophical Quarterly. 62 (1981) 305-22.

4565. Harding, Sandra. "The Inconsistent Scientific Realist." Phil. Studies. 30 (1976) 203-05.

4566. Hauptli, Bruce W. "Quinean Relativism: Beyond Metaphysical Realism and Idealism." Southern Journal of Philosophy. 18 (1980) 393-410.

4567. Heelan, S.J., Patrick A. "A Realist Theory of Physical Science." Continuum. 2 (1964) 334-42.

4568. _____. "Horizon, Objectivity and Reality in the Physical Sciences." IPQ 7 (1967) 375-412.

4569. Hennemann, G. "Die philosophische Problematik der physikalischen Wirklichkeit." Studium Generale. 18 (1965) 601-08.

4570. Hesse, Mary. "Models of Theory-Change." LMPS-4 (1971) 379-91.

4571. _____. "The Hunt for Scientific Reason." PSA 1980. II, 3-22.

4572. Hochberg, Herbert. "On Being and Being Presented." PS 32 (1965) 123-36.

4573. Hooker, C. A. "Craigian Transcriptionism." Am. Phil. Quart. 5 (1968) 152-63.

4574. _____. "Systematic Realism." Synthese. 26 (1974) 409-97. "Correction," 27 (1974) 535-36.

4575. _____. "Philosophy and Meta-Philosophy of Science: Empiricism, Popperianism and Realism." Synthese. 32 (1975) 177-32.

4576. Hübner, Adolf. "On the Logic of Being." International Logic Review. 12 (1981) 27-45.

4577. Janich, P. "Die Sprache der Physik und die Wirklichkeit der Naturwissenschaften." Dialectica. 31 (1977) 301-12.

4578. Joja, Crizantema. "Classique et moderne dans la théorie des entités abstraites." Noesis. 3 (1975) 261-73.

4579. Joseph, Geoffrey. "The Many Sciences and the One World." J. Phil. 77 (1980) 773-91.

4580. Joy, Glenn C. "Instrumentalism: A Duhemian Reply to Popper." Modern Schoolman. 52 (1975) 194-99.

4581. Juhos, Béla. "Two Concepts of Physical Reality." Ratio 12 (1970) 65-78.

4582. Kaeser, E. "Physical Laws, Physical Entities, and Ontology." Dialectica. 31 (1977) 273-300.

4583. Koethe, John. "Putnam's Argument Against Realism." Phil. Review. 88 (1979) 92-99.

4584. Kolb, David. "Sellars and the Measure of All Things." Phil. Studies. 34 (1978) 381-400.

4585. Laudan, Larry. "A Confutation of Convergent Realism." PS 48 (1981) 19-49.

4586. Leeds, Stephen. "A Note on Craigian Instrumentalism." J. Phil. 72 (1975) 177-84.

4587. Lenzen, Victor F. "The Concept of Reality in Physical Theory." Phil. Review. 54 (1945) 321-44.

4588. Leplin, Jarrett. "Reference and Scientific Realism." SHPS 10 (1979) 265-84.

4589. Levine, Vicki Choy. "Sellars' Argument for Extreme Scientific Realism." Pacific Philosophical Quarterly. 61 (1980) 463-68.

4590. MacKinnon, Edward A. "Atomic Physics and Reality." Modern Schoolman. 38 (1960) 37-59. Reprinted in The Problem of Scientific Realism. E. A. MacKinnon, ed. New York: Appleton-Century-Crofts, 1972. 209-31.

4591. _____. "Thomism and Atomism." Modern Schoolman. 38 (1961) 121-41.

4592. _____. The Problem of Scientific Realism. New York: Appleton-Century-Crofts, 1972. 301 pp.

4593. _____. "Theoretical Entities and Metatheories." SHPS 3 (1972) 105-17.

4594. _____. "Scientific Realism: The New Debates." PS 46 (1979) 501-32.

4595. McMullin, Ernan. "Realism in Modern Cosmology." Proc. Am. Cath. Phil. Assn. 29 (1955) 137-50.

4596. Mandelbaum, Maurice. "Philosophy, Science, and Sense-Perception." Proc. & Add. Am. Phil. Assn. 36 (1962-63) 5-20.

4597. _____. Philosophy, Science, and Sense Perception. Baltimore: The Johns Hopkins Press, 1964. 262 pp.

    a. Review essay, "Mandelbaum's Critical Realism," by J. W. N. Watkins. BJPS 16 (1965) 249-52.

4598. Marcus, Solomon. "Conceptual and Contextual Hypostasises of Abstract Entities." Noesis. 1 (1973) 77-83.

4599. Margenau, H. "Einstein's Conception of Reality," in Albert Einstein: Philosopher-Scientist. P. A. Schilpp, ed. New York: Tudor, 1951. 243-68.

4600. _____. "Physical versus Historical Reality." PS 19 (1952) 193-213.

4601. Margolis, Joseph. "Scientific Realism, Ontology, and the Sensory Modes." PS 37 (1970) 114-20.

4602. _____. "Realism's Superiority Over Instrumentalism and Idealism: A Defective Argument." Southern Journal of Philosophy. 17 (1979) 473-79.

4603. Marquis, Donald B. "In Defense of Sir Arthur Eddington." Southwestern Journal of Philosophy. 7, No. 1 (1976) 137-43.

4604. Maxwell, Grover. "The Ontological Status of Theoretical Entities." MSPS 3 (1962) 3-27.

4605. _____. "Theories, Frameworks, and Ontology." PS 29 (1962) 132-38.

4606. _____. "Structural Realism and the Meaning of Theoretical Terms." MSPS 4 (1970) 181-192.

4607. _____. "Theories, Perception, and Structural Realism," in The Nature and Function of Scientific Theories: Essays in Contemporary Science and Philosophy. Robert G. Colodny, ed. Pittsburgh: University of Pittsburgh Press, 1970. 3-34.

4608. Mellor, D. H. "Physics and Furniture," in Studies in the Philosophy of Science. N. Rescher, ed. Oxford: Basil Blackwell, 1969. 171-87.

4609. Mercier, André. "Knowledge and Physical Reality," in Physics, Logic, and History. W. Yourgrau and A. D. Breck, eds. New York: Plenum Press, 1970. 39-50.

4610. _____. "Does Science Coincide with Our Knowledge about Nature?" Dialectics and Humanism. 3, No. 1 (1976) 43-50.

4611. Merrill, G. H. "The Model-Theoretic Argument Against Realism." PS 47 (1980) 69-81.

4612. _____. "Three Forms of Realism." Am. Phil. Quart. 17 (1980) 229-35.

4613. Minogue, Brendan P. "Realism and Intensional Reference." PS 45 (1978) 445-55.

    a. "Minogue on Intensional Reference," by Mary Hesse. 47 (1980) 617-25.

4614. Moore, Harold. "Dewey and the Philosophy of Science." Man and World. 5 (1972) 158-68.

4615. _____. "Scientific Realism and the Compatibilist Thesis: A Defense." Proc. Am. Cath. Phil. Assn. 50 (1976) 24-31.

4616. Morgenbesser, Sidney. "The Realist-Instrumentalist Controversy," in Philosophy, Science, and Method. S. Morgenbesser et al. eds. New York: St. Martin's Press, 1969. 200-18.

4617. Mostepanenko, A. M., and V. M. Mostepanenko. "The Problem of Existence of Microreality and the New Types of Physical Relativity." International Logic Review. 7 (1976) 160-66.

4618. Musgrave, Alan E. "Explanation, Description and Scientific Realism." Scientia. 112 (1977) 727-41. Italian Tr.: 743-55.

4619. _____. "Wittgensteinian Instrumentalism." Theoria. 46 (1980) 65-105.

4620. Nelson, J. O. "Can Systems of Imperceptible Particles Appear to Perceivers?" Mind. 82 (1973) 253-57.

4621. Neumann, Michael. "Fictionalism and Realism." Canadian Journal of Philosophy. 8 (1978) 533-42.

4622. Nola, Robert. "'Paradigms Lost, or the World Regained' -- An Excursion into Realism and Idealism in Science." Synthese. 45 (1980) 317-50.

4623. Noren, Stephen J. "Science, Common Sense, and a Problem for Scientific Realism." Philosophica. 15 (1975) 57-66.

4624. Omel'janovskij, M. E. "The Problem of Reality in Contemporary Physics." Scientia. 110 (1975) 737-49. Italian Tr.: 751-60.

4625. Os, Ch. H. van. "What is Reality?" Synthese. 7 (1948-49) 213-18.

4626. Pauli, W. "Phänomen und physikalische Realitat." Dialectica. 11 (1957) 36-48.

4627. Pearson, C. I. "The Status of Inferred Entities." Phil. Quart. 11 (1961) 158-64.

4628. Polanyi, Michael. "Science and Reality." Synthese. 5 (1946) 137-50. Reprinted in BJPS 18 (1967) 177-96.

4629. Putnman, Hilary. "What is Realism?" Proc. Arist. Soc. 76 n.s. (1975-76) 177-94.

4630. _____. "Realism and Reason." Proc. & Add. Am. Phil. Assn. 50 (1977) 483-98.

4631. _____. Meaning and the Moral Sciences. London and Boston: Routledge and Kegan Paul, 1978. 156 pp.

4632. Quine, Willard V. "On What There Is." Review of Metaphysics. 2 (1948) 21-38.

4633. _____. "Speaking of Objects." Proc. & Add. Am. Phil. Assn. 31 (1957-58) 5-22.

4634. _____. Ontological Relativity and Other Essays. New York: Columbia University Press, 1969. 165 pp.

4635. _____. "Existence," in Physics, Logic, and History. W. Yourgrau and A. D. Breck, eds. New York: Plenum Press, 1970. 89-98.

4636. _____. "Whither Physical Objects?" BSPS 39 (1976) 497-504.

4637. Rădulet, Remus. "Le statut des concepts et des relations abstraites specifiques aux sciences techniques." Noesis. 1 (1973) 65-75.

4638. Robert, Jean-Dominique. "Pensée et 'realites' scientifiques." Laval Théologique et Philosophique. 29 (1973) 165-86, 291-306.

4639. Rosenberg, Jay F. One World and Our Knowledge of It: The Problem of Realism in Post Kantian Perspective. Dordrecht: Reidel, 1980.

4640. Rottschaefer, William A. "Wilfred Sellars and the Demise of the Manifest Image." Modern Schoolman. 53 (1976) 398-404.

4641. Rouse, Joseph. "Kuhn, Heidegger, and Scientific Realism." Man and World. 14 (1981) 269-90.

4642. Schlegel, Richard H. "Contextualistic Realism." PPR 41 (1981) 437-51.

4643. Seely, Charles S. The Philosophy of Science: Essays in Contemporary Realism. New York: Philosophical Library, 1964. xii + 140 pp.

4644. Seigfried, Hans. "Scientific Realism and Phenomenology."
ZPF 34 (1980) 395-404.

4645. Sellars, Wilfrid. "The Language of Theories," in Current
Issues in the Philosophy of Science. H. Feigl and G. Maxwell, eds.
New York: Holt, Rinehart and Winston, 1961. 57-77.

    a. "Comments," by Norwood Russell Hanson. 77-82; by Paul K.
    Feyerabend, 82-83; by Richard Rudner, 84-89.

4646. _____. "Empiricism and Abstract Entities," in The
Philosophy of Rudolf Carnap. P. A. Schilpp, ed. La Salle, Illinois:
Open Court, 1963. 431-68.

4647. _____. Science, Perception and Reality. London: Routledge & Kegan Paul; New York: Humanities Press, 1963. 366 pp.

    a. Review essay by Keith Lehrer. J. Phil. 63 (1966) 266-77.

4648. _____. "Scientific Realism or Irenic Instrumentalism.'
BSPS 2 (1965) 171-204.

4649. _____. Science and Metaphysics. London: Routledge &
Kegan Paul, 1968. 246 pp.

    a. Review essay by Edward MacKinnon. Philosophical Forum.
    1 (1969) 509-45.

    b. Review essay, "Sellars' Semantics," by Gilbert Harman.
    Phil. Review. 79 (1970) 404-19.

    c. Review essay by Richard Rorty. Philosophy. 45 (1970)
    66-70.

    d. Review essay, "The Foundations of Scientific Realism: A
    Critical Review of Wilfrid Sellars' Science and Metaphysics,"
    by Bernard Gendron. IPQ 10 (1970) 129-51.

4650. _____. "Is Scientific Realism Tenable?" PSA 1976. II,
307-34.

4651. _____. Naturalism and Ontology. Reseda, California:
Ridgeview, 1979. 182 pp.

4652. Shapere, Dudley. "Natural Science and the Future of Metaphysics." BSPS 14, SL 60 (1974) 161-71.

4653. Sibley, W. M. "The Pragmatic Theory of Scientific Objects."
Phil. Review. 57 (1948) 248-50.

4654. Sinks, John D. "Fictionalism and the Elimination of Theoretical Terms." PS 39 (1972) 285-90.

4655. Smart, J. J. C. "The Reality of Theoretical Entities." AJP
34 (1956) 1-12.

a. "The Status of Theoretical Entities," by J. P. McKinney. 207-13.

4656. _____. Philosophy and Scientific Realism. London: Routledge & Kegan Paul, 1963. viii + 160 pp.

a. Review essay, "Some Prospects of Synoptic Philosophy," by Joseph Margolis. Ratio. 9 (1967) 105-21.

b. Review essay by M. C. Bradley. AJP 42 (1964) 262-83.

4657. Smith, Peter. Realism and the Progress of Science. New York: Cambridge University Press, 1981. 135 pp.

4658. Strombach, W. "Möglichkeiten und Grenzen des erkenntnistheoretischen Realismus in der heutigen Naturphilosophie." PN 17 (1978-79) 306-26.

4659. Theobald, W. D. "Observation and Reality." Mind. 76 (1967) 198-207.

4660. Thornton, J. B. "Scientific Entities." AJP 31 (1953) 1-21; 73-100.

4661. Tibbetts, P. "Observable versus Inferred Entities: Pragmatic and Phenomenological Considerations." Studium Generale. 24 (1971) 1067-78.

4662. Tuomela, Raimo. "Putnam's Realisms." Theoria. 45 (1979) 114-26.

4663. Wallace, O.P., William A. "The Reality of Elementary Particles." Proc. Am. Cath. Phil. Assn. 38 (1964) 154-66.

4664. _____. "Elementarity and Reality in Particle Physics (with an exchange of letters between E. K. Gora and W. Heisenberg)." BSPS 3 (1967) 236-63.

a. "Comment," by Mary B. Miller. 264-71.

4665. _____. From a Realist Point of View: Essays on the Philosophy of Science. Washington, D.C.: University Press of America, 1979. 375 pp.

4666. Werkmeister, W. H. "The Problem of Physical Reality." PS 19 (1952) 214-24.

4667. Wigner, Eugene P. "Two Kinds of Reality." Monist. 48 (1964) 248-64.

4668. Will, Frederick L. "Reason, Social Practice, and Scientific Realism." PS 48 (1981) 1-18.

4669. Wojciechowski, Jerzy A. "Science and the Notion of Reality." Revue de l'Université d'Ottawa. 31 (1961) Section Speciale, 25-38.

4670. _____. "Do Scientific Laws Give a True Image of Reality?"
Proc. Am. Cath. Phil. Assn. 37 (1963) 206-11.

4671. Yanase, M. M. "Hidden Realism." AJAPS 5, No. 5 (1980)
13-32.

4672. Zinkernagel, P. "On the Problem of Objective Reality as Con-
ceived in the Empiricist Tradition." Mind. 64 (1955) 501-12.

5.6 Scientific Change and Growth

4673. Agassi, Joseph. "Revolutions in Science, Occasional or Permanent?" Organon. 3 (1966) 47-61. Reprinted in BSPS 65 (1981)
104-18.

4674. _____. "Continuity and Discontinuity in the History of
Science." Journal of the History of Ideas. 34 (1973) 609-26.

4675. _____. Science in Flux. BSPS 28. SL 80. Dordrecht:
Reidel, 1975. 553 pp.

   a. Review essay by W. Berkson. Erkenntnis. 12 (1978)
   381-98.

   b. Review essay by John Kekes. Philosophia. 10 (1981)
   43-56.

4676. Agassi, Joseph, and John R. Wettersten. "Stegmüller Squared."
ZAW 11 (1980) 86-94.

4677. Almeder, Robert F. "Science and Idealism." PS 40 (1973)
242-54.

4678. Ambartsumian, V. A. "Materialist Dialectics -- The Methodology
and Logic of Development of Contemporary Natural Science." Soviet
Studies in Philosophy. 10 (1971-72) 210-17. [English reprint from
Voprosy filosofii. (1971, No. 3).]

4679. Amsterdamski, Stefan. "The Evolution of Science: Reformation
and Counter-Reformation." Diogenes. 89 (1975) 21-43.

4680. Andersson, Gunnar. "Presuppositions, Problems, Progress,"
BSPS 59. SL 126. (1979) 3-15.

4681. Audretsch, Jürgen. "Quantum Gravity and the Structure of
Scientific Revolutions." ZAW 12 (1981) 322-39.

4682. Austin, William H. "Paradigms, Rationality, and Partial Communication." ZAW 3 (1972) 203-18.

4683. Baillie, Patricia. "Kuhn's Inductivism." AJP 53 (1975)
54-57.

4684. Baldamus, W. "Das exoterische Paradox der Wissenschaftsfors-chung. Ein Beitrag zur Wissenschaftstheorie Ludwig Flecks." ZAW 10 (1979) 213-33.

4685. Balzer, W. "Incommensurability and Reduction." Acta Philo-sophica Fennica. 30, 2-4 (1978) 313-35.

4686. Barker, Peter. "Can Scientific History Repeat?" PSA 1980. I, 20-28.

4687. Bechler, Zev. "What Have They Done to Kuhn? An Ideological Introduction in Chiaroscuro." SL 145 (1981) 63-86.

    a. "Comment on Zev Bechler's Paper 'What Have They Done to Kuhn?'," by Robert E. Butts. 87-91.

    b. "Comments on Bechler, Niiniluoto, and Sadovsky," by Joseph D. Sneed. 93-104.

4688. Benjamin, A. Cornelius. "Some Theories of the Development of Science." PS 20 (1953) 167-76.

4689. Berkson, William. "Some Practical Issues in the Recent Con-troversy on the Nature of Scientific Revolutions." BSPS 14, SL 60 (1974) 197-210.

4690. Bibler, V. S., B. S. Griaznov, and S. R. Mikulinskij, eds. Očerki istorii i teorii razvitija nauki [Studies on the History and Theory of the Development of Science]. Moscow: Nauka, 1969. 421 pp.

4691. Binkley, Robert W. "Change of Belief or Change of Meaning?" SL 52 (1973) 55-76.

4692. Biser, Erwin, and Enos E. Witmer. "Methodology of Research and Progress in Science." PS 14 (1947) 275-88.

4693. Blachowicz, James A. "Systems Theory and Evolutionary Models of the Development of Science." PS 38 (1971) 178-99.

4694. Blackmore, John T. "Is Planck's 'Principle' True?" BJPS 29 (1978) 347-49.

4695. Böhler, Dietrich. "Paradigmawechsel in analytischer Wissensch-aftstheorie? Wissenschaftsgeschichte und Wissenschaftstheorie Aufgaben der Philosophie." ZAW 3 (1972) 219-42.

4696. Böhme, G. "Die Bedeutung praktischer Argumente für die Ent-wicklung der Wissenschaft." PN 15 (1974-75) 133-51.

4697. _____. "On the Probability of 'Closed Theories'." SHPS 11 (1980) 163-72.

4698. Bronowski, J. "Humanism and the Growth of Knowledge," in The Philosophy of Karl Popper. P. A. Schilpp, ed. La Salle, Illinois: Open Court, 1974. 606-31.

4699. Brown, Harold I. "Problem Changes in Science and Philosophy." Metaphilosophy. 6 (1975) 177-92.

4700. _____. "Paradigmatic Propositions." Am. Phil. Quart. 12 (1975) 85-90.

4701. _____. "Reduction and Scientific Revolutions." Erkenntnis. 10 (1976) 381-85.

4702. _____. "For a Modest Historicism." Monist. 60 (1977) 540-55.

4703. Burian, Richard M. "Conceptual Change, Cross-Theoretical Explanation, and the Unity of Science." Synthese. 32 (1975) 1-28.

4704. Cedarbaum, Daniel. "Relativism, Realism, and Rationality: A Defense of Kuhn's Structure of Scientific Revolutions." Synthesis. 4, No. 4 (1979) 38-49.

4705. Chalmers, Alan. "Towards an Objectivist Account of Theory Change." BJPS 30 (1979) 227-33.

4706. Changeux, Jean-Pierre, et al. "Les progrès des sciences du système nerveaux concernent-ils les philosophes?" Bulletin de la Société Française de Philosophie. 75 (1981) 73-105.

4707. Clay, J. "Combination of Methods." Synthese. 8 (1950-51) 471-79. Reprinted in 10 (n.d.) 111-19.

4708. Cohen, I. Bernard. "The Eighteenth-Century Origins of the Concept of Scientific Revolution." Journal of the History of Ideas. 37 (1976) 257-88.

4709. Cohen, Robert S., and Joseph Agassi. "Dinosaurs and Horses, or: Ways with Nature." Synthese. 32 (1975) 233-48.

4710. Cole, Richard. "Theoretical Becoming." The Philosophy Forum. 11 (1972) 265-84.

4711. Crombie, A. C., ed. Scientific Change: Historical Studies in the Intellectual, Social and Technical Conditions for Scientific Discovery and Technical Invention, from Antiquity to the Present. London: Heinemann, 1963. 896 pp.

4712. Cunningham, F. "Kuhn on Scientific Creativity: An Engelsian Critique." Dialectics and Humanism. 5, No. 3 (1978) 73-80.

4713. Darden, Lindley. "Reasoning in Scientific Change: Charles Darwin, Hugo de Vries, and the Discovery of Segregation." SHPS 7 (1976) 127-69.

4714. Deakin, Michael A. B. "The Impact of Catastrophe Theory on the Philosophy of Science." Nature and System. 2 (1980) 173-88.

4715. Detel, Wolfgang. "Methode and Erkenntnisfortschritt. Kritische Bemerkungen zum Verhältnis von Wissenschaftstheorie und Wissenschaftsgeschichte." ZAW 8 (1977) 237-56.

    a. "Zu 'Methode und Erkenntnisfortschritt' von Wolfgang Detel," by Werner Kutschmann. ZAW 11 (1980) 108-14.

4716. Devitt, Michael. "Against Incommensurability." AJP 57 (1979) 29-50.

4717. Diemer, Alwin, ed. Die Struktur wissenschaftlicher Revolutionen and die Geschichte der Wissenschaften. Meisenheim am Glan: Hain, 1977. 119 pp.

4718. Dilworth, Craig. "On the Nature of the Relation between Successive Scientific Theories." Epistemologia. 1 (1978) 43-76.

4719. _____. Scientific Progress: A Study Concerning the Nature of the Relation between Successive Scientific Theories. SL 153. Dordrecht: Reidel, 1981. 155 pp.

4720. Dolby, R. G. A. "Controversy and Consensus in the Growth of Scientific Knowledge." Nature and System. 2 (1980) 199-218.

4721. Domotor, Zoltan. "Probability Kinematics and Representation of Belief Change." PS 47 (1980) 384-403.

4722. Doppelt, Gerald. "Kuhn's Epistemological Relativism: An Interpretation and Defense." Inquiry. 21 (1978) 33-86.

    a. "Epistemological Relativism in Its Latest Form," by Harvey Siegel. 23 (1980) 107-17.

    b. "A Reply to Siegel on Kuhnian Relativism," by Gerald Doppelt. 117-23.

4723. Düsberg, Klaus Jürgen. "Stegmuller über 'wissenschaftliche Revolutionen'." ZAW 8 (1977) 331-41.

4724. Erpenbeck, J., and U. Röseberg. "Wissenschaftsentwicklung, Theorienentwicklung und Entwicklungstheorie." DZP 25 (1977) 133-49.

4725. Feyerabend, P. K. "Consolations for the Specialist," in Criticism and the Growth of Knowledge. I. Lakatos and A. Musgrave, eds. Cambridge: Cambridge University Press, 1970. 197-230.

4726. Field, Hartry. "Theory Change and Indeterminacy of Reference." J. Phil. 70 (1973) 462-81.

    a. "Against Indeterminacy," by John Earman and Arthur Fine. 74 (1977) 535-38.

4727. Fine, Arthur. "How to Compare Theories: Reference and Change." Nous. 9 (1975) 17-32.

4728. Foster, Stephen. "Historiography and Epistemology in Kuhn." Kinesis. 10 (1979) 3-13.

4729. Frankel, Henry. "The Non-Kuhnian Nature of the Recent Revolution in the Earth Sciences." PSA 1978. II, 197-214.

4730. _____. "The Career of Continental Drift Theory: An Application of Imre Lakatos' Analysis of Scientific Growth to the Rise of Drift Theory." SHPS 10 (1979) 21-66.

4731. _____. "Problem-Solving, Research Traditions, and the Development of Scientific Fields." PSA 1980. I, 29-40.

4732. Gaa, James. "The Replacement of Scientific Theories: Reduction and Explication." PS 42 (1975) 349-72.

4733. Gagnon, Maurice. "Piaget et Kuhn sur l'evolution de la connaissance: une comparaison." Dialogue. 17 (1978) 35-55.

4734. Germain, Paul. "Sur quelques caracteristiques des disciplines scientifiques et sur la portée de la science." Les Etudes philosophiques. 33 (1978) 157-70.

4735. Ginzburg, V. L. "Notes on the Methodology and Evolution of Physics and Astrophysics." Soviet Studies in Philosophy. 20 (1981) 40-82. [English reprint from Voprosy filosofii. (1980, No. 12).]

4736. Glas, E. "Methodology and the Emergence of Physiological Chemistry." SHPS 9 (1978) 291-312.

4737. Goudge, Thomas A. "Peirce and Rescher on Scientific Progress and Economy of Research." Dialogue. 20 (1981) 357-64.

4738. Graham, Loren R. Between Science and Values. New York: Columbia University Press, 1981. 448 pp.

4739. Groenewold, H. J. "Non-Scientific Elements in the Development of Science." Synthese. 10 (n.d.) 293-311.

4740. Grunfeld, Joseph. "Progress in Science." Logique et analyse. 22 (1979) 208-21.

4741. Gutting, Gary. "Conceptual Structures and Scientific Change." SHPS 4 (1973) 209-30.

4742. _____. Paradigms and Revolutions: Applications and Appraisals of Thomas Kuhn's Philosophy of Science. Notre Dame: University of Notre Dame Press, 1980. 339 pp.

4743. Hanson, Norwood Russell. "A Note on Kuhn's Method." Dialogue. 4 (1965) 371-75.

4744. Harper, William L. "Rational Belief Change, Popper Functions and Counterfactuals." WOSPS 6-I (1976) 73-112. Reprinted from Synthese. 30 (1975) 221-62.

a. "Letter" by Robert Stalnaker to W. L. Harper. WOSPS 6-I
(1976) 113-15.

b. "Ramsey Test Conditionals and Iterated Belief Change (A
Response to Stalnaker)," by William L. Harper. 117-35.

4745. _____. "Rational Conceptual Change." PSA 1976. II,
462-94.

4746. _____. "Conceptual Change, Incommensurability and
Special Relativity Kinematics." Acta Philosophica Fennica. 30, 2-4
(1978) 430-61.

4747. Harré, Rom, ed. Problems of Scientific Revolution: Progress
and Obstacles in the Sciences. Oxford: Oxford University Press,
1975. viii + 104 pp.

a. Review essay by Michael Ruse. Erkenntnis. 13 (1978)
407-16.

4748. Hattiangadi, J. N. "Alternatives and Incommensurables: The
Case of Darwin and Kelvin." PS 38 (1971) 502-07.

4749. Heelan, Patrick A. "Heisenberg and Radical Theoretic Change."
ZAW 6 (1975) 113-36.

4750. _____. "The Lattice of Growth in Knowledge." BSPS 59.
SL 136. (1979) 205-11.

4751. Heidelberger, Michael. "Towards a Logical Reconstruction of
Revolutionary Change: The Case of Ohm as an Example." SHPS (1980)
103-21.

4752. Heisenberg, W. "Änderungen der Denkstruktur im Fortschritt
der Wissenschaft." Studium Generale. 23 (1970) 808-16.

4753. Hendrick, R. E., and Anthony Murphy. "Atomism and the Illusion
of Crisis: The Danger of Applying Kuhnian Categories to Current
Particle Physics." PS 48 (1981) 454-68.

4754. Hertzberg, L. "Criteria and the Philosophy of Science."
Acta Philosophica Fennica. 30, 2-4 (1978) 31-50.

4755. Hesse, Mary. Revolutions and Reconstructions in the Philosophy
of Science. Bloomington, IN: Indiana University Press, 1980. xxvi
+ 271 pp.

4756. Hockney, Donald. "Conceptual Structures." SL 52 (1973)
141-66.

4757. Holton, Gerald. "A Heuristic Model for the Growth Process of
Modern Physical Science." Synthese. 10 (n.d.) 190-202.

4758. Hooker, C. A. "Empiricism, Perception and Conceptual Change."
Canadian Journal of Philosophy. 3 (1973) 59-75.

4759. Hosiasson-Lindenbaum, Janina. "Theoretical Aspects of the Advancement of Knowledge." Synthese. 7 (1948-49) 253-61.

4760. Hoyer, U. "Theoriewandel und Strukturerhaltung-das Beispiel der Thermodynamik." PN 16 (1976-77) 421-36.

4761. Hübner, Kurt. "Zur Frage des Relativismus und des Fortschritts in den Wissenschaften." ZAW 5 (1974) 285-303. English Tr.: Man and World. 7 (1974) 394-413.

4762. Hucklenbroich, Peter. Theorie des Erkenntnisfortschritts: zum Verhältnis von Erfahrung and Methoden in den Naturwissenschaften. Meissenheim am Glan: Hain, 1978. 524 pp.

4763. Jones, Gary E. "Kuhn, Popper, and Theory Comparison." Dialectica. 35 (1981) 389-97.

4764. Kalikow, Theo J. "History of Konrad Lorenz's Ethological Theory, 1927-1939: The Role of Meta-Theory, Theory, Anomaly and New Discoveries in a Scientific 'Evolution'." SHPS 6 (1975) 331-41.

4765. Kannegiesser, K., R. Rockhausen, and A. Thom. "Entwicklungsprobleme einer marxistisch-leninistischen philosophischen Wissenschaftstheorie." DZP 17 (1969) 1054-75.

4766. Kantorovich, Aharon. "Towards a Dynamic Methodology of Science." Erkenntnis. 14 (1979) 251-73.

4767. Katz, Jerrold J. "Semantics and Conceptual Change." Phil. Review. 88 (1979) 327-65.

4768. Kearney, H. Science and Change. London: Weidenfeld and Nicolson, 1971. 256 pp.

4769. Kedrov, B. M. "V. I. Lenin on the Dialectics of the Development of Natural Science." Soviet Studies in Philosophy. 10 (1971-72) 231-39. [English reprint from Voprosy filosofii. (1971, No. 3).]

4770. _____. "On Scientific Revolutions and Their Typology." Scientia. I: 114 (1979) 675-92. Italian Tr.: 693-706. II: 115 (1980) 5-21. Italian Tr.: 23-36.

4771. King, M. D. "Reason, Tradition, and the Progressiveness of Science." History and Theory. 10 (1971) 3-32.

4772. Kitchener, Richard F. "Piaget's Genetic Epistemology." IPQ 20 (1980) 377-405.

4773. Kitcher, Philip. "Theories, Theorists and Theoretical Change." Phil. Review. 85 (1978) 519-47. Reprinted in The Philosopher's Annual. 2 (1979) 128-54.

4774. Klaus, Georg. Rationalität, Integration, Information: Entwicklungsgesetze der Wissenschaft in unserer Zeit. München, W. Fink, 1974. 301 pp.

4775. Kleiner, Scott A. "Erotetic Logic and the Structure of Scientific Revolution." BJPS 21 (1970) 149-65.

4776. Klüver, Jürgen, and Wilfried Müller. "Wissenschaftstheorie und Wissenschaftsgeschichte: Die Entdeckung der Benzolformel." ZAW 3 (1972) 243-66.

4777. Kmita, Jerzy. "The Controversy about the Principles of the Development of Science." Poznań Studies. 1, No. 2 (1975) 65-74.

4778. Kneale, William. "Scientific Revolution for Ever?" BJPS 19 (1968) 27-42.

4779. Knorr, Karin D., Herman Strasser, and Hans Georg Zilian, eds. Determinants and Controls of Scientific Developments. Dordrecht: Reidel, 1975. 459 pp.

4780. Kockelmans, Joseph J. "On the Meaning of Scientific Revolutions." The Philosophy Forum. 11 (1972) 243-64.

4781. Koertge, Noretta. "Theory Change in Science." SL 52 (1973) 167-98.

4782. Koeze, J. "Enkele basisbegrippen van Kuhn." Algemeen Nederlands Tijdschrift voor Wijsbegeerte. 67 (1975) 85-95.

4783. Kordig, Carl R. "On Prescribing Description." Synthese. 18 (1968) 459-61.

4784. _____. The Justification of Scientific Change. SL 36 Dordrecht: Reidel, 1971. 119 pp.

      a. "Essay Review: Understanding Scientific Change," by Peter K. Machamer. SHPS 5 (1975) 373-81.

4785. _____. "Objectivity, Scientific Change, and Self-Reference." BSPS 8, SL 39 (1971) 519-23.

4786. _____. "Scientific Transitions, Meaning Invariance, and Derivability." Southern Journal of Philosophy. 9 (1971) 119-25.

      a. "A Note on Meaning Invariance," by Kathryn Pyne Parsons. 126-30.

4787. Körner, Stephan. Categorial Frameworks. New York: Barnes & Noble, 1970. 85 pp.

4788. _____. "Logic and Conceptual Change." SL 52 (1973) 123-36.

      a. "Some Comments on Professor Körner's Poper," by Joseph S. Ullian. 137-40.

4789. Kourany, Janet A. "The Nonhistorical Basis of Kuhn's Theory of Science." Nature and System. 1 (1979) 46-59.

4790. Krah, Wolfgang. "Zum exponentiellen Wachstum der naturwissenschaftlichen Erkenntnisse." DZP 12 (1964) 65-75.

4791. _____. "Zur Forderung nach theoretischem Pluralismus in Permanenz." ZAW 11 (1980) 321-31.

4792. Krajewski, Władysław. Correspondence Principle and Growth of Science. Dordrecht: Reidel, 1977. 138 pp.

4793. Krausser, Peter. "A Cybernetic Systemstheoretical Approach to Rational Understanding and Explanation, Especially of Scientific Revolutions with Radical Meaning Change." Ratio 15 (1973) 221-46.

4794. Krige, John. Science, Revolution, and Discontinuity. Brighton, Sussex: Harvester Press, 1980. 231 pp.

4795. Krüger, Lorenz. "Falsification, Revolution and Continuity in the Development of Science." LMPS-4 (1971) 333-43.

4796. _____. "Wissenschaftliche Revolutionen und Kontinuität der Erfahrung." Neue Hefte für Philosophie. 6/7 (1974) 1-26.

4797. _____. "Intertheoretic Relations as a Tool for the Rational Reconstruction of Scientific Development." SHPS 11 (1980) 89-101.

4798. Kuhn, Thomas S. The Copernican Revolution: Planetary Astronomy in the Development of Western Thought. Cambridge, Mass.: Harvard University Press, 1957. xx + 297 pp.

4799. _____. The Structure of Scientific Revolutions. Chicago: University of Chicago Press, 1962. 172 pp. Second Edition: Enlarged: 1970. 210 pp. Vol. II, No. 2 of International Encyclodia of Unified Science.

    a. Review essay by Dudley Shapere. Phil. Review. 73 (1964) 383-94.

    b. Review essay by Gavin Ardley. Philosophical Studies. (Ireland) 13 (1964) 183-92.

    c. Review essay, "Kuhn on Scientific Revolutions," by Richard L. Purtill. PS 34 (1967) 53-58.

    d. Review essay by Kurt Hübner. Philosophische Rundschau. 15 (1968) 185-95.

    e. Review essay, "Kuhn's Second Thoughts," by A. E. Musgrave. BJPS 22 (1971) 287-97.

    f. "A Note on Thomas S. Kuhn's The Structure of Scientific Revolutions," by Maurice Mandelbaum. Monist. 60 (1977) 445-52.

4800. _____. "The Function of Dogma in Scientific Research," in Scientific Change. A. C. Crombie, ed. London: Heinemann, 1963. 347-69.

4801. _____. "Reflections on my Critics," in Criticism and the Growth of Knowledge. I. Lakatos and A. Musgrave, eds. Cambridge: Cambridge University Press, 1970. 231-78.

4802. _____. "Second Thoughts on Paradigms," in The Structure of Scientific Theories. F. Suppe, ed. Urbana, Illinois: University of Illinois Press, 1974. 459-82.

    a. "Exemplars, Theories and Disciplinary Matrixes," by Frederick Suppe. 483-99.

    b. "Second Thoughts on Paradigms" is reprinted in T. S. Kuhn, The Essential Tension. Chicago: University of Chicago Press, 1977. 293-319.

4803. _____. "Logic of Discovery or Psychology of Research," in The Philosophy of Karl Popper. P. A. Schilpp, ed. La Salle, Illinois: Open Court, 1974. 798-819. Pre-printed in Criticism and the Growth of Knowledge. I. Lakatos and A. Musgrave, eds. Cambridge: Cambridge University Press, 1970. 1-23. Reprinted in T. S. Kuhn, The Essential Tension. Chicago: University of Chicago Press, 1977. 266-92.

4804. _____. "Theory Change as Structure-Change: Comments on the Sneed Formalism." Erkenntnis. 10 (1976) 179-99. Also in WOSPS 12 (1977) 289-309.

4805. _____. "Mathematical versus Experimental Traditions in the Development of Physical Science." The Journal of Interdisciplinary History. 7 (1976) 1-31. Reprinted in T. S. Kuhn, The Essential Tension. Chicago: University of Chicago Press, 1977. 31-65.

4806. _____. The Essential Tension: Selected Studies in Scientific Tradition and Change. Chicago: University of Chicago Press, 1977. 366 pp.

    a. Review essay, "Objectivity, Rationality, Incommensurability, and More," by Harvey Siegel. BJPS 31 (1980) 359-75.

    b. Review essay by Ian Hacking. History and Theory. 18 (1979) 223-36.

    c. Review essay, "Thomas Kuhn e il libro della natura," by E. Bellone. Scientia. 113 (1978) 675-88.

4807. _____. Die Entstehung des Neuen: Studien zur Struktur der Wissenschaftsgebiete. Lorenz Krüger, hrsg. Frankfurt am Main: Suhrkamp, 1977. 472 pp.

4808. Lacharité, Normand. "Le développement des sciences est-il un procès normé? Faut-il choisir entre Kuhn, Feyerabend et Popper?" Dialogue. 17 (1978) 616-33.

4809. Laudan, Larry. "Two Dogmas of Methodology." PS 43 (1976) 585-97.

    a. "Progress Requires Invariance," by Carl R. Kordig. 47 (1980) 141.

4810. _____. Progress and Its Problems: Towards a Theory of Scientific Growth. Berkeley: University of California Press, 1977. x + 257 pp.

    a. "Towards a Richer Model of Man: A Critique of Laudan's Progress and Its Problems," by Robert S. Westman. PSA 1978. II, 493-504.

    b. "In Praise of Truth and Substantive Rationality: Comments on Lauden's Progress and Its Problems," by N. Koertge. PSA 1978, II, 505-21.

    c. "Some Problems about Solving Problems," by D. H. Mellor. PSA 1978. II, 522-29.

    d. "The Philosophy of Progress. . .," by L. Laudan. PSA 1978. II, 530-47.

    e. Review essay by Ernan McMullin. PS 46 (1979) 623-44.

    f. Review essay by John Lossee. SHPS 9 (1978) 333-40.

    g. "Some Problems for Progress and Its Problems," by H. Krips. PS 47 (1980) 601-16.

    h. "Discussion: Anomalous Anomalies," by Larry Laudan. PS 48 (1981) 618-19. [Reply to Krips]

    i. "Laudan's Progress and Its Problems," by David L. Hull. Philosophy of the Social Sciences. 9 (1979) 457-65.

    j. "Laudan and the Problem-Solving Approach to Scientific Progress and Rationality," by Andrew Lugg. 466-74.

    k. "Scientific Progress: The Laudan Manifesto," by Robert E. Butts. 475-83.

    1. "Laudan's Problematic Progress and the Social Sciences," by I. C. Jarvie. 484-97.

    m. "Views of Progress: Separating the Pilgrims from the Rakes," by Larry Laudan. 10 (1980) 273-86.

    n. Review essay by Gary Gutting. Erkenntnis. 15 (1980) 91-103.

    o. Review essay, "Laudan's Problems," by B. Baigrie and J. N. Hattiangadi. Metaphilosophy. 12 (1981) 85-95.

p. Review essay, "Problems with Progress," by Alan Musgrave. Synthese. 42 (1979) 443-64.

q. Review essay, "More Clothes from the Emperor's Bargain Basement," by Paul Fitzgerald. BJPS 32 (1981) 57-71.

r. Review essay, "Laudan's Pragmatic Alternative to Positivist and Historicist Theories of Science," by Gerald Doppelt. Inquiry. 24 (1981) 253-81.

4811. Laudan, Rachel. "The Recent Revolution in Geology and Kuhn's Theory of Scientific Change." PSA 1978. II, 227-39.

4812. Laszlo, E. "A General Systems Model of the Evolution of Science." Scientia. 107 (1972) 379-95.

4813. _____. "The Ideal Scientific Theory: A Thought Experiment." PS 40 (1973) 75-87.

4814. Lehrer, Keith. "Evidence and Conceptual Change." SL 51 (1973) 100-07.

4815. _____. "Evidence, Meaning and Conceptual Change: A Subjective Approach." SL 52 (1973) 94-122.

4816. Leplin, Jarrett. "Truth and Scientific Progress." Studies in History and Philosophy of Science. 12 (1981) 269-91.

4817. Levin, Michael E. "On Theory-Change and Meaning-Change." PS 46 (1979) 407-24.

4818. Lorenz, Kuno. "About Limits of Growth for Scientific Theories." Grazer Philosophische Studien. 12/13 (1981) 79-83.

4819. Lugg, Andrew. "Theory Choice and Resistance to Change." PS 47 (1980) 227-43.

4820. MacIntyre, Alisdair. "Epistemological Crises, Dramatic Narrative and the Philosophy of Science." Monist. 60 (1977) 453-72.

4821. McEvoy, John G. "A 'Revolutionary' Philosophy of Science: Feyerabend and the Degeneration of Critical Rationalism into Sceptical Fallibilism." PS 42 (1975) 49-66.

4822. Machan, Tibor R. "Kuhn, Paradigm Choice and the Arbitrariness of Aesthetic Criteria in Science." Theory and Decision. 8 (1977) 361-62.

4823. Mamchur, E. A. Problems of Theory Selection. [In Russian] Moscow: Nauka, 1975.

4824. Mare, Călina. "Contemporary Epistemology about the Development of Scientific Knowledge." Noesis. 6 (1980) 169-77.

4825. Masterman, Margaret. "The Nature of a Paradigm," in Criticism
and the Growth of Knowledge. I. Lakatos and Alan Musgrave, eds.
Cambridge: Cambridge University Press, 1970. 59-89.

4826. Mertz, Donald W. "Analisi delle Weltanschauungen di Kuhn:
Il problema dell'incommensurabilita Teoria-Natura," in La Scommessa
della verita. Milano: Spirali Edizioni, 1981. 96-111.

4827. Michalos, Alex C. "Theory Appraisal and the Growth of Scientific Knowledge." SHPS 1 (1970-71) 353-61.

4828. Mittelstrass, Jurgen. "Towards a Normative Conception of the
Growth of Knowledge." Nature and System. 2 (1980) 231-44.

4829. Motycka, A. "Kuhn's Sociological Principle of Demarcation."
Poznan Studies. 4 (1978) 264-71.

4830. Moulines, C. Ulises. "Theory-Nets and the Evolution of
Theories: The Example of Newtonian Mechanics." Synthese. 41
(1979) 417-39.

4831. Musgrave, A. "How to Avoid Incommensurability?" Acta Philosophica Fennica. 30, 2-4 (1978) 336-46.

4832. Newton-Smith, William H. "On the Rational Explanation of
Scientific Change." Grazer Philosophische Studien. 12/13 (1981)
47-77.

4833. Niiniluoto, Ilkka. "Verisimilitude, Theory-Change, and
Scientific Progress." Acta Philosophica Fennica. 30, 2-4 (1978)
243-64.

4834. _____. "Scientific Progress." Synthese. 45 (1980)
427-62.

4835. _____. "The Growth of Theories: Comments on the Structuralist Approach." SL 145 (1981) 3-47.

4836. Niiniluoto, Ilkka, and Raimo Tuomela, eds. The Logic and
Epistemology of Scientific Change. Amsterdam: North-Holland Publishing Co., 1979. 461 pp.

4837. Nitschke, August. Revolutionen in Naturwissenschaft und
Gesellschaft. Stuttgart-Bad Cannstatt: Frommann-Holzboog, 1979.
207 pp.

4838. Novakovic, Stanisa. "Is the Transition from an Old Theory to
a New One of a Sudden and Unexpected Character?" BSPS 14, SL 60
(1974) 173-96.

4839. Nowakowa, Izabella. "On the Notion of Correspondence."
Poznan Studies. 1, No. 2 (1975) 75-80.

        a. "A Note on the Implicational Concept of Correspondence,"
        by Lech Witkowski. 2, No. 3 (1976) 120-21.

b. "When do Two Theorems Correspond?" by W. Patryas. 3 (1977) 253-55.

4840. Parvu, Ilie. "Riemann vs. Kant: A Case Study for a New Historiography of the Philosophy of the Sciences." Noesis. 6 (1980) 179-91.

4841. Paterson, Antoinette M. "Velikovsky Versus Academic Lag (The Problem of Hypothesis)." BSPS 32 (1976) 487-98.

4842. Pearce, Glenn and Patrick Maynard, eds. Conceptual Change. SL 52. Dordrecht: Reidel, 1973. 282 pp.

4843. Pedersen, S. A. "Logic and Ontology in the Study of Theory Change." Poznan Studies. 3 (1977) 42-92.

4844. Pietruska-Madej, E. Methodological Problems of Chemical Revolutions. [In Polish] Warsaw: Polish Scientific Publishers, 1975.

4845. _____. "Anomalies and the Dynamics of Scientific Theories." Poznan Studies. 4 (1978) 252-63.

4846. Pitt, Joseph C. "Conceptual Change and Conceptual Tension." Methodology and Science. 14 (1981) 132-38.

4847. Pitt, Joseph C., and Morton Tavel. "Revolutions in Science and Refinements in the Analysis of Causation." ZAW 8 (1977) 48-62.

4848. Pittioni, Veit. "Kritische Bemerkungen zu Kuhns Wissenschaftsauffassung." Conceptus. 8, No. 25 (1974) 91-94.

4849. Poldrack, H. "Kritische Bemerkungen zu Th. S. Kuhns Theorie der Wissenschaftsentwicklung." DZP 29 (1981) 231-41.

4850. Popper, K. R. "Normal Science and Its Dangers," in Criticism and the Growth of Knowledge. I. Lakatos and A. Musgrave, eds. Cambridge: Cambridge University Press, 1970. 51-58.

4851. Przełęcki, M. "Commensurable Referents of Incommensurable Theories." Acta Philosophica Fennica. 30, 2-4 (1978) 347-65.

4852. Quay, S.J., Paul M. "Progress as a Demarcation Criterion for the Sciences." PS 41 (1974) 154-70.

4853. Rabb, J. Douglas. "Incommensurable Paradigms and Critical Idealism." SHPS 6 (1975) 343-46.

4854. Radnitzky, Gérard. "Das Problem der Theorienbewertung. Begrundungsphilosophischer, skeptischer und fallibilistischer Denkstil in der Wissenschaftstheorie." ZAW 10 (1979) 67-97.

4855. Radnitzky, Gérard and Gunnar Andersson, eds. Progress and Rationality in Science. BSPS 58, SL 125. Dordrecht: Reidel, 1978.

400 pp. Revised German version: <u>Fortschritt und Rationalitat der Wissenschaft</u>. Tübingen: J. C. B. Mohr, 1980. 482 pp.

4856. _____. <u>The Structure and Development of Science</u>. SL 136. BSPS 59. Dordrecht: Reidel, 1978. x + 416 pp. Revised German version: <u>Voraussetzungen und Grenzen der Wissenschaft</u>. Tubingen: J. C. B. Mohr, 1981.

4857. Rapp, Friedrich. "Observational Data and Scientific Progress." SHPS 11 (1980) 153-62.

4858. Rescher, Nicholas. <u>Scientific Progress: A Philosophical Essay on the Economics of Research in Natural Science</u>. Pittsburgh: University of Pittsburgh Press, 1978. xiv + 278 pp.

    a. Review essay by Michael Ruse. <u>Nous</u>. 15 (1981) 418-23.

    b. Review essay by Jeff Foss. <u>Canadian Journal of Philosophy</u>. 11 (1981) 761-73.

4859. _____. "Some Issues Regarding the Completeness of Science and the Limits of Scientific Knowledge." BSPS 59. SL 136. (1979) 19-40.

4860. Richards, Robert J. "The Natural Selection Model of Conceptual Evolution." PS 44 (1977) 494-501.

    a. "Comment on 'The Natural Selection Model of Conceptual Evolution'," by Donald T. Campbell. 502-07.

4861. Rochhausen, R. "Die dialektische Einheit 'innerer' und 'ausserer' wissenschaftlicher Determinanten in der Entwicklung naturwissenschaftlicher Theorien." DZP 25 (1977) 552-63.

4862. _____. "Zu einigen Fragen des Erkenntnisfortschritts in der Wissenschaft." DZP 28 (1980) 1048-55.

4863. Rorty, R. "From Epistemology to Hermeneutics." <u>Acta Philosophica Fennica</u>. 30, 2-4 (1978) 11-30.

4864. Rosenberg, Jay F. "Coupling, Retheorizing, and the Correspondence Principle." <u>Synthese</u>. 45 (1980) 351-85.

4865. Ruse, Michael. "What Kind of Revolution Occurred in Geology?" <u>PSA 1978</u>. II, 240-73.

4866. _____. "The Revolution in Biology." <u>Theoria</u>. 36 (1970) 1-22.

4867. Sadovsky, V. N. "Logic and the Theory of Scientific Change." SL 145 (1981) 49-61.

4868. Sagal, Paul T. "Incommensurability Then and Now." ZAW 3 (1972) 298-301.

4869. Savary, Claude. "La conception kuhnienne de la science et le concept d'idéologie." Dialogue. 17 (1978) 266-85.

4870. Schagrin, Morton L. "On Being Unreasonable." PS 40 (1973) 1-9.

    a. "Back to Being Reasonable." by Tibor R. Machan and M. L. Zupan. 42 (1975) 307-10.

    b. "A Response to Machan and Zupan," by Morton Schagrin. 311.

4871. Scheffler, Israel. "Vision and Revolution: A Postscript on Kuhn." PS 39 (1972) 366-74.

4872. Scheibe, Erhard. "Conditions of Progress and the Comparability of Theories." BSPS 39 (1976) 547-68.

4873. Scheurer, Paul. Révolutions de la science et permanence du réel. Paris: Presses Universitaires de France, 1979. 366 pp.

4874. Schopman, Joop. "The History of Semiconductor Electronics -- A Kuhnian Story?" ZAW 12 (1981) 297-302.

4875. Schramm, Alfred. Theorienwandel oder Theorienfortschritt? Wien: VWGO Verb. d. Wissenschaftl. Gesellschaften Oesterreichs, 1975. 187 pp.

4876. Sellars, Wilfrid. "Conceptual Change." SL 52 (1973) 77-93.

4877. Shapere, Dudley. "Meaning and Scientific Change," in Mind and Cosmos. Robert G. Colodny, ed. Pittsburgh: University of Pittsburgh Press, 1966. 41-85.

4878. _____. "Plausibility and Justification in the Development of Science." J. Phil. 63 (1966) 611-21.

    a. "Plausibility of New Hypotheses," by Thomas A. Goudge. 621-24.

    b. "The Plausibility of Theories," by Stephen Toulmin. 624-27.

4879. _____. "What Can the Theory of Knowledge Learn from the History of Knowledge?" Monist. 60 (1977) 488-508.

4880. _____. "The Character of Scientific Change." BSPS 56 (1980) 61-101.

4881. Shimony, Abner. "Comments on Two Epistemological Theses of Thomas Kuhn." BSPS 39 (1976) 569-88.

4882. Short, T. L. "An Analysis of Conceptual Change." Am. Phil. Quart. 17 (1980) 301-09.

4883. Shrader, Douglas. "The Evolutionary Development of Science." Review of Metaphysics. 34 (1980) 273-96.

4884. Siemens, Warren D. "A Logical Empiricist Theory of Scientific Change?" BSPS 8, SL 39 (1971) 524-35.

4885. Skagestad, Peter. "C. S. Peirce on Biological Evolution and Scientific Progress." Synthese. 41 (1979) 85-114.

4886. Sklar, Lawrence. "Do Unborn Hypotheses Have Rights?" Pacific Philosophical Quarterly. 62 (1981) 17-29.

4887. Smirnov, S. N. "External Diversity and Internal Uniformity of Scientific Growth." Acta Philosophica Fennica. 30, 2-4 (1978) 91-110.

4888. Smolicz, J. J. "Kuhn Revisited: Science, Education and Values." Organon. 10 (1974) 45-59.

4889. Sneed, Joseph D. "Philosophical Problems in the Empirical Science of Science: A Formal Approach." Erkenntnis. 10 (1976) 115-46.

4890. _____. "Describing Revolutionary Scientific Change: A Formal Approach." WOSPS 12 (1977) 245-68.

4891. Sparkes, J. J. "Pattern Recognition and Scientific Progress." Mind. 81 (1972) 29-41.

4892. Stegmüller, W. "Structures and Dynamics of Theories: Some Reflections on J. D. Sneed and T. S. Kuhn." Erkenntnis. 9 (1975) 75-100. Spanish Tr.: Dianoia. 21 (1975) 60-84.

4893. _____. "Accidental ('Non-substantial') Theory Change and Theory Dislodgement: To What Extent Logic Can Contribute to a Better Understanding of Certain Phenomena in the Dynamics of Theories." Erkenntnis. 10 (1976) 147-78. Also in WOSPS 12 (1977) 269-88.

4894. _____. "A Combined Approach to the Dynamics of Theories. How to Improve Historical Interpretations of Theory Change by Applying Set Theoretical Structures." BSPS 59. SL 136. (1979) 151-86. Reprinted from Theory and Decision. 9 (1978) 39-75.

4895. _____. "The Structuralist View: Survey, Recent Developments and Answers to Some Criticisms." Acta Philosophica Fennica. 30, 2-4 (1978) 113-29.

4896. Strauss, M. "Logical, Ontological and Methodological Aspects of Scientific Revolutions." WOSPS 12 (1977) 31-49.

4897. Ströker, Elisabeth. "Geschichte als Herausforderung." Neue Hefte für Philosophie. 6/7 (1974) 27-66.

4898. Szumilewicz, Irena. "Incommensurability and the Rationality of the Development of Science." BJPS 28 (1977) 345-50.

4899. Thagard, Paul. "Against Evolutionary Epistemology." PSA 1980. I, 187-96.

4900. Törnebohm, Håkan. "The Growth of a Theoretical Model: A Simple Case Study," in Physics, Logic, and History. W. Yourgrau and A. D. Breck, eds. New York: Plenum Press, 1970. 79-86. "Discussion," 86-88.

4901. _____. "Paradigms in Fields of Research." Acta Philosophica Fennica. 30, 2-4 (1978) 62-90.

4902. Toulmin, Stephen. "Conceptual Revolutions in Science." Synthese. 17 (1967) 75-91. Also in BSPS 3 (1967) 331-47.

   a. "Comment," by L. O. Mink. BSPS 3 (1967) 348-55.

4903. _____. "Does the Distinction Between Normal and Revolutionary Science Hold Water?" in Criticism and the Growth of Knowledge. I. Lakatos and A. Musgrave, eds. Cambridge: Cambridge University Press, 1970. 39-47.

4904. _____. "Scientific Strategies and Historical Change." BSPS 11, SL 58 (1974) 401-14.

4905. Trainor, Paul. "Collingwood on the Possibility of Progress in Metaphysics and the Sciences." Modern Schoolman. 58 (1980) 36-46.

4906. Tuomela, Raimo. "On the Structuralist Approach to the Dynamics of Theories." Synthese. 39 (1978) 211-32.

4907. _____. "Scientific Change and Approximation." Acta Philosophica Fennica. 30, 2-4 (1978) 265-97.

4908. Vandamme, F. J. "Theory Change, Incompatibility and Non-Deducibility." Poznan Studies. 2, No. 2 (1976) 95-98.

4909. Vickers, J. M. "Rules for Reasonable Belief Change." SL 51 (1973) 129-42.

4910. Wagner, Michael. "Modeling of Scientific Revolutions." Nature and System. 3 (1981) 153-71.

4911. Watanabe, Satosi. "Needed: A Historico-Dynamic View of Theory Change." Synthese. 32 (1975) 113-34.

4912. Watkins, J. W. N. "Metaphysics and the Advancement of Science." BJPS 26 (1975) 91-121.

4913. _____. "Against 'Normal Science'," in Criticism and the Growth of Knowledge. I. Lakatos and A. Musgrave, eds. Cambridge: Cambridge University Press, 1970. 25-37.

4914. Wesley, Peter. "Een nieuwe weg in de wetenschapstheorie?" Algemeen Nederlands Tijdschrift voor Wijsbegeerte. 68 (1976) 155-80.

4915. Williams, L. Pearce. "Normal Science, Scientific Revolutions and the History of Science," in Criticism and the Growth of Knowledge. I. Lakatos and A. Musgrave, eds. Cambridge: Cambridge University Press, 1970. 49-50.

4916. Wisdom, J. O. "The Incommensurability Thesis." Phil. Studies. 25 (1974) 299-301.

4917. Wittich, D. "Eine aufschlussreiche Quelle für das Verständnis der gesellschaftlichen Rolle des Denkens von Thomas S. Kuhn." DZP 26 (1978) 105-13.

4918. Yudin, B. G. "The Sociological and the Methodological in the Study of Changes in Science." SL 145 (1981) 105-09.

4919. Ziff, Paul. "Something About Conceptual Schemes." SL 52 (1973) 31-41.

4920. Zilsel, Edgar. "The Genesis of the Concept of Scientific Progress." Journal of the History of Ideas. 6 (1945) 325-49.

5.7 Objectivity and Rationality

4921. Agassi, Joseph. "Rationality and the Tu Quoque Argument." Inquiry. 16 (1973) 395-406. Reprinted in BSPS 65 (1981) 465-76.

   a. "Agassi on Rationality," by Robert E. Innis. Inquiry. 18 (1975) 97-101.

4922. _____. "On Pursuing the Unattainable." BSPS 11, SL 58 (1974) 431-44.

4923. _____. "Between Clarity and Rationality." [In Hebrew] Iyyun. 27 (1976-77) 147-52.

4924. Agassi, Joseph, and I. C. Jarvie. "The Rationality of Irrationalism." Metaphilosophy. 11 (1980) 127-33.

4925. Agazzi, E. "Subjectivity, Objectivity and Ontological Commitment in the Empirical Sciences." WOSPS 12 (1977) 159-71.

4926. _____. "Is Scientific Objectivity Possible Without Measurements?" Diogenes. 104 (1978) 93-111.

4927. _____. "Eine Deutung der wissenschaftlichen Objektivität." Allgemeine Zeitschrift für Philosophie. 3, No. 3 (1978) 20-47.

4928. Ajdukiewicz, K. "The Problem of the Rationality of Fallible Methods of Inference." SL 87 (1977) 13-30. [First published in Studia Filozoficzne. 4 (7) (1958).]

4929. Baran, Bogdan. "Rationality in Objectivistic Epistemology." Reports on Philosophy. 2 (1978) 81-91.

4930. Bar-Hillel, Y. "A Prerequisite for Rational Philosophical Discussion." Synthese. 12 (1960) 328-32.

4931. Bartley, III, William W. The Retreat to Commitment. London: Chatto & Windus, 1964. xii + 233 + iv pp.

4932. Batens, Diderik. "Rationality and Justification." Philosophica. 14 (1974) 83-103.

4933. Bernays, Paul. "Concerning Rationality," in The Philosophy of Karl Popper. P. A. Schilpp, ed. La Salle, Illinois: Open Court, 1974. 597-605.

4934. Blackwell, Richard J. "Science, Objectivity, and Human Values." Proc. Am. Cath. Phil. Assn. 51 (1977) 153-61.

4935. Briskman, Larry. "Historicist Relativism and Bootstrap Rationality." Monist. 60 (1977) 509-39.

4936. Brown, Harold I. "On Being Rational." Am. Phil. Quart. 15 (1978) 241-48.

4937. _____. "Observation and the Foundations of Objectivity." Monist. 62 (1979) 470-81.

4938. Buchdahl, Gerd. "History of Science and Criteria of Choice." MSPS 5 (1970) 204-30.

4939. Cohen, Robert S. "Constraints on Science." BSPS 39 (1976) 79-86.

4940. Costa de Beauregard, O., et al. "Le dilemme objectivité-subjectivité de la mécanique statistique et l'équivalence cybernétique entre information et entropie." Bulletin de la Société Française de Philosophie. 55 (1961) 156-239.

4941. Derksen, A. A. "The Failure of Comprehensively Critical Rationalism." Philosophy of the Social Sciences. 10 (1980) 51-66.

    a. "On the Criticizability of Logic -- A Reply to A. A. Derksen," by W. W. Bartley, III. 67-77.

4942. Destouches, Jean-Louis. "Le rôle de l'activité subjective dans l'élaboration des notions de la physique moderne." Synthese. 7 (1948-49) 75-78.

4943. Destouches-Fevrier, Paulette. "Les notions d'objectivité et de subjectivité en physique atomique." Dialectica. 1 (1947) 127-46.

4944. Earle, William. Objectivity. New York: Noonday Press, 1955. 157 pp.

4945. Echarri, S.J., Jaime. "Racionalidad propria de las ciencias."
Pensamiento. 7 (1951) 147-67.

4946. Esposito, Joseph L. "Criticizing Rationality as Criticism."
Journal of Critical Analysis. 4 (1972-73) 89-96.

4947. _____. "Science and Conceptual Relativism." Phil.
Studies. 31 (1977) 269-77.

4948. Feyerabend, Paul. "On the Critique of Scientific Reason," in
Method and Appraisal in the Physical Sciences. C. Howson, ed.
Cambridge: Cambridge University Press, 1976. 309-39. Also in BSPS
39 (1976) 109-44.

4949. Finocchiaro, Maurice. "Rhetoric and Scientific Rationality."
PSA 1978. I, 235-46.

4950. Freeman, Eugene. "Objectivity as 'Intersubjective Agreement'."
Monist. 57 (1973) 168-75.

4951. Freeman, Eugene, and Henryk Skolimowski. "The Search for
Objectivity in Peirce and Popper," in The Philosophy of Karl Popper.
P. A. Schilpp, ed. La Salle, Illinois: Open Court, 1974. 464-519.

4952. Gaa, James C. "Moral Autonomy and the Rationality of Science."
PS 44 (1977) 513-41.

4953. Gagnebin, S. "Quelques remarques subjectives sur l'objectivité
en science et en philosophie." Synthese. 10 (n.d.) 327-34.

4954. Geraëts, Théodore F., ed. Visages de la rationalité, a propos
de Rationality To-day/La rationalité aujourd'hui. Ottawa: Editions
de l'Université d'Ottawa/The University of Ottawa Press, 1979. 501
pp.

4955. Gerholm, Tor Ragnar. "The Meaning of Scientific Objectivity."
Danish Yearbook of Philosophy. 14 (1977) 97-105.

4956. Geymonat, Ludovio. "Sul concetto di 'crisi' della razionalità
scientifica." Scientia. 110 (1975) 325-41. English Tr.: 343-55.

4957. Grimal, E. "Sur la notion de l'objectivité." Revue Phil.
135 (1945) 236-55.

4958. Grünfeld, J. "Rationality and Scientific Method." Science
et Esprit. 28 (1976) 309-21.

4959. Hansen, Troels Eggers. "Confrontation and Objectivity."
Danish Yearbook of Philosophy. 7 (1970) 13-72.

4960. Harris, Errol E. "Objectivity and Reason." Philosophy. 31
(1956) 55-73.

4961. Harsanyi, J. C. "Advances in Understanding Rational Behavior."
WOSPS 10 (1977) 315-43.

4962. Hattiangadi, J. N. "The Importance of Auxiliary Hypotheses."
Ratio. 16 (1974) 115-20.

4963. _____ . "Rationality and the Problem of Scientific Traditions." Dialectica. 32 (1978) 3-28.

4964. Heelan, Patrick A. "The Role of Subjectivity in Natural
Science." Proc. Am. Cath. Phil. Assn. 43 (1969) 185-94.

4965. _____ . "Scientific Objectivity and Framework Transposi-
tions." Philosophical Studies. 19 (1970) 55-70.

4966. Hempel, Carl G. "Rational Action." Proc. & Add. Am. Phil.
Assn. 35 (1961-62) 5-23.

4967. Hiley, David R. "Relativism, Dogmatism, and Rationality."
IPQ 19 (1979) 133-50.

4968. Hilpinen, Risto, ed. Rationality in Science. Dordrecht:
Reidel, 1980. 256 pp.

4969. Hoering, Walter. "On Judging Rationality." SHPS 11 (1980)
123-36.

4970. Jarvie, I. C. "Toulmin and the Rationality of Science."
BSPS 39 (1976) 311-34.

4971. Jeffrey, Richard C. "Dracula meets Wolfman: Acceptance vs.
Partial Belief." SL 26 (1970) 157-85.

4972. Klackar, Jørgen. "The Objectivity of Physical Description."
Danish Yearbook of Philosophy. 14 (1977) 112-22.

4973. Kekes, John. "Fallibilism and Rationality." Am. Phil.
Quart. 9 (1972) 301-10.

4974. _____ . "Towards a Theory of Rationality." Philosophy
of the Social Sciences. 3 (1973) 275-88.

4975. _____ . A Justification of Rationality. Albany, NY:
State University of New York Press, 1976. 275 pp.

    a. Review essay by Stephen L. Nathanson. IPQ. 19 (1979)
    227-36.

4976. _____ . "Rationality and Problem-Solving." Philosophy
of the Social Sciences. 7 (1977) 351-66.

4977. Keuth, Herbert. "Objectivität und Parteillichkeit in der
Wissenschaft." ZAW 6 (1975) 19-33.

4978. Kockelmans, J. J. G. "L'objectivité des sciences positives
d'apres le point de vue de la phénoménologie." Archives de Philosophie.
27 (1964) 339-55.

4979. Koertge, Noretta. "Bartley's Theory of Rationality." Philosophy of the Social Sciences. 4 (1974) 75-81.

4980. Krausz, Michael. "Relativism and Rationality." Am. Phil. Quart. 10 (1973) 307-12.

4981. Kuhn, Thomas S. "Objectivity, Value Judgment, and Theory Choice," in T. S. Kuhn, The Essential Tension. Chicago: University of Chicago Press, 1977. 320-39.

4982. Kyburg, Jr., Henry E. "Conjunctivitis." SL 26 (1970) 55-82.

4983. Lanczos, C. "Rationalism and the Physical World." BSPS 3 (1967) 181-92.

4984. Leach, J. J. "The Dual Function of Rationality." WOSPS 10 (1977) 393-421.

4985. Lenk, Hans. "Rationalität in den Erfahrungswissenschaften." Philosophische Perspektiven. 5 (1973) 188-99. Part II: Perspektiven der Philosophie. 1 (1975) 85-110.

4986. Litt, Theodor. "Science and Objectivity." Science and Freedom. Boston: Beacon Press, 1955. 255-57.

4987. Lugg, Andrew. "Feyerabend's Rationalism." Canadian Journal of Philosophy. 7 (1977) 755-75.

4988. _____. "Disagreement in Science." ZAW 9 (1978) 276-92.

4989. McMullin, Ernan. "Logicality and Rationality: A Comment on Toulmin's Theory of Science." BSPS 11, SL 58 (1974) 415-30.

4990. _____. "The Rational and the Social." Grazer Philosophische Studien. 12/13 (1981) 13-33.

4991. Machan, Tibor R. "Kuhn's Impossibility Proof and the Moral Element in Scientific Explanations." Theory and Decision. 5 (1974) 355-74.

4992. Marković, Mihailo. "Rationality of Methodological Rules." Grazer Philosophische Studien. 12/13 (1981) 3-11.

4993. Martin, Michael. "Referential Variance and Scientific Objectivity." BJPS 22 (1971) 17-26.

4994. _____. "Ontological Variance and Scientific Objectivity." BJPS 23 (1972) 252-56.

4995. _____. "The Objectivity of a Methodology." PS 40 (1973) 447-50.

4996. Meiland, Jack W. "Kuhn, Scheffler, and Objectivity in Science." PS 41 (1974) 179-87.

a. "Meiland on Scheffler, Kuhn, and Objectivity in Science," by Harvey Siegel. 43 (1976) 441-48.

4997. Metz, A. "Science et subjectivite." Archives de Philosophie. 25 (1962) 35-50.

4998. Meyer, Michel. "Science as a Questioning-Process: A Prospect for a New Type of Rationality." RIP 131-132 (1980) 49-89.

4999. Michalos, Alex C. "Postulates of Rational Preference." PS 34 (1967) 18-22.

a. "A. C. Michalos' 'Postulates of Rational Preference," by John D. Mullen. 37 (1970) 618-19.

5000. _____. "Rationality between the Maximizers and the Satisfiers." BSPS 20, SL 64 (1974) 423-45.

5001. Newton-Smith, W. H. The Rationality of Science. London: Routledge & Kegan Paul, 1981. 312 pp.

5002. Nickles, Thomas. "Introduction: Rationality and Social Context." BSPS 60 (1980) xiii - xxv.

5003. Nicholas, John M. "Scientific Rationality and Local Progress." Nature and System. 2 (1980) 219-30.

5004. Petrie, Hugh G. "Metaphorical Models of Mastery: Or, How to Learn to Do the Problems at the End of the Chapter of the Physics Textbook." BSPS 32 (1976) 301-12.

5005. Pihl, Mogens. "Objectivity in Physics." Danish Yearbook of Philosophy. 14 (1977) 106-11.

5006. Popper, Karl R. "Towards a Rational Theory of Tradition," in The Rationalist Annual for the Year 1949. F. Watts, ed. London: Watts & Co., 1949. 36-55. Reprinted in Conjectures and Refutations. New York: Basic Books; London: Routledge & Kegan Paul, 1962; New York: Harper & Row, 1968. 120-35.

5007. Puligandla, R. "Max Born and the Problem of Objectivity." Scientia. 109 (1974) 499-508.

5008. Putnam, Hilary. "The Impact of Science on Modern Conceptions of Rationality." Synthese. 46 (1981) 359-82.

a. "A Bayesian Marriage of Science and Politics," by Thomas W. Simon. 383-87.

5009. Quintelier, Guy. "Ideal Objectivity, Modern Biology, and Technical Innovation." Man and World. 14 (1981) 369-85.

5010. Rasch, Georg. "On Specific Objectivity -- An Attempt at Formalizing the Request for Generality and Validity of Scientific Statements." Danish Yearbook of Philosophy. 14 (1977) 58-94.

5011. Ross, Jacob Joshua. "Rationality and Common Sense." Philosophy.
53 (1978) 374-81.

5012. Rudner, Richard. "The Scientist qua Scientist Makes Value
Judgments." PS 20 (1953) 1-6.

5013. Scheffler, Israel. Science and Subjectivity. Indianapolis:
Bobbs-Merrill, 1967. 132 pp.

5014. Settle, Tom. "The Rationality of Science versus the Rationality
of Magic." Philosophy of the Social Sciences. 1 (1971) 173-94.

5015. Shapere, Dudley. "Discovery, Rationality, and Progress in
Science: A Perspective in the Philosophy of Science." BSPS 20, SL
64 (1974) 407-19.

5016. Silva, Herman, and George C. Vayonis. "Objectivity and Subjectivity in Scientific Research." PS 20 (1953) 332-34.

5017. Simon, Herbert A. "The Logic of Rational Decision." BJPS 16
(1965) 169-86.

5018. Skolimowski, Henryk. "Knowledge, Language and Rationality:
Statement of the Problem." BSPS 4 (1969) 174-98.

        a. "Comments," by Stephen Toulmin. 199-207.

5019. _____. "Problems of Rationality in Biology," in Studies
in the Philosophy of Biology. F. J. Ayala and T. Dobzhansky, eds.
Berkeley: University of California Press, 1974. 205-24.

5020. _____. "Evolutionary Rationality." BSPS 32 (1976)
191-214.

5021. Smith, John E. "History of Science and the Ideal of Scientific Objectivity." RIP 26 (1972) 172-86.

5022. Sober, Elliott. "The Evolution of Rationality." Synthese.
46 (1981) 95-120.

5023. Stefansen, Niels Christian. "Rationality and Argument."
Danish Yearbook of Philosophy. 9 (1972) 34-54.

5024. Suppes, Patrick. "The Limits of Rationality." Grazer Philosophische Studien. 12/13 (1981) 85-101.

        a. "On 'The Limits of Rationality'," by Karel Lambert.
        103-04.

5025. Swain, Marshall. "The Consistency of Rational Belief." SL
26 (1970) 27-54.

5026. Trangøy, Knut Erik. "Three Thoughts About Objectivity as a
Methodological Norm." Danish Yearbook of Philosophy. 14 (1977)
41-52. Discussion: 53-57.

5027. Toulmin, Stephen E. The Uses of Argument. New York: Cambridge University Press, 1958. 264 pp.

5028. _____. "Rationality and the Changing Aims of Inquiry." LMPS-4 (1971) 885-903.

5029. _____. "From Logical Systems to Conceptual Populations." BSPS 8, SL 39 (1971) 552-64.

5030. _____. "Rationality and Scientific Discovery." BSPS 20, SL 64 (1974) 387-406.

5031. Ullmo, Jean, et al. "La science moderne et la raison." Bulletin de la Société Française de Philosophie. 52 (1958) 47-97.

5032. Verbruggen, F. "Op zoek naar een bepaling van rationaliteit in de geschiedenis van de wetenschappen." Philosophica. 14 (1974) 38-72.

5033. Vermeersch, Etienne. "Can Science be Made Rational?" Philosophica. 17 (1976) 151-68.

5034. Watanabe, Satosi. "Les éléments humains arationnels dans la connaissance scientifique." Archives de Philosophie. 34 (1971) 609-22.

5035. Watkins, J. W. N. "Towards a Unified Decision Theory: A Non-Bayesian Approach." WOSPS 10 (1977) 345-79.

5036. _____. "Comprehensively Critical Rationalism." Philosophy. 44 (1969) 57-62.

      a. "The Grounds of Reason," by Joseph Agassi, I.C. Jarvie, and Tom Settle. 46 (1971) 43-50.

      b. "Watkins on Rationalism," by John Kekes. 51-53.

      c. "Can a Rationalist be Rational about His Rationalism?", by Sheldon Richmond. 54-55.

      d. "CCR: A Reflection," by J. W. N. Watkins. 56-61.

5037. Weiler, Gershon. "Rationality and Criticism." [In Hebrew] Iyyun. 25 (1974) 312-24.

5038. Wojciechowski, Jerzy A. "The Epistemological Problems of De-Anthropomorphization of Modern Science." Proc. Am. Cath. Phil. Assn. 33 (1959) 58-64.

5.8 Truth in Science

5039. Albert, Hans. "Science and the Search for Truth." BSPS 58, SL 125 (1978) 203-20.

5040. Axinn, Sidney. "Fallacy of the Single Risk." PS 33 (1966) 154-62.

5041. Bavink, Bernhard. Was ist Wahrheit in den Naturwissenschaften? Wiesbaden: Brockhaus, 1947. 88 pp.

5042. Bloch, Kurt. "Ueber ein Kriterium der Wahrheit naturwissenschaftlicker Theorienbildung." PN 2 (1952) 110-16.

5043. Boldrini, Marcello. Scientific Truth and Statistical Method. Ruth Kendall, tr. New York: Hafner, 1972. xiv + 264 pp.

5044. Bonjour, Lawrence A. "Sellars on Truth and Picturing." IPQ 13 (1973) 243-66.

5045. Buchanan, Scott. Truth in the Sciences. Charlottesville: University of Virginia Press, 1972. xix + 177 pp.

5046. Christian, Curt. "Inhaltiche und formale Wahrheit." PN 14 (1973) 173-96.

5047. Clark, Romane. "Prima Facie Generalizations." SL 52 (1973) 42-54.

5048. Cohen, L. Jonathan. "What Has Science to do with Truth?" Synthese. 45 (1980) 489-510.

5049. Cousin, D. R. "Carnap's Theories of Truth." Mind. 59 (1950) 1-22.

5050. Dummett, Michael. "Truth." Proc. Arist. Soc. 59 n.s. (1958-59) 141-62.

5051. _____. Truth and Other Enigmas. Cambridge: Harvard University Press, 1978. 470 pp.

5052. Edwards, Rem B. "The Truth and Falsity of Definitions." PS 33 (1966) 76-79.

5053. Frank, Philipp. "The Rôle of Authority in the Interpretation of Science." Synthese. 10 (n.d.) 335-38.

5054. Freundlieb, Dieter. "Zur Problematik einer Diskurstheorie der Wahrheit." ZAW 6 (1975) 82-107.

5055. Friedman, Michael. "Truth and Confirmation." J. Phil. 76 (1979) 361-82.

5056. Fuchs, K. "Ueber das Wahrheitsprinzip in der Physik." DZP 9 (1961) 548-62.

5057. Haack, Susan. "The Pragmatist Theory of Truth." BJPS 27 (1976) 231-49.

5058. _____. "Fallibilism and Necessity." Synthese. 41 (1979) 37-64.

5059. Hanson, Norwood Russell. "Uncertainty." Phil. Review. 64 (1954) 65-73.

5060. Heitler, Walter. Wahrheit und Richtigkeit in den exakten Wissenschaften. Mainz: Verlag der Akademie der Wissenschaften und der Literatur, 1972. 22 pp.

5061. Hempel, Carl G., Roderick Firth, Wilfrid Sellars, Roderick M. Chisholm, and Paul Weiss. "Some Theses on Empirical Certainty." Review of Metaphysics. 5 (1952) 621-29.

5062. Henken, Leon. "Completeness," in Philosophy of Science Today. S. Morgenbesser, ed. New York: Basic Books, 1967. 23-35.

5063. _____. "Truth and Provability," in Philosophy of Science Today. S. Morgenbesser, ed. New York: Basic Books, 1967. 14-22.

5064. Herzberger, Hans G. "Dimensions of Truth." WOSPS 4 (1975) 71-92.

5065. Hesse, Mary. "Truth and the Growth of Scientific Knowledge." PSA 1976. II, 261-80.

   a. "Comments on Shapere and Hesse," by Fred I. Dretske. 299-303.

5066. Hinshaw, Jr., Virgil G. "Epistemological Relativism and the Sociology of Knowledge." PS 15 (1948) 4-10.

5067. Hübner, Kurt. "The Concept of Truth in a Historistic Theory of Science." SHPS 11 (1980) 145-51.

5068. Jaki, Stanley L. "The Role of Faith in Physics." Zygon. 2 (1967) 187-202.

5069. Johnson, Martin C. Science and the Meanings of Truth. London: Faber and Faber, 1946. 179 pp.

5070. Kaplan, Mark, and Lawrence Sklar. "Rationality and Truth." Phil. Studies. 30 (1976) 197-201.

5071. Keuth, Herbert. "Tarski's Definition of Truth and the Correspondence Theory." PS 45 (1978) 420-30.

5072. King-Farlow, John. "Truth Preference and Neuter Propositions." PS 30 (1963) 53-59.

5073. Kleene, S. C. "Computability." in Philosophy of Science Today. S. Morgenbesser, ed. New York: Basic Books, 1967. 36-45.

5074. Krajewski, Władysław. "Approximative Truth of Fact-Statements, Laws, and Theories." Synthese. 38 (1978) 275-80.

5075. Kuznetsov, Boris. "The Value of Scientific Error and the Irreversibility of Science." Diogenes. 97 (1977) 103-23.

5076. Ladrière, Jean. "Vérité et praxis dans la démarche scientifique." Revue Philosophique de Louvain. 72 (1974) 284-310.

5077. Lee, Donald S. "Truth in Empirical Science." Tulane Studies in Philosophy. 14 (1965) 45-92.

5078. Leist, Anton. "Ein Plädoyer für die Beendigung der Suche nach Wahrheitskriterien." ZAW 6 (1976) 217-34.

5079. Loewer, Barry. "The Truth Pays." Synthese. 43 (1980) 369-80.

5080. Lorenzen, Hans-Peter. "Bemerkung über eine Möglichkeit der Definierbarkeit von Wahrheit." ZAW 2 (1971) 64-65.

5081. MacKinnon, Edward M. "The Role of Conceptual and Linguistic Frameworks." Proc. Am. Cath. Phil. Assn. 43 (1969) 24-43.

5082. _____. Truth and Expression. New York: Newman Press, 1971. 212 pp.

5083. Marković, Mihailo. "The Problem of Truth." BSPS 5 (1969) 341-62.

5084. Martin, R. M. Pragmatics, Truth, and Language. BSPS 38. Dordrecht: Reidel, 1979.

5085. Mellor, D. H. "Special Relativity and Truth." Analysis. 34 (1973-74) 74-77.

5086. Meynell, H. "Science, the Truth, and Thomas Kuhn." Mind. 84 (1975) 79-93.

5087. Miller, David. "The Distance between Constituents." Synthese. 38 (1978) 197-212.

5088. Niiniluoto, Ilkka. "Truthlikeness: Comments on Recent Discussions." Synthese. 38 (1978) 281-330.

5089. Nowak, Leszek. "Relative Truth, the Correspondence Principle and Absolute Truth." PS 42 (1975) 187-202.

5090. Pap, Arthur. "Propositions, Sentences, and the Semantic
Definition of Truth." Theoria. 20 (1954) 23-35.

5091. Popper, Karl R. "On the Sources of Knowledge and of Ignorance."
Proceedings of the British Academy. 46 (1960) 39-71. Published
separately by Oxford University Press, 1961. Abbreviated versions
in The Indian Journal of Philosophy 1. (1959) 3-7, and in Encounter
19 (1962) 42-57. Reprinted in Karl R. Popper, Conjectures and Refutations. New York: Basic Books; London: Routledge & Kegan Paul,
1962; New York: Harper & Row, 1968. 3-30.

5092. Quine, W. V. "Carnap and Logical Truth," in The Philosophy of
Rudolf Carnap. P. A. Schilpp, ed. La Salle, Illinois: Open Court,
1963. 385-406. Also in Synthese. 12 (1960) 350-74.

5093. Rescher, Nicholas. The Coherence Theory of Truth. Oxford:
Clarendon Press, 1973. 372 pp.

5094. _____. "Scientific Truth and the Arbitrament of Praxis."
Nous. 14 (1980) 59-74.

5095. Robert, J.-D. "Le second postulat de l'acte scientifique et
le problème du fondement ultime du vrai scientifique." Archives de
Philosophie. 25 (1962) 51-108.

5096. Rosenkrantz, R. D. "Measuring Truthlikeness." Synthese. 45
(1980) 463-88.

5097. Ross, Stephen D. "Truth in Science: Unrestricted Validity."
Transactions of the Charles S. Peirce Society. 6 (1970) 46-57.

5098. Stegmüller, Wolfgang. Das Wahrheitsproblem und die Idee der
Semantik. Eine Einführung in die Theorien von A. Tarski and R. Carnap.
Wien: Springer-Verlag, 1957. x + 328 pp.

    a. Review essay by Ernst Tugendhat. Philosophische Rundschau.
    8 (1960) 131-59.

5099. Stern, Alfred. "Was ist Wahrheit?" Wiener Jahrbuch für Philosophie. 7 (1974) 60-79.

5100. Stroll, Avrum, and Henry Alexander. "'True' and Truth." PS
42 (1975) 384-410.

5101. Ulmer, Karl, ed. Die Wissenschaften and die Wahrheit.
Stuttgart: Kohlhammer, 1966. 204 pp.

5102. Ushenko, A. P. "Truth in Science and in Philosophy." PS 21
(1954) 101-17.

5103. Wedberg, Anders. "Decision and Belief in Science. Comments
on Rudolf Carnap's Views in 'Empiricism, Semantics, and Ontology'."
SL 73 (1975) 161-82. Earlier version printed in Danish Yearbook in
Philosophy. 1 (1964) 139-58.

5104. Weingartner, Paul. "Vier Fragen zum Wahrheitsbegriff."
Salzburger Jahrbuch für Philosophie. 8 (1964) 31-74.

5105. Whyte, L. L. "Some Thoughts on Certainty in Physical Science."
BJPS 14 (1963) 32-38.

5106. Williams, Gardner. "Absolute Truth and the Shadow of Doubt."
PS 15 (1948) 211-24.

5107. Young, J. L. Doubt and Certainty in Science. Oxford: The
Clarendon Press, 1951. viii + 168 pp.

5.9 Determinism and Indeterminism

5108. Adolphe, Lydie. "La crise du déterminisme dans la physique
contemporaine." Les Etudes philosophiques. 12 (1957) 3-11.

5109. Auger, Pierre. "Le microfinalisme." Revue Phil. 143 (1953)
599-619.

5110. Ayers, M. R. The Refutation of Determinism. London: Methuen,
1968. vii + 179 pp.

5111. Barbour, Ian. "Indeterminacy and Freedom: A Reappraisal."
PS 22 (1955) 8-20.

5112. Bendall, Kurt. "Laplacian Determinism and Omnitemporal
Determinateness." J. Phil. 68 (1971) 751-61.

5113. Bennett, Jonathan. "The Status of Determinism." BJPS 14
(1963) 196-119. Erratum, 264.

5114. Blondel, S. "Essai d'une fusion des conceptions déterministes
et indéterministes." RMM 58 (1953) 396-412.

5115. Boyd, Richard N. "Determinism, Laws, and Predictability in
Principle." PS 39 (1972) 431-50.

5116. Bradley, R. D. "Determinism or Indeterminism in Microphysics."
BJPS 13 (1962) 193-215.

5117. Brush, Stephen G. "Irreversibility and Indeterminism:
Fourier to Heisenberg." Journal of the History of Ideas. 37 (1976)
603-30.

5118. Caldirola, P. "Determinismo, indeterminismo, oggettivismo
nella fisica." Scientia. 109 (1974) 607-11. English Tr.: 612-15.

5119. Čapek, Milič. "The Doctrine of Necessity Re-examined."
Review of Metaphysics. 5 (1951) 11-54.

5120. Cassirer, Ernst. Determinism and Indeterminism in Modern
Physics. Historical and Systematic Studies of the Problem of Causality.
New Haven, CT: Yale University Press, 1957. xxiv + 213 pp.

   a. Review essay, "A Hegelian View of Complementarity," by J.
   Agassi. BJPS 9 (1958) 57-63.

5121. Dalla Chiara, M. L., and G. Toraldo de Francia. "The Logical
Dividing Line between Deterministic and Indeterministic Theories."
Studia Logica. 35 (1976) 1-5.

5122. Dear, G. F. "Determinism in Classical Physics." BJPS 11
(1961) 289-304.

5123. Dumitriu, A. "Les degrés de liberté du déterminisme."
Scientia. 101 (1966) 339-46.

5124. Earman, John. "Laplacian Determinism, or Is This Any Way to
Run a Universe?" J. Phil. 68 (1971) 729-44.

5125. Février, Paulette. Déterminisme et indéterminisme. Paris:
Presses Universitaires de France, 1955. xii + 250.

5126. Glymour, Clark. "Determinism, Ignorance, and Quantum Mechanics."
J. Phil. 68 (1971) 744-51.

5127. Hasker, William. "The Transcendental Refutation of Determinism."
Southern Journal of Philosophy. 11 (1973) 175-83.

5128. Hinshaw, Jr., Virgil. "Determinism versus Continuity." PS
26 (1959) 310-24.

5129. Hoering, Walter. "Indeterminism in Classical Physics." BJPS
20 (1969) 247-55.

5130. Holz, H. "Ueber Determinismus und Indeterminismus." PN 16
(1976-77) 344-62.

5131. Hörz, H. "Zum Verhältnis von Kausalität und Determinismus."
DZP 11 (1963) 151-70.

5132. Humphreys, Paul W. "Is 'Physical Randomness' Just Indeterminism
in Disguise?" PSA 1978. II, 98-113.

5133. Jauch, J. M. "Determinism in Classical and Quantal Physics."
Dialectica. 27 (1973) 13-26.

5134. Jørgensen, Jørgen. "A Note on Determinism, Predictability,
and Indeterminism in Atomic Physics." Danish Yearbook of Philosophy.
2 (1965) 60-62.

5135. Kanthack, L. "The So-Called Indeterminism in Physics and the
Freedom of Man's Consciousness." PN 15 (1974-75) 446-60.

5136. Kedrov, B. M. "On Determinism." Soviet Studies in Philosophy.
7, No. 4 (1969) 46-53. [English reprint from Filosofskie Nauki.
(1968, No. 1).]

5137. Kirschenmann, Peter. "Two Forms of Determinism." BSPS 32
(1976) 393-422.

5138. Korch, H. "Bemerkungen zum Begriff des Determinismus." DZP
9 (1961) 796-810.

5139. Kronfli, N. S. "Atomicity and Determinism in Boolean Systems."
International Journal of Theoretical Physics. 4 (1971) 141-3.
Reprinted in WOSPS 5 (1975) 509-12.

5140. Kukla, Andy. "On the Empirical Significance of Pure Determinism." PS 45 (1978) 141-44.

        a. "On the Empirical Content of Determinism," by D. Dieks.
        47 (1980) 124-30.

        b. "Determinism and Predictability: Reply to Dieks," by A.
        Kukla. 131-33.

5141. Laer, P. H. van. "Causalité, déterminisme, prévisibilité et
science moderne." Revue Philosophique de Louvain. 48 (1950) 510-26.

5142. Landé, Alfred. "Determinism versus Continuity in Modern
Science." Mind. 67 (1958) 174-81.

5143. Largeault, Jean. "Ce qui est déterminé ou indéterminé." RMM
85 (1980) 217-23.

5144. _____. "Cause, causalité, déterminisme." Archives de
Philosophie. 44 (1981) 383-402.

5145. Margenau, Henry. "Quantum Mechanics, Free Will, and Determinism." J. Phil. 64 (1967) 714-25.

5146. Mayo, Bernard. "The Incoherence of Determinism." Philosophy.
44 (1969) 89-100.

5147. Meehl, Paul E. "Psychological Determinism and Human Rationality:
A Psychologist's Reactions to Professor Karl Popper's 'Of Clouds and
Clocks'," in MSPS. 4 (1970) 310-72.

5148. Montefiore, A. "Determinism and Causal Order." Proc. Arist.
Soc. 58 n.s. (1957-58) 125-42.

5149. Moreau, Jacques. "Le déterminisme." RIP 24 (1970) 481-93.

5150. Nielsen, Kai. "Is to Abandon Determinism to Withdraw from
the Enterprise of Science?" PPR 28 (1967) 117-21.

5151. O'Connor, D. J. "Determinism and Predictability." BJPS 7
(1957) 310-15.

5152. Perrin, Francis, et al. "L'abandon du déterminisme scienti-fique fondamental." Bulletin de la Société Français de Philosophie. 43 (1949) 145-85.

5153. Polikarov, Azaria. "Determinism in Physics." Soviet Studies in Philosophy. 13 (1974) 67-85. [English reprint from Voprosy filosofii. (1973, No. 7).]

5154. Popper, Karl R. Of Clouds and Clocks: An Approach to the Problem of Rationality and the Freedom of Man. St. Louis: Washington University Press, 1966. 38 pp. Reprinted in K. R. Popper, Objective Knowledge. Oxford: Clarendon Press, 1972. 206-55.

5155. Rietdijk, C. W. "A Rigorous Proof of Determinism derived from the Special Theory of Relativity." PS 33 (1966) 341-44.

5156. Schock, Rolf. "On Determinism, the Universe, and Related Concepts." Synthese. 14 (1962) 255-76.

5157. Scriven, Michael. "The Present Status of Determinism in Physics." J. Phil. 54 (1957) 727-41.

5158. Seifullaev, R. S., and V. I. Ukolova. "Soviet Conference on Problems Concerning the Scientific Concept of Determinism." Soviet Studies in Philosophy. 11 (1972-73) 301-09. [English reprint from Vestnik Moskovskogo universiteta. (1971), No. 6).]

5159. Settle, Tom. "Human Freedom and 1568 Versions of Determinism and Indeterminism," in The Methodological Unity of Science. M. Bunge, ed. Dordrecht: Reidel, 1973. 245-64.

5160. Titze, Hans. "Logik and Determinismus." ZPF 17 (1963) 476-82.

5161. Tonini, V. "Déterminisme et indéterminisme." Scientia. 83 (1948) 39-49.

5162. Vigier, J. P. "Determinism and Indeterminism in a New 'Level' Conception of Matter." LMPS-1 (1960) 262-64.

5163. Weinberger, Ota. "Determinismus und Verantwortung." ZPF 34 (1980) 607-20.

5164. Wójcicki, Ryszard. "Deterministic Systems." Erkenntnis. 9 (1975) 219-27.

5165. Wright, Georg Henrik von. "Determinism and the Study of Man." SL 72 (1976) 415-35.

6. Special Topics in the Philosophy of the Physical Sciences

6.1 Physical Sciences—General

5166. Albritton, Jr., Claude C., ed. The Fabric of Geology. San Francisco: Freeman, Cooper, 1963. 374 pp.

    a. Review essay: "Is Geology Different? A Critical Discussion of The Fabric of Geology," by Richard A. Watson. PS 33 (1966) 172-85.

5167. Bavink, Bernhard. "Die Bedeutung des Konvergenzprinzips für die Erkenntnistheorie der Naturwissenschaften." ZPF 2 (1947) 111-30.

5168. Blair, G. W. Scott. "Some Aspects of the Search for Invariants." BJPS 1 (1950) 230-44.

5169. Bloch, Walter. Polarität, ihre Bedeutung für die Philosophie der modernen Physik, Biologie und Psychologie. Berlin: Duncker & Humblot, 1972. 282 pp.

5170. Bondi, H. Assumption and Myth in Physical Theory. Cambridge: Cambridge University Press, v + 1967. 88 pp.

5171. Caldin, E. F. "Theories and the Development of Chemistry." BJPS 10 (1959) 209-22.

    a. "Physics and Chemistry: Comments on Caldin's View of Chemistry," by D. K. C. MacDonald. BJPS 11 (1960) 222-23.

5172. _____. The Structure of Chemistry. Sheed and Ward, Ltd., 1961. 49 pp.

5173. Čapek, Milič. "Two Types of Continuity." BSPS 13 (1974) 361-75.

5174. Conrad-Martius, Hedwig. Naturwissenschaftlich-Metaphysische
Perspektiven. Heidelberg: F. H. Kerle, 1949. 90 pp.

    a. Review essay by Walter Bohm. Philosophisches Jahrbuch.
    61 (1951) 241-44.

5175. Dagognet, François. "Un oubli certain, un probable retour."
Les Etudes philosophiques. 32 (1977) 31-39.

5176. Feuillée, Pierre. "Les sciences de la terre, des sciences
pour l'homme?" Les Etudes philosophiques. 32 (1977) 3-16.

5177. Fierz, M. "Does a Physical Theory Comprehend an 'Objective,
Real, Single Process'?" in Observation and Interpretation. S.
Körner, ed. New York: Academic Press; London: Butterworths, 1957.
93-96.

5178. Fliegel, G. "Ueber das Verhältnis von Hypothese und Theorie,
dargestellt an einem geowissenschaftlichen Beispiel." DZP 28 (1980)
1125-32.

5179. Frey, Gerhard. "Der Verknüpfungszusammenhang in den mathematischen Naturwissenschaften." PN 3 (1955) 238-51.

5180. Graham, A., G. W. Scott Blair, and R. F. J. Withers. "A
Methodological Problem in Rheology." BJPS 11 (1971) 265-88.

5181. Häberlin, Paul. "Physikalische Theorien in philosophischer
Sicht." PN 3 (1954) 1-40, 279-317.

5182. Hartmann, Max. "Prozess und Gesetz in Physik und Biologie."
PN 2 (1953) 277-92.

5183. Harvey, David. Explanation in Geography. London: Edward
Arnold, 1969. xx + 521 pp.

5184. Hooykaas, Reijer. Natural Law and Divine Miracle: A His-
torical-Critical Study of the Principle of Uniformity in Geology,
Biology, and Theology. Leiden: Brill, 1959. 237 pp.

5185. Kitts, David B. The Structure of Geology. Dallas: Southern
Methodist University Press, 1977. xix + 180 pp.

5186. Koch, R. A. "Zum Wesen des Grundpostulats der Geologie." PN
10 (1967-68) 102-06.

5187. _____. "Zum Problem der Struktur und des Prozesses in
der Erdgeschichte." PN 11 (1969) 446-53.

5188. _____. "Die formale Struktur der Grundgesetze der
Kristallmorphologie." PN 12 (1970) 123-28.

5189. Magnus, A. "Mathematik und ihre Anwendung in der Chemie."
Studium Generale. 6 (1953) 629-37.

5190. Mulckhuyse, J. J. "Molecules and Models." Synthese. 12 (1960) 257-75. Reprinted in SL 3 (1961) 133-51.

5191. Niggli, P. "Zum methodischen Vorgehen in der mineralogischen Wissenschaft." Dialectica. 4 (1950) 271-86.

5192. O'Brien, James F. "Structural and Operational Approaches to the Physical World." Thomist. 22 (1959) 389-400.

5193. Omeljanowski, M. E. "Experimentelle Beobachtung, Theorie und Dialektik in der physikalischen Wissenschaft." DZP 25 (1977) 150-63.

5194. Paneth, F. A. "The Epistemological Status of the Chemical Concept of Element." BJPS 13 (1962) Part I: 1-14. Part II: 144-60.

    a. "Discussion of Professor Paneth's Article," by John Bradley. 13 (1963) 316-17.

    b. "F. A. Paneth's Works and Translations," by Eva Paneth. 317-18.

    c. "Discussion of Professor Paneth's Second Article," by John Bradley. 14 (1963) 39-40.

5195. Prigogine, I. "Irréversibilité et corrélations." RMM 67 (1962) 228-36.

5196. Prigogine, I. et I. Stengers. La nouvelle alliance. Métamorphose des sciences. Paris: Editions Gallimard, 1979. 302 pp.

5197. Rat, Pierre. "La géologie et ses méthodes parmi les sciences de la terre." Les Etudes philosophiques. 32 (1977) 17-30.

5198. Richter, K., and H. Laitko. "Zur Gegenstandsbestimmung der Chemie." DZP 10 (1962) 1278-93.

5199. Röhler, G. "Zur erkenntnistheoretischen Bedeutung von Hypothese, Modellvorstellung und Theorie in der Chemie." DZP 10 (1962) 1294-1307.

5200. Schleichert, H. Elemente der physikalischen Semantik. München: R. Oldenbourg, 1966. 156 pp.

5201. Schüler, W. "Aspekte des Erkenntnisprozesses in der Geophysik." DZP 21 (1973) 207-19.

5202. Simon, R. "Zu einigen Fragen des Verhältnisses von Empirischem und Theoretischem in der chemischen Erkenntnis." DZP 25 (1977) 201-11.

5203. Strombach, Werner. "Die Frage nach Wirklichkeit und Wesenheit des Anorganischen." PN 7 (1961-62) 37-60.

5204. Ullmo, J. "Les sciences de la nature aujourd'hui: méthode
et objet." Les Etudes philosophiques. 16 (1961) 3-10.

5205. Watson, Patty Jo, Steven A. Leblanc, and Charles L. Redman.
Explanation in Archeology: An Explicitly Scientific Approach. New
York: Columbia University Press, xviii + 191 pp.

5206. Weiss, Paul. "The Contemporary World." Review of Metaphysics
6 (1953) 525-38.

        a. "Relativity, Causality and Weiss's Theory of Relations,"
        by Adolf Grünbaum. 7 (1953) 115-23.

        b. "Grünbaum's Relativity and Ontology," by Paul Weiss.
        123-25.

5207. Zinzen, Arthur. Praktische Naturphilosophie. Beihefte zur
Philosophia Naturalis. 1. Meisenheim/Glan: Westkulturverlag Anton
Hain, 1953. 53 pp.

5208. _____. Die ontologische Betrachtungsweise. Beihefte
zur Philosophia Naturalis. 2. Meisenheim am Glan: Verlag Anton
Hain, 1963. 59 pp.

6.2 Physics—General

5209. Abelé, J. "La crise de l'unité dans la physique contemporaine."
Archives de Philosophie. 17, No. 2 (1948) 17-39.

5210. _____. "La physique moderne et la notion de substance."
Archives de Philosophie 18, No. 2 (1952) 42-55.

5211. Agazzi, Evandro. "Physics as Philosophy and as the Paradigm
of Science." Epistemologia. 3 (1980) Fascicolo speciale. 135-48.

5212. Aktchourine, I. A. "Les apories de Zénon, la topologie et la
physique contemporaine." RIP 25 (1971) 565-74.

5213. Akchurin, I. A. "The Methodology of Physics and Topology."
WOSPS 10 (1977) 35-45.

5214. Armstrong, D. M. "Absolute and Relative Motion." Mind. 72
(1963) 209-23.

        a. "Note on Armstrong's 'Absolute and Relative Motion'," by
        Vera Peetz. 79 (1970) 427-30.

5215. Augustynek, Z. "On the Correspondence Principle." Dialectics
and Humanism. 1, No. 3 (1974) 41-43.

5216. Bachelerd, Suzanne. La conscience de rationalité: étude
phénoménologique sur la physique mathématique. Paris: Presses
Universitaires de France, 1958. 215 pp.

5217. Becker, Oskar. "Die Rolle der euklidischen Geometrie in der Protophysik." PN 8 (1964) 49-64.

5218. Berenda, Carlton W. "Phonons--The Quantization of Sound." PS 35 (1968) 179-84.

5219. Bergmann, Gustav. "Physics and Ontology." PS 28 (1961) 1-14.

5220. Bernardini, Carlo. "Le argomentazione non rigose in fisica." Scientia. 111 (1976) 635-44. English Tr.: 645-51.

5221. Bochner, Salomon. "The Role of Mathematics in the Rise of Mechanics." American Scientist. 50 (1962) 294-311.

5222. _____. "The Significance of Some Basic Mathematical Conceptions for Physics." Isis. 54 (1963) 179-205.

5223. Bode, Roy R. "Creationism in Physics and Philosophy." Proc. Am. Cath. Phil. Assn. 29 (1955) 133-37.

5224. Bohn, David. "The Implicate Order: A New Order for Physics." Process Studies. 8 (1978) 73-102.

5225. Bohr, Niels. "Discussion with Einstein on Epistemological Problems in Atomic Physics," in Albert Einstein: Philosopher-Scientist. P. A. Schilpp, ed. New York: Tudor, 1951. 199-242.

5226. _____. "Mathematics and Natural Philosophy." Scientific Monthly. 82 (1956) 80-88.

5227. _____. Atomic Physics and Human Knowledge. New York John Wiley, 1958. 101 pp.

5228. Born, Max. The Restless Universe. Second edition, revised. New York: Dover, 1951. 352 pp.

5229. _____. Experiment and Theory in Physics. New York: Dover, 1956. 44 pp.

5230. _____. Physics in my Generation. New York: Pergamon Press, 1956. vii + 232 pp.

5231. _____. "Physics and Metaphysics." Scientific Monthly. 82 (1956) 229-35.

5232. Bowdery, George J. "The Concept of 'Field' in Electical Theory." PS 13 (1946) 307-24.

5233. Bressan, Aldo. "On the Usefulness of Modal Logic in Axioma-tizations of Physics." BSPS 20, SL 64 (1974) 285-303.

5234. Bridgman, P. W. The Nature of Thermodynamics. Cambridge, MA: Harvard University Press, 1941. Reprinted, with an appendix "Reflections on Thermodynamics," New York: Harper & Row, 1961. 239 pp.

5235. Bright, O. P. Laurence. Whitehead's Philosophy of Physics. London: Sheed and Ward, 1958. 40 pp.

5236. Brillouin, L. "Transformations et avatars de la notion de champ." RMM 67 (1962) 206-13.

5237. Broglie, Louis de. "Un nouveau venu en physique: le champ nucléaire." RMM 56 (1951) 117-27.

5238. _____. Physics and Microphysics. Martin Davidson, tr. New York: Pantheon Books, 1955. Reprint: New York: Harper & Row, 1960. 286 pp.

5239. Buchdahl, G. "Sources of Scepticism in Atomic Theory." BJPS 10 (1959) 120-34.

　　　a. "Epistemology as an Aid to Science: Comments on Dr. Buchdahl's Paper," by J. Agassi. 135-46.

5240. Büchel, Wolfgang. "Messung, Näherung und Zeitrichtung." PN 11 (1969) 162-88.

5241. Bunge, Mario. Foundations of Physics. Berlin, New York: Springer-Verlag, 1967. xii + 311 pp.

　　　a. Review essay, "Corrections to Bunge's Foundations of Physics (1967)," by M. Strauss. Synthese. 19 (1969) 433-42.

　　　b. "Corrections to Foundations of Physics: Correct and Incorrect," by Mario Bunge. 443-52.

　　　c. Review essay, "What About Foundations of Physics?" by Hans Freudenthal. Synthese. 21 (1970) 93-106.

　　　d. "Physical Axiomatics: Freudenthal vs. Bunge," by David Salt. FP 1 (1970-71) 307-13.

　　　e. "More About Foundations of Physics," by Hans Freudenthal. 315-23.

5242. _____. Delaware Seminar in the Foundations of Physics. Berlin: Springer-Verlag, 1967. xii + 193.

5243. _____. "The Physicist and Philosophy." ZAW 1 (1970) 196-208.

5244. _____. Problems in the Foundations of Physics. Berlin: Springer, 1971. ii + 162 pp.

5245. _____. Philosophy of Physics. SL 45. Dordrecht: Reidel, 1973. 248 pp.

5246. Caldirola, P. Dalla microfisica alla macrofisica. Milano: Bibl. Est. Mondadori, 1974.

a. Review essay by E. Bellone. Scientia. 109 (1974) 321-34.

5247. Caldirola, Piero, and Angelo Loinger. "Storie e filosofie della fisica." Epistemologia. 1 (1978) 77-92.

5248. Callen, Herbert. "Thermodynamics as a Science of Symmetry." FP 4 (1974) 423-43.

5249. Čapek, M. "La théorie bergsonienne et la physique moderne." Revue Phil. 143 (1953) 28-59.

5250. _____. The Philosophical Impact of Contemporary Physics. Princeton, NJ: D. Van Nostrand, 1961. 414 pp.

5251. _____. Bergson and Modern Physics. BSPS 7, SL 37. Dordrecht: Reidel, 1971. 414 pp.

a. Review essay by David A. Sipfle. Process Studies. 2 (1972) 306-16.

b. Review essay by F. Heidsieck. RMM 80 (1975) 528-40.

c. Review essay by Pete A. Y. Gunter. Southwestern Journal of Philosophy. 6, No. 1 (1975) 155-66.

d. "Bergson, Nominalism, and Relativity," by M. Čapek. 9, No. 3 (1978) 127-33.

e. "Bergson, Conceptualism, and Indeterminacy: A Rejoinder to Čapek," by Pete A. Y. Gunter. 135-37.

5252. Carnap, Rudolf. Two Essays on Entropy. Abner Shimony, ed. Berkeley: University of California Press, 1977. 130 pp.

a. "Carnap on Entropy," by Abner Shimony [editor's Introduction to Two Essays on Entropy] reprinted in SL 73 (1975) 381-96.

5253. Cartwright, Nancy. "Philosophy of Physics," in Current Research in Philosophy of Science. P. D. Asquith and H. E. Kyburg, Jr., eds. East Lansing, MI: Philosophy of Science Assn., 1979. 381-85.

5254. Chari, C. T. K. "Philosophic Issues about Irreversibility in Classical and Quantum Physics." Methodos. 15 (1963) 205-17.

5255. Clark, S. J., Joseph T. "The Physiognomy of Physics," in Mind and Cosmos. Robert G. Colodny, ed. Pittsburgh: University of Pittsburgh Press, 1966. 276-91.

5256. Clarke, Chris, et al. "Discussion: Is Mathematics Leading Physics by the Nose?" Theoria to Theory. 14 (1980) 5-15; 247-57.

5257. Cohen, Robert S. and Marx W. Wartofsky, eds. Logic and Epistemological Studies in Contemporary Physics. BSPS 13. SL 59. Dordrecht: Reidel, 1973. 462 pp.

5258. Costa de Beauregard, O. "Relation intime entre le principe de Bayes, le principe de Carnot et le principe de retardation des ondes quantifiées." RMM 67 (1962) 214-27.

5259. Daub, Edward E. "Probability and Thermodynamics: The Reduction of the Second Law." Isis. 60 (1969) 318-30.

5260. Day, Michael A. "An Axiomatic Approach to First Law Thermodynamics." Journal of Philosophical Logic. 6 (1977) 119-34.

5261. Del-Negro, Walter von. "Zur Begegnung von Physik und Philosophie." ZPF 17 (1963) 639-54.

5262. _____. Konvergenzen in der Gegenwartsphilosophie und die moderne Physik. Berlin: Duncker und Humblot, 1970. 166 pp.

5263. d'Espagnat, Bernard, et al. "La physique et le réel." Bulletin de la Société Française de Philosophie. 74 (1980) 1-42.

5264. Destouches, J.-L., et al. "La théorie physique et ses principes fondementaux." Bulletin de la Société Française de Philosophie. 42 (1947) 1-32.

5265. _____. "Notions descriptives et notions construites en physique modern." Theoria. 15 (1949) 71-77.

5266. _____. "Problèmes psycho-linguistiques en physique moderne." Synthese. 8 (1950-51) 155-66.

5267. _____. "Aspect dialectique de la notion de système physique." Dialectica. 11 (1957) 57-69.

5268. Destouches-Février, Paulette. "Monde sensible et monde atomique." Theoria. 15 (1949) 78-89.

5269. _____. "Systèmes microscopiques et language microphysique." Synthese. 8 (1950-51) 113-19.

5270. Deutsch, Martin. "Evidence and Inference in Nuclear Research." Daedalus. 87 (1958) 88-98.

5271. d'Haëne, R. "La notion scientifique de l'énergie, son origine et ses limites." RMM 72 (1967) 35-67.

5272. Diederich, Werner. Konventionalität in der Physik. Berlin: Duncker und Humblot, 1974. 265 pp.

5273. Dingle, Herbert. "Die neuen Anschauungen in der Physik." PN 1 (1950) 76-83.

5274. _____. "Particle and Field Theories of Gravitation." BJPS 18 (1967) 57-64.

5275. Dingler, Hugo. "Das physikalische Weltbild." Beihefte zur Zeitschrift für Philosophische Forschung. 4 (1951) 56 pp.

5276. Dorling, Jon. "Einstein's Introduction of Photons: Argument by Analogy or Deduction from the Phenomena?" BJPS 22 (1971) 1-8.

5277. Earman, John. "Combining Statistical-Thermodynamics and Relativity Theory: Methodological and Foundations Problems." PSA 1978. II, 157-85.

   a. "Comments," by Lawrence Sklar. 186-93.

5278. Echarri, S.J., Jaime. "Dualismo de la experiencia y teoría en la física." Pensamiento. 9 (1953) 29-45.

5279. Einstein, Albert, and Leopold Infeld. The Evolution of Physics: The Growth of Ideas from Early Concepts to Relativity and Quanta. London: Cambridge University Press, 1971. 302 pp.

5280. Escat, Gérard. "Sur quelques aspects philosophiques du concret en physique." Les Etudes philosophiques. 22 (1967) 3-16.

5281. Faggiani, Dalberto. "Sulla struttura della fisica." Methodos. 3 (1951) 191-95. English Tr.: 198-201.

5282. _____. "Sulle proposizioni primitive della fisica." Methodos. 4 (1952) 41-59. English Tr.: 59-69.

5283. Farre, George L. "Remarks on the Linguistic Foundations of Physics." Notre Dame Journal of Formal Logic. 6 (1965) 110-22.

5284. Feinberg, Gerald. "Physics and the Thales Problem." J. Phil. 63 (1966) 5-17.

5285. Fertig, H. "Wie objektiv ist die Physik? -- Zur Discussion um die Protophysik." PN 17 (1978-79) 31-55.

5286. Feyerabend, P. K. "Problems of Microphysics," in Frontiers of Science and Philosophy. Robert G. Cologny, ed. Pittsburgh: University of Pittsburgh Press, 1962. 189-283.

5287. _____. "In Defense of Classical Physics." SHPS 1 (1970-71) 59-85.

5288. Fierz, Markus. "Ueber das Wesen der theoretischen Physik." Studia Philosophica. 16 (1956) 130-41.

5289. Finkelstein, David. "Matter, Space and Logic." BSPS 5 (1969) 199-215.

5290. Fisk, Milton. "Cause and Time in Physical Theory." Review of Metaphysics. 16 (1963) 522-49.

   a. "Light Velocity in the Interaction Interpretation of Relativity Theory," by Richard Schlegel. 17 (1963) 286-88.

5291. Frank, Philipp. "Foundations of Physics," in International Encyclopedia of Unified Science. Vol. I., No. 7. O. Neurath, R.

Carnap, and C. W. Morris, eds. Chicago: University of Chicago Press, 1938-55. 423-504.

5292. Froda, Alexandre. "Analyse mathématique du 'Principe de Continuité' en physique." LMPS-1 (1960) 340-47.

5293. Gavroglu, K. "Research Guiding Principles in Modern Physics: Case Studies in Elementary Particle Physics." ZAW 7 (1976) 223-48.

5294. Gent, Werner. "Die Kernphysik und ihre weltanschaulichen Grenzen." PN 6 (1960-61) 202-60.

5295. Giles, Robin. "A Non-Classical Logic for Physics." Studia Logica. 33 (1974) 397-416.

5296. _____. "The Concept of a Proposition in Classical and Quantum Physics." Studia Logica. 38 (1979) 337-53.

5297. Gorgé, Viktor. Philosophie und Physik. Die Wandlung zur heutigen erkenntnis theoretischen Grundhaltung in der Physik. Berlin: Duncker & Humblot, 1960. 137 pp.

5298. Gower, Barry. "Speculation in Physics: The History and Practice of 'Naturphilosophie'." SHPS 3 (1973) 301-56.

5299. Gröbner, W. "Der Begriff 'Struktur' in Mathematik und Physik." Scientia. 92 (1957) 1-7. French Tr.: Supplement, 1-5.

5300. Hanson, Norwood Russell. "Philosophic Problems of Nuclear Science." Cambridge Journal. 7 (1954) 249-51.

5301. Hartley, R. V. L. "A Mechanistic Theory of Extra-Atomic Physics." PS 26 (1959) 295-309.

5302. Havemann, Robert, et al. "Ueber philosophische Fragen der modernen Physik." DZP 1 (1953) 378-405, 629-62; 2 (1954) 188-234, 476-97, 677-94, 928-33; 3 (1955) 106-26, 242-46, 358-84, 736-57; 4 (1956) 82-99, 467-96; 5 (1957) 91-112, 734-35.

5303. Heerden, Pieter J. van. The Foundations of Physics with a Proposal for a Fundamental Theory of Physics. Wassenaar, The Netherlands: Uitgever J. Wistik, 1967. 100 pp.

5304. Heitler, Walter. "The Departure from Classical Thought in Modern Physics," in Albert Einstein: Philosopher-Scientist. P. A. Schilpp, ed. New York: Tudor, 1951. 179-98.

5305. Hennemann, Gerhard. "Philosophie, Religion und moderne Physik." Studia Philosophica. 12 (1952) 18-53.

5306. _____. "Die Bedeutung der Erkenntnistheorie für die Physik." ZPF 7 (1953) 351-67.

5307. Hölling, Joachim. "Zur Kategorialanalyse des physikalischen Feldbegriffes." PN 10 (1967-68) 343-56.

5308. Hutten, E. H. "Symmetry Physics and Information Theory."
Diogenes. 72 (1970) 1-21.

5309. Jaki, Stanley L. The Relevance of Physics. Chicago: University of Chicago Press, 1966. 604 pp.

5310. Jammer, Max. "A Consideration of the Philosophical Implications of the New Physics." BSPS 59. SL 136. (1979) 41-61.

5311. Janich, P. "Wie empirisch ist die Physik?" PN 11 (1969)
291-303.

5312. Juhos, Béla. "Die neuen Formen der empirischen Erkenntnis."
Archiv für Philosoophie. 8 (1958) 255-73.

5313. _____. "Schlüsselbegriffe physikalischer Theorien."
Studium Generale. 20 (1967) 785-95.

5314. Kanitscheider, Bernulf. "Die Rolle der Geometrie innerhalb
physikalischer Theorien." ZPF 26 (1972) 42-55.

5315. Kapp, R. O. "A New Interpretation of Gravitation." BJPS 5
(1955) 331-32.

5316. Kaulbach, Friedrich. "Die Anschauung in der klassischen und
modernen Physik." PN 5 (1958-59) 66-95.

5317. Knauss, G. "Die erkenntnistheoretische Situation der Physik."
Studium Generale. 24 (1971) 1474-1521.

5318. Kolman, E. "The Philosophical Interpretation of Contemporary
Physics." Studies in Soviet Thought. 21 (1980) 1-14.

5319. König, G. "Philosophische Probleme der Physik." PN 18
(1980-81) 87-102.

5320. König, Hans. "Ueber die Methoden der Physik." Synthese. 6
(1947) 25-43.

5321. Koslow, Arnold, ed. The Changeless Order--The Physics of
Space, Time and Motion. New York: George Braziller, 1967. viii +
328.

5322. Körner, S. "On Philosophical Arguments in Physics," in
Observation and Interpretation. S. Körner, ed. New York: Academic
Press; London: Butterworths, 1957. 97-101.

5323. Krajewski, W. "The Role of Correspondence Principle in the
Development of Physics." Dialectics and Humanism. 1, No. 3 (1974)
61-63.

5324. Kratzer, A. "Physik und Mathematik." Studium Generale. 6
(1953) 619-28.

5325. Kulakov, Yu. I., and T. I. Protasiewicz. "Phenomenological Symmetry and the Foundations of Physics." International Logic Review. 4 (1973) 98-101.

5326. Lean, Martin E. "Physics and Metaphysics." Personalist. 53 (1972) 365-94.

5327. Lindsay, R. B. "Physics, Ethics, and the Thermodynamic Imperative," in Philosophy of Science: The Delaware Seminar, II. New York: Interscience Publishers, 1963. 411-48.

5328. _____. The Nature of Physics: A Physicist's Views on the History and Philosophy of His Science. Providence, RI: Brown University Press, 1968. 212 pp.

5329. McGuire, J. E. "Forces, Powers, Aethers, and Fields." BSPS 14, SL 60 (1974) 119-59.

5330. Magyar, George. "On the Dual Nature of Light." BJPS 16 (1965) 44-49.

5331. Margenau, Henry. The Nature of Physical Reality. New York: McGraw-Hill, 1950. xiii + 479 pp.

    a. Review essay, "On Professor Margenau's Kantianism," by Lewis White Beck. PPR 11 (1951) 568-73.

    b. "Reply to Professor Beck," by H. Margenau. 574-78.

    c. Review essay, "Professor Margenau and the Problem of Physical Reality," by W. H. Werkmeister. PS 18 (1951) 183-92.

    d. "Discussion: Physics and Ontology," by Henry Margenau. 19 (1952) 342-45.

    e. "Discussion: The Relativity of Reality," by A. Bachem. 20 (1953) 75-78.

5332. _____. Thomas and the Physics of 1958: A Confrontation. The Aquinas Lecture, 1958. Milwaukee: Marquette University Press, 1958. 61 pp.

    a. Review essay by Pierre H. Conway, O. P. Thomist. 22 (1959) 68-118.

5333. _____. Open Vistas: Philosophical Perspectives of Modern Science. New Haven: Yale University Press, 1961. 256 pp.

5334. _____. Physics and Philosophy: Selected Essays. Dordrecht: Reidel, 1978. xxxvi + 404 pp.

5335. Mattick, Paul. "Marxism and the New Physics." PS 29 (1962) 350-64.

a. "A Reply to Mr. Mattick's Article on Marxism and the New Physics," by Manfred S. Frings. 31 (1964) 289-93.

5336. Maurin, K., and K. Michalski. "Mathematik als Sprache der Physik." PN 16 (1976-77) 363-82.

5337. Maxwell, Nicholas. "Physics and Common Sense." BJPS 16 (1966) 295-311.

5338. Mehra, Jagdish, ed. The Physicist's Conception of Nature. Dordrecht: Reidel, 1973. 839 pp.

a. Review essay by M. M. Yanase. AJAPS 5, No. 2 (1977) 35-47.

5339. Mejbaum, W. "Two Remarks on the Relation of Correspondence in Physics." Dialectics and Humanism. 1, No. 3 (1974) 57-60.

5340. Mercier, A. "Remarks on Physics in General and Relativity in Particular." Dialectics and Humanism. 2, No. 3 (1975) 125-31.

5341. Miller, David L. "Metaphysics in Physics." PS 13 (1946) 281-86.

5342. Misiek, Jozef. "The Extreme Empiricism and Methodology of Physics." Reports on Philosophy. 1 (1977) 51-58.

5343. Mittelstaedt, Peter. Philosophical Problems of Modern Physics. BSPS 18. SL 95. Dordrecht: Reidel, 1976. 211 pp.

5344. Moulines, C. Ulises. "A Logical Reconstruction of Simple Equilibrium Thermodynamics." Erkenntnis. 9 (1975) 101-30.

5345. _____. "An Example of a Theory-Frame: Equilibrium Thermodynamics." SL 146 (1981) 211-38.

5346. Moulines, Carlos-Ulises, and Joseph D. Sneed. "Suppes' Philosophy of Physics," in Patrick Suppes. Radu J. Bogdan, ed. Dordrecht: Reidel, 1979. 59-91.

5347. Myhill, John. "Remarks on the Language of Physics." PS 30 (1963) 305-06.

a. "Remarks on Myhill's Remarks on Coordinate Languages," by H. G. Bohnert. 307-08.

5348. Nowak, I. "The Concept of Dialectical Correspondence." Dialectics and Humanism. 1, No. 3 (1974) 51-55.

5349. _____. "Of Some Modifications of the Concept of Dialectical Correspondence." Dialectics and Humanism. 1, No. 3 (1974) 73-78.

5350. Nye, Mary Jo. Molecular Reality. New York: American Elsevier, 1972. xi + 201 pp.

5351. Omel'ianovskii, M. E. "V. I. Lenin and Problems of Dialectics in Contemporary Physics." Soviet Studies in Philosophy. 10 (1971-72) 240-51. [English reprint from Voprosy filosofii. (1971, No. 3).]

5352. _____. "Axiomatics and the Search for the Foundations of Physics." WOSPS 10 (1977) 47-65.

5353. Oppenheimer, J. Robert. The Flying Trapeze: Three Crises for Physicists. London: Oxford University Press, 1964. Reprint: New York: Harper & Row, 1969. 69 pp.

5354. Parker-Rhodes, A. F. The Theory of Indistinguishables: A Search for Explanatory Principles below the Level of Physics. SL 150. Dordrecht: Reidel, 1981. 248 pp.

5355. Peierls, Rudolf. Surprise in Theoretical Physics. Princeton: Princeton University Press, 1979. viii + 166 pp.

5356. Pham Xuân Yêm. "Modèle d'interaction entre corpuscules en théorie fonctionnelle." SL 3 (1961) 152-54.

5357. Pietschmann, Herbert. "Moderne Physik und Naturphilosophie." PN 12 (1970) 80-86.

5358. Pikler, Andrew G. "Utility Theories in Field Physics and Mathematical Economics." BJPS 5 (1954) 47-58, 303-18.

5359. Planck, Max. A Survey of Physical Theory (Formerly Titled: A Survey of Physics). R. Jones and D. H. Williams, trs. New York: Dover, n.d.

5360. _____. The Philosophy of Physics. New York: W. W. Norton, 1963. 128 pp.

5361. Popper, K. R. "Irreversibility; or, Entropy since 1905." BJPS 8 (1957) 151-55.

5362. Post, E. J. "On People, Topics, and Cults in Physics." Scientia. 115 (1980) 55-68. Italian Tr.: 69-79.

5363. Inadvertently skipped in numbering the entries.

5364. Prigogine, I. "Quelques remarques sur la structure de la physique." RIP 19 (1965) 335-41.

5365. Rantala, V. "Correspondence and Non-Standard Models: A Case Study." Acta Philosophica Fennica. 30, 2-4 (1978) 366-78.

5366. Rietzler, Kurt. Fisica e realtà. Venezia: Neri Pozza Editore, 1955. 160 pp.

5367. Rifkin, Jeremy, with Ted Howard. Entropy: A New World View. New York: Viking, 1980. 305 pp.

5368. Roman, P. "Symmetry in Physics." BSPS 5 (1969) 363-69.

5369. Rosen, Robert. "The Gibbs' Paradox and the Disinguishability of Physical Systems." PS 31 (1964) 232-36.

   a. "Solution of the Gibbs Entropy Paradox," by Alfred Landé. 32 (1965) 192-93.

5370. Rossel, J. "Caractéristiques, tendances et implications de la recherche atomique actuelle." Dialectica. 11 (1957) 49-56.

5371. Rothstein, Jerome. "Information, Logic, and Physics." PS 23 (1956) 31-35.

5372. _____. "Physical Demonology." Methodos. 11 (1959) 99-121.

5373. _____. "Thermodynamics and Some Undecidable Physical Questions." PS 31 (1964) 40-48.

5374. Russell, Bertrand. "Physik und Erfahrung." ZPF 1 (1946) 445-64.

5375. Ruyer, Raymond. "L'activité rationaliste de la physique contemporaine." RMM 57 (1952) 82-92.

5376. Sachs, Mendel. The Field Concept in Contemporary Science. Springfield, Illinois: Charles S. Thomas. 120 pp.

5377. Samuel, Herbert L. Essay in Physics. New York: Harcourt Brace, 1952.

5378. Samuel, Viscount. Essays in Physics. Oxford: Basil Blackwell, 1951. vi + 154 pp.

5379. _____. "A Criticism of Present-Day Physics." Philosophy. 27 (1952) 51-57.

5380. Sapper, K. "Das Erkenntnisproblem in der modernen Physik." Scientia. 84 (1949) 229-36. French Tr.: Supplement, 115-22.

5381. Schilling, Kurt. "Physik und Erkenntnis." PN 5 (1958-59) 459-80.

5382. Schlegel, Richard. "Atemporal Processes in Physics." PS 15 (1948) 25-35.

5383. Seelig, W. "Ueber die erkenntnistheoretische Definition der Kraft als gemeinsame Grundlage der Relativität und der Quantelung." Studium Generale. 24 (1971) 1150-59.

5384. Selvaggi, S.J., Filippo. Problemi della fisica moderna. Brescia: 'La Scuola' Editrice, 1953.

a.   Review essay by F. Kröner.   Dialectica.   7 (1953) 349-70.

5385. Shrader-Frechette, K. "Atomism in Crisis: An Analysis of the Current High Energy Paradigm." PS 44 (1977) 409-40.

5386. _____. "High-Energy Models and the Ontological Status of the Quark." Synthese. 42 (1979) 173-89.

5387. Siciński, M. "Correspondence and Concretization." Dialectics and Humanism. 1, No. 3 (1974) 45-49.

5388. Smith, Vincent Edward. The Philosophical Frontiers of Physics. Washington, D.C.: The Catholic University of America Press, 1947. xii + 210.

5389. _____. The Philosophy of Physics. Jamaica, NY: St. John's University Press, 1961. 85 pp.

5390. _____. "Mathematical Physics in Theory and Practice." Proc. Am. Cath. Phil. Assn. 38 (1964) 74-85.

5391. Somenzi, V. "Fisica e metodologia." Methodos. 1 (1949) 278-83. English Tr.: 284-87.

5392. _____. "Fisica e geometria." Methodos. 3 (1951) 47-52. English Tr.: 52-56.

5393. _____. "Fisica e logica." Methodos. 5 (1953) 273-78.

5394. Strauss, Martin. "Entwicklungsgesetze der Physik." DZP 15 (1967) 217-22.

5395. _____. Modern Physics and Its Philosophy: Selected Papers in the Logic, History, and Philosophy of Science. SL 43. Dordrecht: Reidel, 1972. 297 pp.

5396. Strauss und Torney, Lothar von. "Der Substanzbegriff in der neueren Physik und seine Grenzen." Philosophisches Jahrbuch. 64 (1956) 99-111.

5397. Strombach, Werner. "Der Kraftbegriff." PN 8 (1964) 307-48.

5398. Stuewer, Roger H. "Non-Einsteinian Interpretations of the Photoelectric Effect." MSPS 5 (1970) 246-63.

5399. _____. "On Compton's Research Program." BSPS 39 (1976) 617-33.

5400. Such, Jan. "Controversies over the Correspondence Principle in Physics." Dialectics and Humanism. 1, No. 3 (1974) 65-72.

5401. Taylor, John G. The New Physics. New York: Basic Books, 1972. x + 224 pp.

5402. Tetze, Hans. "Logik, Existenz und Physik." ZPF 19 (1965) 278-305.

5403. Thomasma, O.P., David C. "The Electron Reviewed." New Scholasticism. 41 (1967) 159-90.

5404. _____. "The New Attitude of Contemporary Physics." Revue de l'Université d'Ottawa. 38 (1968) 210-24.

5405. Thüring, Bruno. "Die Axiome der Wirkfähigkeit oder das Gravitations-Axiom." PN 8 (1964) 164-90.

5406. _____. "Das Problem der allegemeinen Gravitation. Gravitation und Trägheit, Masse und Kraft." Scientia. 100 (1965) 259-66. French Tr.: Supplement, 146-52.

5407. _____. Die Gravitation und die philosophischen Grundlagen der Physik. Berlin: Duncker und Humblot Verlag, 1967. 266 pp.

5408. _____. "Zu einer Protophysik der Welle." PN 12 (1970) 421-39.

5409. Tisza, Laszlo. "The Logical Structure of Physics." BSPS 1 (1963) 55-71. Reprinted in Synthese. 14 (1962) 110-31.

5410. Tonini, V. "Principe de correspondance et concept d'action réelle." Scientia. 91. (1956) 297-303.

5411. Tonnelat, M.-A. "Les représentations corpusculaires et l'évolution de la physique." RMM 50 (1945) 203-22.

5412. _____. "Le part d'idéalisme dans la physique contemporaine." RMM 67 (1962) 163-73. English Tr.: Philosophy Today. 7 (1963) 53-61.

5413. Toraldo di Francia, G. "The Concept of Progress in Physics." BSPS 47 (1981) 323-39.

5414. Toulmin, Stephen, ed. Physical Reality: Philosophical Essays on Twentieth-Century Physics. New York: Harper & Row, 1970.

5415. Ullmo, J. "Physique et axiomatique." RMM 54 (1949) 126-38.

5416. Watson, W. H. Understanding Physics Today. Cambridge University Press, 1963. xiii + 219 pp.

5417. Weisskopf, Victor F. Physics in the Twentieth Century. Cambridge, MA: MIT Press, 1972. xv + 368 pp.

5418. Wheeler, John Archibald. "Science and Survival," in Philosophy of Science: The Delaware Seminar, II. New York: Interscience Publishers, 1963. 483-523.

5419. _____. "Physics as Geometry." Epistemologia. 3 (1980) Fascicolo speciale. 59-98.

5420. Whittaker, Sir Edmund. Eddington's Principle in the Philosophy of Science. Cambridge: Cambridge University Press, 1951. vi + 35 pp.

5421. Whyte, Lancelot Law. "Fundamental Physical Theory." BJPS 1 (1951) 303-27.

5422. _____. "Angles in Fundamental Physics." BJPS 3 (1952) 256-58.

5423. _____. "A Dimensionless Physics?" BJPS 5 (1954) 1-17.

5424. _____. "One-Way Processes in Physics and Biophysics." BJPS 6 (1955) 107-21.

5425. _____. The Atomic Problem: A Challenge to Physicists and Mathematicians. London: George Allen & Unwin, 1961. 56 pp.

5426. _____. Essay on Atomism: From Democritus to 1960. Middletown, CT: Wesleyan University Press, 1961. Reprint: New York: Harper & Row, 1963. 108 pp.

5427. Wigner, Eugene P. "Physics and the Explanation of Life." FP 1 (1970-71) 35-45. Also in BSPS 11, SL 58 (1974) 119-32.

5428. Woltjer, H. R. "Physik und Natur." Studium Generale. 7 (1954) 299-307.

5429. Yanase, Mutsuo M. "Reversibilität und Irreversibilität in der Physik." AJAPS 1, No. 2 (1957) 131-50.

5430. Yourgrau, Wolfgang. "Some Problems Concerning Fundamental Constants in Physics," in Current Issues in the Philosophy of Science. H. Feigl and G. Maxwell, eds. New York: Holt, Rinehart and Winston, 1961. 319-42.

5431. Zukav, Gary. The Dancing Wu Li Masters: An Overview of the New Physics. New York: William Morrow and Co., 1979. 352 pp.

6.3 Matter

5432. Adler, Norbert. "Zum Problem der Materie in der modernen Physik." PN 9 (1965-66) 458-84.

5433. Ambacher, Michael. La matière dans les sciences et en philosophie. Paris: Aubier Montaigne, 1972. 156 pp.

5434. Balz, Albert G. A. "Prime Matter and Physical Science." Proc. & Add. Am. Phil. Assn. 29 (1955-56) 5-25.

5435. Barashenkov, V. S. "The Leninist Concept of the Inexhaustibility of Matter in Contemporary Physics." Soviet Studies in Philosophy.

10 (1971-72) 263-68. [English reprint from Voprosy filosofii. (1971, No. 3).]

5436. Bealer, G. "Prediction and Matter." Synthese. 31 (1975) 493-508.

5437. Berghuys, J. J. W. "Materie en energie." Tijdschrift voor Filosofie. 18 (1956) 327-46.

5438. Broglie, Louis de. Matter and Light: The New Physics. W. H. Johnston, tr. New York: W. W. Norton, 1939. Reprint: New York: Dover, n.d. 300 pp.

5439. Brush, S. G. "Mach and Atomism." Synthese. 18 (1968) 192-215.

5440. Büchel, Wolfgang. "Der Materiebegriff der modernen Physik." Philosophisches Jahrbuch. 58 (1948) 55-64.

5441. _____. "Hylemorphismus und Atomphysik." PN 3 (1955) 318-38.

5442. Carella, Michael Jerome. "Heisenberg's Concept of Matter as Potency." Diogenes. 96 (1976) 25-37.

5443. Cook, Kathleen C. "On the Usefulness of Quantities." Synthese. 31 (1975) 443-58.

5444. Delattre, Pierre. L'évolution des systèmes moléculaires. Paris: Maloine-Doin, 1971. 194 pp.

5445. Galperin, F. "Recent Views on the Mass and Extension of the Electron." PPR 7 (1947) 376-90.

5446. Hanson, Norwood Russell. "On Elementary Particle Theory." Scientia. 91 (1956) 81-86. French Tr.: Supplement, 53-58. Also in PS 23 (1956) 142-48.

5447. _____. "The Dematerialization of Matter." PS 29 (1962) 27-38.

    a. "Matter Still Largely Material (A Response to N. R. Hanson's 'The Dematerialization of Matter')," by Herbert Feigl. 39-46.

5448. _____. The Concept of the Positron: A Philosophical Analysis. Cambridge: Cambridge University Press, 1963. ix + 236 pp.

5449. _____. "Dematerialization," in The Concept of Matter. Ernan McMullin, ed. South Bend: University of Notre Dame Press, 1963. 549-61.

5450. Heerden, P. J. van. "What is Matter?" PS 20 (1953) 276-85.

5451. Hill, E. L. "Particles and Fields in Modern Physics," in Philosophy of Science: The Delaware Seminar, II. New York: Interscience Publishers, 1963. 259-89.

5452. Hooker, C. A. "The Non-Necessity of Qualitative Content." Dialogue. 12 (1973) 447-53.

5453. Jammer, Max. Concepts of Mass in Classical and Modern Physics. Cambridge, MA: Harvard University Press, 1961. 230 pp.

5454. Kantorovich, Aharon. "Structure of Hadron Matter: Hierarchy, Democracy, or Potentiality?" FP 3 (1973) 335-49.

5455. Kapp, R. O. "Hypotheses About the Origin and Disappearance of Matter." BJPS 6 (1955) 177-85.

    a. "A Discussion of Professor Kapp's Views," by H. Bondi. 239-41.

    b. "A Reply to Professor Bondi," by R. O. Kapp. 241-43.

    c. "Remarks upon Professor Bondi's Views," by the Rt. Hon. The Earl of Halsbury. 243-44.

    d. "Comment on Lord Halsbury's Remarks," by H. Bondi. 244.

5456. Kedrov, B. M. "Evolution of the Concept of Matter in Science and Philosophy." WOSPS 12 (1977) 187-208.

5457. Kisiel, Theodore J. "The Reality of the Electron." Philosophy Today. 8 (1964) 56-65.

5458. Koslow, A. "Mach's Concept of Mass: Program and Definition." Synthese. 18 (1968) 216-33.

5459. Krąpiec, O.P., Albert M. "Die Theorie der Materie in physikalischer und philosophischer Sicht." Philosophisches Jahrbuch. 69 (1961) 134-76.

5460. Laycock, Henry. "Theories of Matter." Synthese. 31 (1975) 411-42.

5461. Luyten, O.P., Norbert A. "Philosophisches zum Materiebegriff." Philosophisches Jahrbuch. 69 (1961) 125-33. English Tr.: Philosophy Today. 6 (1962) 25-32.

5462. McMullin, Ernan, ed. The Concept of Matter. Notre Dame, Indiana: University of Notre Dame Press, 1963. 624 pp. First half reprinted under title: The Concept of Matter in Greek and Medieval Philosophy. 1965. 343 pp. Second half revised and reprinted under title: The Concept of Matter in Modern Philosophy. 1978. 304 pp.

5463. _____. "From Matter to Mass." BSPS 2 (1965) 25-45.

    a. "Comments," by Marx W. Wartofsky, 46-53.

5464. _____. "Material Causality." WOSPS 12 (1977) 209-41.

5465. Machlup, Fritz. "If Matter Could Talk," in Philosophy, Science, and Method. S. Morgenbesser et al. eds. New York: St. Martin's Press, 1969. 286-305.

5466. Meigne, Maurice. Structure de la matière. Paris: Presses Universitaires de France, 1963. 96 pp.

5467. Mellor, D. P. The Evolution of Atomic Theory. New York: American Elsevier, 1972. vi + 171 pp.

5468. Melsen, Andrew G. van. From Atomos to Atom. Pittsburgh Duquesne University Press, 1952. xii. + 240 pp.

5469. Moretti, J. "A la recherche de l'unité de la matière." Archives de Philosophie. 17, No. 2 (1948) 40-61.

5470. Neuhäusler, Anton. "Die Elementarteilchen und die zweite Kantische Antinomie." PN 3 (1956) 484-94.

5471. Njegovan, Vladimir. "Wie soll die Materie definiert?" PN 7 (1961-62) 442-50.

5472. O'Brien, James F. "Antimatter: A Study of Generation and Corruption in the Subatomic Realm." Philosophical Studies. (Ireland) 23 (1974) 152-65.

5473. Pais, A. "The Structure of Matter," in Philosophy of Science: The Delaware Seminar, II. New York: Interscience Publishers, 1963. 291-318.

5474. Pham Xuân Yêm. "Modèle d'interaction entre corpuscules en théorie fonctionelle." Synthese. 12 (1960) 276-78.

5475. Post, Heinz. "The Problem of Atomism." BJPS 26 (1975) 19-26.

5476. Reignier, J. "Qu'est-ce qu'une particule élémentaire?" Dialectica. 25 (1971) 153-65.

5477. Reiser, Oliver L. "Matter, Anti-Matter, and Cosmic Symmetry." PS 24 (1957) 271-74.

5478. Reyna, Ruth. The Philosophy of Matter in the Atomic Era: A New Approach to the Philosophy of Science. Bombay and New York: Asia Pub. House, 1962. 255 pp.

5479. Ritchie, A. D. "The Atomic Theory as Metaphysics and as Science." Proc. Arist. Soc. 45 n.s. (1944-45) 71-88.

5480. Sachs, Mendel. The Search for a Theory of Matter. New York:
McGraw Hill, 1972. 221 pp.

5481 _____ . "On the Nature of Light and the Problem of
Matter." WOSPS 2 (1973) 346-68.

5482. _____ . Ideas of Matter: From Ancient Times to Bohr
and Einstein. Washington, D.C.: University Press of America, 1981.
334 pp.

5483. Saini, Hugo. "Les particules fondamentales de la physique
contemporaine." Studia Philosophica. 12 (1952) 70-102.

5484. Schlossberger, Eugene. "Aristotelian Matter, Potentiality
and Quarks." Southern Journal of Philosophy. 17 (1979) 507-22.

5485. Schneider, Friedrich. "Das Problem der Materie in der gegenwärtigen Naturphilosophie." PN 5 (1958-59) 322-37.

5486. Shrader-Frechette, K. S. "Recent Changes in the Concept of
Matter: How Does 'Elementary Particle' Mean?" PSA 1980. I, 302-16.

5487. Speakman, J. C. "The Molecule -- The Evolution of a Concept."
Philosophical Journal. 2 (1965) 55-74.

5488. Suppes, Patrick. "Aristotle's Concept of Matter and Its
Relation to Modern Concepts of Matter." Synthese. 28 (1974) 27-50.

5489. Toulmin, Stephen, and June Goodfield. The Architecture of
Matter. New York: Harper & Row, 1962. 398 pp.

5490. Wallace, William A. "Immateriality and Its Surrogates in
Modern Science." Proc. Am. Cath. Phil. Assn. 52 (1978) 16-27.

6.4 Cosmology

5491. Abramenko, B. "The Age of the Universe." BJPS 5 (1954)
237-52.

5492. Ambartsumian, V. A. "Astronomy and Microphysics." Soviet
Studies in Philosophy. 2, No. 4 (1964) 23-30. [English reprint
from Voprosy filosofii. (1963, No. 6).]

5493. Bennett, Jonathan. "The Age and Size of the World." Synthese.
23 (1971) 127-46.

5494. Berenda, Carlton W. "Notes on Cosmology." J. Phil. 42
(1945) 545-48.

5495. _____ . "On the Cosmological Indeterminacy Principle of
McCrae." PS 31 (1964) 265-70.

5496. Bergmann, Peter G. "Cosmology as a Science." FP 1 (1970-71) 17-22. Also in BSPS 11, SL 58 (1974) 181-88.

5497. Bertiau, F. C. "The Science of Cosmogony: Its Principles and Problems." IPQ 3 (1963) 80-93.

5498. Bird, J. H. "The Beginning of the Universe." Arist. Soc. Supp. Vol. 40 (1966) 139-50.

5499. Bondi, H. Cosmology. Cambridge: Cambridge University Press, 1952. 179 pp.

5500. Borzeszkowski, H.-H. von, and R. Wahsner. "Kosmologie-Physik oder Metaphysik?" DZP 26 (1978) 242-46.

5501. Bunge, Mario. "Cosmology and Magic." Monist. 47 (1962) 116-41.

5502. Cancienne, Donald. "The Age of the Universe." Laval Théologique et Philosophique. 24 (1968) 9-38.

5503. Chari, C. T. K. "Information, Cosmology and Time." Dialectica. 17 (1963) 369-80.

5504. Chaudhury, Probas Jiban. "The Problem of Form and Content in Physical Science." PPR 10 (1949) 229-37.

5505. Clarke, C. J. S. "Quantum Theory and Cosmology." PS 41 (1974) 317-32.

        a. "Many Worlds are Better than None," by Stanley Kerr. 43 (1976) 578-82.

        b. "Reply to Stanley Kerr," by C. J. S. Clarke. 583-84.

5506. Coffey, Brian. "Notes on Modern Cosmological Speculation." Modern Schoolman. 29 (1952) 183-96.

5507. Davies, J. T. "The Age of the Universe." BJPS 5 (1954) 191-202.

5508. _____. "On Extrapolation, with Special Reference to the 'Age of the Universe'." BJPS 7 (1956) 129-38.

5509. Davidson, W. "Philosophical Aspects of Cosmology." BJPS 13 (1962) 120-29.

5510. Denbigh, K. G. An Inventive Universe. New York: G. Braziller, 1975. 219 pp.

5511. Dingle, Herbert. "Philosophical Aspects of Cosmology." Vistas in Astronomy. 1 (1955) 162-66.

5512. Earman, John. "The Closed Universe." Nous. 4 (1970) 261-69.

5513. Ellis, B. "Has the Universe a Beginning in Time?" AJP 33 (1955) 32-37.

5514. Finlay-Freundlich, E. "Cosmology," in International Encyclopedia of Unified Science. Vol. I, No. 8. O. Neurath, R. Carnap, and C. W. Morris, eds. Chicago: University of Chicago Press, 1938-55. 505-65.

5515. Friedmann, Hermann. Wissenschaft und Symbol. München: Biederstein-Verlag, 1949. 502 pp.

     a. Review essay by Walter Böhm. Philosophisches Jahrbuch. 61 (1951) 229-40.

5516. Gal-Or, B. "On Some Basic Interrelationships between Thermodynamics, Relativity and Cosmology." Scientia. 105 (1970) 361-69.

5517. Good, I. J. "Black and White Hole Hierarchical Universes." Theoria to Theory. 10 (1977) 191-201.

5518. Grünbaum, Adolf. "Some Highlights of Modern Cosmology and Cosmogony." Review of Metaphysics. 5 (1952) 481-98.

5519. Goudge, T. A. "Physical Cosmology and Philosophical Physics." Review of Metaphysics. 7 (1954) 444-51.

5520. Hanson, Norwood Russell. "Some Philosophical Aspects of Contemporary Cosmologies," in Philosophy of Science: The Delaware Seminar, II. New York: Interscience Publishers, 1963. 465-82.

5521. _____. "Conservation 'Beyond the Edge'." AJP 42 (1964) 227-31.

5522. Harré, R. "Philosophical Aspects of Cosmology." BJPS 13 (1962) 104-19.

5523. Heckmann, Otto. "Theorie und Erfahrung in der Kosmologie." Philosophisches Jahrbuch. 61 (1951) 13-17.

5524. _____. "Weltmodelle." Studium Generale. 18 (1965) 183-93.

5525. Heidmann, Jean, Jacques Merleau-Ponty, et al. "La Connaissance de l'univers: L'acquis et l'incertain." Bulletin de la Société Française de Philosophie. 70 (1976) 85-124.

5526. Hosinski, Thomas E. "Creation and the Origin of the Universe." Thought 48 (1973) 213-39; 386-403.

5527. Hübner, Kurt. "Ist das Universum nur eine Idee? Eine Analyse der relativistischen Kosmologie." Allgemeine Zeitschrift für Philosophie. 2, No. 2 (1977) 1-20.

5528. Hutten, Ernest H. "A Note on 'The Age of the Universe'."
BJPS 6 (1955) 58-61.

5529. _____. "Methodological Remarks Concerning Cosmology."
Monist. 47 (1962) 104-15.

5530. Ivanenko, Dmitri D. "The Problems of Unifying Cosmology with
Microphysics," in Physics, Logic, and History. W. Yourgrau and A.
D. Breck, eds. New York: Plenum Press, 1970. 105-111.

5531. Järnefelt, G. "Some Cosmological Points of Controversy."
Ajatus. 26 (1964) 99-124.

5532. Jordan, Pascual. "Neuere Gesichtspunkte der kosmologischen
Theorienbildung." Philosophisches Jahrbuch. 61 (1951) 8-12.

5533. Kapp, Reginald O. Towards a Unified Cosmology. London:
Hutchison & Co., 1960. 303 pp.

5534. Layzer, David. "Cosmic Evolution." BSPS 11, SL 58 (1974)
203-14.

5535. Lovell, Bernard. Emerging Cosmology. New York: Columbia
University Press, 1981. 208 pp.

5536. McCrea, W. H. "Modern Cosmology and the Concept of God."
Philosophy. 28 (1953) 160-63.

5537. _____. "Information and Prediction in Cosmology."
Monist. 47 (1962) 94-103.

5538. McMullin, Ernan. "Is Philosophy Relevant to Cosmology?" Am.
Phil. Quart. 18 (1981) 177-89.

5539. Merleau-Ponty, Jacques. "Réflexions sur la cosmologie contemporaine." RMM 63 (1958) 428-67.

5540. _____. Cosmologie du XXe siècle: Étude épistémologie
et historique des théories de la cosmologie contemporaine. Paris:
Gallimard, 1965. 533 pp.

5541. _____. "La cosmologie et la science contemporaine."
Dialogue. 7 (1968) 194-202.

5542. _____. "Logique, mathematiques et cosmologie." Les
Etudes philosophiques. 24 (1969) 499-511.

5543. _____. "Les hypothèses en cosmologie." RIP 25 (1971)
32-43.

5544. Meurers, Joseph. Das Alter des Universums. Meisenheim am
Glan: Westkulturverlag Anton Hain, 1954. 103 pp.

5545. _____. Weltallforschung. Meisenheim a. Glan: Hain,
1971. 112 pp.

5546. _____. "Philosophische Probleme der Kosmologie." PN 18 (1980-81) 116-45.

5547. Milne, E. A. Modern Cosmology and the Christian Idea of God. Oxford: Clarendon Press, 1952. 160 pp.

5548. Misner, Charles W. "Infinity in Physics and Cosmology." Proc. Am. Cath. Phil. Assn. 55 (1981) 59-72.

5549. Morrison, Philip. "The Physics of the Large," in Beyond the Edge of Certainty: Essays in Contemporary Science and Philosophy. Robert G. Colodny, ed. Englewood Cliffs, NJ: Prentice-Hall, 1965. 118-44.

5550. _____. "Open or Closed?" BSPS 11, SL 58 (1974) 189-202.

5551. Munitz, Milton K. "Kantian Dialectic and Modern Scientific Cosmology." J. Phil. 48 (1951) 325-38.

5552. _____. "One Universe or Many?" Journal of the History of Ideas. 12 (1951) 231-55.

5553. _____. "Scientific Method in Cosmology." PS 19 (1952) 108-30.

5554. _____. "Creation and the 'New' Cosmology." BJPS 5 (1954) 32-46.

5555. _____. Space, Time, and Creation: Philosophical Aspects of Scientific Cosmology. Glencoe, IL: Free Press and Falcon's Wing Press, 1957. xi + 182 pp. Reprint: New York: Collier Books, 1961.

5556. _____. "The Logic of Cosmology." BJPS 13 (1962) 34-50.

5557. _____. "On the Use of 'Universe' in Cosmology." Monist. 48 (1964) 185-94.

5558. Narlikar, Jayant. The Structure of the Universe. Oxford University Press, 1977. 264 pp.

5559. Nogar, O.P., Raymond J. "The Mystery of Cosmic Epigenesis." Proc. Am. Cath. Phil. Assn. 39 (1965) 112-24.

5560. North, J. D. The Measure of the Universe. Oxford: 1965. xxviii + 436 pp.

5561. Öpik, Ernst J. "The Age of the Universe." BJPS 5 (1954) 203-14.

5562. Orlicki, W. "The Methodological Analysis of the Steady State Theory of the Universe." Reports on Philosophy. 5 (1981) 109-20.

5563. Polikarow, A. "Zum Problem der Thermodynamik des Universums." DZP 11 (1963) 197-211.

5564. Poole, H. W. "A Close-Clipped View of the Universe, with a Note on its 'Origin' and 'Age', and our Position in it." BJPS 6 (1955) 43-50.

5565. Prokhovnik, S. J. "Cosmology versus Relativity -- The Reference Frame Paradox." FP 3 (1973) 351-58.

5566. Rohmer, Jean. "Astronomy and Philosophy." Philosophy Today. 1 (1957) 279-89.

5567. Russo, F. "Cosmologie du XXe siecle." Archives de Philosophie. 30 (1967) 398-410.

5568. Schlegel, Richard. "The Age of the Universe." BJPS 5 (1954) 226-36.

5569. _____. "The Problem of Infinite Matter in Steady-State Cosmology." PS 32 (1965) 21-31.

      a. "On the Self Consistency of the Steady-State Cosmology," by David Hawkins. 38 (1971) 273-79.

      b. "Reply to David Hawkins," by Richard Schlegel. 280-81.

5570. Schrödinger, E. Expanding Universes. London: Cambridge University Press, 1956. 93 pp.

5571. Sciama, D. W. "The Origin of the Universe." LMPS-3 (1967) 397-400.

5572. _____. Modern Cosmology. Cambridge: Cambridge University Press, 1971. 212 pp.

5573. _____. "The Universe as a Whole." Epistemologia. 3 (1980) Fascicolo speciale. 99-122.

5574. Scriven, Michael. "The Age of the Universe." BJPS 5 (1954) 181-90.

      a. "The Age of the Universe," by G. J. Whitrow. 334.

      b. "Some Comments on 'The Age of the Universe'," by C. K. Grant. 6 (1955) 248-51.

5575. Sinha, Ajit Kumar. "Philosophical Implications of Scientific Cosmology." Philosophical Quarterly (India). 37 (1964) 55-68.

5576. Smart, W. M. The Origin of the Earth. Cambridge University Press, 1951. viii + 239 pp.

5577. Strombach, Werner. "Das Problem der kosmischen Gesetzmässigkeiten." PN 7 (1961-62) 391-405.

5578. Swinburne, R. G. "Cosmological Horizons." PS 33 (1966) 210-14.

5579. _____. "The Beginning of the Universe." Arist. Soc.
Supp. Vol. 40 (1966) 125-38.

5580. Thüring, Bruno. "Methodologische Untersuchungen zur Kosmologie."
Methodos. 6 (1954) 95-114.

5581. Toulmin, Stephen. "Historical Inference in Science: Geology
as a Model for Cosmology." Monist. 47 (1962) 142-58.

5582. Velikovsky, Immanuel. Worlds in Collision. New York:
Doubleday, 1950. Reprint: New York: Dell, 1967. 400 pp.

5583. Vogt, Heinrich. "Der gegenwärtige Stand der Kosmologie." PN
9 (1965-66) 266-84.

5584. Weingard, Robert. "Some Philosophical Aspects of Black
Holes." Synthese. 42 (1979) 191-219.

5585. Wheeler, J. A. "Genesis and Observership." WOSPS 10 (1977)
3-33.

5586. Whitrow, G. J. The Structure of the Universe. London:
Library, 1949. 171 pp.

5587. _____. "The Limits of the Physical Universe." Studium
Generale. 5 (1952) 329-37.

5588. _____. "The Age of the Universe." BJPS 5 (1954) 215-25.

5589. _____. The Structure and Evolution of the Universe:
An Introduction to Cosmology. New York: Harper & Row, 1959. 212
pp.

5590. _____. "Is the Physical Universe a Self-Contained
System?" Monist. 47 (1962) 77-93.

5591. Whitrow, G. J., and Bondi, H. "Is Physical Cosmology a
Science?" BJPS 4 (1954) 271-83.

5592. Yourgrau, Wolfgang. "Cosmology and Logic -- An Intractable
Issue?" BSPS 39 (1976) 739-53.

5593. Zeldovich, Ia. B., and I. D. Novikov. "Contemporary Trends
in Cosmology." Soviet Studies in Philosophy. 14 (1976) 28-49.
[English reprint from Voprosy filosofii. (1975, No. 6).]

6.5 Classical Mechanics

5594. Adler, Carl G. "Why is Mechanics Based on Acceleration?" PS
47 (1980) 146-52.

5595. Agassi, J. "Between Micro and Macro." BJPS 14 (1963) 26-31.

5596. Bălan, Stefan. "Two Important Entities from the Past of
Classical Mechanics." Noesis. 2 (1974) 47-58.

5597. Balzer, W. "Die epistemologische Rolle des zweiten Newtonshen
Axioms." PN 17 (1978-79) 131-49.

5598. Bopp, Erich. "Starrer Körper und euklidische Geräte." PN 3
(1955) 383-91.

5599. Brush, Stephen G. "Statistical Mechanics and the Philosophy
of Science: Some Historical Notes." PSA 1976. II, 551-84.

5600. Buchdahl, G. "Science and Logic: Some Thoughts on Newton's
Second Law of Motion in Classical Mechanics." BJPS 2 (1951) 217-35.

5601. Bunge, Mario. "Machs Beitrag zur Grundlegung der Mechanik."
PN 11 (1969) 189-203.

5602. Christensen, Ferrel. "The Problem of Inertia." PS 48 (1981)
232-47.

5603. Cohen, I. Bernard. "Hypotheses in Newton's Philosophy."
BSPS 5 (1969) 304-26.

5604. Corsiglia, L. "Elementary Concepts of Space, Time and Matter."
Scientia. 108 (1973) 535-42, 863-66.

5605. Dorling, Jon. "The Eliminability of Masses and Forces in
Newtonian Particle Mechanics: Suppes Reconsidered." BJPS 28 (1977)
55-57.

5606. Earman, John, and Michael Friedman. "The Meaning and Status
of Newton's Law of Inertia and the Nature of Gravitational Forces."
PS 40 (1973) 329-59.

5607. Ellis, Brian. "The Origin and Nature of Newton's Laws of
Motion," in Beyond the Edge of Certainty: Essays in Contemporary
Science and Philosophy. Robert G. Colodny, ed. Englewood Cliffs,
NJ: Prentice-Hall, 1965. 29-68.

        a. "A Response to Ellis's Conception of Newton's First Law,"
        by Norwood Russell Hanson. 69-74.

5608. _____. "The Existence of Forces." SHPS 7 (1976) 171-85.

5609. Flaschner, Ludwig. "Zur ontologischen Begründung der Mass-
begriffe und Gesetze in Mechanik und Elektrizitätslehre." PN 2
(1952) 137-77.

5610. Galgani, Luigi. "Meccanica classica e meccanica quantistica."
Scientia. 110 (1975) 469-82. English Tr.: 483-91.

5611. Gamow, George. "The Three Kings of Physics," in Physics, Logic, and History. W. Yourgrau and A. D. Breck, eds. New York: Plenum Press, 1970. 203-07.

5612. Giedymin, Jerzy. "Logical Comparability and Conceptual Disparity between Newtonian and Relativistic Mechanics." BJPS 24 (1973) 270-76.

5613. Grigoryan, A. T. "Appreciation de la mechanique newtonienne et de l'autobiographie d'Einstein." Scientia. 96 (1961) 356-63.

5614. Hager, N., and U. Röseberg. "Philosophisch-weltanschauliche Aspekte des Weltbildes der klassischen Physik." DZP 25 (1977) 577-86.

5615. Hanson, Norwood Russell. "Waves, Particles, and Newton's 'Fits'." Journal of the History of Ideas. 21 (1960) 370-91.

5616. _____. "The Law of Inertia: A Philosopher's Touchstone." PS 30 (1963) 107-21.

5617. _____. "Newton's First Law: A Philosopher's Door into Natural Philosophy," in Beyond the Edge of Certainty: Essays in Contemporary Science and Philosophy. R. Colodny, ed. Englewood Cliffs, NJ: Prentice-Hall, 1965. 6-28.

5618. Hay, William H. "On the Nature of Newton's First Law of Motion." Phil. Review. 65 (1956) 95-102.

5619. Hertz, P. "Gibbs' Theory: Its Foundations and Applications." Dialectica. 10 (1956) 368-83.

5620. Hobson, Arthur. Concepts in Statistical Mechanics. London: Gordon and Breach, 1971. 172 pp.

5621. Hooker, C. A. "Defense of a Non-Conventionalist Interpretation of Classical Mechanics." BSPS 13 (1974) 123-91.

a. "Comments on C. A. Hooker: Systematic Realism," by Laszlo Tisza. 192-95.

5622. Hoyer, Ulrich. "Ist das zweite Newtonsche Bewegungsaxiom ein Naturgesetz?" ZAW 8 (1977) 292-301.

5623. Hunt, I. E., and W. E. Suchting. "Force and 'Natural Motion'." PS 36 (1969) 233-50.

5624. Hutten, E. H. "On the Principle of Action by Contact." BJPS 2 (1951) 45-51.

5625. Jammer, Max. Concepts of Force: A Study in the Foundations of Dynamics. Cambridge, MA: Harvard University Press, 1957. Reprint: New York: Harper and Brothers, 1962. x + 269 pp.

5626. Janich, Peter. "Ist Masse ein 'theoretischer Begriff'?" ZAW 8 (1977) 302-14.

5627. Juhos, Béla. "Das Prinzip der virtuellen Geschwindigkeiten." PN 9 (1965-66) 55-113.

5628. Kannegiesser, K.-H. "Zum zweiten Hauptsatz der Thermodynamik." DZP 9 (1961) 841-59.

5629. Klein, Martin J. "Order, Organization and Entropy." BJPS 4 (1953) 158-60.

    a. "Entropy and Disorder," by J. M. Burgers. 5 (1954) 70-71.

5630. Koslow, Arnold. "The Law of Inertia: Some Remarks on Its Structure and Significance," in Philosophy, Science, and Method. S. Morgenbesser et al. eds. New York: St. Martin's Press, 1969. 549-67.

5631. Krbek, Fr. von. "Grundlegung der Mechanik." Scientia. 91 (1956) 265-70. French Tr.: Supplement, 161-66.

5632. Kronfli, N. "Probabilistic Formulation of Classical Mechanics." International Journal of Theoretical Physics. 3 (1970) 395-400. Reprinted in WOSPS 5 (1975) 503-08.

5633. Kurth, Rudolf. "Masse als Menge der Materie." PN 7 (1961-62) 356-64.

5634. Lavis, David. "The Role of Statistical Mechanics in Classical Physics." BJPS 28 (1977) 255-79.

5635. McCrea, W. H. "Action at a Distance." Philosophy. 27 (1952) 70-76.

5636. McKinsey, J. C. C., and Suppes, Patrick. "On the Notion of Invariance in Classical Mechanics." BJPS 5 (1955) 290-302.

5637. McMullin, Ernan. Newton on Matter and Activity. Notre Dame, Indiana: University of Notre Dame Press, 1978. 160 pp.

5638. Malament, David B., and Sandy L. Zabell. "Why Gibbs Phase Averages Work -- The Role of Ergodic Theory." PS 47 (1980) 339-49.

5639. Quay, S.J., Paul M. "A Philosophical Explanation of the Explanatory Functions of Ergodic Theory." PS 45 (1978) 47-59.

5640. Rapoport, Anatol. "Verbal Difficulties in the Application of Newtonian Physics." Synthese. 7 (1948-49) 79-92.

5641. Santirocco, Raymond A. "Note on the Axiomatic Formulation of Electrostatics." BJPS 12 (1961) 132-45.

5642. Schulman, L. S., R. G. Newton, and R. Shtokhamer. "Model of Implication in Statistical Mechanics." PS 42 (1975) 503-11.

5643. Simon, Herbert A. "The Axiomatization of Classical Mechanics." PS 21 (1954) 340-43.

5644. Sklar, Lawrence. "Statistical Explanation and Ergodic Theory." PS 40 (1973) 194-212.

    a. "A Partial Vindication of Ergodic Theory," by K. S. Friedman. 43 (1976) 151-62.

5645. _____. "Inertia, Gravitation and Metaphysics." PS 43 (1976) 1-23.

5646. Tisza, Laszlo. "The Foundations of Statistical Mechanics." PSA 1976. 585-608.

    a. "Comments on the Papers of Brush and Tisza," by Abner Shimony. 609-16.

    b. "Response to Comments," by L. Tisza. 617-18.

5647. _____. "Classical Statistical Mechanics Versus Quantal Statistical Thermodynamics: A Study in Contrasts." WOSPS 6-III (1976) 207-16.

5648. Thüring, B. "Logische Betrachtungen zum Energieerhaltungssatz und zum Perpetuum Mobile." Scientia. 86 (1951) 253-58. French Tr.: Supplement, 103-08.

5649. _____. "Operative oder analytische Definition des Begriffs Inertialsystem?" PN 18 (1980-81) 255-42.

5650. Tonnelat, Marie-Antoinette. "La relativité avant Einstein." Organon 2 (1965) 79-103.

5651. Toulmin, Stephen. "Criticism in the History of Science: Newton on Absolute Space, Time, and Motion." Phil. Review. 68 (1959) 1-29; 203-27.

    a. "Mathematical Ideals and Metaphysical Concepts," by Dudley Shapere. 69 (1960) 376-85.

5652. Whitrow, G. J. "On the Foundations of Dynamics." BJPS 1 (1950) 92-107.

    a. "The Law of Inertia," by L. L. Whyte and G. J. Whitrow. 2 (1951) 58.

5653. Wicken, Jeffrey S. "Causal Explanations in Classical and Statistical Thermodynamics." PS 48 (1981) 65-77.

5654. Wolf, Rudolf. "Ueber die Konsequenzen der Mehrfachdefinition von Grundbegriffen in der Theorie der klassischen Mechanik." PN 2 (1952) 238-50.

6.6 Quantum Mechanics

5655. Albertson, James. "The Statistical Nature of Quantum Mechanics."
BJPS 13 (1962) 229-33.

5656. Almog, J. "Perhaps (?), New Logical Foundations are Needed
for Quantum Mechanics." Logique et analyse. 21 (1978) 253-77.

5657. Angelidis, Th. D. "Momentum Conservation Decides Heisenberg's
Interpretation of the Uncertainty Formulas." FP 7 (1977) 431-49.

5658. Aronov, R. A. "Toward a Logic of the Microworld." Soviet
Studies in Philosophy. 9 (1970-71) 212-17. [English reprint from
Voprosy filosofii. (1970, No. 2).]

5659. Audi, Michael. The Interpretation of Quantum Mechanics.
Chicago: University of Chicago Press, 1973. xiv + 200 pp.

    a. Review essay, "Wave-Particle Duality," by M. L. G. Redhead.
    BJPS 28 (1977) 65-74.

5660. Barreau, H. "La métaphysique de Bernard d'Espagnat." RMM 86
(1981) 364-78.

    a. "Response à Hervé Barreau," by B. d'Espagnat. 379-88.

5661. Bastin, Ted, ed. Quantum Theory and Beyond. Cambridge:
Cambridge University Press, 1971. viii + 345 pp.

    a. Review essay by J. Bub. BJPS 24 (1973) 78-90.

    b. Review essay by Ian Drummond and Jon Dorling. Theoria
    to Theory. 5, No. 4 (1971) 77-90.

5662. _____. "Probability in a Discrete Model of Particles
and Observations." SL 78 (1976) 195-220. Reprinted from Synthese.
29 (1974) 203-28.

5663. Bedau, Hugo, and Paul Oppenheim. "Complementarity in Quantum
Mechanics: A Logical Analysis." Synthese. 13 (1961) 201-32.

5664. Belinfante, F. J. A Survey of Hidden-Variable Theories.
Oxford: Pergamon Press, 1973. 243 pp.

5665. _____. Measurements and Time Reversal in Objective
Quantum Theory. New York: Pergamon Press, 1975. xix + 116 pp.

5666. Beltrametti, E. G., and G. Cassinelli. "On State Transformations Induced by Yes-No Experiments, in the Context of Quantum
Logic." Journal of Philosophical Logic. 6 (1977) 369-79.

5667. _____. "Problems of the Proposition-State Structure of Quantum Mechanics." BSPS 47 (1981) 215-35.

5668. Benoist, Rodney W., Jean-Paul Marchand, and Wolfgang Yourgrau. "Statistical Inference and Quantum Mechanical Measurement." FP 7 (1977) 827-33.

5669. Berenda, Carlton W. "De Broglie and Bohr Yet Once More." Southwestern Journal of Philosophy. 3, No. 2 (1972) 137-39.

5670. Bergstein, T. Quantum Physics and Ordinary Language. London: Macmillan, 1972. xii + 61 pp.

5671. Bernays, P. "Ueber den Unterschied zwischen realisticher und konservater Tendenz in der heutigen theoretischen Physik." RMM 67 (1962) 142-46.

5672. Beth, E. W. "Semantics of Physical Theories." SL 3 (1961) 48-51.

5673. Bigelow, John C. "Quantum Probability in Logical Space." PS 46 (1979) 223-43.

5674. Birkhoff, G. and J. von Neumann. "The Logic of Quantum Mechanics." Annals of Mathematics. 37 (1936) 823-43. Reprinted in WOSPS 5 (1975) 1-26.

5675. Bjørnestad, Øistein. "A Note on the So-Called Yes-No Experiments and the Foundations of Quantum Mechanics." SL 78 (1976) 235-45. Reprinted from Synthese. 29 (1974) 243-54.

5676. Blokhintsev, D. I. The Philosophy of Quantum Mechanics. New York: Humanities Press, 1968. viii + 132 pp.

5677. Bohm, David. Causality and Chance in Modern Physics. Foreward by Louis de Broglie. London: Routledge & Kegan Paul; New York: Van Nostrand, 1957. Reprint: New York: Harper & Row, 1961. xi + 170 pp.

   a. Review essay by P. K. Feyerabend. BJPS 10 (1960) 321-38.

5678. _____. "A Proposed Explanation of Quantum Theory in terms of Hidden Variables at a Sub-quantum-mechanical Level," in Observation and Interpretation. S. Körner, ed. New York: Academic Press; London: Butterworths, 1957. 33-40.

5679. _____. "Classical and Non-Classical Concepts in the Quantum Theory." BJPS 12 (1961-62) 265-80. Reprinted in Physical Reality. S. Toulmin, ed. New York: Harper & Row, 1970. 197-216.

5680. _____. "Quantum Theory as an Indication of a New Order in Physics. Part A. The Development of New Orders as Shown Through the History of Physics." FP 1 (1970-71) 359-81. "B. Implicate and Explicate Order in Physical Law." 3 (1973) 139-68.

5681. Bohm, D. J., and B. J. Hiley. "On the Intuitive Understanding of Nonlocality as Implied by Quantum Theory." FP 5 (1975) 93-109.

5682. Bohr, Niels. "On the Notions of Causality and Complementarity." Dialectica. 2 (1948) 312-19.

5683. Bonsack, François. "Indeterminismus und Freiheit in unvollständig beschriebenen Systemen." Studia Philosophica. 20 (1960) 18-36.

5684. Bopp, F. "The Principles of the Statistical Equations of Motion in Quantum Theory," in Observation and Interpretation. S. Körner, ed. New York: Academic Press; London: Butterworths, 1957. 189-96.

5685. Born, Max. Natural Philosophy of Cause and Chance. Oxford: Clarendon Press, 1949. viii + 215 pp.

5686. _____. "The Interpretation of Quantum Mechanics." BJPS 4 (1953) 95-106.

5687. Born, M., and W. Biem. "Dualismus in der Quantentheorie." PN 10 (1967-68) 411-18.

    a. "Dualismus in der Quantentheorie (Eine Entgegnung)," by A. Landé. 11 (1969) 395-96.

    b. "Zu Alfred Landés Entgegnung," by W. Biem. 397.

5688. Broglie, Louis de. "Hasard et contingence en physique quantique." RMM 50 (1945) 241-52.

5689. _____. "Le microscope électronique et la dualité des ondes et des corpuscules." RMM 52 (1947) 3-14.

5690. _____. "Au delà des mouvantes limites de la science." RMM 52 (1947) 277-89.

5691. _____. "Sur la complémentarité des idées d'individu et de système." Dialectica. 2 (1948) 325-30.

5692. _____. "Sur la relation d'incertitude de la seconde quantification." RIP 3 (1949) 166-75.

5693. _____. "La physique quantique restera-t-elle indéterministe?" Bulletin de la Société Française de Philosophie. 46-47 (1952/53) 135-73.

5694. _____. Éléments de théorie des quanta et de mécanique ondulatoire. Paris: Gauthier-Villars, 1953. viii + 302 pp.

    a. Review essay, "Methodological Aspects of Physics and Geometry," by E. H. Hutten. BJPS 5 (1954) 256-61.

5695. _____ . La physique quantique restera-t-elle indéterministe? Exposé du problème suivi de la reproduction de certains
documents et d'une contribution de M. J-P. Vigier. Paris: Gauthier-
Villars, 1953. vii + 113 pp.

    a. Review essay by E. H. Hutten. BJPS 5 (1954) 159-64.

5696. _____ . Une Tentative d'interprétation causale et non
linéaire de la mécanique ondulatoire (la théorie de la double solution).
Paris: Gauthier-Villars, 1956. vii + 297 pp.

5697. _____ . The Current Interpretation of Wave Mechanics: A
Critical Study. Amsterdam: Elsevier, 1964. ix + 94 pp.

5698. _____ . "The Reinterpretation of Wave Mechanics." FP 1
(1970-71) 5-15.

5699. Brown, Harvey R., and M. L. G. Redhead. "A Critique of the
Disturbance Theory of Indeterminacy in Quantum Mechanics." FP 11
(1981) 1-20.

5700. Bub, Jeffrey. "Hidden Variables and the Copenhagen Interpretation." BJPS 19 (1968) 185-210.

5701. _____ . "On the Completeness of Quantum Mechanics."
WOSPS 2 (1973) 1-65.

5702. _____ . The Interpretation of Quantum Mechanics. WOSPS
3. Dordrecht: Reidel, 1974. 155 pp.

    a. Review essay by Clark Glymour. Canadian Journal of
    Philosophy. 6 (1976) 161-75.

    b. Review essay by G. J. Smith. IPQ 17 (1977) 339-46.

    c. Review essay by Edward MacKinnon. Philosophia. 10
    (1981) 89-124.

5703. _____ . "Popper's Propensity Interpretation of Probability
and Quantum Mechanics." MSPS 6 (1975) 416-29.

5704. _____ . "Randomness and Locality in Quantum Mechanics."
SL 78 (1976) 397-420.

5705. _____ . "Hidden Variables and Locality." FP 6 (1976)
511-25.

5706. _____ . "The Statistics of Non-Boolean Event Structures."
WOSPS 6-III (1976) 1-16.

5707. _____ . "What is Philosophically Interesting about
Quantum Mechanics?" WOSPS 10 (1977) 69-79.

5708. _____ . "Von Neumann's Projection Postulate as a Probability Conditionalization Rule in Quantum Mechanics." Journal of
Philosophical Logic. 6 (1977) 381-90.

5709. _____. "Some Reflections on Quantum Logic and Schrödinger's Cat." BJPS 30 (1979) 27-39.

5710. _____. "Hidden Variables and Quantum Logic-A Sceptical Review." Erkenntnis. 16 (1981) 275-93.

    a. "Invited Comment on Professor Bub's Paper," by R. L. Hudson. 295-97.

5711. Bub, Jeffrey and William Demopoulos. "The Interpretation of Quantum Mechanics." BSPS 13 (1974) 92-122.

5712. Bub, Jeffrey, and Vandana Shiva. "Non-local Hidden Variable Theories and Bell's Inequality." PSA 1978. I, 45-53.

5713. Büchel, Wolfgang. "Quantenphysik und kritischer Realismus." PN 5 (1958-59) 3-54.

5714. _____. "Das H-Theorem und seine Umkehrung." PN 6 (1960-61) 167-201.

5715. _____. "Der Bellsche Beweis. Eine Fallstudie." ZAW 8 (1977) 221-36.

5716. Bugajski, Sławomir. "Probability Implication in the Logics of Classical and Quantum Mechanics." Journal of Philosophical Logic. 7 (1978) 95-106.

5717. _____. "Only if 'Acrobatic Logic' is Non-Boolean." PSA 1980. I, 264-71.

5718. Bunge, Mario. "Strife About Complementarity." BJPS 6 (1955) 1-12, 141-54.

5719. _____. "The Philosophy of the Space-Time Approach to the Quantum Theory." Methodos. 7 (1955) 295-308.

5720. _____. Quantum Theory and Reality: Studies in the Foundations, Methodology and Philosophy of Science. Vol. II. New York: Springer-Verlag, 1967. 117 pp.

    a. "Discussion Review: The Interface of Philosophy and Physics," by Joseph Agassi. PS 39 (1972) 263-65.

5721. _____. "Analogy in Quantum Theory: From Insight to Nonsense." BJPS 18 (1968) 265-86.

5722. Caldirola, P. "Teoria della misurazione e teoremi ergodici nella meccanica quantistica." Scientia. 99 (1964) 219-31. French Tr.: Supplement, 129-40.

5723. _____. "Problemi vecchi e nuovi nella interpretazione della teoria quantistica." Scientia. 110 (1975) 653-99. English Tr.: 701-35.

5724. Cantore, S.J., Enrico. "Philosophy in Atomic Physics: Complementarity." Modern Schoolman. 34 (1957) 79-104.

5725. _____ ____. Atomic Order: An Introduction to the Philosophy
of Microphysics. Cambridge, MA: MIT Press, 1969. xi + 334 pp.

5726. Cartwright, Nancy Delaney. "Superposition and Macroscopic
Observation." SL 78 (1976) 221-34. Reprinted from Synthese. 29
(1974) 229-42.

5727. _____. "A Dilemma for the Traditional Interpretation of
Quantum Mixtures." BSPS 20, SL 64. (1974) 251-58.

5728. _____. "The Sum Rule Has Not Been Tested." PS 44
(1977) 107-12.

5729. _____. "The Only Real Probabilities in Quantum Mechanics."
PSA 1978. I, 54-59.

5730. Chapman, Tobias. "Quantum Logic and Modality." Logique et
analyse. 24 (1981) 99-111.

5731. Chari, C. T. K. "Quantum Mechanical Perplexities." Philosophical Quarterly (India). 26 (1953) 177-86.

5732. _____. "Philosophical Exaggerations of Quantum Field
Theory." Philosophical Quarterly (India). 37 (1965) 227-32.

5733. _____. "Information Theory, Quantum Mechanics, and
'Linguistic Duality'." Dialectica. 20 (1966) 66-87.

5734. _____. "Towards Generalized Probabilities in Quantum
Mechanics." Synthese. 22 (1971) 438-47.

5735. Chernavska, Ariadna. "The Impossibility of Bivalent, Truth-
Functional Semantics for the Non-Boolean Propositional Structures of
Quantum Mechanics." Philosophia. 10 (1981) 1-18.

5736. Chernoff, Paul R., and Jerrold E. Marsden. "Some Remarks on
Hamiltonian Systems and Quantum Mechanics." WOSPS 6-III (1976)
35-54.

5737. Cohen, Leon. "Can Quantum Mechanics be Reformulated as a
Classical Probability Theory?" PS 33 (1966) 317-22.

5738. _____. "Joint Probability Distributions in Quantum
Mechanics." WOSPS 2 (1973) 66-79.

5739. Colodny, Robert G., ed. Paradigms and Paradoxes: The Philosophical Challenge of the Quantum Domain. University of Pittsburgh
Series in the Philosophy of Science, Vol. V. Pittsburgh: University
of Pittsburgh Press, 1972. 446 pp.

     a. Review essay, "The Trouble with Quanta," by Heinz Post.
     BJPS 24 (1973) 277-82.

b. Review essay by Jeffrey Bub and William Demopoulos.
Philosophia. 6 (1976) 333-44.

5740. Cortes, Alberto. "Leibniz's Principle of the Identity of
Indiscernibles: A False Principle." PS 43 (1976) 491-505.

a. "Does Quantum Mechanics Disprove the Principle of the
Identity of Indiscernibles?" by R. L. Barnette. 45 (1978)
466-70.

b. "Quantum Theory and the Identity of Indiscernibles Revisited," by Allen Ginsberg. 48 (1981) 487-91.

5741. Corvez, M. "Positivisme ou réalisme de la physique contemporaine." RMM 76 (1971) 362-76.

5742. Costa de Beauregard, O. "Complémentarité et relativité."
Revue Phil. 145 (1955) 385-409.

5743. _____. "Le 'paradoxe' des corrélations d'Einstien et de
Schrödinger et l'épaisseur temporelle de la transition quantique."
Dialectica. 19 (1965) 280-89.

5744. _____. "Time Symmetry and Interpretation of Quantum
Mechanics." FP 6 (1976) 539-59.

5745. _____. "Le paradoxe d'Einstein, Podolsky et Rosen."
Bulletin de la Société Française de Philosophie. 71 (1977) 1-30.

5746. _____. "Le paradoxe d'Einstein (1927) ou d'Einstein-
Podolsky-Rosen (1935)." Logique et analyse. 22 (1979) 425-33.

5747. _____. "A Burning Question: Einstein's Paradox of
Correlations." Diogenes. 110 (1980) 83-97.

5748. Cyranski, John F. "Quantum Measurement as a Communication
with Nature." FP 8 (1978) 805-22.

5749. Dalla Chiara, Maria Luisa. "Logical Self Reference, Set
Theoretical Paradoxes and the Measurement Problem in Quantum Mechanics."
Journal of Philosophical Logic. 6 (1977) 331-47.

5750. _____. "Quantum Logic and Physical Modalities." Journal
of Philosophical Logic. 6 (1977) 391-404.

5751. _____. "Is There a Logic of Empirical Sciences?" BSPS
47 (1981) 187-96.

5752. Darachschani, Homayun. "Wissenschaftstheoretische Bemerkungen
zur Axiomatik der Quantenmechanik." Erkenntnis. 10 (1976) 361-70.

5753. Del-Negro, Walter. "Statistische Gesetze und Determination."
ZPF 7 (1953) 65-71.

5754. Demopoulos, William. "What is the Logical Interpretation of Quantum Mechanics?" BSPS 32 (1976) 721-28.

5755. _____. "Fundamental Statistical Theories." SL 78 (1976) 421-32.

5756. _____. "The Possibility Structure of Physical Systems?" WOSPS 6-III (1976) 55-80.

5757. _____. "Completeness and Realism in Quantum Mechanics." WOSPS 10 (1977) 81-88.

5758. _____. "Boolean Representations of Physical Magnitudes and Locality." Synthese. 42 (1979) 101-20.

5759. Denecke, Heinz-Martin. "Quantum Logic of Quantifiers." Journal of Philosophical Logic. 6 (1977) 405-13.

5760. D'Espagnat, B. "Things, Structures and Phenomena in Quantum Physics." LMPS-3 (1967) 377-84.

5761. _____. "Nonseparability and Quantum Logic." AJAPS 5, No. 2 (1977) 11-16.

5762. Destouches, Jean-Louis. "Quelques aspects théoretique de la notion de complémentarité." Dialectica. 2 (1948) 351-82.

5763. _____. "Déterminisme et indéterminisme en physique moderne." Actualités scientifiques et industrielles. 1066 (1949) 7-60.

5764. _____. "Sur le débat actuel du déterminisme et de l'indéterminisme dans les théories quantiques." Revue Phil. 143 (1953) 1-27.

5765. _____. "Théories prévisionnelles et théories réalistes en microphysique." RMM 67 (1962) 174-205.

5766. _____. "Aspects philosophiques récents liés aux théories quantiques." RIP 17 (1963) 190-96.

5767. _____. "General Mathematical Physics and Schemes, Application to the Theory of Particles." Dialectica. 19 (1965) 345-48.

5768. _____. "Confrontation entre réalisme et positivisme à propos du paradoxe d'Einstein-Podolsky-Rosen." Logique et analyse. 22 (1979) 381-406.

5769. _____. "The Notion of Object in Microphysics." Epistemologia. 2 (1979) 39-48.

5770. Destouches, Paulette. "Manifestations et sens de la notion de complémentarité." Dialectica. 2 (1948) 383-412.

5771. Destouches-Février, P. "Contradiction et complémentarité." Synthese. 7 (1948-49) 173-82.

5772. _____. "La structure ouverte des théories quantiques." Revue Phil. 142 (1952) 369-91.

5773. Dingle, Herbert. "Causality and Statistics in Modern Physics." BJPS 21 (1970) 223-46. Correction, 410.

5774. Dishkant, H. "Semantics of the Minimal Logic of Quantum Mechanics." Studia Logica. 30 (1972) 23-32.

5775. Domotor, Zoltan. "The Probability Structure of Quantum-Mechanical Systems." SL 78 (1976) 147-78. Reprinted from Synthese. 29 (1974) 155-86.

5776. Doty, Patricia J. "Complementarity and Its Analogies." J. Phil. 55 (1958) 1089-1104.

5777. Drieschner, M. "Is (Quantum) Logic Empirical?" Journal of Philosophical Logic. 6 (1977) 415-23.

5778. Dunn, J. Michael. "Quantum Mathematics." PSA 1980. II, 512-31.

5779. Edlin, Gregor. "Dialektik und Komplementarität." Studia Philosophica. 24 (1964) 66-89. English Tr.: Philosophy Today. 10 (1966) 75-87.

5780. Edwards, David A. "The Mathematical Foundations of Quantum Mechanics." Synthese. 42 (1979) 1-70.

5781. Ehrenberg, W. Dice of the Gods. Dundee: G. C. Stevenson, 1977. 110 pp.

5782. Einstein, Albert. "Quanten-Mechanik und Wirklichkeit." Dialectica. 2 (1948) 320-24.

5783. Elsasser, Walter M. "Quantum Mechanics, Amplifying Processes, and Living Matter." PS 18 (1951) 300-26.

5784. _____. "A Natural Philosophy of Quantum Mechanics Based on Induction." FP 3 (1973) 117-37.

5785. Emch, G., and J. M. Jauch. "Structures logiques et mathématiques en physique quantique." Dialectica. 19 (1965) 259-79.

5786. Erwin, Edward. "Quantum Logic and the Status of Classical Logic." Logique et analyse. 21 (1978) 279-92.

5787. Faggiani, D. "Fisica quantistica e tradizione filosofica." Scientia. 106 (1971) 993-1003. English Tr.: 1004-13.

5788. Faye, Jan. "The Influence of Harald Høffding's Philosophy on Niels Bohr's Interpretation of Quantum Mechanics." Danish Yearbook of Philosophy. 16 (1979) 37-72.

a. "On Hoffding and Bohr: A Reply to Jan Faye," by David Favrholdt. 73-78.

5789. Feibleman, J. K. "The Principle of Indeterminacy Re-examined." Ratio. 3 (1961) 133-51.

5790. Feinberg, Gerald. "Philosophical Implications of Contemporary Particle Physics," in Paradigms and Paradoxes: The Philosophical Challenge of the Quantum Domain. Robert G. Colodny, ed. Pittsburgh: University of Pittsburgh Press, 1972. 33-46.

5791. Février, P. L'Interprétation physique de la mécanique ondulatoire et des théories quantiques. Paris: Gauthier-Villars, 1956. vii + 216 pp.

5792. Feyerabend, Paul. "Reichenbach's Interpretation of Quantum-Mechanics." Phil. Studies. 9 (1958) 49-59. Reprinted in WOSPS 5 (1975) 109-22.

5793. _____. "Complementarity." Arist. Soc. Supp. Vol. 32 (1958) 75-104.

5794. _____. "Professor Bohm's Philosophy of Nature." BJPS 10 (1959-60) 321-38. Reprinted in Physical Reality. S. Toulmin, ed. New York: Harper & Row, 1970. 173-96.

5795. _____. "Niels Bohr's Interpretation of the Quantum Theory," in Current Issues in the Philosophy of Science. H. Feigl and G. Maxwell, eds. New York: Holt, Rinehart and Winston, 1961. 371-90.

a. "Comments," by Norwood Russell Hanson. 390-98.

b. "Rejoinder," by Paul K. Feyerabend. 398-400.

5796. _____. "On a Recent Critique of Complementarity." PS 35 (1968) 309-31; 36 (1969) 82-105.

5797. Filippe, U. "Les principes de la mécanique quantique. Leur interprétation par une logique trivalente." Les Etudes philosophiques. 9 (1946) 121-27.

5798. Finch, P. D. "On the Structure of Quantum Logic." The Journal of Symbolic Logic. 34 (1969) 275-82. Reprinted in WOSPS 5 (1975) 415-26.

5799. _____. "Quantum Mechanical Physical Quantities as Random Variables." WOSPS 6-III (1976) 81-104.

5800. Fine, Arthur. "Logic, Probability, and Quantum Theory." PS 35 (1968) 101-11.

5801. _____. "Realism in Quantum Measurements." Methodology and Science. 1 (1968) 210-20.

5802. _____. "Some Conceptual Problems of Quantum Theory," in
Paradigms and Paradoxes: The Philosophical Challenge of the Quantum
Domain. Robert G. Colodny, ed. Pittsburgh: University of Pitts-
burgh Press, 1972. 3-31.

5803. _____. "Probability and the Interpretation of Quantum
Mechanics." BJPS 24 (1973) 1-37.

5804. _____. "On the Completeness of Quantum Theory." SL 78
(1976) 249-82. Reprinted from Synthese. 29 (1974) 257-90.

5805. _____. "Conservation, the Sum Rule and Confirmation."
PS 44 (1977) 95-106.

5806. _____. "How to Count Frequencies: A Primer for Quantum
Realists." Synthese. 42 (1979) 145-54.

5807. _____. "Correlations and Physical Locality." PSA 1980.
II, 535-62.

     a. "Critique of the Papers of Fine and Suppes," by Abner
     Shimony. 572-80.

5808. Fine, Arthur, and Paul Teller. "Algebraic Constraints on
Hidden Variables." FP 8 (1978) 629-36.

5809. Fine, Terrence L. "Towards a Revised Probabilistic Basis for
Quantum Mechanics." SL 78 (1976) 179-94. Reprinted from Synthese.
29 (1974) 187-202.

5810. Finkelstein, David. "The Physics of Logic," in Paradigms
and Paradoxes: The Philosophical Challenge of the Quantum Domain.
Robert G. Colodny, ed. Pittsburgh: University of Pittsburgh Press,
1972. 47-66.

5811. _____. "The Leibniz Project." Journal of Philosophical
Logic. 6 (1977) 425-39.

5812. Fleming, John J. "Sub-Quantum Entities." PS 31 (1964)
271-74.

5813. Fock, V. "La physique quantique et les idéalisations classiques."
Dialectica. 19 (1965) 223-45.

5814. _____. "Quantum Physics and Philosophical Problems."
FP 1 (1970-71) 293-306.

5815. Fogarasi, Béla. "Ist der Komplementaritätsgedanke widerspruch-
frei?" DZP 3 (1955) 190-94.

5816. Fok, V. A. "Quantum Physics and Philosophical Problems."
Soviet Studies in Philosophy. 10 (1971-72) 252-56. [English reprint
from Voprosy filosofii. (1971, No. 3).]

5817. Folse, Henry J. "The Copenhagen Interpretation of Quantum Theory and Whitehead's Philosophy of Organism." Tulane Studies in Philosophy. 23 (1974) 32-47.

5818. _____. "The Formal Objectivity of Quantum Mechanical Systems." Dialectica. 29 (1975) 127-44.

5819. _____. "Complementarity and the Description of Experience." IPQ 17 (1977) 377-92.

5820. _____. "Quantum Theory and Atomism: A Possible Ontological Resolution of the Quantum Paradox." Southern Journal of Philosophy. 16 (1978) 629-40.

5821. Foulis, D. J., and C. H. Randall. "Empirical Logic and Quantum Mechanics." SL 78 (1976) 73-104. Reprint from Synthese. 29 (1974) 81-112.

5822. Fraassen, Bas C. van. "A Formal Approach to the Philosophy of Science," in Paradigms and Paradoxes: The Philosophical Challenge of the Quantum Domain. Robert G. Colodny, ed. Pittsburgh: University of Pittsburgh Press, 1972. 303-66.

    a. "Van Fraassen's Modal Model of Quantum Mechanics," by Nancy Cartwright. PS 41 (1974) 199-202.

5823. _____. "Semantic Analysis of Quantum Logic." WOSPS 2 (1973) 80-113.

5824. _____. "The Einstein-Podolsky-Rosen Paradox." SL 78 (1976) 283-302. Reprinted from Synthese. 29 (1974) 291-310.

5825. _____. "The Labyrinth of Quantum Logics." BSPS 13 (1974) 224-54. Reprinted in WOSPS 5 (1975) 577-607.

5826. _____. "The Formal Representation of Physical Quantities." BSPS 13 (1974) 196-209.

    a. "Comments," by E. J. Post, 210-13; by John Stachel, 214-23.

5827. _____. "Hidden Variables and the Modal Interpretation of Quantum Theory." Synthese. 42 (1979) 155-65.

5828. Fraassen, Bas C. van, and C. A. Hooker. "A Semantic Analysis of Niels Bohr's Philosophy of Quantum Theory." WOSPS 6-III (1976) 221-41.

5829. Fraïssé, Roland. "Essai sur la logique d l'indéterminisme et la ramification de l'espace-temps." SL 78 (1976) 19-46. Reprinted from Synthese. 29 (1974) 27-54.

5830. Freistadt, Hans. "Dialectical Materialism: A Further Discussion." PS 24 (1957) 25-40.

5831. Freundlich, Yehudah. "Mind, Matter, and Physicists." FP 2 (1972) 129-48.

5832. _____. "Two Views of an Objective Quantum Theory." FP 7 (1977) 279-300.

5833. _____. "In Defense of Copenhagenism." SHPS 9 (1978) 151-79.

5834. _____. "Copenhagenism and Popperism." BJPS 29 (1978) 145-77.

5835. Frey, Gerhard. "Subjektive und objektive Unbestimmtheit." Methodos. 4 (1952) 301-16.

5836. Friedman, Michael, and Clark Glymour. "If Quanta Had Logic." Journal of Philosophical Logic. 1 (1972) 16-28.

5837. Friedman, M., and H. Putnam. "Quantum Logic, Conditional Probability, and Interference." Dialectica. 32 (1978) 305-16.

5838. Friedrichs, K. O. "Remarks on the Notion of State in Quantum Mechanics." FP 9 (1979) 515-24.

5839. Fries, Horace. "Is the psi-Function Description 'Complete'? A Layman's Question." PS 19 (1952) 166-69.

5840. Fuchs, Walter R. "Ansätze zu einer Quantenlogik." Theoria. 30 (1964) 137-40.

5841. Fujiwara, Izuru. "Quantum Theory of State Reduction and Measurement." FP 2 (1972) 83-110.

5842. Gardner, Michael R. "Is Quantum Logic Really Logic." PS 38 (1971) 508-29.

5843. _____. "Quantum-Theoretical Realism: Popper and Einstein v. Kochen and Specker." BJPS 23 (1972) 13-23.

5844. _____. "Two Deviant Logics for Quantum Theory: Bohr and Reichenbach." BJPS 23 (1972) 89-109.

5845. Gauthier, Yvon. "The Use of the Axiomatic Method in Quantum Physics." PS 38 (1971) 429-37.

5846. Gerharz, Reinhold. "Ansätze zu einer kausalistischen Strukturtheorie des Elektrons." PN 13 (1971-72) 443-57.

5847. _____. "Ist das Photon eine Kombination aus Wirkung und Energie?" PN 14 (1973) 225-30.

5848. Gerjuoy, E. "Is the Principle of Superposition Really Necessary?" WOSPS 2 (1973) 114-42.

5849. Gerlich, G. "Some Remarks on Classical Probability Theory in Quantum Mechanics." Erkenntnis. 16 (1981) 335-38.

5850. Gibbins, Peter. "A Note on Quantum Logic and the Uncertainty Principle." PS 48 (1981) 122-26.

5851. Glymour, Clark. "The Sum Rule is Well-Confirmed." PS 44 (1977) 86-94.

5852. Gonseth, F. "Remarque sur l'idée de complémentarité." Dialectica. 2 (1948) 413-20.

5853. Greechie, Richard J. "Some Results from the Combinatorial Approach to Quantum Logic." SL 78 (1976) 105-19. Reprinted from Synthese. 29 (1974) 113-27.

5854. Greechie, R. J., and Stanley P. Gudder. "Quantum Logics." WOSPS 2 (1973) 143-73. Reprinted in WOSPS 5 (1975) 545-76.

5855. Groenewold, H. J. "Quantum Mechanics and Its Models." Synthese. 9 (n.d.) 97-103. Reprinted in 10 (n.d.) 203-09.

5856. _____. "Objective and Subjective Aspects of Statistics in Quantum Description," in Observation and Interpretation. S. Körner, ed. New York: Academic Press; London: Butterworths, 1957. 197-203.

5857. _____. "Foundations of Quantum Theory; Statistical Interpretation." SL 31 (1970) 180-99.

5858. Grossman, Neal. "Quantum Mechanics and Interpretations of Probability Theory." PS 39 (1972) 451-60.

5859. _____. "The Ignorance Interpretation Defended." PS 41 (1974) 333-44.

5860. _____. "Metaphysical Implications of the Quantum Theory." Synthese. 35 (1977) 79-98. Reprinted in SL 132 (1979) 605-24.

5861. Grünbaum, Adolf. "Complementarity in Quantum Physics and Its Philosophical Generalization." J. Phil. 54 (1957) 713-27.

5862. Guccione, S. "Quantum Logic and the Two-Slit Experiment." BSPS 47 (1981) 237-47.

5863. Hanson, Norwood R. "Uncertainty." Phil. Review. 63 (1954) 65-73.

5864. _____. "The Logic of the Correspondence Principle." Scientia. 93 (1958) 63-70. French Tr.: Supplement, 43-51.

5865. _____. "The Copenhagen Interpretation of Quantum Theory." American Journal of Physics. 27 (1959) 1-15. Reprinted in Physical Reality. S. Toulmin, ed. New York: Harper & Row, 1970. 143-72.

5866. _____. "Five Cautions for the Copenhagen Interpretation's Critics." PS 26 (1959) 325-37.

5867. _____. "Are Wave Mechanics and Matrix Mechanics Equivalent Theories?" in Current Issues in the Philosophy of Science. H. Feigl and G. Maxwell, eds. New York: Holt, Rinehart and Winston, 1961. 401-25.

    a. "Comments," by E. L. Hill. 425-28.

    b. "Rejoinder" by Norwood Russell Hanson. 428.

5868. _____. "Wave Mechancis and Matrix Mechanics." Czechoslovakian Journal of Theoretical Physics. B 11 (1961).

5869. _____. "Heisenberg and Schrödinger." Journal of Philosophy. 59 (1962) 320-22.

5870. Hardegree, Gary M. "The Conditional in Quantum Logic." SL 78 (1976) 55-72. Reprinted from Synthese. 29 (1974) 63-80.

5871. _____. "Stalnaker Conditionals and Quantum Logic." Journal of Philosophical Logic. 4 (1975) 399-421.

5872. _____. "The Modal Interpretation of Quantum Mechanics." PSA 1976. I, 82-103.

5873. _____. "Relative Compatibility in Conventional Quantum Mechanics." FP 7 (1977) 495-510.

5874. _____. "Reichenbach and the Logic of Quantum Mechanics." Synthese. 35 (1977) 3-40. Reprinted in SL 132 (1979) 475-512.

5875. _____. "Reichenbach and the Interpretation of Quantum Mechanics." SL 132 (1979) 513-66.

5876. _____. "Micro-States in the Interpretation of Quantum Theory." PSA 1980. I, 43-54.

5877. _____. "An Axiom System for Orthomodular Quantum Logic." Studia Logica. 40 (1981) 1-12.

5878. Harré, R. "Philosophy and Quantum Physics." Philosophy. 35 (1960) 341-43.

5879. Hartshorne, Charles. "Charles Peirce and Quantum Mechanics." Transactions of the Charles S. Peirce Society. 9 (1973) 191-201.

5880. _____. "Bell's Theorem and Stapp's Revised View of Space-Time." Process Studies. 7 (1977) 183-91.

5881. Healey, Richard. "Comments on Kochen's Specification of Measurement Interactions." PSA 1978. II, 277-94.

5882. _____ . "Quantum Realism: Naïvete is No Excuse."
Synthese. 42 (1979) 121-44.

5883. Heelan, Patrick. "Quantum and Classical Logic: Their Respective Roles." Synthese. 21 (1970) 2-33. Reprinted in BSPS 13
(1974) 318-49.

5884. _____ . "Complementarity, Context Dependence, and Quantum
Logic." FP 1 (1970-71) 95-110.

5885. Heisenberg, Werner. Philosophic Problems of Nuclear Science.
London: Faber and Faber, 1952. 126 pp.

   a. Review essay by Ernan McMullin. Philosophical Studies
   (Ireland). 3 (1953) 106-21.

5886. _____ . Nuclear Physics. New York: Philosophical
Library, 1953. 225 pp.

5887. _____ . Physics and Philosophy: The Revolution in
Modern Science. With an Introduction by F. S. C. Northrop. New
York: Harper & Brothers, 1958.

   a. Review essay by Jean Wahl. RMM 66 (1961) 326-33.

5888. _____ . The Physicist's Conception of Nature. London:
Hutchinson, 1958. 192 pp.

5889. _____ . Introduction to the Unified Field Theory of
Elementary Particles. London: Interscience Publishers, 1966. 177
pp.

   a. Review essay by J. M. Jauch. FP 1 (1970-71) 183-89.

5890. _____ . Der Teil und das Ganze - Gespräche im Umkreis
der Atomphysik. München: R. Piper & Co. Verlag, 1969.

5891. _____ . Physics and Beyond: Encounters and Conversa-
tions. Arnold J. Pomerans, tr. New York: Harper & Row, 1971. 257
pp.

5892. _____ . Fysica in Perspectief. Utrecht, Antwerpen: Het
Spectrum, 1974. 160 pp.

5893. _____ . "The Philosophical Background of Modern Physics."
[In Hebrew] Iyyun. 26 (1975) 1-16.

5894. Hellman, Geoffrey. "Quantum Logic and Meaning." PSA 1980.
II, 493-51.

5895. _____ . "A Probabilistic Version of the Kochen-Specker
No-Hidden-Variable Proof." Synthese. 44 (1980) 495-500.

5896. _____ . "Quantum Logic and the Projection Postulate."
PS 48 (1981) 469-86.

5897. Hennemann, Gerhard. "Zur erkenntnistheoretisch-ontologischen Problematik der modernen Atomphysik." Studia Philosophica. 17 (1957) 73-86.

5898. _____. "Zur Frage nach dem ontologischen Hintergrund der modernen Atomphysik." PN 6 (1960-61) 32-54.

5899. _____. "Die philosophische Problematik des physikalischen Wirklichkeitsbegriffes." ZPF 15 (1961) 415-43.

5900. Hermann, Armin. The Genesis of Quantum Theory. W. N. Claude, tr. Cambridge, MA: MIT Press, 1971. viii + 164 pp.

5901. Hill, E. L. "Quantum Physics and the Relativity Theory," in Current Issues in the Philosophy of Science. H. Feigl and G. Maxwell, eds. New York: Holt, Rinehart and Winston, 1961. 429-41.

   a. "Comments," by Paul K. Feyerabend. 441-44.

   b. "Rejoinder" by E. L. Hill. 444-45.

5902. Hooker, C. A. "Energy and the Interpretation of Quantum Mechanics." AJP 49 (1971) 262-70.

5903. _____. "The Nature of Quantum Mechanical Reality: Einstein Versus Bohr," in Paradigms and Paradoxes: The Philosophical Challenge of the Quantum Domain. Robert G. Colodny, ed. Pittsburgh: University of Pittsburgh Press, 1972. 67-302.

5904. _____. "Metaphysics and Modern Physics: A Prolegomenon to the Understanding of Quantum Theory." WOSPS 2 (1973) 174-304.

5905. _____. Contemporary Research in the Foundations and Philosophy of Quantum Theory. WOSPS 2. Dordrecht: Reidel, 1973. xx + 385 pp.

5906. _____. The Logico-Algebraic Approach to Quantum Mechanics. WOSPS 5. Dordrecht: Reidel. Vol. I: Historical Evolution, 1975, 607 pp.; Vol. II: Contemporary Consolidation, 1979, 466 pp.

5907. Holdsworth, David G. "Category Theory and Quantum Mechanics (Kinematics)." Journal of Philosophical Logic. 6 (1977) 441-53.

5908. _____. "A Role for Categories in the Foundations of Quantum Theory." PSA 1978. I, 257-67.

5909. Hörz, Herbert. "Die philosophische Bedeutung der Heisenberg-schen Unbestimmtheitsrelation." DZP 8 (1960) 702-09.

5910. _____. "Bemerkungen zum Begriff des Faktischen in der Kopenhagener Deutung der Quantenmechanik." DZP 10 (1962) 75-84.

5911. _____. "Zu einigen philosophischen Problemen der Theorie der Elementarteilchen." DZP 13 (1965) 828-50.

5912. Hoyer, U. "Ueber eine rationale statistische Grundlegung der
Wellenmechanik." PN 18 (1980-81) 356-67.

5913. Hübner, Kurt. "Zur gegenwärtigen philosophischen Diskussion
der Quantenmechanik." PN 9 (1965-66) 3-21.

5914. _____. "Ueber die Philosophie der Wirklichkeit in der
Quantenmechanik." PN 14 (1973) 3-24.

5915. _____. "The Philosophical Background of Hidden Variables
in Quantum Mechanics." Man and World. 6 (1973) 421-40.

5916. Hughes, R. I. G. "Quantum Logic and the Interpretation of
Quantum Mechanics." PSA 1980. I, 55-67.

5917. Jammer, Max. The Conceptual Development of Quantum Mechanics.
New York: McGraw-Hill, 1966. 400 pp.

5918. _____. The Philosophy of Quantum Mechanics. New York:
John Wiley, 1974. xi + 536 pp.

5919. Jasselette, P. "Le paradoxe EPR et les idées d'Einstein sur
la réalité." Logique et analyse. 22 (1979) 407-23.

5920. _____. "La non séparabilité et la notion de système en
physique quantique." Epistemologia. 3 (1980) 3-12.

5921. Jauch, J. M. The Foundations of Quantum Mechanics. Reading,
MA: Addison-Wesley, 1968. 299 pp.

5922. _____. Are Quanta Real? Bloomington and London: Indiana University Press, 1973. xii + 106 pp.

        a. Review essay, "Wave-Particle Duality," by M. L. G. Redhead.
        BJPS 28 (1977) 65-74.

5923. _____. "The Quantum Probability Calculus." SL 78
(1976) 123-46. Reprinted from Synthese. 29 (1974) 131-54.

5924. Jauch, J. M., and C. Piron. "On the Structure of Quantal
Proposition Systems." Helvetica Physica Acta. 43 (1969) 842-48.
Reprinted in WOSPS 5 (1975) 427-36.

5925. Jones, Roger. "Causal Anomalies and the Completeness of
Quantum Theory." Synthese. 35 (1977) 41-78. Reprinted in SL 132
(1979) 567-604.

5926. Jones, William B. "Bell's Theorem, H. P. Stapp, and Process
Theism." Process Studies. 7 (1977) 250-61.

5927. Juhos, Béla von. "Mögliche Gesetzformen in der Quantenphysik."
PN 3 (1955) 211-37.

5928. Kägi-Romano, U. "Quantum Logic and Generalized Probability
Theory." Journal of Philosophical Logic. 6 (1977) 455-62.

5929. Kaila, Eino. "Zur Metatheorie der Quantenmechanik." Acta Philosophica Fennica. 5 (1950) 1-136.

5930. Kamlah, Andreas. "The Connection between Reichenbach's Three-Valued and V. Neumann's Lattice-Theoretical Quantum Logic." Erkenntnis. 16 (1981) 315-25.

5931. Kanthak, Lothar, and Ursula Wegener. "Zum Zusammenhang zwischen Projektionsoperatoren und Eigensschaften." ZAW 7 (1976) 249-57.

5932. Kellett, B. H. "The Physics of the Einstein-Podolsky-Rosen Paradox." FP 7 (1977) 735-57.

5933. Kemble, Edwin C. "Reality, Measurement, and the State of the System in Quantum Mechanics." 18 (1951) 273-99.

    a. "Comment on a Paper by Professor Kemble," by J. P. McKinney. 20 (1953) 227-31.

    b. "Reply to Mr. McKinney," by E. C. Kemble. 232-35.

5934. Kirschenmann, Peter. "Reciprocity in the Uncertainty Relations." PS 40 (1973) 52-58.

5935. _____. "Stability Implies Chance." AJAPS 5, No. 2 (1977) 27-34.

5936. Kochen, S., and E. P. Specker. "Logical Structures Arising in Quantum Theory," in The Theory of Models. J. Addison, L. Henkin, and A. Tarski, eds. Amsterdam: North-Holland Pub. Co., 1965. Reprinted in WOSPS 5 (1975) 263-76.

5937. _____. "The Calculus of Partial Propositional Functions," in Logic, Methodology and Philosophy of Science. U. Bar-Hillel, ed. Amsterdam: North-Holland Pub. Co., 1965. Reprinted in WOSPS 5 (1975) 277-92.

5938. _____. "The Problem of Hidden Variables in Quantum Mechanics." Journal of Mathematics and Mechanics. 17 (1967) 59-67. Reprinted in WOSPS 5 (1975) 293-328.

5939. Komar, Arthur. "The Quantitative Epistemological Content of Bohr's Correspondence Principle." Synthese. 21 (1970) 83-92.

5940. _____. "The General Relativistic Quantization Program." WOSPS 2 (1973) 305-27.

5941. Konczewska, H. "Le crise de la physique contemporaine et la conversion idéaliste." Revue Phil. 145 (1955) 410-31.

5942. Konrad, W. "Dialektik der Elementarteilchenprozesse." DZP 15 (1967) 1069-85.

5943. Kouznetsov, Boris. "Einstein et Bohr." Organon. 2 (1965) 105-21.

5944. Kracklauer, A. F. "On the Imaginable Content of de Broglie Waves." Scientia. 109 (1974) 111-19.

5945. Krips, Henry P. "Two Paradoxes in Quantum Mechanics." PS 36 (1969) 145-52.

a. "Against Krips' Resolution of Two Paradoxes in Quantum Mechanics," by C. A. Hooker. 38 (1971) 418-28.

b. "The Einstein-Podolski-Rosen Paradox," by Herman Erlichson. 39 (1972) 83-85.

5946. _____. "Quantal Quandaries." AJP 52 (1974) 133-45.

5947. _____. "Foundations of Quantum Theory." FP 4 (1974) I: 181-93; II: 381-94; III: 6 (1976) 639-59.

5948. Kropp, Gerhard. "Zum Begriff der Komplementarität." PN 1 (1951) 446-62.

5949. Kumakura, K. "On the Concept of Phase in Quantum Mechanics." AJAPS 5, No. 1 (1976) 37-45.

5950. Kundt, U., and K. Lanius. "Der moderne Atomismus." DZP 27 (1979) 203-12.

5951. Kunsemüller, H. "Zur Axiomatik der Quantenlogik." PN 8 (1964) 363-76.

5952. Kurth, R. "Zur Deutung der Heisenberg-Ungleichung." PN 16 (1976-77) 10-39.

5953. Kuznetsov, B. G. "On Quantum-Relativistic Logic." Soviet Studies in Philosophy. 9 (1970-71) 203-11. [English reprint from Voprosy filosofii. (1970, No. 2).]

5954. Laitko, H. "Zur Dialektik von Kontinuität und Diskontinuität und einigen physikalischen Problemen." DZP 12 (1964) 54-64.

5955. Landé, Alfred. "Continuity, A Key to Quantum Mechanics." PS 20 (1953) 101-09.

5956. _____. "Quantum Indeterminacy, A Consequence of Cause-Effect Continuity." Dialectica. 8 (1954) 199-209.

5957. _____. Foundations of Quantum-Mechanics: A Study in Continuity and Symmetry. Yale University Press, 1955. 106 pp.

5958. _____. "The Logic of Quanta." BJPS 6 (1956) 300-20.

5959. _____. "Non-Quantal Foundations of Quantum Theory." PS 24 (1957) 309-20. Reprinted in Dialectica, 19 (1965) 349-57, and in

Physics, Logic, and History. W. Yourgrau and A. D. Breck, eds. New York: Plenum Press, 1970. 297-306.

5960. _____. "Ist die Dualität in der Quantentheorie ein Erkenntnisproblem?" PN 5 (1958-59) 498-502.

5961. _____. "From Dualism to Unity in Quantum Mechanics." BJPS 10 (1959) 16-24.

5962. _____. _From Dualism to Unity in Quantum Physics_. Cambridge: Cambridge University Press, 1960. xvi + 114 pp.

    a. Review essay by W. Yourgrau. BJPS 12 (1961) 158-66.

5963. _____. "From Duality to Unity in Quantum Mechanics," in _Current Issues in the Philosophy of Science_. H. Feigl and G. Maxwell, eds. New York: Holt, Rinehart and Winston, 1961. 350-60.

    a. "Comments," by Henryk Mehlberg. 360-70.

5964. _____. "The Case Against Quantum Duality." PS 29 (1962) 1-6.

5965. _____. "Causality and Dualism on Trial," in _Philosophy of Science: The Delaware Seminar_, I. New York: Interscience Publishers, 1963. 327-48.

5966. _____. "Vom Dualismus zur einheitlichen Quantentheorie." PN 8 (1964) 232-41.

5967. _____. _New Foundations of Quantum Mechanics_. London: Cambridge University Press, 1965. 171 pp.

5968. _____. "Why Do Quantum Theorists Ignore the Quantum Theory?" BJPS 15 (1965) 307-13.

5969. _____. "Quantum Theory without Dualism." _Scientia_. 101 (1966) 208-12. French Tr.: Supplement, 91-95.

5970. _____. "Unity in Quantum Theory." FP 1 (1970-71) 191-202.

5971. _____. "The Decline and Fall of Quantum Dualism." PS 38 (1971) 221-23.

5972. _____. "Einheit in der Quantenwelt (gegen den Bohr-Heisenberg'schen Positivismus.)" _Dialectica_. 26 (1972) 115-30.

5973. _____. _Quantum Mechanics in a New Key_. New York: Exposition Press, 1973. vii + 131 pp.

5974. _____. "The Laws behind the Quantum Laws." BJPS 27 (1976) 43-50.

a. "Comments on Landé," by Jon Dorling. 160.

5975. _____. "Why the World is a Quantum World." SL 78
(1976) 433-44.

5976. Landsberg, P. T. "The Uncertainty Principle as a Problem in
Philosophy." Mind. 56 (1947) 250-56.

5977. Lanz, Lodovico. "Il principio di indeterminazione." Scientia.
111 (1976) 315-24. English Tr.: 325-32.

5978. Latzer, Robert W. "Errors in the No Hidden Variable Proof of
Kochen and Specker." SL 78 (1976) 323-64. Reprinted from Synthese.
29 (1974) 331-72.

5979. Leininger, C. W. "Concerning Some Proposals for Quantum
Logic." Notre Dame Journal of Formal Logic. 10 (1969) 95-96.

5980. Lenk, Hans. "Philosophische Kritik an Begründungen von
Quantenlogiken." PN 11 (1969) 413-25.

5981. Lenzen, Victor F. "Concepts and Reality in Quantum Mechanics."
PS 16 (1949) 279-86.

5982. Lestienne, R. "Faut-il reformuler la physique? Scientia.
108 (1973) 69-86. English Tr.: 87-101.

5983. Lévy-Leblond, Jean-Marc. "Towards a Proper Quantum Theory."
Dialectica. 30 (1976) 161-96.

5984. Lindenberg, Siegwart, and Paul Oppenheim. "Generalization of
Complementarity." Synthese. 28 (1974) 117-40.

5985. Lopes, J. L., and M. Paty, eds. Quantum Mechanics, A Half
Century Later. Dordrecht: Reidel, 1977. x + 310 pp.

5986. Lorente, Miguel. "Cincuenta años de mecanica cuántica: un
reto de la física a la filosofía." Pensamiento. 31 (1975) 407-28.

5987. _____. "La metodología científica en la mecánica cuántica:
tema obligado de la filosofía de la ciencia." Pensamiento. 32
(1976) 405-32.

5988. Lossee, Jr., John P. "The Use of Philosophic Arguments in
Quantum Physics." PS 31 (1964) 10-17.

5989. Ludwig, G. "Quantum Theory as a Theory of Interactions between Macroscopic Systems Which can be Described Objectively."
Erkenntnis. 16 (1981) 359-87.

a. "The Structure of Stern-Gerlach Experiments and Ludwig's
Approach to Quantum Theory," by H.-J. Schmidt. 389-95.

5990. MacKinnon, Edward. "Ontic Commitments of Quantum Mechanics."
BSPS 13 (1974) 255-308.

a. "Comments," bu John Stachel. 309-17.

5991. _____. Scientific Explanation and Atomic Physics.
Chicago: University of Chicago Press, 1982. 450 pp.

5992. McGrath, James. "Only If Quanta Had Logic." PSA 1978. I,
268-76.

5993. _____. "A Formal Statement of Schrödinger's Cat Paradox." PSA 1980. I, 251-63.

5994. Mackay, D. M. "Complementarity." Arist. Soc. Supp. Vol. 32
(1958) 105-22.

5995. Mackey, G. W. The Mathematical Foundations of Quantum Mechanics.
New York: W. A. Benjamin, 1963. 137 pp.

5996. March, A. "Der Raum in der Mikrophysik." Studium Generale.
5 (1952) 338-42.

5997. Margenau, Henry. "Reality in Quantum Mechanics." PS 16
(1949) 287-302.

a. "Realism and Neo-Kantianism in Professor Margenau's
Philosophy of Quantum Mechanics," by Adolf Grünbaum. 17
(1950) 26-34.

5998. _____. "Causality in Quantum Electrodynamics." Diogenes.
6 (1954) 74-84.

5999. _____. "The Philosophical Legacy of the Quantum Theory,"
in Mind and Cosmos. Robert G. Colodny, ed. Pittsburgh: University
of Pittsburgh Press, 1966. 330-56.

6000. Margenau, Henry, and John Compton. "Report on Recent Developments in the Philosophy of Quantum Mechanics." Synthese. 8 (1950-51)
260-71.

6001. Margenau, Henry, and James L. Park. "The Physics and Semantics
of Quantum Measurement." FP 3 (1973) 19-28.

6002. Marioni, Cesare, and Emilio Montaldi. "La teoria unitaria
delle particelle elementari." Scientia. 111 (1976) 333-47. English
Tr.: 349-59.

6003. Marlow, A. R. "Implications of a New Axiom Set for Quantum
Logic." BSPS 13 (1974) 350-60.

6004. Maxwell, Nicholas. "Does the Minimal Statistical Interpretation of Quantum Mechanics Resolve the Measurement Problem?" Methodology and Science. 8 (1975) 84-101.

6005. _____. "Toward a Micro Realistic Version of Quantum
Mechanics." FP 6 (1976) I: 275-92; II: 661-76.

6006. Mehlberg, Henry. Time, Causality, and the Quantum Theory. 2 Vols. BSPS 19. Dordrecht: Reidel, 1980.

6007. Mehra, Jagdish. "The Quantum Principle: Its Interpretation and Epistemology." Dialectica. 27 (1973) 75-158.

6008. Menzel, Donald H., and David Layzer. "The Physical Principles of the Quantum Theory." PS 16 (1949) 303-24.

6009. Mercier, André, and Gottfried Ruttimann. "Le problème posé à la logique par l'apparition des théories quantiques en physique moderne." Les Etudes philosophiques. 24 (1969) 461-73.

6010. Mermin, N. David. "Quantum Mysteries for Anyone." J. Phil. 78 (1981) 397-408.

6011. Metz, A. "La physique des quanta et l'épistémologie meyerson-ienne." RMM 54 (1949) 139-62.

6012. Mittelstaedt, Peter. "Quantum Logic." BSPS 32 (1976) 501-14.

6013. _____. "On the Applicability of the Probability Concept to Quantum Theory." WOSPS 6-III (1976) 155-65.

6014. _____. "Time Dependent Propositions and Quantum Logic." Journal of Philosophical Logic. 6 (1977) 463-72.

6015. _____. Quantum Logic. SL 126. Dordrecht: Reidel, 1978. 149 pp.

6016. _____. "The Metalogic of Quantum Logic." PSA 1978. I, 249-56.

6017. _____. "The Modal Logic of Quantum Logic." Journal of Philosophical Logic. 8 (1979) 479-504.

6018. Mittelstaedt, P., and E. W. Stachow. "Operational Foundation of Quantum Logic." FP 4 (1974) 355-65.

6019. _____. "The Principle of Excluded Middle in Quantum Logic." Journal of Philosophical Logic. 7 (1978) 181-208.

6020. Moldauer, P. A. "Reexamination of the Arguments of Einstein, Podolsky, and Rosen." FP 4 (1974) 195-205.

6021. Müller-Markus, S. "Zum Anschaulichkeitsproblem in der Quan-tenmechanik." Studies in Soviet Thought. 2 (1962) 289-300.

6022. _____. "V. P. Branskij: Die philosophische Bedeutung des Anschaulichkeitsproblems in der modernen Physik." Studies in Soviet Thought. 3 (1963) 191-201.

6023. _____. Die Komplementarität in der Sowjetphilosophie." Studies in Soviet Thought. 4 (1964) 33-47.

6024. Muynck, Willem M. de, Peter A. E. M. Janssen, and Alexander
Santman. "Simultaneous Measurement and Joint Probability Distributions in Quantum Mechanics." FP 9 (1979) 71-122.

6025. Nagasaka, Gen-Ichiro. "The Einstein-Podolsky-Rosen Paradox
Reexamined." BSPS 8, SL 39 (1971) 437-45.

6026. Nagel, E. "La compétence de la raison." Les Etudes philosophiques. 10 (1964) 181-90.

6027. Nartonis, D. K. "Quantum Fact and Fiction." BJPS 25 (1974)
329-32.

6028. Naumann, Hans. "Zur erkenntnistheoretischen Bedeutung der
Diskussion zwischen Albert Einstein und Niels Bohr." DZP 7 (1959)
389-411.

6029. Neumann, John von. Mathematical Foundations of Quantum
Mechanics. R. T. Beyer, tr. Princeton University Press, 1955. 472
pp.

6030. Nilson, Donald Richard. "Hans Reichenbach on the Logic of
Quantum Mechanics." Synthese. 34 (1977) 313-60. Reprinted in SL
132 (1979) 427-75.

6031. Nordin, Ingemar. "Determinism and Locality in Quantum Mechanics."
Synthese. 42 (1979) 71-90.

6032. Ochs, W. "On the Strong Law of Large Numbers in Quantum Probability Theory." Journal of Philosophical Logic. 6 (1977) 473-80.

6033. _____. "Some Comments on the Concept of State in Quantum
Mechanics." Erkenntnis. 16 (1981) 339-56.

        a. "Comment on the Contribution by W. Ochs about the Ignorance
        Interpretation of States," by K.-E. Hellwig. 357-58.

6034. Omeljanowski, M. "Das Realitätsproblem in der Quantenphysik."
DZP 8 (1969) 280-90.

6035. _____. "Das Problem der Anschaulichkeit in der Physik."
DZP 10 (1962) 1008-18.

6036. _____. "The Concept of Dialectical Contradictions in
Quantum Physics." Soviet Studies in Philosophy. 2, No. 3 (1963-64)
17-30.

6037. _____. "Die Idee des dialektischen Widerspruchs in der
Quantenphysik." DZP 13 (1965) 47-63.

6038. _____. "On the Concepts of Elementary and Complex in
Microphysics." Soviet Studies in Philosophy. 4, No. 4 (1966) 3-14.
[English reprint from Voprosy filosofii. (1965, No. 10).]

6039. _____. "Lenin und die Dialektik in der modernen Physik."
DZP 18 (1970) 5-19.

6040. Pagels, Heinz R. The Cosmic Code: Quantum Physics as the
Language of Nature. New York: Simon and Schuster, 1982. 370 pp.

6041. Park, D. "The Idea of a Particle in Microphysics." Dialectica.
19 (1965) 246-58.

6042. Park, J. L. "Quantum Physics and the Macrocosmos." Scientia.
103 (1968) 569-84. French Tr.: Supplement, 281-95.

6043. _____. "Quantum Systems." Scientia. 105 (1970) 269-79,
370-81. French Tr.: Supplement, 107-16.

6044. _____. "The Concept of Transition in Quantum Mechanics."
FP 1 (1970-71) 23-33.

6045. Paty, M. "L'univers des particules élementaires." Scientia.
109 (1974) 729-48. English Tr.: 749-65.

6046. Pauli, W. "Die philosophische Bedeutung der Idee der Komplementarität." Experientia. 5 (1950) 72-75.

6047. _____. "Einstein's Contributions to Quantum Theory," in
Albert Einstein: Philosopher-Scientist. P. A. Schilpp, ed. New
York: Tudor, 1951. 147-60.

6048. Peat, F. David. "Quantum Physics and General Relativity:
the Search for a Deeper Theory." WOSPS 2 (1973) 328-45.

6049. Peschke, J. von. "Eine Deutung der Quantentheorie." PN 15
(1974-75) 232-39.

6050. Petersen, Aage. Quantum Physics and the Philosophical Tradition. Cambridge, MA: MIT Press, 1968. ix + 200 pp.

6051. _____. "On the Philosophical Significance of the Correspondence Argument." BSPS 5 (1969) 242-52.

6052. Petry, Günther. "Das Problem des Korpuskel-Welle-Dualismus."
PN 5 (1958-59) 338-47.

6053. Phipps, Jr., T. E. "Time Asymmetry and Quantum Equations of
Motion." FP 3 (1973) 435-55.

6054. Piron, C. "Survey of General Quantum Physics." Foundations
of Physics. 2 (1972) 287-314. Reprinted in WOSPS 5 (1975) 513-44.

6055. _____. "On the Logic of Quantum Logic." Journal of
Philosophical Logic. 6 (1977) 481-84.

6056. _____. "Ideal Measurement and Probability in Quantum
Mechanics." Erkenntnis. 16 (1981) 397-401.

a. "Piron's Foundation of Quantum Mechanics," by W. Balzer. 403-06.

b. "Piron's Approach to the Foundations of Quantum Mechanics," by W. K. Essler and G. Zoubek. 411-18.

6057. Pitt, Axel. "Die dialektische Begründung der quantenmechanischen Statistik durch die metaphysik Hegels." PN 13 (1971-72) 371-93.

6058. Polikarov, Azaria, ed. Philosophische Probleme der Physik der Elementarteilchen. Munich: Manz, 1966. 116 pp.

6059. Pool, J. C. T. "Semimodularity and the Logic of Quantum Mechanics." Communications on Mathematical Physics. 9 (1968) 212-28. Reprinted in WOSPS 5 (1975) 395-414.

6060. _____. "Baer*-Semigroups and the Logic of Quantum Mechanics." Communications on Mathematical Physics. 9 (1968) 118-41. Reprinted in WOSPS 5 (1975) 365-94.

6061. Popper, K. R. "Indeterminism in Quantum Physics and in Classical Physics." BJPS 1 (1950) 117-33, 173-95.

6062. _____. "The Propensity Interpretation of the Calculus of Probability, and the Quantum Theory," in Observation and Interpretation. S. Körner, ed. New York: Academic Press; London: Butterworths, 1957. 65-70.

6063. _____. "Birkhoff and von Neumann's Interpretation of Quantum Mechanics." Nature. 219 (1968) 682-85.

a. "Popper and Quantum Logic," by Erhard Scheibe. BJPS 25 (1974) 319-28.

6064. Post, E. J. "Uncertainty and Metric Structure." Scientia. 112 (1977) 81-88. Italian Tr.: 89-94.

6065. Price, W., and S. Chissick, eds. The Uncertainty Principle and Foundations of Quantum Mechanics: A Fifty Year's Survey. New York: John Wiley, 1977. xvii + 572 pp.

6066. Puigrefagut, Ramón. "Del determinismo clásico a la indeterminación cuantista." Pensamiento. 1 (1945) 413-46.

6067. _____. "¿Crisis del determinismo en la fisica contemporánea?" Pensamiento. 5 (1949) 435-54; 6 (1950) 63-77.

6068. Putnam, Hilary. "A Philosopher Looks at Quantum Mechanics," in Beyond the Edge of Certainty: Essays in Contemporary Science and Philosophy. Robert G. Colodny, ed. Englewood Cliffs, NJ: Prentice-Hall, 1965. 75-101.

6069. _____. "Is Logic Empirical?" BSPS 5 (1969) 216-41.

a. "Putnam and the Two-Slit Experiment," by Peter Gibbins.
Erkenntnis. 16 (1981) 235-41.

6070. _____. "How to Think Quantum-Logically." SL 78 (1976)
47-54. Reprinted from Synthese. 29 (1974) 55-62.

6071. _____. "Quantum Mechanics and the Observer." Erkenntnis.
16 (1981) 193-219.

a. "Comment on Putnam's 'Quantum Mechanics' and the Observer',"
by D. Mayr. 221-25.

b. "On Understanding Quantum Logic," by W. Hoering. 227-33.

c. "Answer to a Question from Nancy Cartwright," by H.
Putnam. 407-10.

6072. Radžabov, U. A. "The Correspondence Principle: History and
Present State." Danish Yearbook of Philosophy. 17 (1980) 59-82.

6073. Rayski, Jerzy. "The Probability of a More Realistic Interpretation of Quantum Mechanics." FP 3 (1973) 89-100.

6074. _____. "A Realistic Interpretation of Quantum Mechanics."
Dialectics and Humanism. 1, No. 3 (1074) 137-52.

6075. _____. "Epistemological and Mathematical Foundations of
Quantum Mechanics." FP 7 (1977) 151-64.

6076. _____. "Controversial Problems of Measurements within
Quantum Mechanics." FP 9 (1979) 217-36.

6077. Redhead, M. L. G. "Some Philosophical Aspects of Particle
Physics." SHPS 11 (1980) 279-304.

6078. _____. "Experimental Tests of the Sum Rule." PS 48
(1981) 50-64.

6079. Reichenbach, Hans. Philosophic Foundations of Quantum Mechanics.
Berkeley: University of California Press, 1944. x + 182 pp.

a. Review essay by Ernest Nagel. J. Phil. 42 (1945) 437-44.

b. "Reply to Ernest Nagel's Criticism of My Views on Quantum
Mechanics." by Hans Reichenbach. 43 (1946) 239-47.

c. "Professor Reichenbach on Quantum Mechanics: A Rejoinder,"
by Ernest Nagel. 247-50.

d. Review essay by W. Pauli. Dialectica. 1 (1947) 176-78.

6080. _____. "The Principle of Anomaly in Quantum Mechanics."
Dialectica. 2 (1948) 337-50.

6081. Richter, Ewald. "Bermerkungen zur 'Quantenlogik'." PN 8 (1964) 225-31.

6082. Roig Gironella, J. "El determinismo de la moderna fisica cuántica examinado a la luz de la noción filosófica de causalidad." Pensamiento. 9 (1953) 47-75.

6083. Röseberg, U. "Objektive Gesetze und Subjekt-Objekt-Dialektik." DZP 27 (1979) 180-90.

6084. Rosenfeld, L. "Misunderstandings About the Foundations of Quantum Theory," in Observation and Interpretation. S. Körner, ed. New York: Academic Press; London: Butterworths, 1957. 41-45.

6085. _____. "Le conflit épistémologique entre Einstein et Bohr." RMM 67 (1962) 147-51. English Tr.: Philosophy Today. 7 (1963) 74-77.

6086. Ross, David J. "Operator-observable Correspondence." SL 78 (1976) 365-96. Reprinted from Synthese. 29 (1974) 373-404.

6087. Rossteutscher, Helmut. "Ueber Kausalität und Komplementarität." PN 10 (1967-68) 357-68.

6088. Russo, F. "Conceptions de la physique contemporaine." Archives de Philosophie. 30 (1967) 106-13.

6089. Sachs, Mendel. "A New Approach to the Theory of Fundamental Processes." BJPS 15 (1964) 213-43.

6090. _____. "Is Quantization Really Necessary?" BJPS 21 (1970) 359-70.

6091. Sačkov, Ju. V. "Sulla struttura delle teorie statistiche in fisica." Scientia. 113 (1978) 739-56. English Tr.: 757-70.

6092. Schatzmann, Evry. "Quantenphysik und Realität." DZP 2 (1954) 621-41.

6093. Scheibe, E. The Logical Analysis of Quantum Mechanics. Oxford: Pergamon Press, 1973. viii + 204 pp.

   a. Review essay, "Orthodoxy in Quantum Mechanics," by M. L. G. Redhead. BJPS 25 (1974) 352-58.

6094. Schiller, Ralph. "Interpretations of the Quantum Theory." Synthese. 14 (1962) 5-16.

   a. "Comments on the Papers of Prof. S. Schiller and Prof. A Siegel," by Abner Shimony. 14 (1962) 189-92.

6095. _____. "Deterministic Interpretations of the Quantum Theory." BSPS 1 (1963) 144-55.

6096. Schlegel, Richard. "Statistical Explanation in Physics: The Copenhagen Interpretation." Synthese. 21 (1970) 65-82.

6097. _____. "Quantum Physics and Human Purpose." Zygon. 8 (1973) 200-20.

6098. _____. "Quantum Physics and the Divine Postulate." Zygon. 14 (1979) 163-85.

6099. _____. Superposition and Interaction: Coherence in Physics. Chicago: University of Chicago Press, 1980. 302 pp.

6100. Schrödinger, E. "Are There Quantum Jumps?" BJPS 3 (1952) 109-23; 233-42.

    a. "Are There Quantum Jumps?" by Edmund Whittaker. 348-49.

6101. _____. "Relativistic Quantum Theory." BJPS 4 (1954) 328-29.

6102. Scoledes, G. M. "The Determinism of Quantum-Mechanical Probability Statements." PS 39 (1972) 195-203.

6103. Sharp, D. H. "The Einstein-Podolsky-Rosen Paradox Re-examined." PS 28 (1961) 225-33.

    a. "Comments on the Paper of David Sharp," by Hilary Putnam. 234-34.

    b. "Comments on Professor Putnam's Comments," by H. Margenau and E. P. Wigner. 29 (1962) 292-93.

    c. "Comments on Comments on Comments: A Reply to Margenau and Wigner," by Hilary Putnam. 31 (1964) 1-6.

    d. "Reply to Professor Putnam," by Henry Margenau and Eugene P. Wigner. 7-9.

    e. "Sharp and the Refutation of the Einstein, Podolsky, Rosen Paradox," by C. A. Hooker. 38 (1971) 224-33.

6104. Shimony, Abner. "Quantum Physics and the Philosophy of Whitehead." BSPS 2 (1965) 307-30.

    a. "Comments on Shimony's Paper," by J. M. Burgers, 331-42.

6105. _____. "The Status of Hidden-Variable Theories." LMPS-4 (1971) 593-601.

6106. _____. "Metaphysical Problems in the Foundations of Quantum Mechanics." IPQ 18 (1978) 3-17.

6107. Shiva, Vandana. "Are Quantum Mechanical Transition Probabilities Classical? A Critique of Cartwright's Interpretation of Quantum Theory." Synthese. 44 (1980) 501-08.

6108. Siegel, Armand. "Operational Aspects of Hidden-Variable Quantum Theories with a Postscript on the Impact of Scientific Trends on Art." Synthese. 14 (1962) 171-88. Reprinted in BSPS 1 (1963) 156-73.

   a. "Comments," by A. Shimony. BSPS 1 (1963) 174-77.

6109. Sofonea, Liviu. "Possibilités et implications d'une mécanique fondée sur le concept d'invariant." Noesis. 2 (1974) 71-81.

6110. Stapp, Henry Pierce. "Theory of Reality." FP 7 (1977) 313-23.

6111. _____. "Quantum Mechanics, Local Causality, and Process Philosophy." W. B. Jones, ed. Process Studies. 7 (1977) 173-82.

6112. _____. "Whiteheadian Approach to Quantum Theory and the Generalized Bell's Theorem." FP 9 (1979) 1-25.

6113. _____. "Locality and Reality." FP 10 (1980) 767-95.

6114. Smith, Vincent E. "Cognitive Aspects of the Heisenberg Principle." Thomist. 12 (1949) 474-99.

6115. Sneed, Joseph D. "Von Neumann's Argument for the Projection Postulate." PS 33 (1966) 22-39.

6116. _____. "Quantum Mechanics and Classical Probability Theory." Synthese. 21 (1970) 34-64.

6117. Specker, E. P. "The Logic of Propositions Which are not Simultaneously Decidable." Dialectica. 14 (1960) 239-46. Reprinted in WOSPS 5 (1975) 135-40.

6118. Sperber, Gunnar. "On Measurement and Irreversible Processes." FP 4 (1974) 163-79.

6119. Stachel, John. "The 'Logic' of 'Quantum Logic'." BSPS 32 (1976) 515-26.

6120. Stachow, E.-W. "Completeness of Quantum Logic." Journal of Philosophical Logic. 5 (1976) 237-80.

6121. _____. "How does Quantum Logic Correspond to Physical Reality?" Journal of Philosophical Logic. 6 (1977) 485-96.

6122. _____. "Quantum Logical Calculi and Lattice Structures." Journal of Philosophical Logic. 7 (1978) 347-86.

6123. _____. "A Model-Theoretic Semantics for Quantum Logic." PSA 1980. I, 272-80.

6124. Stairs, Allen. "On Arthur Fine's Interpretation of Quantum Mechanics." Synthese. 42 (1979) 91-100.

6125. Stakhanov, I. P. "The Logic of 'Possibility'." Soviet Studies in Philosophy. 9 (1970-71) 218-21. [English reprint from Voprosy filosofii. (1970, No. 2).]

6126. Stein, Howard. "Is There a Problem of Interpreting Quantum Mechanics." Nous. 4 (1970) 93-103.

6127. _____. "On the Conceptual Structure of Quantum Mechanics," in Paradigms and Paradoxes: The Philosophical Challenge of the Quantum Domain. Robert G. Colodny, ed. Pittsburgh: University of Pittsburgh Press, 1972. 367-438.

6128. Straubinger, H. "Quantenphysik und Metaphysik." Philosophisches Jahrbuch. 60 (1950) 306-22.

6129. Strauss, M. D. H. "Quantum Theory and Logic." Bulletin of the British Society for the History of Science. 1 (1950) 99-101.

6130. Strauss, Martin. "Two Concepts of Probability in Physics." LMPS-4 (1971) 603-15.

6131. _____. "The Logic of Complementarity and the Foundation of Quantum Theory," in Modern Physics and Its Philosophy. M. Strauss, ed. Dordrecht: Reidel, 1972. Reprinted in WOSPS 5 (1975) 27-44.

6132. _____. "Foundations of Quantum Mechanics," in Modern Physics and Its Philosophy. M. Strauss, ed. Dordrecht: Reidel, 1972. Reprinted in WOSPS 5 (1975) 351-64.

6133. _____. "Logics for Quantum Mechanics." FP 3 (1973) 265-76.

6134. Strauss und Torney, Lothar von. "Das Komplementaritätsprinzip der Physik in philosophischer Analyse." ZPF 10 (1956) 109-29.

6135. Stehr, G. "Zum Problem der objektiven Realität im quantenmechanischen Formalismus." DZP 9 (1961) 2021-39,

6136. Suppes, Patrick. "Probability Concepts in Quantum Mechanics." PS 28 (1961) 378-89.

6137. _____. "The Role of Probability in Quantum Mechanics," in Philosophy of Science: The Delaware Seminar, II. New York: Interscience Publishers, 1963. 319-37.

6138. _____. "Logics Appropriate to Empirical Theories," in The Theory of Models. J. W. Addison, L. Henkin, and A. Tarski, eds. Amsterdam: North-Holland Pub. Co., 1965. Reprinted in WOSPS 5 (1975) 329-40.

6139. _____. "The Probabilistic Argument for a Non-Classical Logic of Quantum Mechanics." PS 33 (1966) 14-21. Reprinted in WOSPS 5 (1975) 329-40. French Tr.: Synthese. 16 (1976) 74-85.

6140. _____. "Popper's Analysis of Probability in Quantum Mechanics," in The Philosophy of Karl Popper. P. A. Schilpp, ed. La Salle, Illinois: Open Court, 1974. 760-74.

6141. _____. Logic and Probability in Quantum Mechanics. SL 78. Dordrecht: Reidel, 1976. 541 pp.

6142. _____. "Causal Analysis of Hidden Variables." PSA 1980. II, 563-71.

6143. _____. "Probability in Relativistic Particle Theory." Erkenntnis. 16 (1981) 299-305.

    a. "Some Remarks on a Paper by P. Suppes," by Andreas Kamlah. 327-33.

6144. _____. "Some Remarks on Hidden Variables and the EPR Paradox." Erkenntnis. 16 (1981) 311-14.

6145. Suppes, Patrick, and Mario Zanotti. "On the Determinism of Hidden Variable Theories with Strict Correlation and Conditional Statistical Independence of Variables." SL 78 (1976) 445-55.

6146. _____. "Stochastic Incompleteness of Quantum Mechanics." SL 78 (1976) 303-22. Reprinted from Synthese. 29 (1974) 311-30.

6147. Tarozzi, Gino. "The Principle of Empiricism and Quantum Theory." Epistemologia. 3 (1980) 13-28.

6148. _____. "On the Essential Role of the Realist Hypothesis in All Derivations of EPR-Type Paradoxes." Epistemologia. 4 (1981) 407-22.

6149. Teller, Paul. "On the Problem of Hidden Variables for Quantum Mechanical Observables with Continuous Spectra." PS 44 (1977) 475-77.

6150. _____. "Quantum Mechanics and the Nature of Continuous Physical Quantities." J. Phil. 76 (1979) 345-61.

6151. _____. "The Projection Postulate and Bohr's Interpretation of Quantum Mechanics." PSA 1980. II, 201-23.

6152. Törnebohm, Håkan. "On Two Logical Systems Proposed in the Philosophy of Quantum-Mechanics." Theoria. 23 (1957) 84-101.

6153. Toro, T. "Symmetries in Subnuclear Physics and Some Philosophical Implications of Their Violation." Noesis. 5 (1979) 169-74.

6154. _____. "On the Einstein-Podolsky-Rosen Paradox and the Model of Hidden Parameters of Quantum Mechanics." Noesis. 7 (1981) 141-46.

6155. Uhlmann, A. "Neuere Vorstellungen aus dem Bereich der Elementarteilchenphysik." DZP 13 (1965) 315-20.

6156. Ullmo, J. "La mécanique quantique et la causalité." Revue
Phil. 139 (1949) 257-87.

6157. _____. "Le théorème de von Neumann et la causalité."
RMM 56 (1951) 143-70.

6158. Verde, Mario. "Werner Heisenberg e la fisica del xx secolo."
Scientia. 111 (1976) 305-09. English Tr.: 311-14.

6159. Vigier, J.-. "The Concept of Probability in the Frame of the
Probabilistic and the Causal Interpretation of Quantum Mechanics,"
in Observation and Interpretation. S. Körner, ed. New York:
Academic Press; London: Butterworths, 1957. 71-77.

6160. _____. "Possible Internal Subquantum Motions of Elementary Particles," in Physics, Logic, and History. W. Yourgrau and
A. D. Breck, eds. New York: Plenum Press, 1970. 191-97.

6161. Wahsner, R. "J. P. Terletzkis Determinismusauffassung in den
modernen Quantenphysik." DZP 10 (1962) 1019-32.

6162. Wallace, Rachel. "A New Approach to Probabilities in Mechanics."
Erkenntnis. 16 (1981) 243-62.

     a. "Comment on R. Wallace," by E.-W. Stachow. 263-73.

6163. Wegener, Ursula. "Eine Lösung der 'wahrscheinlichkeitstheoreti-
schen Antinomie der Quantenmechanik'." ZAW 9 (1978) 149-56.

6164. _____. "Ein Vergleich der von Ludwig bzw. Popper vor-
geschlagenen Interpretationen der Quantenmechanik." ZAW 11 (1980)
357-66.

6165. Wiesskopf, Victor F. "Quality and Quantity in Quantum Physics."
Daedalus. 88 (1959) 592-605.

6166. Weizsäcker, C. F. von. "Probability and Quantum Mechanics."
BJPS 24 (1973) 321-37.

6167. Wenzlaff, B., and U. Kundt. "Zur Dialektik der Mikrobewegung."
DZP 9 (1961) 828-40.

6168. Werkmeister, W. H. "An Epistemological Basis for Quantum
Physics." PS 17 (1950) 1-25.

6169. Wessels, Linda. "Schrödinger's Route to Wave Mechanics."
SHPS 10 (1979) 311-40.

6170. _____. "What Was Born's Statistical Interpretation?"
PSA 1980. II, 187-200.

6171. Whyte, L. L. "The Scope of Quantum Mechanics." BJPS 9
(1958) 133-34.

6172. Wiegand, F. "Eine nichtstatistische Deutung der Quantentheorie." PN 12 (1970) 181-83.

6173. Wigner, Eugene P. "Epistemological Perspective on Quantum Theory." WOSPS 2 (1973) 369-85.

6174. Wigner, E. P., and M. M. Yanase. "Analysis of the Quantum Mechanical Measurement Process." AJAPS 4, No. 3 (1973) 15-30.

6175. Wilmsen, Arnold. "Zur Ontologie des Korpuskel-Welle-Dualismus." PN 3 (1955) 392-99.

6176. Wonsowski, G. W., and G. A. Kursanow. "Ueber den Zusammenhang der dynamischen und statistischen Gesetzmassigkeiten in den atomaren Erscheinungen." DZP 6 (1958) 40-57.

6177. Workman, Rollin W. "Is Indeterminism Supported by Quantum Theory?" PS 26 (1959) 251-59.

6178. Yanase, Mutsuo. "Some Remarks on Concepts in Theoretical Physics." AJAPS 2, No. 5 (1965) 38-41.

6179. Yourgrau, Wolfgang, and Stanley Mandelstam. Variational Principles in Dynamics and Quantum Theory. London: Sir Isaac Pitman and Sons, 1960.   xi + 180 pp.

6180. Yourgrau, Wolfgang, and Alwyn van der Merwe, eds. Perspectives in Quantum Theory: Essays in Honor of Alfred Landé. Cambridge, MA: MIT Press, 1971. xxxvii + 283 pp.

    a. Review essay, "A Landé Festschrift," by Michael R. Gardner. BJPS 24 (1973) 72-78.

6181. Zeh, H. D. "Toward a Quantum Theory of Observation." FP 3 (1973) 109-16.

6182. _____. "Quantum Theory and Time Asymmetry." FP 9 (1979) 803-18.

6183. Zeman, J. Jay. "Quantum Logic with Implication." Notre Dame Journal of Formal Logic. 20 (1979) 723-28.

6184. Zierler, N. "Axioms for Non-Relativistic Quantum Mechanics." Pacific Journal of Mathematics. 11 (1961) 1151-69. Reprinted in WOSPS 5 (1975) 149-70.

6185. Zierler, N., and M. Schlessinger. "Boolean Embeddings of Orthomodular Sets and Quantum Logic." Duke Mathematical Journal. 32 (1965) 251-62. Reprinted in WOSPS 5 (1975) 247-62.

6186. Zimmerman, E. J. "Time and Quantum Theory," in The Voices of Time. J. T. Fraser, ed. New York: George Braziller, 1966. 479-99.

6187. Zinov'ev, A. A. "On the Logic of Microphysics." Soviet
Studies in Philosophy. 9 (1970-71) 222-36. [English reprint from
Voprosy Filosofii. (1970, No. 2).]

6.7 Relativity Theory

6188. Abelé. J. "La theorie de la relativité et le jugement de
realité en physique." Archives de Philosophie. 19, No. 3 (1956)
3-24.

6189. Adam, E. "The Logical Character of the Relativity of Space
and Time." [In Hebrew] Iyyun. 7 (1956) 99-100.

6190. Aliotta, Antonio. Il relativismo, l'idealismo, e la teoria
di Einstein. Roma: Perrela, 1948. 256 pp.

6191. Angel, Roger B. "Relativity and Covariance," in The Methodological Unity of Science. M. Bunge, ed. Dordrecht: Reidel, 1973.
53-68.

6192. Augustynek, Z. "Past, Present and Future in Relativity."
Studia Logica. 35 (1976) 45-53.

6193. Bachelard, Gaston. "The Philosophic Dialectic of the Concepts
of Relativity," in Albert Einstein: Philosopher-Scientist. P. A.
Schilpp, ed. New York: Tudor, 1951. 563-80.

6194. Bafico, R. "Tachyoni - Eventuali particelle più veloci della
luce." Scientia. 105 (1970) 694-705. English Tr.: Supplement,
210-20.

6195. Barashenkov, V. S. "On the Possibility of Elementary Processes
Exceeding the Speed of Light." Soviet Studies in Philosophy. 15
(1976-77) 25-41. [English reprint from Voprosy filosofii. (1976,
No. 5).]

6196. Barreau, H. "Einstein et les concepts d'espace et de temps."
RMM 85 (1980) 357-69.

6197. Barter, E. G. Relativity and Reality. New York: Philosophical Library, 1953. 131 pp.

6198. Beauregard, Laurent A. "The Sui Generis Conventionality of
Simultaneity." PS 43 (1976) 469-90.

        a. "Is Signal Synchrony Independent of Transport Synchrony?"
        by Brian Ellis. 45 (1978) 309-11.

6199. _____. "Round-Trip Clock Retardation and the Conventionality of Simultaneity." FP 7 (1977) 769-82.

6200. Berenda, Carlton W. "On Birkhoff's and Einstein's Relativity Theory." PS 12 (1945) 116-19.

6201. Bergmann, P. G. "Physics and Geometry." LMPS-2 (1964) 343-46.

6202. Bitsakis, Eftichios. "Les théories relativistes et le monde microphysique." Scientia. 111 (1976) 361-82. Italian Tr.: 383-99. English Tr.: 401-16.

6203. Böhm, Walter. "Realismus und Idealismus in der Einsteinschen Relativitätstheorie." Philosophisches Jahrbuch. 64 (1956) 112-24.

6203A. Bondi, Hermann. "General Relativity as an Open Theory," in Physics, Logic, and History. W. Yourgrau and A. D. Breck, eds. New York: Plenum Press, 1970. 265-71. "Discussion," 271-76.

6204. Borzeszkowski, H.-H. von, and R. Wahsner. "Erkenntnistheoretischer Apriorismus und Einsteins Theorie." DZP 27 (1979) 213-22.

6205. Bowman, Peter A. "Einstein's Second Treatment of Simultaneity." PSA 1976. I, 71-81.

6206. Bridgman, P. W. A Sophisticate's Primer of Relativity. Middletown, CT: Wesleyan University Press, 1962. Reprint: New York: Harper & Row, 1965. 164 pp.

6207. Broglie, Louis de. "The Scientific Work of Albert Einstein," in Albert Einstein: Philosopher-Scientist. P. A. Schilpp, ed. New York: Tudor, 1951. 107-28.

6208. Brühlmann, Otto. "Von der metaphysikalischen Grundlage der Physik." Studia Philosophica. 15 (1955) 10-34.

6209. Büchel, Wolfgang. "Relativitätstheorie und kritischer Realismus." PN 7 (1961-62) 4-36.

6210. _____. "Metrisches Feld, Kraftfeld, Feldquanten - Zur Interpretation der allgemeinen Relativitätstheorie." PN 8 (1964) 22-48.

6211. _____. "Zur Begründung und Deutung der Relativitätstheorie." PN 10 (1967-68) 211-36.

6212. _____. "3°K-Strahlung, Tachyonen und Relativitätstheorie. Ueber Symmetrie und Asymmetrie in der Natur." PN 12 (1970) 51-69.

6213. _____. "Die Struktur wissenschaftlicher Revolutionen und das Uhren-'Paradoxon'." ZAW 5 (1974) 218-25.

6214. Builder, Geoffrey. "The Resolution of the Clock Paradox." PS 26 (1959) 135-44.

6215. Byrne, Patrick H. "Relativity and Indeterminism." FP 11 (1981) 913-32.

6216. Caldirola, P., and E. Recami. "Causality and Tachyons in Relativity." BSPS 47 (1981) 249-98.

6217. Čapek, Milič. "Relativity and the Status of Space." Review of Metaphysics. 9 (1955) 169-99.

6218. _____. "Time in Relativity Theory: Arguments for a Philosophy of Becoming," in The Voices of Time. J. T. Fraser, ed. New York: George Braziller, 1966. 434-54.

6219. _____. "Relativity and the Status of Becoming." FP 5 (1975) 607-17.

6220. Chapman, T. "Special Relativity and Indeterminism." Ratio 15 (1973) 107-10.

6221. Chari, C. T. K. "On Representations of Time as 'The Fourth Dimension' and their Metaphysical Inadequacy." Mind. 58 (1949) 218-21.

6222. Christensen, Ferrel. "Special Relativity and Space-like Time." BJPS 32 (1981) 37-53.

6223. Chudinov, E. M. "The General Theory of Relativity and the Space-Time Structure of the Universe." Soviet Studies in Philosophy. 6, No. 2 (1967) 51-60. [English reprint from Voprosy filosofii. (1967, No. 3).]

6224. Clarke, C. J. S. "Time in General Relativity." MSPS 8 (1977) 94-108.

6225. Cleobury, F. H. "The Bearing of Relativity on the Controversy between Realism and Idealism." Phil. Quart. 4 (1954) 302-09.

6226. Coffa, J. Alberto. "Elective Affinities: Weyl and Reichenbach." SL 132 (1979) 267-304.

6227. Cohen, Robert S. "Epistemology and Cosmology: E. A. Milne's Theory of Relativity." Review of Metaphysics. 3 (1950) 385-405.

6228. Costa de Beauregard, O. "La relativité en microphysique." Archives de philosophie. 19, No. 3 (1956) 25-35.

6229. _____. Théorie synthétique de la relativité restreinte et des quanta. Paris: Gauthiers-Villars, 1957. xii + 200 pp.

6230. _____. "Time in Relativity Theory: Arguments for a Philosophy of Being," in The Voices of Time. J. T. Fraser, ed. New York: George Braziller, 1966. 417-33.

6231. _____. "Le troisième orage du 20e siècle: le paradoxe d'Einstein." Epistemologia. 1 (1978) 305-12.

6232. Cullwick, E. G. "Einstein and Special Relativity: Some Inconsistencies in his Electrodynamics." BJPS 32 (1981) 167-76.

6233. d'Espagnat, Bernard. "Einstein et le principe de la causalité."
Epistemologia. 3 (1980) Fascicolo speciale. 123-34.

6234. _____. "The Concepts of Influences and of Attributes as
Seen in Connection with Bell's Theorem." FP 11 (1981) 205-34.

6235. Dingle, Herbert. "Scientific and Philosophical Implications
of the Special Theory of Relativity," in Albert Einstein: Philosopher-
Scientist. P. A. Schilpp, ed. New York: Tudor, 1951. 535-54.

6236. _____. "The Doppler Effect and the Foundations of
Physics." BJPS 11 (1960) 11-31, 113-29.

    a. "A Point of Professor Dingle's," by P. J. van Heerden.
    BJPS 12 (1961) 70.

    b. "A Reply to Dr. van Heerden," by Herbert Dingle. 71.

6237. _____. "Relativity and Electromagnetism: An Epistemological Appraisal." PS 27 (1960) 233-53.

6238. _____. "Reason and Experiment in Relation to the Special
Relativity Theory." BJPS 15 (1964) 41-61.

    a. "A Note on Professor Dingle's Paper on Relativity," by
    Hubert Schleichert. 15(1965) 331.

    b. "Reply to Dr. Schleichert," by Herbert Dingle. 331-32.

6239. _____. "Time in Relativity Theory: Measurement or
Coordinate?" in The Voices of Time. J. T. Fraser, ed. New York:
George Braziller, 1966. 455-72.

6240. Dorling, Jon. "Did Einstein Need General Relativity to Solve
the Problem of Absolute Space? Or Had the Problem already been
Solved by Special Relativity?" BJPS 29 (1978) 311-23.

6241. Earman, John. "Implications of Causal Progagation outside
the Null Cone." AJP 50 (1972) 222-37.

6242. _____. "Covariance, Invariance, and the Equivalence of
Frames." FP 4 (1974) 267-89.

6243. Earman, John, and Clark Glymour. "Lost in the Tensors:
Einstein's Struggles with Covariance Principles, 1912-1916." SHPS 9
(1978) 251-78.

6244. _____. "The Gravitational Red Shift as a Test of General
Relativity: History and Analysis." SHPS 11 (1980) 175-214.

6245. Einstein, Albert. The Meaning of Relativity. Edwin P.
Adams, tr. London: Methuen; Princeton: Princeton University
Press, 1922. Later editions: 1945, 1950, 1953, 1956. Fifth edition:
166 pp.

6246. _____. Relativity: The Special and the General Theory. 16th Edition. New York: Crown Publishers, 1961. 164 pp.

6247. Ellis, Brian. "On Conventionality and Simultaneity -- A Reply." AJP 49 (1971) 177-203.

6248. Ellis, Brian, and Peter Bowman. "Conventionality in Distant Simultaneity." PS 34 (1967) 116-36.

6249. Evans, M. G. "The Relativity of Simultaneity: A Critical Analysis." Dialectica. 16 (1962) 61-82.

   a. "Résponse à M. Evans et à quelques autres," by F. Bonsack. 83-9.

   b. "Suite et fin de la discussion sur la relativité de la vitesse relative," by M. G. Evans and F. Bonsack. 299-301.

6250. _____. "On the Falsity of the Fitzgerald-Lorentz Contraction Hypothesis." PS 36 (1969) 354-62.

   a. "The Lorentz-Fitzgerald Contraction Hypothesis and the Combined Rod Contraction-Clock Retardation Hypothesis," by Herman Erlichson. 38 (1971) 605-09.

6251. Feenberg, Eugene. "Conventionality in Distant Simultaneity." FP 4 (1974) 121-26.

6252. _____. "Distant Synchrony and the One-Way Velocity of Light." FP 9 (1979) 329-37.

6253. Felt, James W. "Mach's Principle Revisited." Laval Théologique et Philosophique. 20 (1964) 35-49.

6254. Fitzgerald, Paul. "Relativity Physics and the God of Process Philosophy." Process Studies. 2 (1972) 251-76.

6255. Fowler, Dean R. "Whitehead's Theory of Relativity." Process Studies. 5 (1975) 159-74.

6256. Francis, R. "On the Interpretation and Transitivity of Non-standard Synchronisms." BJPS 31 (1980) 165-73.

6257. Fraassen, Bas C. van. "Conventionality in the Axiomatic Foundations of the Special Theory of Relativity." PS 36 (1969) 64-73.

6258. Frank, Philipp. Relativity: A Richer Truth. Boston: Beacon Press, 1950. xvi + 142 pp.

6259. Friedman, Michael. "Relativity Principles, Absolute Objects and Symmetry Groups." SL 56 (1973) 296-320.

6260. Gardner, Michael. "Relationism and Relativity." BJPS 28 (1976) 215-33.

6261. Giannoni, Carlo. "Einstein and the Lorentz-Poincaré Theory of Relativity." BSPS 8, SL 29 (1971) 575-89.

6262. _____. "Special Relativity in Accelerated Systems." PS 40 (1973) 282-92.

6263. _____. "A Philosopher Looks at Relativity Theory." Scientia. 109 (1974) 779-89.

6264. _____. "A Universal Axiomatization of Kinematical Theories." PSA 1978. I, 60-70.

6265. _____. "Relativistic Mechanics and Electrodynamics without One-Way Assumptions." PS 45 (1978) 17-46.

6266. _____. "Clock Retardation, Absolute Space, and Special Relativity." FP 9 (1979) 427-44.

6267. Gehlhar, F. "Relativität und Dialektik." DZP 27 (1979) 223-32.

6268. Gérard, R. "De l'univers de champ à l'universe de mouvement." RMM 71 (1966) 1-14.

6269. Geroch, Robert. "Prediction in General Relativity." MSPS 8 (1977) 81-93.

6270. Gödel, Kurt. "A Remark on the Relationship between Relativity Theory and Idealistic Philosophy." in Albert Einstein: Philosopher-Scientist. P. A. Schilpp, ed. New York: Tudor, 1951. 555-62.

6271. Godfrey-Smith, William. "Special Relativity and the Present." Phil. Studies. 36 (1979) 233-44.

6272. Goenner, Herbert. "Mach's Principle and Einstein's Theory of Relativity." BSPS 6 (1970) 200-16.

6273. Gonzales-Gascon, F. "Some Remarks for a Broadening of Special Relativity." Scientia. 111 (1976) 653-60. Italian Tr.: 661-66.

6274. Graves, J. C. The Conceptual Foundations of Contemporary Relativity Theory. Cambridge, MA: MIT Press, 1971. ix + 361 pp.

　　a. Review essay by George Berger. Erkenntnis. 10 (1976) 413-19.

　　b. Review essay by Allen I. Janis. SHPS 4 (1973) 300-06.

6275. Grieder, Alfons. "Protophysik der Zeit und Relativitätstheorie." Dialectica. 30 (1976) 145-60.

6276. _____. "Relativity, Causality and the 'Substratum'." BJPS 28 (1977) 35-48.

6277. Griese, A. "Einsteins philosophischer Ausgangspunkt bei der Schaffung einer einheitlichen Feldtheorie." DZP 11 (1963) 1203-16.

6278. Grünbaum, Adolf. "Relativity and the Atomicity of Becoming." Review of Metaphysics. 4 (1950) 143-86. Corrigenda, 465.

6279. _____. "The Clock Paradox in the Special Theory of Relativity." PS 21 (1954) 249-53.

    a. "The Clock Paradox in the Special Theory of Relativity," by Boris Leaf. 22 (1955) 45-52.

    b. "Reply to Dr. Leaf," by A. Grünbaum. 53.

    c. "Comments on Professor Grünbaum's 'The Clock Paradox in the Special Theory of Relativity," by Håkan Törnebohm. 231-32.

    d. "Reply to Dr. Törnebohm's Comments on my Article," by A. Grünbaum. 233.

6280. _____. "Philosophical Principles in the Special Theory of Relativity." [In Hebrew] Iyyun. 9 (1958) 249-62.

6281 _____. "The Genesis of the Special Theory of Relativity," in Current Issues in the Philosophy of Science. H. Feigl and G. Maxwell, eds. New York: Holt, Rinehart and Winston, 1961. 43-53.

    a. "Comments," by Michael Polanyi. 53-55.

6282. _____. "Simultaneity by Slow Clock Transport in the Special Theory of Relativity." PS 36 (1969) 5-43.

    a. "The Transitivity of Non-Standard Synchronisms," by Philip L. Quinn. BJPS 25 (1974) 78-82.

    b. "The Intransitivity of Non-Standard Synchronisms," by I. W. Roxburgh. 26 (1975) 47-49.

    c. "Synchronisation Rules and Transitivity," by Jarrett Leplin. 27 (1976) 399-402.

6283. _____. "Geometrodynamics and Ontology." J. Phil. 70 (1973) 775-800.

6284. _____. "Space, Time, and Matter: The Foundations of Geometrodynamics." BSPS 20, SL 64 (1974) 3-5.

6285. Grünbaum, Adolf and Allen I. Janis. "The Geometry of the Rotating Disk in the Special Theory of Relativity." Synthese. 34 (1977) 281-300. Reprinted in SL 132 (1979) 321-40.

6286. Guggenheimer, H. W. "General Relativity and Nuclear Reactions." Dialectica. 14 (1960) 183-87.

6287. Guttenberg, A. Ch. de. "Das neue physikalische Weltbild und Einstein." Philosophisches Jahrbuch. 65 (1957) 375-93.

6288. Gutting, Gary. "Einstein's Discovery of Special Relativity." PS 39 (1972) 51-68.

6289. Hadley, Henry G. "Hindu-Philosophie und die Einsteinsche Theorie." PN 10 (967-68) 107-11.

6290. Hägerström, Axel. Erkenntnistheoretische Voraussetzungen der speziellen Relativitätstheorie Einsteins." Theoria. 12 (1946) 1-68.

6291. Havas, P. "Relativity and Causality." LMPS-2 (1964) 347-62.

6292. _____. "Causality Requirements and the Theory of Relativity." Synthese. 18 (1968) 75-102. Reprinted in BSPS 5 (1969) 151-78.

     a. "Comments," by John Stachel. BSPS 5 (1969) 179-98.

6293. Hof, W. "Die Relativitätstheorie in realistischer Begründung." PN 12 (1970) 173-80.

6294. Hölling, Joachim. Realismus und Relativität. München: Wilhelm Fink Verlag, 1971. 262 pp.

6295. Holton, Gerald. "Einstein, Michelson, and the 'Crucial' Experiment." Isis. 60 (1969) 133-97.

6296. Hörz, H. "Albert Einstein und die Philosophie." DZP 27 (1979) 149-61.

6297. Hoyer, Ulrich. "Kant-Mach-Einstein." Perspektiven der Philosophie. 4 (1978) 103-18.

6298. Hsu, J. P., and T. N. Sherry. "Common Time in a Four-Dimensional Symmetry Framework." FP 10 (1980) 57-76.

6299. Hudgin, Richard H. "Coordinate-Free Relativity." Synthese. 24 (1972) 281-97. Reprinted in SL 56 (1973) 366-82.

6300. Hund, F. "Denkschemata und Modelle in der Physik." Studium Generale. 18 (1965) 174-83.

6301. Illy, József. "Revolutions in a Revolution." Studies in History and Philosophy of Science. 12 (1981) 175-210.

6302. Infeld, Leopold. Albert Einstein: His Work and Its Influence on Our World. Revised edition. New York: Charles Scribner's Sons, 1950. 134 pp.

6303. _____. "General Relativity and the Structure of Our Universe," in Albert Einstein: Philosopher-Scientist. P. A. Schilpp, ed. New York: Tudor, 1951. 475-500.

6304. Jackson, Frank, and Robert Pargetter. "Relative Simultaneity
in the Special Theory of Relativity." PS 44 (1977) 464-74.

    a. "Jackson and Pargetter's Criterion of Distant Simultaneity,"
by Roberto Torretti. 46 (1979) 302-05.

    b. "Comment on 'Relative Simultaneity in the Special Theory
of Relativity,'" by Carlo Giannoni. 306-09.

    c. "A Reply to Torretti and Giannoni," by Frank Jackson and
Robert Pargetter. 310-15.

    d. "Jackson and Pargetter on Distant Simultaneity," by Burke
Townsend. 47 (1980) 646-55.

6305. Janis, Allen I. "Synchronism by Slow Transport of Clocks in
Noninertial Frames of Reference." PS 36 (1969) 74-81.

6306. Jones, Roger. "Is General Relativity Generally Relativistic?"
PSA 1980. II, 363-81.

6307. Joseph, Geoffrey. "Geometry and Special Relativity." PS 46
(1979) 425-38.

6308. Juhos, Béla. "Die 'Metrik' als Bestandteil der empirischen
Bescreibung." Archiv für Philosophie. 7 (1957) 209-28.

6309. _____. "The Characterization of States of Translatory
Motion." Ratio. 6 (1964) 28-49.

6310. _____. "Virtuelle Geschwindigkeiten als verborgene
Parameter." PN 11 (1969) 440-45.

6311. _____. "Logische Analyse des Relativitätsprinzips." PN
11 (1969) 207-17.

6312. Kaila, Eino. "Mistä Einstein-Minkowskin invarianssiteoriassa
on kysymys." Ajatus. 21 (1958) 9-121.

6313. Kar, Robert. "Kausalistische Erklärung der relativistischen
Paradoxa." PN 8 (1964) 250-54.

6314. _____. "Die spezielle Para-Relativitätstheorie." PN 10
(1967-68) 392-406.

6315. _____. "Die allgemeine Para-Relativitätstheorie." PN
13 (1971-72) 221-56.

6316. _____. "Die Ursache der geodätischen Form der freien
Weltlinien im Gravitationsfeld." PN 15 (1974-75) 218-31.

6317. Keswani, G. H. "Origin and Concept of Relativity." BJPS.
Part I: 15 (1965) 286-306. Part II: 16 (1965) 19-32. Part III:
16 (1966) 273-94.

a. "Note on Mr. Keswani's Articles, Origin and Concept of Relativity," by Herbert Dingle. 16 (1965) 242-46.

b. "The Origin and Concept of Relativity," by H. B. Levinson. 246-48.

c. "A Note on the Difference between the Lorentz-Fitzgerald Contraction and the Einstein Contraction," by Karl R. Popper. 332-33.

d. "Origin and Concept of Relativity (Parts I and II): Reply to Professor Dingle and Mr. Levinson," by G. H. Keswani. 17 (1966) 149-52.

e. "Origin and Concept of Relativity: Reply to Professor Popper," by G. H. Keswani. 234-36.

g. "A Note on Relativistic Phenomena in an Ether Theory," by S. J. Prokhovnik. 18 (1968) 322-23.

h. "Length Contraction and Clock Synchronisation: The Empirical Equivalence of the Einsteinian and Lorentzian Theories," by Jon Dorling. 19 (1968) 67-69.

i. "Some New Aspects of Relativity: Remarks on Keswani's Paper," by M. F. Podlaha. 26 (1975) 133-37.

6318. Kingsley, James M. "On the Consistency of the Postulates of Special Relativity." FP 5 (1975) 295-300.

a. "The Consistency of the Postulates of Special Relativity," by Ø. Grøn and M. Nicola. 6 (1976) 677-80.

b. "Consistency in Relativity," by R. E. Chatham. 681-85.

6319. Klotz, A. H. "On Some Philosophical Aspects of the Unified Field Theories." Studium Generale. 22 (1969) 1189-1214.

6320. Korch, H. "Albert Einstein über philosophische Fragen der naturwissenschaftlichen Erkenntnis." DZP 27 (1979) 167-79.

6321. Kouznetsov, B. "Complementarity and Relativity." PS 33 (1966) 199-209.

6322. _____. "Einstein et le principe de Mach." Organon. 6 (1969) 265-77.

6323. Kwassow, A. A. "Materialismus und Relativismus." DZP 8 (1960) 802-19.

6324. Laucks, Irving F. "Was Newton Right After All?" PS 26 (1959) 229-39.

6325. Laue, Max von. "Inertia and Energy," in Albert Einstein: Philosopher-Scientist. P. A. Schilpp, ed. New York: Tudor, 1951. 501-34.

6326. Laymon, Ronald. "Independent Testability: The Michelson-Morley and Kennedy-Thorndike Experiments." PS 47 (1980) 1-37.

    a. "The Michelson-Morley and Kennedy-Thorndike Experiments," by Herman Erlichson. 48 (1981) 620-22.

6327. Leaf, Boris. "Vectorial Composition of Velocities in Relativity." PS 22 (1954) 321-24.

6328. Lamaître, Georges Edward. "The Cosmological Constant," in Albert Einstein: Philosopher-Scientist. P. A. Schilpp, ed. New York: Tudor, 1951. 437-56.

6329. Levin, Michael E. "Length Relativity." J. Phil. 68 (1971) 164-74.

6330. Levy, Edwin. "Competing Radical Translations: Examples, Limitations and Implications." BSPS 8, SL 39 (1971) 590-605.

6331. Loinger, A. "Simmetria e relatività." Scientia. 98 (1963) 201-06. French Tr.: Supplement, 113-18.

6332. Lorenzen, P. "Eine Revision der Einsteinschen Revision." PN 16 (1976-77) 383-91.

6333. _____. "Die allgemeine Relativitätstheorie als eine Revision der Newtonschen Gravitationstheorie." PN 17 (1978-79) 1-9.

6334. Luce, R. Duncan, and Louis Narens. "A Qualitative Equivalent to the Relativistic Addition Law for Velocities." Synthese. 33 (1976) 483-87.

6335. McCrea, W. H. "On the Objective of Einstin's Work." BJPS 8 (1957) 18-29.

6336. McMorris, M. N. "The Second Postulate of Einstein's Theory of Special Relativity." International Logic Review. 9 (1978) 77-83.

6337. McVittie, G. C. "General Relativity and Time in the Solar System." Studium Generale. 23 (1970) 197-202.

6338. Margenau, Henry, and Richard A. Mould. "Relativity: An Epistemological Appraisal." PS 24 (1957) 297-308.

6339. Matchinski, M. "Note sur les sens des notions en philosophie et en science et sur la 'relativité' en physique et en philosophie." RMM 73 (1968) 237-43.

6340. Maund, B. "The Conventionality of Temporal Relations in Relativity Theory." PS 41 (1974) 394-407.

6341. _____. "Tachyons and Causal Paradoxes." FP 9 (1979) 557-74.

6342. Mehlberg, H. "Space, Time, Relativity." LMPS-2 (1964) 363-80.

6343. Menger, Karl. "Modern Geometry and the Theory of Relativity," in Albert Einstein: Philosopher-Scientist. P. A. Schilpp, ed. New York: Tudor, 1951. 457-74.

6344. Metz, A. "Bergson, Einstein et les relativistes." Archives de Philosophie. 22 (1959) 369-84.

6345. _____. "L'interprétation philosophique de la théorie de la relativité." Bulletin de la Société Française de Philosophie. 61 (1967) 33-80.

6346. _____. "La theorie de la relativité en Union Sovietique." Archives de Philosophie. 30 (1967) 262-71.

6347. Miller, Arthur I. Albert Einstein's Special Theory of Relativity. Reading, MA: Addison-Wesley, 1981. 496 pp.

6348. Miller, G. "The Inexactness of Time." FP 3 (1973) 389-98.

6349. Milne, E. A. "Gravitation without General Relativity," in Albert Einstein: Philosopher-Scientist. P. A. Schilpp, ed. New York: Tudor, 1951. 409-36.

6350. Misner, Charles W. "Some Topics for Philosophical Inquiry Concerning the Theories of Mathematical Geometrodynamics and of Physical Geometrodynamics." BSPS 20, SL 64. (1974) 7-29.

6351. Mittelstaedt, Peter. "Conventionalism in Special Relativity." FP 7 (1977) 573-83.

6352. Mohorovičic, Stjepan. "Ueber die Möglichkeit auch anderer spezieller Relativitätstheorien." Methodos. 10 (1958) 267-86.

6353. Molchanov, Iu. B. "The Causality Principle and the Hypothesis of Speeds Exceeding That of Light." Soviet Studies in Philosophy. 15 (1976-77) 42-61. [English reprint from Voprosy filosofii. (1976, No. 5).]

6354. Moon, Parry, and Domina Eberle Spencer. "Retardation in Cosmology." PS 25 (1958) 287-92.

6355. _____. "Mach's Principle." PS 26 (1959) 125-34.

6356. Moreno, O. P., Antonio. "Time and Relativity: Some Philosophical Considerations." Thomist. 45 (1981) 62-79.

6357. Moser, M. "Aethertheorie und Relativitätstheorie." PN 14 (1973) 210-24.

6358. Moser, M., B. Juhos, and H. Schleichert. "Gespräch über das Uhrenparadoxon." PN 10 (1967-68) 23-41.

a. "Zum 'Gespräch über das Uhrenparadoxon'," by D. Laptschinsky.
12 (1970) 90-92.

b. "Nochmals das 'Uhrenparadoxon' der Relativitätstheorie,"
by Karl Saur. 93-100.

6359. Müller-Markus, S. "Einstein and Soviet Philosophy." Studies
in Soviet Thought. 1 (1961) 78-87.

6360. _____. "Erkenntnisprobleme der Relativitätstheorie."
PN 7 (1961-62) 180-206.

6361. _____. "Die Prinzipien der allgemeinen Relativitätstheorie."
Synthese. 15 (1963) 336-78.

6362. _____. "Soviet Discussion on General Relativity Theory."
Studies in Soviet Thought. 5 (1965) 204-22.

6363. _____. Einstein und die Sowjetphilosophie. Dordrecht:
Reidel, 1966. Vol. I: 455 pp.; Vol. II: 509 pp.

6364. Noonan, H. W. "The Four-Dimensional World." Analysis. 37
(1976-77) 32-39.

6365. Nordenson, Harald. Relativity, Time and Reality: A Critical
Investigation of the Einstein Theory of Relativity from a Logical
Point of View. London: Allen and Unwin, 1969. 215 pp.

6366. North, J. D. "The Time Coordinate in Einstein's Restricted
Theory of Relativity." Studium Generale. 23 (1970) 203-23.

6367. Øhrstrøm, Peter. "Conventionality in Distant Simultaneity."
FP 10 (1980) 333-43.

6368. Papapetrou, A. "General Relativity -- Some Puzzling Questions."
BSPS 13 (1974) 376-87.

6369. Petry, Günther. "Ist der 'Aether' als kosmologische Grund-
kategorie haltbar?" Philosophisches Jahrbuch. 67 (1959) 365-88.

6370A. Pfarr, Joachim. "Die Protophysik der Zeit und das Relativitätsprinzip." ZAW 7 (1976) 298-326.

a. "Die Protophysik der Zeit und das Relativitätsprinzip,"
by Peter Janich. 9 (1978) 343-47.

6370. Podlaha, M. F. "Light Signal Synchronisation and Clock
Transport Synchronisation in the Theory of Relativity." BJPS 30
(1979) 376-80.

6371. Prokhovnik, S. J. "The Case for an Aether." BJPS 14 (1963)
195-207.

6372. _____. The Logic of Special Relativity. Cambridge
University Press, 1967. xiv + 128 pp.

6373. Prokop, Walter. "Zur Deutung der Einsteinschen Energie-Masse-Relation." DZP 8 (1960) 52-61.

    a. "Masse und Energie," by A. Schurupow. 880-82.

6374. Puigrefagut, Ramón. "Implicaciones filosóficas de la relatividad especial einsteiniana." Pensamiento. 12 (1956) 411-30.

6375. _____. "La relatividad restringida y los sistemas filosóficas." Pensamiento. 14 (1958) 135-58.

6376. Rabbeno, G. "La pure logique des 'Théories de la relativité'." Scientia. 82 (1947) 1-6.

6377. Reichenbach, Hans. "The Philosophical Significance of the Theory of Relativity," in Albert Einstein: Philosopher-Scientist. P. A. Schilpp, ed. New York: Tudor, 1951. 287-312.

6378. _____. The Theory of Relativity and A Priori Knowledge. Translated and edited, with an Introduction, by Maria Reichenbach. Berkeley: University of California Press, 1965. xliv + 115 pp.

6379. _____. "Die relativistische Zeitlehre." Scientia. 110 (1975) 765-75. French Tr.: 777-85. Reprinted from Scientia (1924).

6380. Rietdijk, C. W. "Special Relativity and Determinism." PS 43 (1976) 598-609.

6381. Robertson, H. P. "Geometry as a Branch of Physics," in Albert Einstein: Philosopher-Scientist. P. A. Schilpp, ed. New York: Tudor, 1951. 313-32.

6382. Rosen, Philip. "The Clock Paradox and Thermodynamics." PS 26 (1959) 145-47.

6383. Rosenthal-Schneider, Ilse. "Presuppositions and Anticipations in Einstein's Physics," in Albert Einstein: Philosopher-Scientist. P. A. Schilpp, ed. New York: Tudor, 1951. 129-46.

6384. Rosser, W. G. V. "The Clock Hypothesis and the Lorentz Transformations." BJPS 29 (1978) 349-53.

6385. Roxburgh, I. W., and R. K. Tavakol. "Conventionalism and General Relativity." FP 8 (1978) 229-37.

6386. Russell, Bertrand. A B C of Relativity. London: George Allen and Unwin, 1958. 139 pp.

6387. Sachs, Mendel. "On the Elementarity of Measurement in General Relativity: Toward a General Theory." Synthese. 17 (1967) 29-53.

6388. _____. "On the Logical Status of Equivalence Principles in General Relativity Theory." BJPS 27 (1976) 225-29.

6389. _____. "Elementary Particle Physics from General Relativity." FP 11 (1981) 329-54.

6390. Salmon, Wesley C. "The Conventionality of Simultaneity." PS 36 (1969) 44-63.

6391. _____. "The Philosophical Significance of the One-Way Speed of Light." Nous. 11 (1977) 253-92.

6392. Sandgathe, Franz. "Ein Vorschlag zu einer Aenderung der speziellen Relativitätstheorie." Archiv für Philosophie. 5 (1955) 241-304.

6393. Sapper, Karl, ed. Kritik und Fortbildung der Relativitätstheorie. Graz: Akademische Druck u. Verlagsantstatlt, 1959.

6394. Schild, A. "The Principle of Equivalence." Monist. 47 (1962) 20-39.

6395. Schlegel, Richard. "An Interaction Interpretation of Special Relativity Theory." FP 3 (1973) I: 169-84; II: 227-95; III: 5 (1975) 197-215.

6396. _____. "A Lorentz-Invariant Clock." FP 7 (1977) 245-53.

6397. _____. "The Clock Paradox: Some New Thoughts." PS 44 (1977) 306-12.

6398. Schleichert, Hubert. "Lösungsversuche für das Uhrenparadoxon, erkenntnislogisch betrachtet." PN 9 (1956-66) 326-40.

6399. _____. "Ueber die logische Stellung der relativistischen Messtheorie." ZAW 1 (1970) 243-51.

6400. Schmutzer, E. "Der Energieerhaltungssatz und die relativische Physik." DZP 14 (1966) 1087-99.

6401. _____. "New Approach to Interpretation Problems of General Relativity by Means of the Splitting-Up-Formalism of Space-Time." SL 31 (1970) 121-36.

    a. "Comments on Schmutzer's Paper," by A. Grünbaum and M. Strauss. 137-39.

6402. Schock, Rolf. "The Inconsistency of the Theory of Relativity." ZAW 12 (1981) 285-96.

6403. Seaman, Francis. "Whitehead and Relativity." PS 22 (1945) 222-26.

6404. Sellars, Roy Wood. "The Philosophy and Physics of Relativity." PS 13 (1946) 177-95.

6405. _____. "A Note on the Theory of Relativity." J. Phil. 43 (1946) 309-17.

a. "Comments Upon Roy Sellars' Views on Relativity," by Carlton W. Berenda. 44 (1947) 15-18.

6406. _____. "Materialism and Relativity." Phil. Review. 55 (1946) 25-51.

6407. _____. "Physical Realism and Relativity: Unfinished Business." PS 23 (1956) 75-81.

6408. _____. "Gestalt and Relativity." PS 23 (1956) 275-79.

6409. _____. "A Note on the Semantics of Relativity." J. Phil. 70 (1973) 537-38.

6410. Severi, F. "I fondamenti logici della relativitá." Scientia. 90 (1955) 277-82. French Tr.: Supplement, 161-66.

6411. Siminel, Gh. "Physics and Relativity." International Logic Review. 7 (1976) 34-49.

6412. Sivadjian, J. "Le temps et le mouvement." Revue Phil. 152 (1962) 345-50; 153 (1963) 311-30.

6413. _____. "Deux mobiles qui se déplacent dans une même direction avec la même vitesse peuvent-ils se rencontrer?" RMM 80 (1975) 363-68.

6414. Sepetys, Jonas. A Critique of Relativity. New York: Philosophical Library, 1968. 64 pp.

6415. Spector, Marshall. Methodological Foundations of Relativistic Mechanics. Notre Dame, IN: University of Notre Dame Press, 1972. xvii + 174 pp.

6416. Stachel, John. "The Rise and Fall of Geometrodynamics." BSPS 20, SL 64 (1974) 31-54.

6417. Stein, Howard. "On the Paradoxical Time-Structures of Gödel." PS 37 (1970) 589-601.

6418. _____. "Graves on the Philosophy of Physics." J. Phil. 69 (1972) 621-34.

a. "Some Aspects of General Relativity and Geometrodynamics," by John Earman. 634-47.

b. "Reply to Stein and Earman," by John C. Graves. 647-49.

6419. _____. "Some Philosophical Prehistory of General Relativity." MSPS 8 (1977) 3-49.

6420. Strauss, M. "Einstein's Theories and the Critics of Newton: An Essay in Logico-Historical Analysis." Synthese. 18 (1968) 251-84.

6421. Titze, Hans. "Versuch einer relativistischen Lösung des Paradoxons der speziellen Relativitatstheorie." PN 13 (1971-72) 279-89.

6422. Tonini, V. "La relativitâ a cinquant'anni dalla prima formulazione einsteiniana." Scientia. 90 (1955) 283-90. French Tr.: Supplement, 167-74.

6423. Tonnelat, Marie Antoinette. "Einstein: mythe et réalité." Scientia. 114 (1979) 297-326. Italian Tr.: 327-47. English Tr.: 349-69.

6424. Törnebohm, Håkan. A Logical Analysis of the Theory of Relativity. Stockholm: Almquist & Wiksell, 1952. 273 pp.

6425. _____. "Epistemological Reflexions over the Special Theory of Relativity and Milne's Conception of Two Times." PS 24 (1957) 57-69.

6426. _____. "The Space-Time Theory within the Special Theory of Relativity: An Essay in Reconstructive Analysis." Methodos. 10 (1958) 243-66.

6427. _____. "Concepts of Velocity in the Special Theory of Relativity: A Study in Concept Formation." Methodos. 12 (1960) 297-309.

6428. _____. "The Lorentz-Formulae and the Metrical Principle." PS 29 (1962) 269-78.

6429. _____. Concepts and Principles in the Space-Time Theory within Einstein's Special Theory of Relativity. Gothobergensis: Acta Universitatis, 1963. 98 pp.

6430. _____. "The Clock Paradox and the Notion of Clock Retardation in the Special Theory of Relativity." Theoria. 29 (1963) 79-90.

6431. _____. "Two Concepts of Simultaneity in the Special Theory of Relativity." Theoria. 29 (1963) 147-53.

6432. _____. "On the Concepts of Distance and Length in the Special Theory of Relativity." Theoria. 29 (1963) 283-89.

6433. _____. "A Foundational Study of Einstein's Special Space-Time Theory." Scientia. 104 (1969) 375-87. French Tr.: Supplement, 219-29.

6434. _____. "Two Studies Concerning the Michelson-Morley Experiment." FP 1 (1970-71) 47-56.

6435. Treder, Hans-Jürgen. "Relativity Theory and Historicity of Physical Systems," in Physics, Logic, and History. W. Yourgrau and A. D. Breck, eds. New York: Plenum Press, 1970. 253-60. "Discussion," 260-64.

6436. _____. "Einstein zu der Beziehung zwischen Experiment
und Theorie." DZP 27 (1979) 162-66.

6437. Tucci, Pasquale. "I contributi di H. Reichenbach ai fondamenti
filosofici della teoria della relatività." Scientia. 110 (1975)
787-820. English Tr.: 821-46.

6438. Ueno, Yoshio. "A Meta-Physical Investigation for the Special
Theory of Relativity - Preliminary Remarks." AJAPS I: 4, No. 2
(1972) 1-14; II: 4, No. 3 (1973) 1-14; III: 4, No. 5 (1975) 37-51.

6439. Ujemow, A. I. "Das heliozentrische System des Kopernikus und
die Relativitätstheorie." DZP 2 (1954) 418-45.

6440. Vogtherr, Karl. "Relativität und Realität." PN 2 (1954)
501-04.

6441. _____. "Die Ermittlung der Gleichzeitigkeit." Methodos.
7 (1955) 309-18. English Tr.: 319-23.

6442. _____. "Die Massenveraenderlichkeit nach der Relativit-
aetstheorie." Methodos. 9 (1957) 183-98. English Tr.: 199-207.

6443. _____. "Die Voraussetzungen der Relativitätstheorie."
PN 6 (1960-61) 55-82.

6444. Weingard, Robert. "Relativity and the Reality of Past and
Future Events." BJPS 23 (1972) 119-21.

6445. _____. "On the Ontological Status of the Metric in
General Relativity." J. Phil. 72 (1975) 426-31.

6446. _____. "General Relativity and the Conceivability of
Time Travel." PS 46 (1979) 328-32.

6447. Wenzl, Aloys. "Einstein's Theory of Relativity Viewed from
the Standpoint of Critical Realism, and Its Significance for Philosophy," in Albert Einstein: Philosopher-Scientist. P. A. Schilpp,
ed. New York: Tudor, 1951. 581-606.

6448. Wheeler, John Archibald. "The Universe in the Light of
General Relativity." Monist. 47 (1962) 40-76.

6449. _____. Einstein's Vision. Berlin: Springer-Verlag,
1968. 108 pp.

6450. Whitbeck, Caroline. "Simultaneity and Distance." J. Phil.
66 (1969) 329-40.

6451. Willer, Jörg. Relativität und Eindeutigkeit: Hugo Dinglers
Beitrag zur Begründungsproblematik. Meisenheim am Glan: A. Hain,
1973. 213 pp.

6452. Williams, L. Pearce. Relativity Theory: Its Origins and
Impact on Modern Thought. New York: Wiley, 1968. viii + 159 pp.

6453. Winnie, John A. "Special Relativity Without One-Way Velocity
Assumptions." PS 37 (1970) 81-99; 223-38.

6454. Woodward, J. F., and W. Yourgrau. "The Incompatibility of
Mach's Principle and the Principle of Equivalence in Current Gravitation Theory." BJPS 23 (1972) 111-16.

  a. "Comments on 'The Incompatibility of Mach's Principle and
  the Principle of Equivalence in Current Gravitation Theory,"
  by Ronald G. Newburgh. 24 (1973) 263-64.

  b. "Mach's Principle, the Equivalence Principle and Gravitation: A Rejoinder to Newburgh," by J. F. Woodward and W.
  Yourgrau. 264-70.

  c. "On the Mach Principle and General Relativity," by Mendel
  Sachs. 26 (1975) 49-51.

  d. "Mach's Principle: Micro -or Macrophysical?" by J.
  Woodward and W. Yourgrau. 137-41.

6455. Zahar, E. G. "Why did Einstein's Programme Supercede Lorentz's?"
BJPS 24 (1973) 95-123, 223-62.

  a. "Zahar on Einstein," by Paul K. Feyerabend. 25 (1974)
  25-28.

  b. "On Lorentz's Methodology," by Arthur I. Miller. 29-45.

  c. "Einstein versus Lorentz: Research Programmes and the
  Logic of Theory Evaluation," by Kenneth F. Schaffner. 45-78.

  d. "Did Einstein's Programme Supercede Lorentz's?" by S. J.
  Prokhovnik. 25 (1974) 336-40.

  e. "On Einstein's Second Postulate," by Steve Wykstra. 27
  (1976) 259-61.

  f. "Some New Aspects of Relativity: Remarks on Zahar's
  Paper," by M. F. Podlaha. 261-67.

  g. "Einstein's Debt to Lorentz: A Reply to Feyerabend and
  Miller," by Elie Zahar. 29 (1978) 49-60.

  h. "A Reply to 'Some New Aspects of Relativity: Remarks on
  Zahar's Paper'," by Arthur I. Miller. 252-56.

  i. "Ad Hocness and the Appraisal of Theories," by M. Redhead.
  355-61.

6456. _____. "Mach, Einstein, and the Rise of Modern Science."
BJPS 28 (1977) 195-213.

  a. "Zahar on Mach, Einstein and Modern Science," by Paul
  Feyerabend. 31 (1980) 273-82.

b. "Second Thoughts about Machian Positivism: A Reply to
Feyerabend," by Elie Zahar. 32 (1981) 267-76.

6457. Zel'dovich, Ia. B. "Einstein's Creative Legacy." Soviet
Studies in Philosophy. 19, No. 3 (1980-81) 11-37. [English reprint
from Voprosy filosofii. (1980, No. 6).]

6.8 Space, Time, Space-Time

6458. Abramenko, B. "On Dimensionality and Continuity of Physical
Space and Time." BJPS 9 (1958) 89-109.

a. "Lattice Structure of Space-Time," by I. J. Good. 9
(1959) 317-19.

6459. Adams, J. Q. "Grünbaum's Solution to Zeno's Paradoxes."
Philosophia. 3 (1973) 43-50.

a. "Reply to J. Q. Adams 'Grünbaum's Solution to Zeno's
Paradoxes'," by Adolf Grünbaum. 51-57.

6460. Ahundov, Murad D. "Lo spazio e il tempo nella struttura
della teoria fisica." Scientia. 113 (1978) 365-78. English Tr.:
379-89.

6461. Aleksandrov, A. D. "Space and Time in Contemporary Physics
in the Light of Lenin's Philosophical Ideas." Soviet Studies in
Philosophy. 10 (1971-72) 257-62. [English reprint from Voprosy
filosofii. (1971, No. 3).]

6462. Augustynek, Zdzislaw. "Three Studies in the Philosophy of
Space and Time." BSPS 3 (1967) 447-65.

6463. Baker, Lynne Rudder. "Temporal Becoming: The Argument from
Physics." Philosophical Forum. 6 (1974) 218-36.

6464. _____. "On the Mind-Dependence of Temporal Becoming."
PPR 39 (1979) 341-57.

6465. Barreau, H. "L'irréversibilité du temps en physique et dans
le langage ordinaire." RMM 84 (1979) 13-31.

6466. _____. "Conception relationnelle et conception absolu-
tiste du temps et de l'espace-temps." Archives de Philosophie. 43
(1980) 57-72.

6467. Basri, Saul A. A Deductive Theory of Space and Time. Amsterdam: North-Holland Pub. Co., 1966. xi + 163 pp.

6468. Berenda, Carlton W. "The Determination of Past by Future
Events: A Discussion of the Wheeler-Feynman Absorption-Radiation
Theory." PS 14 (1947) 13-19.

6469. Berger, George. "The Conceptual Possibility of Time Travel." BJPS 19 (1968) 152-55.

6470. _____. "Temporally Symmetric Causal Relations in Minkowski Space-Time." Synthese. 24 (1972) 58-73. Reprinted in SL 56 (1973) 56-71.

6471. _____. "Elementary Causal Structures in Newtonian and Minkowskian Space-Time." Theoria. 40 (1974) 191-201.

6472. Bergmann, Gustav. "Duration and the Specious Present." PS 27 (1960) 39-47.

6473. Bergmann, Peter G. "Geometry and Observables." MSPS 8 (1977) 275-80.

6474. Biser, Erwin. "Postulates for Physical Time." PS 19 (1952) 50-69.

    a. "A Simple Theory of Time," by Robert Nordberg. 20 (1953) 236-37.

    b. "Time and Events," by Erwin Biser. 238-40.

6475. Black, Max. "The 'Direction' of Time." Analysis. 19 (1959) 54-63.

6476. Borsari, Raffaele. "Spazio, tempo e moto operativo." International Logic Review. 3 (1972) 38-43. English Tr.: 44-49.

6477. Borzeszkowski, H.-H. von, and R. Wahsner. "Zur Beziehung von experimenteller Methode und Raumbegriffe. DZP 28 (1980) 685-96.

6478. Bowman, Peter A. "The Conventionality of Slow-Transport Synchrony." BSPS 32 (1976) 423-34.

6479. _____. "On Conventionality and Simultaneity -- Another Reply." MSPS 8 (1977) 433-47.

6480. Broglie, Louis de. "L'espace et le temps dans la physique quantique." RMM 54 (1949) 113-25.

6481. Brotman, Honor. "Could Space be Four Dimensional?" Mind. 61 (1952) 317-27.

6482. Bruzzaniti, Giuseppe. "L'introduzione del cronone nella teoria classica dell'elettrone. Analisi di alcuni aspetti storici ed epistemologici." Epistemologia. 4 (1981) 381-406.

6483. Büchel, Wolfgang. "Zur 'Protophysik' von Raum und Zeit." PN 12 (1970) 261-81.

6484. Bunge, Mario. "On Multi-dimensional Time." BJPS 9 (1958) 39.

a. "Multidimensional Time," by H. A. C. Dobbs. 225-27.

6485. _____. "Physical Time: The Objective and Relational
Theory." PS 35 (1968) 355-88.

a. "Bunge on Time," by Michael Ruse. 39 (1972) 82.

6486. _____. "Time Asymmetry, Time Reversal, and Irreversibility." Studium Generale. 23 (1970) 562-70.

6487. Caldirola, Piero, and Erasmo Recami. "The Concept of Time in
Physics." Epistemologia. 1 (1978) 263-304.

6488. Čapek, Milič. "The Theory of Eternal Recurrence in Modern
Philosophy of Science, with Special Reference to C. S. Peirce."
J. Phil. 57 (1960) 289-96.

a. "Čapek on Eternal Recurrence," by Bas C. van Fraassen.
59 (1962) 371-76.

6489. _____. "The Myth of Frozen Passage: The Status of
Becoming in the Physical World." BSPS 2 (1965) 441-63.

a. "Physics and Flux: Comments on Professor Čapek's Essay,"
by Donald Williams, 464-75.

6490. _____. "The Fiction of Instants." Studium Generale.
24 (1971) 31-43.

6491. _____. The Concepts of Space and Time. Their Structure
and Their Development. BSPS 22. SL 74. Dordrecht: Reidel, 1976.
570 pp.

6492. Carnap, R. "Les avantages d'une géométrie des espaces physiques
non-euclidienne." Revue Phil. 163 (1973) 317-32.

6493. Casagrande, Federico. "L'introduzione del cronone nella
teoria classica e quantistica dell'elettrone." Scientia. 112
(1977) 401-15. English Tr.: 417-27.

6494. Caws, Peter. "On Being in the Same Place at the Same Time."
Am. Phil. Quart. 2 (1965) 63-66.

6495. Chari, C. T. K. "J. W. Dunne and the Misconception About
Relativistic Time." Philosophical Quarterly (India). 23 (1950)
1-11.

6496. _____. "Is Time Metrizable?" Philosophical Quarterly
(India). 24 (1951) 203-15.

6497. _____. "A Note on Time and Space." Philosophical
Quarterly (India). 25 (1952) 135-38.

6498. _____. "A Note on Multi-Dimensional Time." BJPS 8
(1957) 155-58.

6499. _____. "On the Metaphysical Status of 'Past-Present-Future'." Philosophical Quarterly (India). 30 (1957) 173-82.

6500. Christensen, Ferrel. "How to Establish Non-Conventional Isochrony." BJPS 28 (1977) 49-54.

    a. "The Conventionality of Uniform Time," by Ian W. Roxburgh. 172-77.

6501. Clatterbaugh, Kenneth C. "A Note on Newtonian Time." PS 40 (1973) 281-84.

6502. Cole, Richard. "A Curious Consequence of Conventionalism in Geometry." Southwestern Journal of Philosophy. 1, No. 1 (1970) 121-24.

6503. Costa de Beauregard, O. "Deux problèmes en épistémologie du temps de la physique." RIP 16 (1962) 358-77.

6504. _____. La notion de temps. Équivalence avec l'espace. Paris: Hermann, 1963. 207 pp.

    a. Review essay by P. Lévy. RMM 70 (1965) 96-102.

6505. _____. Le second principe de la science du temps. Entropie, information, irreversibilité. Paris: Éditions du Seuil, 1963. 158 pp.

6506. _____. "Irreversibility Problems." LMPS-2 (1964) 313-42.

6507. _____. "Autofondation." Dialectica. 24 (1970) 247-54.

6508. _____. "On Time, Information and Life." Dialectica. 22 (1968) 187-205.

6509. _____. "No Paradox in the Theory of Time Anisotropy." Studium Generale. 24 (1971) 10-18.

6510. _____. "Two Lectures on the Direction of Time." Synthese. 35 (1977) 129-54. Reprinted in SL 132 (1979) 341-66.

6511. Craig, E. J. "Phenomenal Geometry." BJPS 20 (1969) 121-34.

6512. Dako, M. "The Direction of Time." Studium Generale. 22 (1969) 965-84.

6513. Davies, P. C. W. The Physics of Time Asymmetry. Berkeley: University of California Press, 1974. 232 pp.

6514. _____. Space and Time in the Modern Universe. Cambridge: Cambridge University Press, 1977. 232 pp.

6515. Deledalle, Gérard. "Le temps physique." Les Etudes philosophique. 17 (1962) 237-40.

6516. Demopoulos, William. "On the Relation of Topological to Metrical Structure." MSPS 4 (1970) 263-72.

6517. Denbigh, K. G. "Thermodynamics and the Subjective Sense of Time." BJPS 4 (1953) 183-91.

6518. _____. "In Defence of the Direction of Time." Studium Generale. 23 (1970) 234-44.

6519. Deppert, Wolfgang. "Grundlagen einer Theorie der Systemzeiten." Allgemeine Zeitschrift für Philosophie. 6, No. 2 (1981) 1-25.

6520. Dingle, Herbert. "The Philosophical Significance of Space-Time." Proc. Arist. Soc. 48 n.s. (1947-48) 153-64.

6521. _____. "Time in Philosophy and in Physics." Philosophy. 54 (1979) 99-104.

6522. Dingler, H. "Geometrie und Wirklichkeit." Dialectica. 9 (1955) 341-62. 10 (1956) 80-93.

6523. Dobbs, H. A. C. "The Relation Between the Time of Psychology and the Time of Physics." BJPS 2 (1951) 122-41, 177-92.

    a. "Time of Psychology and of Physics," by G. W. Scott Blair. 3 (1952) 82-85.

    b. "Time of Psychology and of Physics," by G. W. Scott Blair. 3 (1953) 359.

    c. "The Time of Psychology and of Physics," by H. A. C. Dobbs. 4 (1953) 161-64.

    d. "E. A. Milne's Scales of Time," by Adolf Grünbaum. 4 (1954) 329-31.

    e. "Mr. Dobbs' Two-Dimensional Theory of Time," by C. W. K. Mundle. 331-37.

    f. "E. A. Milne's Scales of Time," by G. J. Whitrow. 5 (1954) 151.

    g. "The Time of Physics and Psychology," by H. A. C. Dobbs. 7 (1956) 156-60.

6524. _____. "The 'Present' in Physics." BJPS 19 (1969) 317-24.

    a. "Are Present Events Themselves Transiently Past, Present and Future? A Reply to H. A. C. Dobbs," by Adolf Grünbaum. 20 (1969) 145-53.

    b. "Reply to Professor Grünbaum," by H. A. C. Dobbs. 21 (1970) 275-78.

c. "Grünbaum vs. Dobbs: The Need for Physical Transiency," by Frederick Ferre. 278-80.

6525. Domotor, Zoltan. "Causal Models and Space-Time Geometries." Synthese. 24 (1972) 5-57. Reprinted in SL 56 (1973) 3-55.

6526. Dongorozi, C.-S. "Sur l'asymétrie de l'espace-temps." Organon. 7 (1970) 291-94.

6527. Dretske, Fred I. "Moving Backward in Time." Phil. Review. 71 (1962) 94-98.

6528. Düsberg, K. J. Zur Messung von Raum und Zeit. Eine Kritik der sogenannten Protophysik. Königstein/Ts.: Athenäum-Hain-Skriptor-Hanstein, 1980. 117 pp.

6529. Earman, John. "Irreversibility and Temporal Asymmetry." J. Phil. 64 (1967) 543-49.

a. "Earman on Temporal Anistropy," by George Berger. 68 (1971) 132-37.

b. "Berger on Earman on Temporal Anisotropy," by Robert Weingard. 69 (1972) 786-90.

6530. _____. "On Going Backwards in Time." PS 34 (1967) 211-22.

6531. _____. "The Anisotropy of Time." AJP 47 (1969) 273-95.

a. "The Asymmetry of Time," by H. Krips. 49 (1971) 204-10.

6532. _____. "Space-Time, or How to Solve Philosophical Problems and Dissolve Philosophical Muddles without Really Trying." J. Phil. 67 (1970) 259-77.

6533. _____. "Till the End of Time." Methodology and Science. 4 (1971) 2-40. Also in MSPS 8 (1977) 109-33.

6534. _____. "Kant, Incongruous Counterparts and the Nature of Space and Space-Time." Ratio. 13 (1971) 1-18.

6535. _____. "Notes on the Causal Theory of Time." Synthese. 24 (1972) 74-86. Reprinted in SL 56 (1973) 72-84.

a. "Earman on the Causal Theory of Time," by Bas C. van Fraassen. Synthese. 24 (1972) 87-95. Reprinted in SL 56 (1973) 85-93.

6536. _____. "An Attempt to Add a Little Direction to 'The Problem of the Direction of Time'." PS 41 (1974) 15-47.

6537. _____. "How to Talk About the Topology of Time." Nous. 11 (1977) 211-26.

6538. Ehrlich, Walter. "Das Zeitproblem." ZPF 13 (1959) 369-84.

6539. Eichinger, B. E. "Projective Spacetime." FP 7 (1977) 673-703.

6540. Ellis, Brian. "Universal and Differential Forces." BJPS 14 (1963) 177-94.

6541. Ferré, Frederick. "Grünbaum on Temporal Becoming." IPQ 12 (1972) 426-45.

6542. Fields, H. "On the Status of 'The Direction of Time'." Methodology and Science. 12 (1979) 213-35.

6543. Fine, Arthur. "Reflections on a Relational Theory of Space." SL 56 (1973) 234-67. Reprinted from Synthese. 22 (1971) 448-81.

6544. Fischer, Roland. "Biological Time," in The Voices of Time. J. T. Fraser, ed. New York: George Braziller, 1966. 357-82.

6545. Fisk, Milton. "The Epistemological Status of Time's Arrow." Proc. Am. Cath. Phil. Assn. 38 (1964) 166-77.

6546. Fitzgerald, Paul. "Nowness and the Understanding of Time." BSPS 20, SL 64 (1974) 259-81.

6547. _____. "Is Temporality Mind Dependent?" PSA 1980. I, 283-91.

6548. Fliedner, D. "Zum Problem des vierdimensionalen Raumes." PN 18 (1980-81) 388-412.

6549. Fraassen, Bas C. van. An Introduction to the Philosophy of Time and Space. New York: Random House, 1970. 224 pp.

6550. _____. "Time: Physical and Experienced." Epistemologia. 1 (1978) 323-36.

6551. Fraser, J. T., ed. The Voices of Time: A Cooperative Survey of Man's Views of Time as Expressed by the Sciences and by the Humanities. New York: George Braziller, 1966. 710 pp.

6552. Fraser, J. T. Of Time, Passion and Knowledge. New York: Braziller, 1975. xiii + 529 pp.

6553. Fraser, J. T., and N. Lawrence, eds. The Study of Time II. New York: Springer-Verlag, 1975. 486 pp.

6554. Fraser, J. T., N. Lawrence, and D. Park, eds. The Study of Time III. New York, Heidelberg, Berlin: 1978. 727 pp.

6555. Freeman, Eugene, and Wilfrid Sellars, ed. Basic Issues in the Philosophy of Time. La Salle, Illinois: Open Court, 1971. 241 pp.

6556. Freundlich, Yehudah. "'Becoming' and the Asymmetries of Time." PS 40 (1973) 496-517.

6557. Frey, Gerhard. "Zum Problem des Konventionalismus in der mathematischen und physikalischen Begriffsbildung." ZPF 9 (1957) 385-89.

6558. _____. "Der Zeitbegriff in den Naturwissenschaften." Studium Generale. 14 (1961) 420-28.

6559. Friedman, Michael. "Grünbaum on the Conventionality of Geometry." Synthese. 24 (1972) 219-35. Reprinted in SL 56 (1973) 217-33.

6560. _____. "Simultaneity in Newtonian Mechanics and Special Relativity." MSPS 8 (1977) 403-32.

6561. Frye, Royal M. "Impacts of Modern Physics on the Problem of Time." Philosophical Forum. 6 (1948) 2-8.

6562. Gagnebin, S. "Structure et substructure de la géometrie." Dialectica. 11 (1957) 405-33.

6563. Gale, Richard M. "McTaggert's Analysis of Time." Am. Phil. Quart. 3 (1966) 145-52.

6564. _____. The Philosophy of Time: A Collection of Essays. Garden City, NY: Doubleday, 1967. 507 pp.

6565. _____. The Language of Time. New York: Humanities Press, 1968. viii + 248 pp.

   a. "Gale's Analysis of the Concept of Time," by Richard J. Blackwell. Modern Schoolman. 47 (1970) 346-50.

   b. Review essay by Paul Fitzgerald. Philosophical Forum. 5 (1974) 424-40.

   c. Review essay by A. N. Prior. Mind. 78 (1969) 453-60.

6566. Gardner, Michael R. "The Unintelligibility of 'Observational Equivalence'." PSA 1976. I, 104-16.

6567. Gehlhar, F. "Raum und Zeit als Existenzformen der Materie." DZP 23 (1975) 898-912.

6568. Giedymin, Jerzy. "On the Origin and Significance of Poincaré's Conventionalism." SHPS 8 (1977) 271-301.

6569. _____. Science and Convention: Essays on Henri Poincaré's Philosophy of Science and the Conventionalist Tradition. New York: Pergamon Press, 1982. xvii + 229 pp.

6570. Glymour, Clark. "Topology, Cosmology and Convention." Synthese. 24 (1972) 195-218. Reprinted in SL 56 (1973) 193-216.

6571. _____. "Physics by Convention." PS 39 (1972) 322-40.

6572. _____. "The Epistemology of Geometry." Nous. 11 (1977) 227-51.

6573. _____. "Indistinguishable Space-Times and the Fundamental Group." MSPS 8 (1977) 50-60.

    a. "Observationally Indistinguishable Space-Times: Comments on Glymour's Paper," by David Malament. 61-80.

6574. Gold, Thomas. "Cosmic Processes and the Nature of Time," in Mind and Cosmos. Robert G. Colodny, ed. Pittsburgh: University of Pittsburgh Press, 1966. 311-29.

6575. Gold, T., and D. L. Schumacher, eds. The Nature of Time. Ithaca, NY: Cornell University Press, 1967. ix + 225 pp.

6576. Goldblatt, Robert. "Diodorean Modality in Minkowski Space-time." Studia Logica. 39 (1980) 219-36.

6577. Gonseth, F. La géométrie et le problème de l'espace. Neuchâtel: Editions du Griffon, 1955.

6578. _____. "La géométrie et le probleme de l'espace." Studium Generale. 11 (1958) 108-15.

6579. _____. "Temps et Syntaxe." Dialectica. 22 (1968) 206-13.

6580. Gössmann, G. "Recent Discussions of the Micro-Structure of Space and Time in Soviet Philosophy." Studies in Soviet Thought. 4 (1964) 296-304.

6581. Greenaway, Frank, ed. Time and the Sciences. Paris: Unesco, 1979. 182 pp.

6582. Griese, A. "Zeit-Bewegung-Entwicklung." DZP 27 (1979) 191-202.

6583. Griese, A., and R. Wahsner. "Zur Ausarbeitung einer philosophischen Raum-Zeit-Theorie." DZP 15 (1967) 691-704.

6584. Grünbaum, Adolf. "A Consistent Conception of the Extended Linear Continuum as an Aggregate of Unextended Elements." PS 19 (1952) 288-306.

6585. _____. "Das Zeitproblem." Archiv für Philosophie. 7 (1957) 165-208.

6586. _____. "Conventionalism in Geometry," in The Axiomatic Method. L. Henkin, P. Suppes, and A. Tarski, eds. Amsterdam: North-Holland, 1959. 216-22.

a. "Physical Geometry and Physical Laws," by Arthur Fine. PS 31 (1964) 156-62.

6587. _____. "The Nature of Time," in Frontiers of Science and Philosophy. Robert G. Coloduy, ed. Pittsburgh: University of Pittsburgh Press, 1962. 147-88.

6588. _____. "Geometry, Chronometry, and Empiricism." MSPS 3 (1962) 405-526.

6589. _____. Philosophical Problems of Space and Time. New York: A. A. Knopf, 1963. 448 pp. Second, enlarged edition: BSPS 12. SL 55. Dordrecht: Reidel, 1973. 884 pp.

   a. Review essay by Roberto Torretti. Dialogos. 10, No. 27 (1974) 89-117.

   b. Review essay by Arthur I. Miller. Isis. 66 (1975) 590-94.

   c. "Remarks on Miller's Review of Philosphical Problems of Space and Time," by Adolf Grünbaum. 68 (1977) 447-48.

   d. "Reply," by Arthur I. Miller. 449-50.

6590. _____. "Carnap's Views on the Foundations of Geometry," in The Philosophy of Rudolf Carnap. P. A. Schilpp, ed. La Salle, Illinois: Open Court, 1963. 599-684.

6591. _____. "The Anisotropy of Time." Monist. 48 (1964) 219-47.

6592. _____. "Geometrie, Zeitmessung und Empirismus." Archiv für Philosophie. 12 (1964) 179-303.

6593. _____. "Space and Time," in Philosophy of Science Today. S. Morgenbesser, ed. New York: Basic Books, 1967. 125-35.

6594. _____. Geometry and Chronometry in Philosophical Perspective. Minneapolis: University of Minnesota Press, 1968. 378 pp.

   a. Review essay, "Geometry and Convention: A Critical Discussion," by Milton Fisk. Thomist. 33 (1969) 343-51.

   b. Review essay by J. Merleau-Ponty. Revue Phil. 160 (1970) 471-75.

   c. Review essay, "Reflections on a Relational Theory of Space," by Arthur Fine. Synthese. 22 (1971) 448-81.

6595. _____. Modern Science and Zeno's Paradoxes. Middletown: Wesleyan University Press, 1967. x + 148 pp. 2nd Edition: London: Allen & Unwin, 1968.

6596. _____. "Can an Infinitude of Operations be Performed in a Finite Time?" BJPS 20 (1969) 203-18.

6597. _____. "Space, Time and Falsifiability: Critical Exposition and Reply to 'A Panel Discussion of Grünbaum's Philosophy of Science' [PS 36 (1979)]." Part I: PS 37 (1970) 469-588. Reprinted as Chapter 16 of A. Grünbaum, Philosophical Problems of Space and Time. Second Edition. Dordrecht: Reidel, 1973.

> a. "Intrinsic Metrics on Continuous Spatial Manifold," by Philip L. Quinn. PS 43 (1976) 396-414.

6598. _____. "Simultaneity by Slow Clock Transport in the Special Theory of Relativity." SL 31 (1970) 140-66. [Abridged version of an essay in PS 36 (1969) 5-43.]

6599. _____. "The Meaning of Time." SL 24 (1970) 147-77. [Revised version of Chapter I of Modern Science and Zeno's Paradoxes.] Reprinted in Basic Issues in the Philosophy of Time, Eugene Freeman and Wilfrid Sellars, eds. La Salle, Illinois: Open Court, 1971. 195-228.

6600. _____. "The Ontology of the Curvature of Empty Space in the Geometrodynamics of Clifford and Wheeler." SL 56 (1973) 268-95.

6601. _____. "Popper's Views on the Arrow of Time," in The Philosophy of Karl Popper. P. A. Schilpp, ed. La Salle, Illinois: Open Court, 1974. 775-97.

6602. _____. "Absolute and Relational Theories of Space and Space-Time." MSPS 8 (1977) 303-73.

6603. Hacking, Ian. "The Identity of Indiscernibles." J. Phil. 72 (1975) 249-56.

> a. "On Cracking That Nut, Absolute Space," by Robert Weingard. PS 44 (1977) 288-91.

6604. Hanson, Norwood Russell. "On Being in Two Places at Once." Review of Metaphysics. 12 (1958) 3-18.

6605. Harré, R. "More Points on Discontinuity," Synthese. 17 (1967) 104-06.

6606. Harrison, Craig. "On the Structure of Space-Time." Synthese. 24 (1972) 180-94. Reprinted in SL 56 (1973) 178-92.

6607. Hellpach, Willy. "Dimensionen in Raum and Zeit." PN 1 (1950) 179-88.

6608. Henmueller, Frank, and Karl Menger. "What is Length?" PS 28 (1961) 172-77.

6609. Hinckfuss, Ian. The Existence of Space and Time. Oxford: Clarendon Press, 1975. xii + 153 pp.

a. Review essay by Chris Mortensen. AJP 55 (1977) 149-57.

6610. Holt, Dennis Charles. "Time Travel: The Time Discrepancy Paradox." Philosophical Investigations. 4, No. 4 (1981) 1-16.

6611. Hönigswald, Richard. "Gleichzeitigkeit und Raum." Archiv für Philosophie. 2 (1948) 67-95.

6612. Hooker, Clifford A. "The Relational Doctrines of Space and Time." BJPS 22 (1971) 97-130.

6613. Horwich, Paul. "Grünbaum on the Metric of Space and Time." BJPS 26 (1975) 199-211.

6614. _____. "On the Existence of Time, Space, and Space-Time." Nous. 12 (1978) 397-419.

6615. Hoy, Ronald C. "The Role of Genidentity in the Causal Theory of Time." PS 42 (1975) 11-18.

6616. Hsu, J. P. "The Analysis of Time: Is the Relativistic Time Unique?" FP 9 (1979) 55-69.

6617. Hübner, Kurt. "A Philosophical Discussion of the Concept of Time in Physics." Epistemologia. 3 (1980) Fascicolo speciale. 149-60.

6618. Hughes, Martin. "Absolute Rotation." BJPS 32 (1981) 359-66.

6619. Hund, F. "Die Zeit in der Begriffswelt des Physikers." Studium Generale. 8 (1955) 469-76.

6620. _____. "Zeit als physikalisher Begriff." Studium Generale. 23 (1970) 1088-1101.

6621. Huron, Roger. "Notes sur le temps des physiciens et des mathématiciens." Les Etudes Philosophiques. 17 (1962) 19-46.

6622. Hwang, Sun-Tak. "A New Interpretation of Time Reversal." FP 2 (1972) 315-26.

6623. Jaeglé, Pierre. Essai sur l'espace et le temps ou propos sur la dialectique de la nature. Paris: Editions sociales, 1976. 128 pp.

6624. Jammer, Max. Concepts of Space: The History of Theories of Space in Physics. Cambridge: Mass.: Harvard University Press, 1954. xviii + 196 pp. Reprint: New York: Harper, 1960.

a. Review essay, "The Philosophical Retention of Absolute Space in Einstein's General Theory of Relativity," by Adolf Grünbaum. Phil. Review. 66 (1957) 525-34.

6625. Janich, Peter. Die Protophysik der Zeit. Mannheim: Biblio-graphisches Institut; 1969. 177 pp.

a. Review essay by Gernot Böhme. Philosophische Rundschau. 20 (1973) 94-111.

b. Review essay, "Zur konstruktivistischen Protophysik der Zeit: Eine immanete Kritik," by R. Aschelberg, G. Herrgott, and P. Krausser. ZAW 9 (1978) 112-33.

c. Review essay (on second edition), "Methode oder Dogma?" by Andreas Kamlah. ZAW 12 (1981) 138-62.

6626. Jobe, Evan K. "Nature's Choice of Time." AJP 58 (1980) 347-59.

6627. Joseph, Geoffrey. "Riemannian Geometry and Philosophical Conventionalism." AJP 57 (1979) 225-36.

6628. Juhos, Béla. "Die zweidimensionale Zeit." Archiv für Philosophie. 11 (1961) 3-27.

6629. Kalmus, Hans. "Organic Evolution and Time," in The Voices of Time. J. T. Fraser, ed. New York: George Braziller, 1966. 330-52.

6630. Kamlah, Andreas. "Hans Reichenbach's Relativity of Geometry." Synthese. 34 (1977) 249-64. Reprinted in SL 132 (1979) 251-66.

6631. Kamp, Hans. "Formal Properties of 'Now'." Theoria. 37 (1971) 227-73.

6632. Kanitscheider, Bernulf. "Geochronometrie und Geometrodynamik: Zum Problem der Konventionalismus." ZAW 4 (1973) 261-302.

6633. _____. Vom absoluten Raum zur dynamischen Geometrie. Mannheim: Bibliographisches Institut, 1976. 139 pp.

a. Review essay by J. Pfarr. Grazer Philosophische Studien. 6 (1978) 179-91.

6634. _____. "Singularitäten, Horizonte und das Ende der Zeit." PN 16 (1976-77) 480-511.

6635. Kapp, R. O. "Space." BJPS 10 (1959) 1-15.

6636. Krbek, F. von. "Raumformen." Scientia. 95 (1960) 275-78. French Tr.: Supplement, 145-48.

6637. Krimsky, Sheldon. "The Multiple-World Thought Experiment and Absolute Space." Nous. 6 (1972) 266-73.

6638. Kuchowicz, B. "Ueber eine Synthese der klassischen Raum-und Zeitauffassungen." PN 12 (1970) 87-89.

6639. Kurth, Rudolf. "Ueber Zeit und Zeitmessung." PN 8 (1964) 65-90.

6640. _____. "Die topologische Struktur der Zeit." PN 15
(1974-75) 359-74.

6641   Lacey, Hugh M.   "The Causal Theory of Time: A Critique of
Grünbaum's Version." PS 35 (1968) 332-54.

6642. _____. "The Scientific Intelligibility of Absolute
Space: A Study of Newtonian Argument." BJPS 21 (1970) 317-42.

6643. Landsberg, P. T. "Time in Statistical Physics and Special
Relativity." Studium Generale. 23 (1970) 1108-58.

6644. Langevin, Paul. "L'évolution de l'espace et du temps."
Scientia. 108 (1973) 221-39. English Tr.: 285-300.

6645. Latzer, Robert W. "Nondirected Light Signals and the Structure of Time." Synthese. 24 (1972) 236-80. Reprinted in SL 56
(1973) 321-65.

6646. Lawrence, Nathaniel. "Time Represented as Space." The Monist
53 (1969) 447-56. Reprinted in Basic Issues in the Philosophy of
Time. E. Freeman and W. Sellars, eds. La Salle, Illinois: Open
Court, 1971. 123-32.

6647. Leite-Lopes, J. "L'évolution des notions d'espace et du
temps." Scientia. 107 (1972) 411-33. English Tr.: 434-53.

6648. Lestienne, R. "Caractères de la durée physique." Scientia.
107 (1972) 77-89, 279-93. English Tr.: 90-100, 294-306.

6649. _____. "Entropie, temps mécanique et flèche cosmologique."
Scientia. 115 (1980) 337-58. Italian Tr.: 359-72. English Tr.:
373-86.

6650. Lorenzen, Paul. "Die Eindeutigkeit der Zeitmessung." ZAW 7
(1976) 359-61.

6651. Lucas, J. R. A Treatise on Time and Space. London: Methuen,
1973. 321 pp.

6652. MacKinnon, Edward. "Time in Contemporary Physics." IPQ 2
(1962) 428-57.

6653. McGilvray, James A. "A Defense of Physical Becoming."
Erkenntnis. 14 (1979) 275-99.

6654. McGonigle, T. G. "Euclidean Space: A Lasting Philosophical
Obsession." BJPS 21 (1970) 185-91.

6655. Machamer, Peter K., and Robert G. Turnbull, eds. Motion and
Time, Space and Matter. Columbus: Ohio State University Press,
1976. 559 pp.

6656. Malament, David. "Causal Theories of Time and the Conventionality of Simultaneity." Nous. 11 (1977) 293-300.

6657. Margenau, Henry. "Can Time Flow Backwards?" PS 21 (1954) 79-92.

6658. Masani, Alberto. "La freccia del tempo, le costanti di natura e la cosmologia." Epistemologia. 2 (1979) 3-38.

6659. Massey, Gerald J. "Toward a Clarification of Grünbaum's Concept of Intrinsic Metric." PS 36 (1969) 331-45.

   a. "On Massey's Explication of Grünbaum's Conception of Metric," by Bas C. van Fraassen. 346-53.

6660. _____. "Is 'Congruence' a Peculiar Predicate?" BSPS 8, SL 39 (1971) 606-15.

6661. Matthews, Geoffrey. "Time's Arrow and the Structure of Spacetime." PS 46 (1979) 82-97.

6662. Mead, George H. "Relative Space-Time and Simultaneity." Review of Metaphysics. 17 (1974) 514-35. [Previously unpublished paper by Mead, first discovered in 1931 by Charles W. Morris.]

6663. Mehlberg, Henryk. "Physical Laws and Time's Arrow," in Current Issues in the Philosophy of Science. H. Feigl and G. Maxwell, eds. New York: Holt, Rinehart and Winston, 1961. 105-38.

6664. _____. "Philosophical Aspects of Physical Time." Monist. 53 (1969) 340-84. Reprinted in Basic Issues in the Philosophy of Time. E. Freeman and W. Sellars, eds. La Salle, Illinois: Open Court, 1971. 16-60.

6665. _____. "Individuality, Reality and Space-Time." Methodology and Science. 10 (1977) 34-63.

6666. Meiland, J. W. "Temporal Parts and Spatio-Temporal Analogies." Am. Phil. Quart. 3 (1966) 64-70.

6667. Mercier, André. "Les conditions physiques et la notion de temps." Studia Philosophica. 10 (1950) 85-114.

6668. _____. "Petits prolégomènes à une étude sur le temps." Theoria. 29 (1963) 277-82.

6669. Merleau-Ponty, Jacques. "Problems of Physical Time." Diogenes. 56 (1966) 115-40.

6670. _____. "Réflexions sur la distinction de l'espace et du temps." Revue Phil. 157 (1967) 453-60.

6671. Metz, A. "Le temps, la physique moderne et la philosophie." Archives de Philosophie. 27 (1964) 592-99.

6672. _____. "De la durée intuitive au temps scientifique." Dialogue. 5 (1966) 184-204.

6673. Millard, Richard M. "Space, Time, and Space-Time." Philosophical Forum. 13 (1955) 29-53.

6674. Miller, Arthur T. "The Myth of Gauss' Experiment on the Euclidean Nature of Physical Space." Isis. 63 (1972) 345-48.

    a. "Comments on Miller's 'The Myth of Gauss' Experiment on the Euclidean Nature of Physical Space," Comment I by George Goe, Comment II by B. L. van der Waerden. 65 (1974) 83-85.

    b. "Reply," by A. I. Miller. 86-87.

6675. Miller, David L. "The Importance of Presents in Contemporary Science." 24 (1957) 19-25.

6676. _____. "The Functions of Pasts in Science." Southern Journal of Philosophy. 3 (1965) 77-82.

6677. Miller, Peter. "A Pluralistic Account of Space." IPQ 11 (1971) 180-212.

6678. Milne, E. A. "A Modern Conception of Time." Philosophy. 25 (1950) 68-72.

6679. Mirman, R. "Comments on the Dimensionality of Time." FP 3 (1973) 321-33.

6680. _____. "The Direction of Time." FP 5 (1975) 491-511.

6681. Montgomery, Hugh. "Space." Philosophical Journal. 6 (1969) 127-40.

6682. Mooij, J. J. A. "La philosophie géométrique de Henri Poincaré." Synthese. 16 (1966) 53-65.

6683. Moon, Parry, and Domina Eberle Spencer. "On the Establishment of a Universal Time." PS 23 (1956) 216-29.

    a. "Remarks Concerning Moon and Spencer's 'On the Establishment of a Universal Time." by A. Grünbaum. 24 (1957) 77-78.

6684. Moreno, Antonio. "Time and Relativity: Some Philosophical Considerations." Thomist. 45 (1981) 62-79.

6685. Mortensen, Chris, and Graham Nerlich. "Physical Topology." Journal of Philosophical Logic. 7 (1978) 209-23.

6686. Mühlhölzer, Felix. "Zur Protophysik der Zeit: Eine erneute Kritik." ZAW 12 (1981) 340-52.

6687. Müller, Georg. "Seinszeit und Gegenwartszeit, Zeitraum und Weltraum." PN 8 (1964) 91-108.

6688. Murakami, Yoichiro. "Basis of the Structure of Time." AJAPS 3, No. 4 (1969) 16-25.

6689. Narliker, J. V. "The Direction of Time." BJPS 15 (1965) 281-85.

6690. Neemann, U. "Philosophische Probleme von Raum und Zeit." PN 18 (1980-81) 146-59.

6691. Nerlich, Graham. The Shape of Space. Cambridge: Cambridge University Press, 1976. xi + 280 pp.

    a. Review essay by Chris Mortensen. AJP 55 (1977) 149-57.

    b. Review essay by C. A. Hooker. Dialogue. 20 (1981) 783-98.

6692. _____. "What Can Geometry Explain?" BJPS 30 (1979) 69-83.

    a. "On Things and Causes in Spacetime," by Hugh Mellor. 31 (1980) 282-88.

6693. _____. "Is Curvature Intrinsic to Physical Space?" PS 46 (1979) 439-58.

6694. Newton-Smith, W. H. The Structure of Time. London: Routledge & Kegan Paul, 1980. 262 pp.

6695. O'Gorman, F. P. "Poincaré's Conventionalism of Applied Geometry." SHPS 8 (1977) 303-50.

    a. "Poincaré's Thesis That any and all Stellar Parallax Findings are Compatible with the Euclideanism of the Pertinent Astronomical 3-Space," by Adolf Grünbaum. 9 (1978) 313-18.

    b. "Poincaré's Retention of Euclid on Apparently Adverse Parallactic Findings: A Reply to A. Grünbaum," by F. P. O'Gorman. 319-21.

6696. Palter, Robert. "On the Significance of Space-Time." Review of Metaphysics. 9 (1955) 149-55.

6697. Park, D. "The Changing Role of Concepts of Space and Time in Physics." Studium Generale. 20 (1967) 10-14.

6698. _____. "Are Space and Time Necessary?" Scientia. 105 (1970) 210-22. French Tr.: Supplement, 84-96.

6699. _____. "The Myth of the Passage of Time." Studium Generale. 24 (1971) 19-30.

6700. _____. The Image of Eternity: Roots of Time in the Physical World. Amherst, MA: The University of Massachusetts Press, 1980. x + 149 pp.

6701. Post, E. J. "The Geometric Substratum of Physics." Scientia. 106 (1971) 391-96.

6702. _____. "The Logic of Time Reversal." FP 9 (1979) 129-61.

6703. _____. "Time Asymmetries in Classical and in Nonclassical Physics." FP 9 (1979) 831-63.

6704. Prokhovnik, Simon J. "Time as a Universal Property of Nature." Epistemologia. 1 (1978) 313-22.

6705. Putnam, Hilary. "An Examination of Grünbaum's Philosophy of Geometry," in Philosophy of Science: The Delaware Seminar, II. New York: Interscience Publishers, 1963. 205-55.

    a. "Reply to Hilary Putnam's 'An Examination of Grünbaum's Philosophy of Geometry'," by Adolf Grunbaum. BSPS 5 (1969) 1-150.

6706. _____. "Time and Physical Geometry." J. Phil. 64 (1967) 240-47.

    a. "Simultaneity and the Future," by Errol E. Harris. BJPS 19 (1968) 254-56.

6707. _____. "The Refutation of Conventionalism." Nous. 8 (1974) 41-52.

    a. "A Defense of Conventionalism," by Henry E. Kyburg, Jr. 11 (1977) 75-95.

6708. Quinn, O.S.A., John M. "The Irreversibility of Time: A Realistic Approach." Proc. Am. Cath. Phil. Assn. 39 (1965) 103-12.

6709. Quinton, Anthony. "Space and Times." Philosophy. 37 (1962) 130-47.

    a. "The Unity of Space-Time: Mathematics versus Myth Making," by J. J. C. Smart. AJP 45 (1967) 214-17.

6710. Reichenbach, Hans. The Direction of Time. Berkeley: University of California Press; London: CUP, 1957. xi + 280 pp.

    a. "Critical Study," by J. J. C. Smart. Phil. Quart. 8 (1958) 72-77.

6711. _____. The Philosophy of Space and Time. M. Reichenbach and J. Freund, trs. New York: Dover, 1958. 295 pp. [Translation of Philosophie der Raum-Zeit-Lehre, 1928.]

6712. Reidemeister, K. "Zur Logik der Lehre vom Raum." Dialectica. 6 (1952) 327-42.

6713. Reinberg, Alain. "La chronobiologie et le renouvellement de la notion de temps dans les sciences de la vie." Epistemologia. 1 (1978) 251-62.

6714. Roberts, George W. "Some Points About Discontinuity." _Synthese_.
17 (1967) 100-03.

6715. Rogers, Ben. "On Discrete Spaces." _Am. Phil. Quart._ 5
(1968) 117-23.

6716. Rosen, Steven M. "Synsymmetry." _Scientia_. 110 (1975)
539-49. Italian Tr.: 551-58.

6717. Sachs, Mendel. "On the Mach Principle and Relative Space-Time."
BJPS 23 (1972) 117-19.

6718. Salmon, Wesley C. _Space, Time and Motion: A Philosophical
Introduction_. Encino, CA: Dickenson Publishing Co., 1975. 147 pp.
Second edition, revised. 1981. 140 pp.

      a. Review essay, "Mathematics and Reality," by Simon Prokhovnik.
      BJPS 28 (1977) 189-94.

6719. _____. "The Curvature of Physical Space." MSPS 8
(1977) 281-302.

6720. Schäfer, K. "Die Ziet und die übrigen Dimensionen." _Studium
Generale_. 20 (1967) 1-10.

6721. Schlegel, Richard. _Time and the Physical World_. East Lansing:
Michigan State University Press, 1961. xii + 211 pp.

6722. _____. "Time and Thermodynamics," in _The Voices of
Time_. J. T. Fraser, ed. New York: George Braziller, 1966. 500-23.

6723. Schlesinger, G. "What Does the Denial of Absolute Space
Mean?" AJP 45 (1967) 44-60.

      a. "The Denial of Absolute Space and the Hypothesis of a
      Universal Nocturnal Expansion: A Rejoinder to George Schle-
      singer," by Adolf Grünbaum. 61-91.

      b. "Who's Afraid of Absolute Space?" by John Earman. 48
      (1970) 287-319.

      c. "Why I am Afraid of Absolute Space," by Adolf Grünbaum.
      49 (1971) 96.

6724. _____. "Two Notions of the Passage of Time." _Nous_. 3
(1969) 1-16.

6725. _____. "The Structure of McTaggart's Argument." _Review
of Metaphysics_. 24 (1971) 668-77.

6726. _____. "The Similarities between Space and Time."
Mind. 84 (1975) 161-76.

      a. "Space and Time Re-assimilated," by Bernard Mayo. 85
      (1976) 576-80.

b. "Comparing Space and Time Once More," by G. Schlesinger. 87 (1978) 264-66.

c. "Space and Time," by J. M. Shorter. 90 (1981) 61-78.

6727. _____. Aspects of Time. Indianapolis: Hackett, 1980. 150 pp.

6728. Sellars, Wilfrid. "Time and the World Order." MSPS 3 (1962) 527-616.

6729. Shackle, G. L. S. "Time and Thought." BJPS 9 (1959) 285-98.

6730. Shapere, Dudley. "Space, Time, and Language - An Examination of Some Problems and Methods of the Philosophy of Science," in Philosophy of Science: The Delaware Seminar, II. New York: Interscience Publishers, 1963. 139-70.

6731. _____. "The Causal Efficacy of Space." PS 31 (1964) 111-21.

6732. Sklar, Lawrence. "The Conventionality of Geometry," in Studies in the Philosophy of Science. N. Rescher, ed. Oxford: Basil Blackwell, 1969. 42-60.

6733. _____. "Absolute Space and the Metaphysics of Theories." Nous. 6 (1972) 289-309.

6734. _____. Space, Time, and Spacetime. Berkeley: University of California Press, 1974. 435 pp.

a. Review essay by Robert Weingard. PS 44 (1977) 167-73.

b. Review essay by David Malament. J. Phil. 73 (1976) 306-23.

6735. _____. "Facts, Conventions, and Assumptions in the Theory of Space-Time." MSPS 8 (1977) 206-74.

6736. _____. "What Might be Right about the Causal Theory of Time." Synthese. 35 (1977) 155-72. Reprinted in SL 132 (1979) 367-84.

6737. _____. "Up and Down, Left and Right, Past and Future." Nous. 15 (1981) 111-29.

6738. Smart, J. J. C. "The Moving 'Now'." AJP 31 (1953) 184-87.

6739. _____. "The Temporal Asymmetry of the World." Analysis. 14 (1954) 79-82.

a. "Professor Smart on Temporal Asymmetry," by Bernard Mayo. AJP 33 (1955) 38-44.

b. "Mr. Mayo on Temporal Asymmetry," by J. J. C. Smart.
124-27.

6740. _____. "Causal Theories of Time." Monist. 53 (1969)
385-95. Reprinted in Basic Issues in the Philosophy of Time. E.
Freeman and W. Sellars, eds. La Salle, Illinois: Open Court, 1971.
61-71.

6741. Snider, Caroline Whitbeck. "The Confusion Concerning Universal
Forces." BJPS 18 (1967) 64-66.

6742. Spisani, Franco. Significato e struttura del tempo. The
Meaning and Structure of Time. Testo bilingue. Bologna: Azzoguidi,
1972. 161 pp.

6743. Stancovici, V., and L. Stanciu. "La Strattura dello spazio e
del tempo nella logica produttiva di Franco Spisani." International
Logic Review. 6 (1975) 23-50. English Tr.: 53-80.

6744. Stein, Howard. "On Einstein-Minkowski Space-Time." J. Phil.
65 (1968) 5-23.

a. "The Logic of Simuntaneity," by John W. Lango. 66 (1969)
340-50.

b. "A Note on Time and Relativity Theory," by Howard Stein.
67 (1970) 289-94.

6745. _____. "On Space-Time Ontology: Extracts from a Letter
to Adolf Grünbaum." MSPS 8 (1977) 374-402.

6746. Ströker, Elisabeth. Philosophische Untersuchungen zum Raum.
Frankfurt/M: Vittorio Klostermann. 1965. 366 pp.

a. Review essay by Joachim Hölling. Philosophische Rundschau.
15 (1968) 209-19.

6747. Suppes, Patrick. "Some Open Problems in the Philosophy of
Space and Time." Synthese. 24 (1972) 298-316. Reprinted in SL 56
(1973) 383-401.

6748. _____. Space, Time, and Geometry. SL 56. Dordrecht:
Reidel, 1973. 424 pp.

6749. _____. "Is Visual Space Euclidean?" Synthese. 35
(1977) 397-422.

6750. Swinburne, R. G. "Times." Analysis. 25 (1964-65) 185-91.

6751. _____. Space and Time. London: Macmillan, 1968. 319
pp.

a. Review essay by Paul Fitzgerald. PS 43 (1976) 618-37.

b. Review essay by Richard M. Gale. <u>J. Phil.</u> 67 (1970) 300-17.

6752. _____. "Conventionalism about Space and Time." BJPS 31 (1980) 255-72.

6753. Tati, Takao. "Concepts of Space-Time in Physical Theories -- Non-Spatio-Temporal Description and Universal Present." AJAPS 5, No. 2 (1977) 17-25.

6754. Taylor, J. G. "Time in Particle Physics." <u>Studium Generale</u>. 23 (1970) 1102-07.

6755. Thom, René. "L'espace et la réalité physique. Réflexions sur Mach et Einstein." <u>Epistemologia</u>. 4 (1981) 313-22.

6756. Thomason, Richmond H. "Indeterminist Time and Truth-Value Gaps." <u>Theoria</u>. 36 (1970) 264-81.

6757. Titze, H. "Gedanken über den absoluten Raum." PN 12 (1970) 184-89.

6758. _____. "Gedanken zur Gerichtetheit der Zeit." PN 18 (1980-81) 368-87.

6759. Törnebohm, Håkan. "Durations and Distances in Time." <u>Theoria</u>. 37 (1971) 209-26.

6760. Toulmin, Stephen, and June Goodfield. <u>The Discovery of Time</u>. New York: Harper & Row, 1965. 280 pp.

6761. Trautman, Andrzej. "Comparison of Newtonian and Relativistic Theories of Space-Time." <u>Organon</u>. 2 (1965) 123-29.

6762. Treder, H.-J. "Die Eigenschaften physikalischer Prozesse und die geometrische Struktur von Raum und Zeit." DZP 14 (1966) 562-65.

6763. Tummers, J. H. "La géométrie et le problème de l'espace." <u>Dialectica</u>. 16 (1962) 56-60.

6764. Vuillemin, Jules. "Poincaré's Philosophy of Space." <u>Synthese</u>. 24 (1972) 161-79. Reprinted in SL 56 (1973) 159-77.

6765. Watanabe, S. "Le concept de temps en physique moderne et la durée pure de Bergson." RMM 56 (1951) 128-42.

6766. _____. "Time and the Probabilistic View of the World," in <u>The Voices of Time</u>. J. T. Fraser, ed. New York: George Braziller, 1966. 527-63.

6767. _____. "Creative Time." <u>Studium Generale</u>. 23 (1970) 1057-87.

6768. Webb, C. W. "Can Space be Time-Like?" <u>J. Phil.</u> 74 (1977) 462-75.

6769. Weingard, Robert. "On Traveling Backward in Time." Synthese. 24 (1972) 117-32. Reprinted in SL 56 (1973) 115-30.

6770. _____. "On the Unity of Space." Phil. Studies. 29 (1976) 215-20.

6771. _____. "Space-Time and the Direction of Time." Nous. 11 (1977) 119-31.

6772. Wenzlaff, B. "Die physikalischen Eigenschaften von Raum und Zeit. DZP 11 (1963) 838-50.

6773. Weyl, Hermann. Space-Time-Matter. Henry L. Brose, tr. New York: Dover Publications, 1951. xviii + 330 pp.

6774. Wheeler, John A. "Curved Empty Space-Time as the Building Material of the Physical World." LMPS-1 (1960) 361-74.

6775. Whiteman, Michael. Philosophy of Space and Time and the Inner Constitution of Nature. London: George Allen & Unwin; New York: Humanities Press, 1967. 436 pp.

6776. Whitrow, G. J. "Why Physical Space Has Three Dimensions." BJPS 6 (1955) 13-31.

6777. _____. The Natural Philosophy of Time. London: Thomas Nelson and Sons, Ltd., 1961. 324 pp. Second Edition: New York: Oxford University Press, 1980. 400 pp.

6778. _____. "On the Nature of Time." RIP 16 (1962) 319-32.

6779. _____. "Time and the Universe," in The Voices of Time. J. T. Fraser, ed. New York: George Braziller, 1966. 564-81.

6780. _____. "Time and Cosmical Physics." Studium Generale. 23 (1970) 224-33.

6781. _____. What is Time? London: Thames & Hudson, 1972; New York: Holt, Rinehart & Winston, 1973. Reprinted under title The Nature of Time. Harmondsworth, Middlesex, England: Penguin Books, 1975. 153 pp.

6782. _____. "On the Impossibility of an Infinite Past." BJPS 29 (1978) 39-45.

    a. "On the Possibility of an Infinite Past: A Reply to Whitrow," by Karl Popper. 47-48.

    b. "The Infinite Past Regained: A Reply to Whitrow," by John Bell. 30 (1979) 161-65.

    c. "Whitrow and Popper on the Impossibility of an Infinite Past," by William Lane Craig. 165-70.

     d. "General Relativity and the Length of the Past," by
Robert Weingard. 170-72.

6783. Whyte, L. L. "The Electric Current. A Study of the Role of
Time in Electron Physics." BJPS 3 (1952) 243-55.

     a. "Has a Single Electron a Transit Time?" by L. L. Whyte
349-50.

6783A. _____. "Geodesics and the Space and Time of Physical
Observations." BJPS 4 (1954) 337-38.

6784. Williams, Donald C. "The Myth of Passage." J. Phil. 48
(1951) 457-72.

6785. Winnie, John A. "The Causal Theory of Space-Time." MSPS 8
(1977) 134-205.

6786. Wolfson, Paul, and James Woodward. "Scientific Explanation
and Sklar's Views of Space and Time." PS 46 (1979) 287-94.

6787. Wright, G. H. von. Time, Change and Contradiction. Cambridge: Cambridge University Press, 1969. 32 pp.

6788. Yanase, Michael M. "The Concept of Space-Time in Modern
Physics." Archives de Philosophie. 34 (1971) 667-72.

6789. _____. "On Aevum - between Time and Eternity." AJAPS
4, No. 5 (1975) 31-35.

6790. Yukawa, Hideki. "Elementary Particles and Space-Time Structure." AJAPS 1, No. 2 (1957) 91-100.

6791. Zaret, David. "Absolute Space and Conventionalism." BJPS 30
(1979) 211-26.

6792. _____. "A Limited Conventionalist Critique of Newtonian
Space-Time." PS 47 (1980) 474-94.

6793. Zeman, Jiři. Time in Science and Philosophy: An International Study of Some Current Problems. Amsterdam: Elsevier, 1971.
305 pp.

6794. Zenzen, Michael. "Popper, Grünbaum and de facto Irreversibility." BJPS 28 (1977) 313-24

6795. Zwart, P. J. "The Flow of Time." Synthese. 24 (1972)
133-60. Reprinted in SL 56 (1973) 131-58.

6796. _____. About Time. Amsterdam: North-Holland; New York:
American Elsevier, 1975. 266 pp.

7. Special Topics in the Philosophy of the Biological Sciences

7.1 Biological Sciences—General

6797. Afanasjew, W. G. "Ueber Bertalanffys 'organismische' Konzeption." DZP 10 (1962) 1033-46.

    a. "Die 'organismische Auffassung' Bertalanffys," by A. Bendmann. 11 (1963) 216-22.

6798. Agar, W. E. "The Wholeness of the Living Organism." PS 15 (1948) 179-91.

6799. Ageno, M. "L'oggetto della biologia: un problema insoluto." Scientia. 105 (1970) 193-209. French Tr.: Supplement, 69-83.

6800. Anokhin, P. K. "Philosophical Aspects of the Theory of a Functional System." Soviet Studies in Philosophy. 10 (1971-72) 269-76. [English reprint from Voprosy filosofii. (1971, No. 3).]

6801. Arber, Agnes. The Natural Philosophy of Plant Form. Cambridge University Press, 1950. xiv + 247 pp.

6802. _____. The Mind and the Eye. Cambridge: The University Press, 1954. 146 pp.

6803. _____. The Manifold and the One. London: John Murray, 1957. xiii + 146 pp.

6804. Ayala, Francisco J. "The Concept of Biological Progress," in Studies in the Philosophy of Biology. F. J. Ayala and T. Dobzhansky, eds. Berkeley: University of California Press, 1974. 339-55.

6805. Ayala, F. J., and T. Dobzhansky, eds. Studies in the Philosophy of Biology. London: Macmillan, 1974. xix + 390 pp.

6806. Bachem, A. "Heisenberg's Indeterminacy Principle and Life."
PS 19 (1952) 261-72.

6807. Bailey, Orville T. "Levels of Research in the Biological
Sciences." PS 12 (1945) 1-7.

6808. Barthélemy-Madaule, Madeleine. L'idéologie du hasard et de
la nécessité. Paris: du Seuil, 1972. 223 pp.

6809. Bawden, H. Heath. "The Psychical as a Biological Directive."
PS 14 (1947) 56-67.

     a. "Biological Directiveness and the Psychical: A Note," by
     Ralph S. Lillie. 266-68.

     b. "The 'Psychical' as Secondary and as Secret," by Ralph
     Gregory. 15 (1948) 76-79.

6810. Beck, William S. Modern Science and the Nature of Life. New
York: Harcourt, Brace, 1957. 302 pp.

6811. Beckner, Morton. The Biological Way of Thought. New York:
Columbia University Press, 1959. viii + 200 pp.

6812. _____. "Metaphysical Presuppositions and the Description
of Biological Systems," in Form and Strategy in Science. J. R.
Gregg and F. T. C. Harris, eds. Dordrecht: Reidel, 1964. 15-29.
Reprinted from Synthese. 15 (1963) 260-74.

6813. _____. "Aspects of Explanation in Biological Theory,"
in Philosophy of Science Today. S. Morgenbesser, ed. New York:
Basic Books, 1967. 148-59.

6814. Beigbeder, Marc. Le contre-Monod. Paris: Grasset, 1970.
351 pp.

6815. Bernier, R., and P. Pirlot. Organe et fonction: Essai de
biophilosophie. Paris: Maloine-Doin-Edisem, 1977. 153 pp.

6816. Bertalanffy, Ludwig von. Das biologische Weltbild. Band I:
Die Stellung des Lebens in Natur und Wissenschaft. Bern: A. Franke
Ag. Verlag, 1949. 202 pp.

6817. _____. "An Outline of General System Theory." BJPS 1
(1950) 134-65.

6818. _____. Theoretische Biologie. Bern: A. Francke Ag.,
1951. 418 pp.

6819. _____. Problems of Life -- An Evaluation of Modern
Biological Thought. London: C. A. Watts & Co., 1952. ix + 216 pp.
Reprint: New York: Harper & Brothers, 1960.

     a. Review essay by T. A. Goudge. Review of Metaphysics. 7
     (1953) 282-89.

6820. _____. Biophysik des Fliessgleichgewichts. Braunschweig: Sammlung Vieweg, Heft 124, 1953. iv + 56 pp.

6821. _____. "The Biophysics of the Steady State of the Organism." Scientia. 89 (1954) 361-65. French Tr.: Supplement, 166-70.

6822. _____. Robots, Men and Minds. New York: George Braziller, 1967. x + 150 pp.

6823. _____. General System Theory. New York: G. Braziller, 1968. xv + 289 pp.

    a. Review essay by M. A. Abell. RMM 83 (1978) 308-32.

6824. _____. "The Model of Open Systems: Beyond Molecular Biology," in Biology, History, and Natural Philosophy. A. D. Breck and W. Yourgrau, eds. New York: Plenum, 1972. 17-30.

6825. Blandino, S.J., G. Theories on the Nature of Life. D. Olsoufieff, tr. New York: Philosophical Library, 1969. xiii + 374 pp.

6826. Blum, H. F. "Complexity and Organization," in Form and Strategy in Science. J. R. Gregg and F. T. C. Harris, eds. Dordrecht: Reidel, 1964. 59-65. Reprinted from Synthese. 15 (1963) 115-21.

6827. Boden, Margaret. "The Case for a Cognitive Biology." Arist. Soc. Supp. Vol. 54 (1980) 25-49.

6828. Böhm, Walter. "Ueber die Anwendbarkeit des Technomorphismus in der Biologie." Philosophisches Jahrbuch. 61 (1951) 201-14.

6829. Bonhoeffer, Karl Friedrich. "Ueber physikalisch-chemische Modelle von Lebensvorgängen." Studium Generale. 1 (1947-48) 137-43.

6830. Bonin, Gerhardt von. "Types and Similitudes: A Enquiry into the Logic of Comparative Anatomy." PS 13 (1946) 196-202.

6831. Bronowski, J. "New Concepts in the Evolution of Complexity, Stratified Stability and Unbounded Plans." Zygon. 5 (1970) 18-35. Also in Synthese. 21 (1970) 228-46; and in BSPS 11, SL 58 (1974) 133-51.

    a. "Commentary," by R. W. Burhoe. Zygon. 5 (1970) 36-40.

6832. Buican, D. "Microphénomène et philosophie de la biologie moderne." Scientia. 109 (1974) 335-54. English Tr.: 355-71.

6833. Bünning, E. "Der Lebensbegriff in der Physiologie." Studium Generale. 12 (1959) 127-33.

6834. Burch, George B. "The Nature of Life." Review of Metaphysics. 5 (1951) 1-10.

6835. Butler, J. A. V. The Life Process. New York: Basic Books,
1972. 256 pp.

6836. Caldirola, Piero. "Ruolo della complessità nella descrizione
fisica dei sistemi biologici." Scientia. 116 (1981) 5-14. English
Tr.: 15-21.

6837. Canguilhem, Georges. "Le tout et la partie dans la pensée
biologique." Les Etudes Philosophiques. 21 (1966) 3-16.

6838. Caplan, Arthur L., ed. The Sociobiology Debate: Readings
on Ethical and Scientific Issues. New York: Harper & Row, 1978.
514 pp.

6839. Carles, S.J., Jules. "Le problème de l'unité de la vie."
Archives de Philosophie. 17, No. 2 (1948) 62-83.

6840. _____. "L'unité du vivant." Les Etudes Philosophiques.
21 (1966) 29-42.

6841. Caspari, Ernst. "On the Conceptual Basis of the Biological
Sciences," in Frontiers of Science and Philosophy. Robert G. Colodny,
ed. Pittsburgh: University of Pittsburgh Press, 1962. 131-45.

6842. Chedd, Graham. The New Biology. New York: Basic Books,
1972. xiv + 306 pp.

6843. Cochran, A. "Life and the Wave Properties of Matter."
Dialectica. 19 (1965) 290-312.

6844. Cramer, F. "Macht und Verantwortung der biologischen Forschung." Studium Generale. 22 (1969) 481-93.

6845. Davison, John. "Animal Organization as a Problem in Cell
Form," in Form and Strategy in Science. J. R. Gregg and F. T. C.
Harris, eds. Dordrecht: Reidel, 1964. 363-77.

6846. Dobzhansky, Theodosius. "On Cartesian and Darwinian Aspects
of Biology," in Philosophy, Science, and Method. S. Morgenbesser et
al, eds. New York: St. Martin's Press, 1969. 165-78.

6847. Dubois, Georges. "Originalité et finalité des êtres vivants."
Studia Philosophica. 16 (1946) 108-29.

6848. Dubouchet, Jeanne. "Information biologique et entropie."
Archives de Philosophie. 38 (1975) 79-119.

6849. Dubuisson, M. "Biologie moderne." Scientia. 84 (1949)
126-31.

6850. Duchesneau, F. "Définition de l'organisation et théorie
cellulaire." Revue Phil. 165 (1975) 401-30.

6851. _____. "Analyse fonctionnelle et causalité biologique."
RIP 131-132 (1980) 229-67.

6852. Echarri, Jaime. "Azar y necesidad en la filosofía de la vida de J. Monod. In Memoriam." Pensamiento. 33 (1977) 5-33.

6853. Edelman, Gerald M. "The Problem of Molecular Recognition by a Selective System," in Studies in the Philosophy of Biology. F. J. Ayala and T. Dobzhansky, eds. Berkeley: University of California Press, 1974. 45-56.

6854. Ehrenberg, Rudolf. Metabiologie. Heidelberg: Lambert Schneider, 1950. 341 pp.

    a. Review essay by W. Böhm. Philosophisches Jahrbuch. 61 (1951) 245-52.

6855. Elsasser, Walter M. Atom and Organism. Princeton, NJ: Princeton University Press, 1966. ix + 143 pp.

6856. _____. The Chief Abstractions of Biology. Amsterdam and New York: North-Holland and American Elsevier, 1975. xiv + 261 pp.

6857. Fantoli, Annibale. "Jacques Monod and the Natural Philosophy of Modern Biology: Some Epistemological Considerations. AJAPS 4, No. 4 (1974) 23-40.

6858. Favarger, C. "Vérité scientifique et compréhension du vivant: Reflexions d'un naturaliste." Dialectica. 12 (1958) 37-67.

    a. "Deux remarques à propos de l'article de M. C. Favarger," by F. Gonseth. 68-69.

6859. Fischer, Hans. "Einheit und Methode in der Biologie." Synthese. 6 (1947-48) 412-25.

6860. Foerster, H. von, ed. Cybernetics: Circular, Causal and Feedback Mechanisms in Biological and Social Systems. New York: Josiah Macy, Jr. Foundation, 1952. 240 pp.

6861. Förster, H. "Allgemeine biologische Gesetzlichkeiten im Vergleich zu physikalischen Gesetzen." Studium Generale. 7 (1954) 40-45.

6862. Fothergill, Philip G. Life and Its Origin. London: Sheed and Ward, 1958. 70 pp.

6863. Fouchet, J. "Le déterminisme peut-il prétendre à une description complète des phenomènes biologiques?" Revue Phil. 148 (1958) 48-70.

6864. Frolow, I. T. "Probleme der Methodologie in der biologischen Forschung." DZP 12 (1964) 1444-59.

6865. _____. "Leben und Erkenntnis." DZP 20 (1972) 564-81.

6866. _____. "The Nature of Contemporary Biological Knowledge." Soviet Studies in Philosophy. 12 (1973-74) 27-49. [English reprint from Voprosy filosofii. (1972, No. 11).]

6867. Fuchs-Kittowski, K., S. M. Rapaport, H.-A. Rosenthal, and G. Wintgen. "Zur Dialektik von Notwendigkeit und Zufall in der Molekularbiologie." DZP 20 (1972) 418-43.

6868. Gierer, A. "The Physical Foundations of Biology and the Problem of Psychophysics." Ratio 12 (1970) 47-64.

6869. Glucksmann, Alfred. "Monod's Conception of 'The Cell' and of 'Teleonomic Systems'." Theoria to Theory. 10 (1977) 203-10.

6870. Goodfield, June. "Theories and Hypotheses in Biology: Theoretical Entities and Functional Explanation." BSPS 5 (1969) 421-49.

   a. "Comments on 'Theories and Hypotheses in Biology'," by Ernst Mayr, 450-56.

   b. "Comments: Theoretical Entities versus Theories," by Joseph Agassi, 457-59.

6871. Gould, R. P. "The Place of Historical Statements in Biology." BJPS 8 (1957) 192-210.

6872. Gould, Stephen Jay. "D'Arcy Thompson and the Science of Form." New Literary History, II, No. 2. (1971). Reprinted in BSPS 27 (1976) 66-97.

6873. Grégoire, Franz. "Note sur la philosophie de l'organisme." Revue Philosophique de Louvain. 46 (1948) 275-334.

6874. Grene, Marjorie. "On Some Distinctions Between Men and Brutes." Ethics. 57 (1947) 121-27. Reprinted in BSPS 23, SL 66 (1974) 243-53.

6875. _____. "Bohm's Metaphysics and Biology," in Towards a Theoretical Biology. II: Sketches. C. H. Waddington, ed. Edinburgh: Edinburgh University Press, 1969. 61-69. Reprinted in BSPS 23, SL 66 (1974) 180-88.

6876. _____. Approaches to a Philosophical Biology. New York: Basic Books, 1969. ix + 295 pp.

6877. _____. "People and Other Animals." Philosophical Forum. 3 (1972) 157-72. Reprinted in PN 14 (1973) 25-38, and in BSPS 23, SL 66 (1974) 346-60.

6878. _____. "Aristotle and Modern Biology." Journal of the History of Ideas. 33 (1972) 395-424. Reprinted in BSPS 23, SL 66 (1974) 74-107, and in BSPS 27 (1976) 3-36.

6879. _____. The Understanding of Nature: Essays in the Philosophy of Biology. BSPS 23. SL 66. Dordrecht: Reidel, 1974. 360 pp.

6880. Grene, Marjorie and Everett Mendelsohn, eds. Topics in the
Philosophy of Biology. BSPS 27. SI 84. Dordrecht: Reidel, 1976.
454 pp.

a. Review essay, "Philosophy of Biology Today: No Grounds
for Complacency," by Michael Ruse. Philosophia. 8 (1979)
785-96.

6881. Grobstein, Clifford. "Organizational Levels and Explanation."
BSPS 27 (1976) 145-52.

6882. Gunter, Pete A. Y. "Biological Time and Biological Mechanism:
Reflections on the 'New Embryology'." Southwestern Journal of Philosophy. 2, No. 1 (1971) 173-83.

6883. Haefner, James W. "Two Metaphors of the Niche." Synthese.
43 (1980) 123-54.

6884. Harris, F. T. C. "A Representation of Animal Growth," in
Form and Strategy in Science. J. R. Gregg and F. T. C. Harris, eds.
Dordrecht: Reidel, 1964. 235-50.

6885. Herrmann, Heinz. "The Unity of the Morphological and Functional Aspects of Living Matter." PS 14 (1947) 254-60.

6886. _____. "An Account of Recent Biological Methodology:
Causal Law and Transplanar Analysis." PS 20 (1953) 149-56.

6887. _____. "Biological Field Phenomena: Facts and Concepts,"
in Form and Strategy in Science. J. R. Gregg and F. T. C. Harris,
eds. Dordrecht: Reidel, 1964. 343-62.

6888. Hocevar, H. E. "Die Seinsstruktur der Pflanzen." PN 15
(1974-75) 15-65.

6889. Hogben, Lancelot. Science in Authority. New York: W. W.
Norton, 1963. 157 pp.

6890. Holmes, S. J. "Micromerism in Biological Theory." Isis. 39
(1948) 145-58.

6891. Hull, David L. "What Philosophy of Biology is Not." Synthese.
20 (1969) 157-84.

6892. _____. Philosophy of Biological Science. Englewood
Cliffs, NJ: Prentice-Hall, 1974. 148 pp.

6893. _____. "Philosophy of Biology," in Current Research
in Philosophy of Science. P. D. Asquith and H. E. Kyburg, Jr., eds.
East Lansing, MI: Philosophy of Science Assn., 1979. 421-35.

6894. Hutten, Ernest H. "Physics and Biology." BJPS 11 (1960)
101-08.

6895. Ingle, Dwight J. Principles of Research in Biology and Medicine. Philadelphia: J. B. Lippincott, 1958. 123 pp.

6896. Jardine, N. "The Concept of Homology in Biology." BJPS 18 (1967) 125-39.

6897. Jonas, Hans. The Phenomenon of Life: Toward a Philosophical Biology. New York: Harper & Row, 1966. x + 303 pp.

   a. "Review essay: "Existentialism and Biology," by T. A. Goudge. Dialogue. 5 (1967) 603-08.

6898. Kahane, E. "La maîtrise du hasard et les sciences de la vie." RMM 73 (1968) 133-48.

6899. Kaiser, H., and W. Voigt. "Probleme einer allgemeinen oder theoretischen Biologie." DZP 15 (1967) 435-45.

6900. Kamarýt, J. "Die Bedeutung der Theorie des offenen Systems in der gegenwärtigen Biologie." DZP 9 (1961) 2040-59.

6901. Kapp, R. O. "Living and Lifeless Machines." BJPS 5 (1954) 91-103.

6902. Kauffman, Stuart A. "Articulation of Parts Explanation in Biology and the Rational Search for Them." BSPS 8 (1971) 257-72. Reprinted in BSPS 27 (1976) 245-63.

6903. _____. "Constraints on the Sociobiologists' Program." PSA 1976. II, 32-47.

6904. Kempermann, Carl Theo. "Zur derzeitigen Situation der Morphologie." PN 2 (1954) 479-500, 3 (1955) 173-93, 339-60.

6905. Klatt, B. "Gedanken zur Zoologie als einer theoretischen Wissenschaft." Studium Generale. 7 (1954) 1-13.

6906. Klein, M. "Sur les résonances de la philosophie de la nature en biologie moderne et contemporaine." Revue Phil. 144 (1954) 514-43.

6907. Kleiner, Scott A. "Essay Review: The Philosophy of Biology." Southern Journal of Philosophy. 13 (1975) 523-42.

6908. Klohr, Olof. "Besonderheiten der Gesetze der Biologie." DZP 5 (1957) 313-26.

6909. Konecsni, Johnemery. Biology and the Philosophy of Science. Washington: University Press of America, 1977. viii + 170 pp.

6910. Körner, U. "Zur Bestimmung des naturwissenschaftlichen Begriffs Leben und Fragen des Begreifens von Entwicklung." DZP 18 (1970) 960-79.

6911. Kremiansky, V. I. "Hyperstructures and 'Infra'-Systems of Organized and Organizing Information in Biology." LMPS-4 (1971) 637-45.

6912. Kruseman, W. M. "Aspecten der moderne Biologie in Nederland." Synthese. 5 (1946) 174-82.

6913. _____. "Gradation of Language in Biological Systematics." Synthese. 8 (1950-51) 175-81.

6914. Kuźnicki, L. "On the Development of the Greatest Unifying Theories in Biology." Organon. 3 (1966) 95-103.

6915. Lagerspetz, Kari Y. H. "Individuality and Creativity: Is Biology Different?" Synthese. 20 (1969) 254-60.

6916. Läsker, L. "Ist die moderne Biologie mechanistisch?" DZP 11 (1963) 184-96.

6917. _____. "Erkenntnistheoretische Probleme der modernen Zellphysiologie." DZP 11 (1963) 1389-1400.

6918. Lehmann, F. E. "Objekt und Methode in der Biologie." Synthese. 6 (1947) 44-56.

6919. Lewontin, R. C. "Sociobiology - A Caricature of Darwinism." PSA 1976. II, 22-31.

6920. Ley, H. "Jacques Monod und die Relevanz von Kategorien." DZP 20 (1972) 681-96.

6921. Lillie, Ralph S. "Philosophy of Organism: A Rejoinder to Professor Werkmeister." PPR 8 (1948) 706-11.

6922. _____. "Some Aspects of Theoretical Biology." PS 15 (1948) 118-34.

6923. Lindenmayer, Aristid. "Life Cycles as Hierarchical Relations," in Form and Strategy in Science. J. R. Gregg and F. T. C. Harris, eds. Dordrecht: Reidel, 1964. 416-70.

6924. _____. "Theories and Observations of Developmental Biology." WOSPS 10 (1977) 103-18.

6925. Löther, R. "Philosophische Probleme der Biologie." DZP 14 (1966) 315-27.

6926. Ludwig, W. "Probleme und Aufgaben der Biomathematik." Studium Generale. 6 (1953) 637-46.

6927. MacIver, Robert M. Life: Its Dimensions and Its Bounds. New York: Harper, 1960. 144 pp.

6927A. McDowell, Sister Margaret Ann. The Rythmic Universe." The Thomist. 24 (1961) 502-18. Reprinted in The Dignity of Science. James A. Weisheipl, ed. Baltimore: The Thomist Press, 1961. 366-82.

6927B. McIntosh, Robert P. "The Background and Some Current Problems of Theoretical Ecology." Synthese. 43 (1980) 195-256.

6928. McShane, S.J., Philip. "Insight and the Strategy of Biology." Continuum. 2 (1964) 374-88.

6929. Macovschi, E. "Les bases philosophiques de la conception biostructurale." Noesis. 3 (1975) 291-99.

6930. Mainx, Felix. Foundations of Biology. (International Encyclopedia of Unified Science. Vol. I, No. 9.) Chicago: The University of Chicago Press, 1955. ii + 86 pp.

6931. Manier, Edward. "'Fitness' and Some Explanatory Patterns in Biology." Synthese. 20 (1969) 206-18.

6932. _____. "Functionalism and the Negative Feedback Model in Biology." BSPS 8, SL 39 (1971) 225-40.

6933. Mannoury, G. "Sociobiology." Synthese. 5 (1947) 522-25.

6934. María de Alejandro, José. "Sobre el azar y la necesidad." Pensamiento. 28 (1972) 387-411.

6935. May, Eduard. "Das Vitalismusproblem und die Erklärung der Lebensphänomene." PN 2 (1952) 251-57.

6936. Mayr, Ernst. "Footnotes on the Philosophy of Biology." PS 36 (1969) 197-202.

6937. Medawar, Peter B. The Future of Man. New York: Basic Books, 1960. 128 pp.

6938. Mendelsohn, Everett. "Physical Models and Physiological Concepts: Explanation in Nineteenth Century Biology." BSPS 2 (1965) 127-50.

    a. "Comments," by Ernst Mayr, 151-55.

6939. _____. "Philosophical Biology versus Experimental Biology," in Actes, Tome I, B, Discours et Conferences Colloques, Xll^e Congres Internationale d'Histoire des Sciences. Paris: 1968. Reprinted in BSPS 27 (1976) 37-65.

6940. Mertz, David B., and David E. McCauley. "The Domain of Laboratory Ecology." Synthese. 43 (1980) 95-110.

6941. Milos, Plamenac. "Biophysical Analysis of Vital Force of Living Matter." PN 12 (1970) 440-45.

6942. Mocek, Reinhard. "Theoretische Biologie und Naturphilosophie im ideologischen Klassenkampf." DZP 8 (1960) 408-32.

6943. Monod, Jacques. Chance and Necessity: An Essay on the Natural Philosophy of Modern Biology. New York: Knopf, 1971. xiv + 198 pp.

a. Review essay, "Peut-on justifier une métaphysique par les résultats des sciences?" by G. Malicot and J. Parain-Vial. Les Etudes philosophiques. 26 (1971) 483-94.

b. Review essay, "Un radicalismo que pudo ser esperanzador," by F. Riaza. Pensamiento. 28 (1972) 5-27.

c. Review essay by Ana Katz. RMM 79 (1974) 118-25.

d. "Quelques notes et questions sur Le hasard et la nécessité de Jacques Monod (Seuil, 1970)," by Germain Dandenault. Dialogue. 13 (1974) 355-61.

6944. _____. "On Chance and Necessity," in Studies in the Philosophy of Biology. F. J. Ayala and T. Dobzhansky, eds. Berkeley: University of California Press, 1974. 357-75.

6945. Moser, Simon. "Der Begriff des Lebens (einführende Bemerkung)." PN 6 (1960-61) 125.

a. "Was ist Leben?," by Max Hartmann. 125-40.

b. "Zur Frage nach dem Ursprung des Lebens," by H. Friedrich-Freksa. 141-44.

c. "Die Historizität als Wesenszug des Lebendigen," by Gerhard Heberer. 145-52.

d. "Zum 'Begriff des Lebens' - Eine Rückbesinnung," by S. Moser. 153-59.

c. "Stellungnahme zu der 'Rückbesinnung' von Simon Moser," by M. Hartmann, 160-61; by H. Friedrich-Freksa, 162-64; by G. Heberer, 165-66.

6946. Munson, Ronald, ed. Man and Nature: Philosophical Issues in Biology. New York: Delta, 1971. xxi + 413 pp.

6947. _____. "Is Biology a Provincial Science?" PS 42 (1975) 428-47.

6948. Needham, J. Order and Life. Cambridge, MA: M. I. T. Press, 1968. xvii + 175 pp.

6949. Nowiński, C. "Jacques Monod: Chance and Necessity." Dialectics and Humanism. 1, No. 4 (1974) 137-49.

6950. _____. "Biologische Gesetze und dialektische Methode." DZP 23 (1975) 926-37.

6951. Núñez de Castro, Ignacio. "Epistemología de la bioquímica y biología molecular." Pensamiento. 35 (1980) 425-35.

6952. Parsegian, V. L. "Biological Trends within Cosmic Processes." Zygon. 8 (1973) 221-43.

6953. Partashnikov, Anatoly. "Soviet Philosophy of Biology Today."
Studies in Soviet Thought. 14 (1974) 1-25.

6954. Pattee, H. H. "Physical Theories of Biological Co-ordination."
BSPS 27 (1976) 153-73.

6955. Patten, Bernard C., and Gregor T. Auble. "Systems Approach
to the Concept of Niche." Synthese. 43 (1980) 155-82.

6956. Peters, Hans M. "Sociomorphic Models in Biology." Ratio. 3
(1960) 26-43.

6957. Peters, Robert Henry. "Useful Concepts for Predictive Ecology." Synthese. 43 (1980) 257-70.

6958. Piaget, Jean. Biologie et connaissance: Essai sur les relations entre les régulations organique et les processus cognitifs.
Paris: Editions Gallimard, 1967. Tr. by B. Walsh: Biology and
Knowledge: An Essay on the Relations between Organic Regulations
and Cognitive Processes. Chicago: University of Chicago Press,
1971. 384 pp.

6959. Piattelli-Palmarini, M. "Equilibria, Crystals, Programs,
Energetic Models, and Organizational Models." BSPS 47 (1981) 341-59.

6960. Pilet, P. E. "L'idonéisme et la recherche en biologie." RIP
24 (1970) 494-507.

6961. _____. "Quelques aspects méthodologiques de l'information biologique." Dialectica. 25 (1971) 3-15.

6962. _____. "La pluridisciplinarité en biologie: recherche
de base et applications." Scientia. 116 (1981) 609-20. Italian
Tr.: 621-28. English Tr.: 629-36.

6963. Pirlot, Paul. "Organicisme en biologie et en psychologie."
Dialogue. 9 (1970) 303-36.

6964. Pirlot, Paul, and Rejane Bernier. "Preliminary Remarks on
the Organ-Function Relation," in The Methodological Unity of Science.
M. Bunge, ed. Dordrecht: Reidel, 1973. 71-83.

6965. Plochmann, George K. "D'Arcy Thompson: His Conception of
the Living Body." PS 20 (1953) 139-48.

6966. Polanyi, Michael. "Life's Irreducible Structure." Science.
160 (1968) 1308-12. Reprinted in Knowing and Being. M. Grene, ed.
Chicago: University of Chicago Press, 1969. 225-39.

6967. Pollard, Ernest C. "Are Life Processes Governed by Physical
Laws?" in Philosophy of Science: The Delaware Seminar, II. New
York: Interscience Publishers, 1963. 395-410.

6968. Portmann, Adolf. "Naturaufassung und Menschenbild in der
modernen Biologie." PN 7 (1961-62) 251-65.

6969. Powers, William T. "Biological Research and Catholic Philosophy." Proc. Am. Cath. Phil. Assn. 22 (1947) 172-76.

6970. Prado, C. G. "Sociobiology and Materialist Theories of Mind." Dialogue. 20 (1981) 247-68.

6971. Prat, Naftali. "Diamat and Contemporary Biology." Studies in Soviet Thought. 21 (1980) 181-209.

6972. Purton, A. C. "Biological Function." Phil. Quart. 29 (1979) 10-24.

6973. Rashevsky, N. "The Devious Roads of Science," in Form and Strategy in Science. J. R. Gregg and F. T. C. Harris, eds. Dordrecht: Reidel, 1964. 51-58.

6974. _____. "A Unified Approach to Biological and Social Organisms." LMPS-3 (1967) 403-12.

6975. Raven, Chr. P. "Formalization of the Fundamental Concepts in Some Fields of Biology." Synthese. 7 (1948-49) 93-99.

6976. _____. "Irrational Elements in Some Theories of Life." Synthese. 10 (n.d.) 359-63.

6977. Rensch, Bernhard. "Polynomistic Determination of Biological Processes," in Studies in the Philosophy of Biology. F. J. Ayala and T. Dobzhansky, eds. Berkeley: University of California Press, 1974. 241-58.

6978. Reutterer, Alois. "Die Auferstehung des Vitalismus." Conceptus. 8, No. 25 (1974) 63-71.

6979. Rochhausen, R. "Einige Probleme der modernen Biologie im Lichte des dialektisch-materialistischen Determinismus." DZP 9 (1961) 66-87.

6980. Rosenberg, Alexander. "The Supervenience of Biological Concepts." PS 45 (1978) 368-86.

6981. _____. Sociobiology and the Preemption of Social Science. Baltimore: The Johns Hopkins University Press, 1980. 176 pp.

6982. Rostand, Jean. Error and Deception in Science: Essays on Biological Aspects of Life. A. J. Pomerans, tr. New York: Basic Books, 1960. 196 pp.

6983. Ruse, Michael E. "Are There Laws in Biolgy?" AJP 48 (1970) 234-46.

6984. _____. "Two Biological Revolutions." Dialectica. 25 (1971) 17-38.

6985. _____. The Philosophy of Biology. London: Hutchinson University Library, 1973. 231 pp.

a. Review essay by Ronald Munson. SHPS 5 (1974) 73-85.

b. "Functionalism and the Possbility of Group Selection," by
Vernon Pratt. 5 (1975) 371-72.

c. Review essay by Nils Roll-Hansen. Inquiry. 17 (1974)
131-42.

d. Review essay by Peter Achinstein. Canadian Journal of
Philosophy. 4 (1975) 745-54.

e. Review essay, "A Logical Empiricist Looks at Biology," by
David Hull. BJPS 28 (1977) 181-89.

6986. _____. "Sociobiology: Sound Science or Muddled Metaphysics?" PSA 1976. II, 48-73.

6987. _____. Sociobiology: Sense or Nonsense? Dordrecht:
Reidel, 1979. xi + 231 pp.

a. Review essay by J. L. Mackie. Erkenntnis. 15 (1980)
189-94.

b. Review essay by Paul Thagard. Canadian Journal of Philosophy. 11 (1981) 751-59.

6988. Russell, E. S. "The 'Drive' Element in Life." BJPS 1 (1950)
108-16.

6989. Ruyer, R. "La psychobiologie et la science." Dialectica.
13 (1959) 103-22.

6990. _____. "Les deux types de dés-information en biologie."
Les Etudes philosophiques. 15 (1960) 341-52.

6991. Săhleanu, Victor. "Structuralismes biologiques et structuralismes 'biologisants'." Noesis. 5 (1979) 187-91.

6992. Schilling, Kurt. "Das Lebensproblem in Philosophie und Biologie." PN 1 (1952) 532-52.

6993. Schubert-Soldern, R. von. "Das chemodynamische Lebensprinzip."
Philosophisches Jahrbuch. 61 (1951) 146-57.

6994. Schurig, V. "Gegenstand und Theorien der Biologie im dialektischen Denken." PN 13 (1971-72) 316-40.

6995. Shapere, Dudley. "On the Relations Between Compositional and
Evolutionary Theories," in Studies in the Philosophy of Biology.
F. J. Ayala and T. Dobzhansky, eds. Berkeley: University of Califor-
nia Press, 1974. 187-204.

6996. Sheldrake, Rupert. "Three Approaches to Biology. I: The
Mechanistic Theory of Life. II: Vitalism. III: The Organismic
Approach." Theoria to Theory. 14 (1980-81) 125-44; 227-40; 301-11.

6997. Sherrington, Sir Charles. Man on his Nature. Harmondsworth, Middlesex: Penguin Books, 1955. 312 pp.

6998. Simberloff, Daniel. "A Succession of Paradigms in Ecology: Essentialism to Materialism and Probabilism." Synthese. 43 (1980) 3-40.

    a. "A Note on Simberloff's 'Succession of Paradigms in Ecology'," by Marjorie Grene. 41-46.

    b. "Dialectics and Reductionism in Ecology," by Richard Levins and Richard Lewontin. 47-78.

    c. "Reply," by Daniel Simberloff. 79-94.

6999. Simon, Michael A. The Matter of Life: Philosophical Problems of Biology. New Haven: Yale University Press, 1971. xi + 258.

7000. Simpson, George Gaylord. "Biology and the Nature of Science." Science. 139 (1963) 81-88.

7001. Sinnott, Edmund Ware. Life and Mind. Yellow Springs, Ohio: Antioch Press, 1957. 29 pp.

    a. "Review essay, "Sinnott's Philosophy of Purpose," by David L. Miller. Review of Metaphysics. 11 (1958) 637-47.

    b. Review essay, "Sinnott's Philosophy of Organism," by Alfred P. Stiernotte. 12 (1959) 654-61.

7002. _____. The Bridge of Life: From Matter to Spirit. New York: Simon and Schuster, 1966. 255 pp.

7003. Skiebe, K. "Ueberlegungen zur Gesetzesproblematik in der Biologie." DZP 23 (1975) 946-54.

7004. Smart, J. J. C. "Can Biology be an Exact Science?" Synthese. 11 (1959) 359-68.

7005. Smith, C. U. M. The Problem of Life: An Essay in the Origins of Biological Thought. New York: John Wiley, 1976. 343 pp.

7006. Smith, Edward T. "The Vitalism of Hans Driesch." Thomist. 18 (1955) 186-227.

7007. Smith, Vincent E., ed. Philosophy of Biology. New York: St. John's University Press, 1962. x + 95 pp.

7008. Soran, Viorel, and Ana P. Fabian. "Des processus quantiques impliqués dans la genèse et la fonctionnalité des systèmes vivants." Noesis. 7 (1981) 147-53.

7009. Southwood, FRS, T.R.E. "Ecology - A Mixture of Pattern and Probablism." Synthese. 43 (1980) 111-22.

7010. Stengers, Isabelle. "La description de Kuhn et son application à la biologie contemporaine." Annales de l'Institut de Philosophie. (Brussels) (1973) 179-226.

7011. Stent, Gunther S. Paradoxes of Progress. San Francisco: W.
H. Freeman, 1978. xii + 231.

7012. Strong, Jr., Donald R. "Null Hypotheses in Ecology." Synthese.
43 (1980) 271-86.

7013. Sussman, Héctor J., and Raphael S. Zahler. "Catastrophe
Theory as Applied to the Social and Biological Sciences: A Critique."
Synthese. 37 (1978) 117-216.

7014. Szarski, H. "The Explanation of Facts in Biological Sciences."
Scientia. 95 (1960) 17-21. French Tr.: Supplement, 9-13.

7015. Szent-Györgyi, Albert. "Electronic Mobility in Biological
Processes," in Biology, History, and Natural Philosophy. A. D.
Breck and W. Yourgrau, eds. New York: Plenum, 1972. 31-36.

7016. Thomas, E. "Zu methodologischen Fragen der Arbeit mit Fer-
mentmodellen." DZP 10 (1962) 1308-17.

7017. Thomas, Lewis. The Lives of a Cell: Notes of a Biology
Watcher. New York: Viking Press, 1974. 153 pp.

7018. Thompson, R. Paul. "Is Sociobiology a Pseudoscience?" PSA
1980. I, 363-70.

7019. Thum, P. G. "Indeterminismo fisico e autonomia della vita."
Scientia. 85 (1948) 199-208. French Tr.: Supplement, 85-94.

7020. Timoféeff-Ressovsky, N. W., and K. G. Zimmer. Das Treffer-
prinzip in der Biologie. Leipzig, 1947.

        a. "Logische Bermerkungen zu gewissen Prinzipien der Quant-
        enbiologie," by Max Bense. ZPF 3 (1948) 106-12.

7021. Tomlin, E. W. F. "The Concept of Life." Heythrop Journal. 18
(1977) 289-304.

7022. Troll, Wilhelm, and Anneliese Meister. "Wesen und Aufgabe
der Biosystematik im ontologischer Betractung." Philosophisches
Jahrbuch. 61 (1951) 105-31.

7023. Urbani, Enrico. "Biologia e Bionica, realtà e prospettive."
Scientia. 114 (1979) 69-91. English Tr.: 93-106.

7024. Vácha, J. "Biology and the Problem of Normality." Scientia.
113 (1978) 823-46. Italian Tr.: 847-65.

7025. Vossius, G. "Die Anwendung der Systemtheorie in der Biologie."
Studium Generale. 18 (1965) 306-13.

7026. Waddington, C. H. The Nature of Life: The Main Problems and Trends of Thought in Modern Biology. London: George Allen & Unwin, 1961; New York: Atheneum, 1962. Reprint: New York: Harper & Row, 1966. 133 pp.

7027. _____. Towards a Theoretical Biology. I: Prolegomena. II: Sketches. Edinburgh: Edinburgh University Press, 1969.

7028. Weber, H. "Grenzen in der Biologie." Studium Generale. 5 (1952) 356-63.

7029. Weill, Robert. "Problèmes d'unité et d'identité en biologie: l'individu, l'individualité, la personnalité." Les Etudes philosophiques. 21 (1966) 17-28.

7030. Weiss, Paul A. "The Living System: Determinism Stratified," in Beyond Reductionism. A. Koestler and J. R. Smythies, eds. New York: Macmillan, 1970. 3-55. Also in Studium Generale. 22 (1969) 361-400.

7031. Werkmeister, W. H. "Some Philosophical Implications of the Life Sciences." Personalist. 35 (1954) 117-27.

7032. Whyte, Lancelot Law. The Unitary Principle in Physics and Biology. New York: Henry Holt, 1949. ix + 182 pp.

    a. Review essay: "Biology and Unitary Principle," by Ralph S. Lillie. PS 18 (1951) 193-207.

7033. _____. Aspects of Form. London: Percy Lund Humphries. 1951. ix + 249 pp.

    a. Review essay, "The World of Form," by W. H. Thorpe. BJPS 2 (1952) 318-22.

7034. _____. "Note on the Structural Philosophy of Organism." BJPS 5 (1954) 332-34.

7035. _____. "On the Relation of Physical Laws to the Processes of Organisms." BJPS 7 (1957) 347-50.

7036. Wilkie, J. S. "Causation and Explanation in Theoretical Biology." BJPS 1 (1951) 273-90.

7037. Williams, Roger J. Biochemical Individuality: The Basis for the Genetotrophic Concept. New York: John Wiley, 1946. 214 pp.

7038. Wilson, Edward O. Sociobiology: The New Synthesis. Cambridge, MA: The Belknap Press of Harvard University Press, 1975.

7039. _____. On Human Nature. Cambridge, MA: Harvard University Press, 1978. 260 pp.

7040. Wimsatt, William C. "Some Problems with the Concept of 'Feedback'." BSPS 8, SL 29 (1971) 241-56.

7041. _____. "Complexity and Organization." BSPS 20, SL 64 (1974) 67-86. Reprinted in BSPS 27 (1976) 174-93.

7042. Withers, R. F. J. "Morphological Correspondence and the Concept of Homology," in Form and Strategy in Science. J. R. Gregg and F. T. C. Harris, eds. Dordrecht: Reidel, 1964. 378-94.

7043. Wolff, Étienne, et al. "Le climat de la découverte en biologie." Bulletin de la Société Française de Philosophie. 60 (1966) 117-49.

7044. Wolvekamp, H. P. "The Concept of the Organism as an Integrated Whole." Dialectica. 20 (1966) 196-214.

7045. Woodger, J. H. Biology and Language: An Introduction to the Methodology of the Biological Sciences, including Medicine. Cambridge: The University Press, 1952. xiv + 364 pp.

    a. Review essay, "Metabiology," by J. O. Wisdom. BJPS 4
       (1954) 339-44.

7046. _____. "From Biology to Mathematics." BJPS 3 (1952) 1-21.

7047. _____. "Formalization in Biology." Logique et analyse. 1 (1958) 97-104.

7048. _____. "Biology and Physics." BJPS 11 (1960) 89-100.

    a. "Chairman's Comments on the Contributions of Woodger and
       Hutten," by C. F. A. Pantin. 109-12.

7049. _____. "Identité et diversité dans la science naturelle spécialement en biologie." Archives de Philosophie. 25 (1962) 260-79.

7050. Wright, Larry. "Functions." Phil. Review. 82 (1973) 139-68.

    a. "Wright on Functions," by Christopher Boorse. 85 (1976)
       70-86.

7051. Wright, Sewall. "Biology and the Philosophy of Science." Monist. 48 (1964) 265-90.

7052. Wuketits, Franz M. Wissenschaftstheoretische Probleme der modernen Biologie. Berlin: Verlag Duncker & Humblot, 1978. 294 pp.

7053. Young, J. Z. Doubt and Certainty in Science: A Biologist's Reflection on the Brain. Oxford: Clarendon Press, 1951. viii + 168 pp.

    a. "Philosophy and Brain Physiology," by C. A. Campbell.
       Phil. Quart. 3 (1953) 51-56.

7054. Zaw, Susan Khin. "The Case for a Cognitive Biology." Arist. Soc. Supp. Vol. 54 (1980) 51-71.

7055. Zöllner, Walter. "Was ist Leben?" ZPF 3 (1948) 399-410.

   a. "Noch Einmal: Was ist Leben?" by Theodor Haering. 4 (1949) 257-68.

   b. "Was ist Leben?" by Hans A. Lindemann. 5 (1950) 234-36.

   c. "Stellungnahme zu den Bermerkungen über die Studie 'Was ist Leben'?" by Walter Zöllner. 236-39.

7056. Zwart, P. J. "Het fundamentele dualisme van de natuur." Algemeen Nederlands Tijdschrift voor Wijsbegeerte. 70 (1978) 229-44.

7.2 Biological Species

7057. André, Hans. "Typologie und Ontologie das Lebendigen im Gegenwartsschrifttum." Philosophisches Jahrbuch. 61 (1951) 220-28.

7058. Ayala, Francisco J. "What is a Species? 1937-1977." Scientia. 111 (1976) 605-06. Italian Tr.: 607-08.

7059. Brainerd, Barron. "Semi-lattices and Taxonomic Systems." Nous. 4 (1970) 189-99.

7060. Caplan, Arthur L. "Have Species Become Déclasse?" PSA 1980. I, 71-82.

7061. Cocchiarella, Nino. "On the Logic of Natural Kinds." PS 43 (1976) 202-22.

7062. Dobzhansky, Theodosius. "What is a Species?" Scientia. 111 (1976) 617-23. Italian Tr.: 625-30. [Reprinted from Scientia (1937).]

7063. Dubois, Georges. "Des attributs statiques de l'espèce aux tendances évolutives des groups taxinomiques." Dialectica. 1 (1947) 264-76.

7064. Dupré, John. "Natural Kinds and Biological Taxa." Phil. Review. 90 (1981) 66-90.

7065. Favarger, Claude. "Réflexions sur l'importance de l'espèce et de la classification en botanique." Dialectica. 1 (1947) 253-63.

7066. Gagnebin, Elie. "La notion d'espèce en biologie." Dialectica. 1 (1947) 229-42.

7067. Ghigi, A. "La specie." Scientia. 83 (1948) 175-81. French Tr.: Supplement, 78-84.

7068. Giray, Erol. "An Integrated Biological Approach to the Species Problem." BJPS 27 (1976) 317-28.

7069. Gregg, John R. The Language of Taxonomy. New York: Columbia University Press, 1954. ix + 70 pp.

7070. Grene, Marjorie. "Is Genus to Species as Matter to Form? Aristotle and Taxonomy." BSPS 23, SL 66 (1974) 108-26. Also in Synthese. 28 (1974) 51-70.

  a. "Matter as Goo: Comments on Grene's Paper," by Richard Rorty. Synthese. 28 (1974) 71-78.

7071. Hull, David L. "The Effect of Essentialism on Taxonomy - Two Thousand Years of Stasis." BJPS 15 (1965) 314-26; 16 (1965) 1-18.

7072. _____. "Are the 'Members' of Biological Species 'Similar' to Each Other?" BJPS 25 (1974) 332-34.

  a. "Professor Hull and the Evolution of Species," by L. Jonathan Cohen. 334-36.

7073. _____. "Are Species Really Individuals?" Systematic Zoology. 25 (1976) 174-91.

7074. _____. "Contemporary Systematic Philosophies." BSPS 27 (1976) 396-440.

7075. _____. "The Ontological Status of Species as Evolutionary Units." WOSPS 10 (1977) 91-102.

7076. _____. "A Matter of Individuality." PS 45 (1978) 335-60.

  a. "Back to Class: A Note on the Ontology of Species," by Arthur L. Caplan. 48 (1981) 130-40.

  b. "Biological Species as Natural Kinds," by David B. Kitts and David J. Kitts. PS 46 (1979) 613-22.

  c. "Kitts and Kitts and Caplan on Species," by David L. Hull. 48 (1981) 141-52.

7077. _____. "The Principle of Biological Classification: The Use and Abuse of Philosophy." PSA 1978. II, 130-53.

7078. Kiester, A. Ross. "Natural Kinds, Natural History and Ecology." Synthese. 43 (1980) 331-42.

7079. Lehman, Hugh. "Are Biological Species Real?" PS 34 (1967) 157-67.

  a. "Biological Species: Mr. Lehman's Thesis," by Ronald Munson. 37 (1970) 121-24.

  b. "Reply to Munson," by Hugh Lehman. 125-30.

7080. Lorch, J. "The Natural System in Biology." PS 28 (1961)
282-95.

7081. Łuszczewska-Romahnowa, S. "Classification as a Kind of Distance Function. Natural Classifications." SL 87 (1977) 341-73.
First published in Studia Logica. 12 (1961).

7082. Mayr, Ernst. "Species Concepts and Definitions." BSPS 27
(1976) 353-71.

7083. Montalenti, Giuseppe. "Che cos'è una specie? 1907-1937."
Scientia. 111 (1976) 609-12. English Tr.: 613-16.

7084. Morchio, Renzo. "Il concetto di specie nella biologia moderna." Epistemologia. 2 (1979) 49-76.

7085. Platnick, Norman I., and Gareth Nelson. "The Purposes of
Biological Classification." PSA 1978. II, 117-29.

7086. Pratt, Vernon. "Biological Classification." BJPS 23 (1972)
305-27. Reprinted in BSPS 27 (1976) 372-95.

7087. _____. "Foucault and the History of Classification
Theory." SHPS 8 (1977) 163-71.

7088. Rosenberg, Alexander. "Species Notions and the Theoretical
Hierarchy of Biology." Nature and System. 2 (1980) 163-72.

7089. Ruse, Michael. "Definitions of Species in Biology." BJPS 20
(1969) 97-119.

       a. "Morphospecies and Biospecies: A Reply to Ruse," by
       David L. Hull. 21 (1970) 280-82.

       b. "The Species Problem: A Reply to Hull," by Michael Ruse.
       369-71.

7090. Sklar, Abe. "On Category Overlapping in Taxonomy," in Form
and Strategy in Science. J. R. Gregg and F. T. C. Harris, eds.
Dordrecht: Reidel, 1964. 395-401.

7091. Thomason, Richmond H. "Species, Determinates and Natural
Kinds." Nous. 3 (1969) 95-101.

       a. "Thomason on Natural Kinds," by Howard Kahane. 409-12.

7092. Valen, Leigh van. "An Analysis of Some Taxonomic Concepts,"
in Form and Strategy in Science. J. R. Gregg and F. T. C. Harris,
eds. Dordrecht: Reidel, 1964. 402-15.

7093. _____. "Individualistic Classes." PS 43 (1976) 539-41.

7094. White, Michael J. D. "Speciation: Is It a Real Problem?"
Scientia. 114 (1979) 455-68. Italian Tr.: 471-80.

7095. Woods, John. "On Species and Determinates." Nous. 1 (1967)

243-54.

7.3 Evolution

7096. Abeloos, M. "La conception scientifique de l'évolution biologique." Revue Phil. 152 (1962) 173-86.

7097. Adrian Marie, O.P., Sister. "Nature's Law: Competition or
Cooperation" New Scholasticism. 33 (1959) 493-513.

7098. Alexander, Richard D. "Evolution, Human Behavior, and Determinism." PSA 1976. II, 3-21.

7099. Anthony, G. F. Penn. "Whither Evolution? Some Questions to
Teilhard de Chardin." IPQ 15 (1975) 71-82.

7100. Aron, Jean-Paul. "The Problem of Evolution." Diogenes. 7
(1954) 90-103.

7101. Ashley, O.P., Benedict M. "Causality and Evolution." Thomist.
36 (1972) 199-230.

7102. Ayala, O.P., Francisco J. "Man in Evolution: A Scientific
Statement and Some Theological and Ethical Implications." Thomist.
31 (1967) 1-20.

7103. _____. "The Evolutionary Thought of Teilhard de Chardin,"
in Biology, History, and Natural Philosophy. A. D. Breck and W.
Yourgrau, eds. New York: Plenum, 1972. 207-16.

7104. Azar, Larry. "Heredity versus Evolution." Philosophical
Studies (Ireland). 20 (1971) 152-65.

7105. Barker, A. D. "An Approach to the Theory of Natural Selection." Philosophy. 44 (1969) 271-90.

7106. Barnett, S. A., ed. A Century of Darwin. Cambridge, MA:
Harvard University Press, 1958. 376 pp.

7107. Barricelli, Nils Aall. "Esempi numerici di processi di evoluzione." Methodos. 6 (1954) 45-68.

7108. Beatty, John. "What's Wrong with the Received View of Evolutionary Theory?" PSA 1980. II, 397-426.

7109. Benl, Gerhard. "Wissenschaft und Glaube." PN 1 (1950)
116-31.

7110. Bertalanffy, Ludwig von. "On the Logical Status of the
Theory of Evolution." Laval Théologique et Philosophique. 8 (1952)
161-68.

7111. Beurton, P. "Zur Dialektik in der biologischen Evolution."
DZP 23 (1975) 913-25.

7112. _____. "Zum Verhältnis von Mikro-und Makroevolution."
DZP 24 (1976) 810-26.

7113. _____. "Biologische Evolution und Subjekt-Objekt-Dialektik." DZP 27 (1979) 558-70.

7114. Blanc, A. C. "De l'emploi inadéquat du terme 'primitif':
Considérations sur l'évolution et la systématique." Dialectica. 11
(1957) 247-75.

7115. Blum, Harold F. Time's Arrow and Evolution. Princeton:
Princeton University Press, 1951. Second edition, 1955. Reprint:
New York: Harper & Row, 1962. x + 220 pp.

7116. Boesiger, Ernest. "Evolutionary Theories After Lamarck and
Darwin," in Studies in the Philosophy of Biology. F. J. Ayala and
T. Dobzhansky, eds. Berkeley: University of California Press,
1974. 21-44.

7117. Bone, E. "Polygénisme et polyphylétisme." Archives de
Philosophie. 23 (1960) 99-141.

7118. Bowler, Peter J. "The Changing Meaning of 'Evolution'."
Journal of the History of Ideas. 36 (1975) 95-114.

    a. "Herbert Spencer and 'Evolution' - An Additional Note,"
    by Peter J. Bowler. 367.

7119. Bradie, Michael, and Mark Gromko. "The Status of the Principle
of Natural Selection." Nature and System. 3 (1981) 3-12.

7120. Braithwaite, R. B., et al. "Discussion: Neo-Darwinism."
Theoria to Theory. 13 (1979) 87-107; 313-17; 14 (1980) 17-25.

7121. Brandon, Robert N. "'Evolution'." PS 45 (1978) 96-109.

7122. _____. "Adaptation and Evolutionary Theory." SHPS 9
(1978) 181-206.

7123. _____. "A Structural Description of Evolutionary Theory."
PSA 1980. 427-39.

7124. Bruna, M. "Beschouwingen van biologische evolutie." Tijdschrift voor Filosofie. 23 (1961) 565-89.

7125. Büchel, W. "Kosmische und biologische Evolution." PN 17
(1978-79) 280-305.

7126. Burhoe, Ralph Wendell. "Natural Selection and God." Zygon.
7 (1972) 30-63.

7127. Byerly, Henry. "Teleology and Evolutionary Theory: Mechanisms and Meanings." Nature and System. 1 (1979) 157-76.

7128. Campbell, Margaret. "The Theory of Natural Selection - Its Status and Adequacy." Methodology and Science. 11 (1978) 129-45.

7129. Caplan, Arthur. "Testability, Disreputability, and the Structure of the Modern Synthetic Theory of Evolution." Erkenntnis. 13 (1978) 261-78.

7130. _____. "Darwinism and Deductivist Models of Theory Structure." SHPS 10 (1979) 341-53.

7131. Casserley, Julian Victor Langmead. "The Evolution of Evolution," in Physics, Logic, and History. W. Yourgrau and A. D. Breck, eds. New York: Plenum Press, 1970. 115-22. "Discussion," 122-27.

7132. Cassidy, John. "Philosophical Aspects of the Group Selection Controversy." PS 45 (1978) 575-94.

7133. _____. "Ambiguities and Pragmatic Factors in the Units of Selection Controversy." PS 48 (1981) 95-111.

7134. Centore, F. F. "Darwin on Evolution: A Re-estimation." Thomist. 33 (1969) 456-96.

7135. _____. "Evolution After Darwin." Thomist. 33 (1969) 718-36.

7136. Chaisson, Eric J. "Cosmic Evolution: A Synthesis of Matter and Life." Zygon. 14 (1979) 23-39.

7137. Cimutta, J. "Die Stellung der evolutionaren Plastizität in evolutionären Prozess." DZP 15 (1967) 1456-69.

7138. _____. "Die Dialektik von Zufall und Notwendigkeit im Evolutionsgeschehen." DZP 17 (1969) 967-84.

7139. Clarke, John R. "The General Adaptation Syndrome in the Study of Animal Populations." BJPS 3 (1953) 350-52.

7140. Clay, J. "The Concept of Evolution." Synthese. 7 (1948-49) 560-74.

7141. D'Armagnac, C. "Épistémologie et philosophie de l'évolution." Archives de Philosophie. 23 (1960) 153-64.

7142. Deely, O.P., John N. "The Emergence of Man: An Inquiry into the Operation of Natural Selection in the Making of Man." New Scholasticism. 40 (1966) 141-76.

7143. _____. "The Philosophical Dimensions of the Origin of Species." Thomist. 33 (1969) 75-149; 251-342.

7144. Deely, John N., and Raymond J. Nogar, eds. The Problem of Evolution: A Study of the Philosophical Repercussions of Evolutionary Science. New York: Appleton-Century-Crofts, 1973. xix + 470 pp.

    a. Review essay by John C. Cahalan. New Scholasticism. 49 (1975) 350-62.

7145. Dobzhansky, Theodosius. Evolution, Genetics, and Man. New York: John Wiley, 1955. 398 pp.

7146. _____. Mankind Evolving: The Evolution of the Human Species. New Haven: Yale University Press, 1962. xiii + 381 pp.

7147. _____. "Scientific Explanation - Chance and Antichance in Organic Evolution," in Philosophy of Science: The Delaware Seminar, I. New York: Interscience Publishers, 1963. 209-22.

7148. _____. "Creative Evolution." Diogenes. 58 (1967) 62-74.

7149. _____. "Teilhard de Chardin and the Orientation of Evolution." Zygon. 3 (1968) 242-58.

7150. _____. "Chance and Creativity in Evolution," in Studies in the Philosophy of Biology. F. J. Ayala and T. Dobzhansky, eds. Berkeley: University of California Press, 1974. 307-38.

7151. Donceel, S.J., Joseph. "Causality and Evolution: A Survey of Some Neo-scholastic Theories." New Scholasticism. 39 (1965) 295-315.

7152. Dowdeswell, W. H. The Mechanism of Evolution. London: William Heinemann, 1955. Second edition, 1958. Reprint: New York: Harper & Row, 1960. x + 115 pp.

7153. Duchesne, Jules. "Reflections on a New Perspective of the Evolution of Living Organisms." Dialectica. 32 (1978) 155-64.

7154. Dufault, O.M.I., Lucien. "The Philosophical and Biological Implications of Evolution." Proc. Am. Cath. Phil. Assn. 26 (1952) 66-80.

7155. Dunham, B., D. Fridshal, R. Fridshal, and J. H. North. "Design by Natural Selection," in Form and Strategy in Science. J. R. Gregg and F. T. C. Harris, eds. Dordrecht: Reidel, 1964. 306-11. Reprinted from Synthese. 15 (1963) 254-59.

7156. Ebersole, F. B., and Shrewsbury, M. M. "Origin Explanations and the Origin of Life." BJPS 10 (1959) 103-119.

7157. Eccles, J. C. "Cultural Evolution versus Biological Evolution." Zygon. 8 (1973) 282-93.

7158. Egami, Fujio. "Perché si studia l'origine della vita?"
Scientia. 113 (1978) 811-15. English Tr.: 817-21.

7159. Eisenstein, I. "Ist die Evolutionstheorie wissenschaftlich
begründet?" PN 15 (1974-75) 241-92, 404-45.

7160. Elliott, Francis. "The Creative Aspect of Evolution." IPQ 6
(1966) 230-47.

7161. Emberger, L. "Regards sur la phylogénèse des végétaux."
Archives de Philosophie. 23 (1960) 79-98.

7162. Emerson, Alfred E. "Dynamic Homeostasis: A Unifying Principle in Organic, Social, and Ethical Evolution." Zygon. 3 (1968)
129-68.

7163. Erbrich, P. "Ist die naturwissenschaftliche Evolutionstheorie
hinreichend?" PN 14 (1973) 156-72.

7164. 'Espinasse, Paul G. "Genetical Semantics and Evolutionary
Theory," in Form and Strategy in Science. J. R. Gregg and F. T. C.
Harris, eds. Dordrecht: Reidel, 1964. 330-42.

7165. Ewing, J. Franklin. "Précis on Evolution." Thought. 25
(1950) 53-78.

7166. Fabrizi, E. "Ein Denkmodell der universellen Evolution." PN
15 (1974-75) 191-27.

7167. Feibleman, James K. "Darwin and Scientific Method." Tulane
Studies in Philosophy. 8 (1959) 3-14.

7168. Fisher, R. A. Creative Aspects of Natural Law. New York:
Cambridge University Press, 1951. 23 pp.

7169. Fletcher, William W. "The Origin of Life." Philosophical
Journal. 1 (1964) 49-61.

7170. Forest, Herman S., and Thomas Morrill. "Biological Expansion - Perspective on Evolution." Monist. 48 (1964) 291-305.

7171. Frisch, S.J., J. E. "On the Study of Regularities in Evolution." AJAPS 3, No. 2 (1967) 33-39.

7172. Gaussen, H. "Le sens de l'evolution." Scientia. 86 (1951)
319-23.

7173. Gellner, Ernest. "Ideal Language and Kinship Structure." PS
24 (1957) 235-42.

      a. "Descent Systems and Ideal Language," by Rodney Needham.
      27 (1960) 96-101.

      b. "The Concept of Kinship, with Special Reference to Mr.
      Needham's 'Descent Systems and Ideal Language," by Ernest
      Gellner. 27 (1960) 187-204.

c. "Physical and Social Kinship," by J. A. Barnes. 28 (1961) 296-99.

7174. Ghiselin, Michael T. "On Semantic Pitfalls of Biological Adaptation." PS 33 (1966) 147-53.

7175. _____. The Triumph of the Darwinian Method. Berkeley: University of California Press, 1968. 287 pp.

    a. Review essay, "Darwin's Method or Methods? by Frank N. Egerton. SHPS 2 (1971) 281-86.

7176. _____. The Economy of Nature and the Evolution of Sex. Berkeley: University of California Press, 1974. xii + 346 pp.

7177. Got, Etienne. Évolution individuelle et évolution collective. Paris: Maloine et Doin, 1976. 309 pp.

7178. Goudge, T. A. "The Concept of Evolution." Mind. 63 (1954) 16-25.

    a. "The Concept of Evolution," by R. J. Spilbury. 544-45.

    b. "On the Logical Geography of Neo-Mendelism," by Paul G. 'Espinasse. 65 (1956) 75-77.

7179. _____. "What is a Population?" PS 22 (1955) 272-79.

7180. _____. The Ascent of Life: A Philosophical Study of the Theory of Evolution. Toronto: University of Toronto Press, 1961. 236 pp.

7181. _____. "Another Look at Emergent Evolutionism." Dialogue. 4 (1965) 273-85.

7182. Gould, Stephen Jay. Ontogeny and Phylogeny. Cambridge, MA: The Belnap Press of Harvard University Press, 1977. ix + 501 pp.

7183. Greenberg, Joseph H. "The Logical Analysis of Kinship." PS 16 (1949) 58-64.

7184. Greene, John C. Science, Ideology, and World View: Essays in the History of Evolutionary Ideas. Berkeley: University of California Press, 1981. 168 pp.

7185. Grégoire, Franz. "L'incidence des règles méthodologiques et des positions philosophiques sur l'hypothèse de l'evolution des espèces." Revue Philosophique de Louvain. 52 (1954) 416-46.

7186. Grene, Marjorie. "Two Evolutionary Theories." BJPS 9 (1958) 110-27, 185-93. Reprinted in BSPS 23, SL 66 (1974) 127-53.

    a. "Two Evolutionary Theories, A Discussion," by Walter J. Bock and Gerd von Wahlert. BJPS 15 (1963) 140-46.

b. "On Evolutionary Theories," by Leigh Van Valen. 146-52.

c. "Two Evolutionary Theories: A Reply," by Marjorie Grene. 152-54.

d. "Two Evolutionary Theories by M. Grene: A Further Discussion," by G. S. Carter. BJPS 14 (1964) 345-49.

e. "Two Evolutionary Theories: Reply to Dr. Carter," by Marjorie Grene. 349-51.

7187. _____. "Statistics and Selection." BJPS 12 (1961) 25-42. Reprinted in BSPS 23, SL 66 (1974) 154-71.

7188. _____. "Darwin and Philosophy." BSPS 23, SL 66 (1974) 189-200.

7189. _____. "Explanation and Evolution." BSPS 23, SL 66 (1974) 207-27.

7190. _____. "Changing Concepts of Darwinian Evolution." Monist. 64 (1981) 195-213.

7191. Gruner, Rolf. "On Evolution and Its Relation to Natural Selection." Dialogue. 16 (1977) 708-14.

7192. Hahn, Lewis E. "Contextualism and Cosmic Evolution-Revolution." The Philosophy Forum. 11 (1972) 3-40.

7193. Hamilton, H. J. "A Thermodynamic Theory of the Origin and Hierarchical Evolution of Living Systems." Zygon. 12 (1977) 289-335.

7194. Hanson, Earl D. "Phylogeny as Description and Explanation." Scientia. 115 (1980) 387-403. Italian Tr.: 405-12.

7195. Heuss, Eugen. "Zur Topik des Entwicklungs (Evolutions-) Begriffes." Studia Philosophica. 9 (1949) 42-79.

7196. Heuts, J.-J. "Discussion sur les bases du néo-darwinisme." Archives de Philosophie. 23 (1960) 59-78.

7197. Hörz, H. "Natürliche Evolution und philosophische Entwicklungstheorie." DZP 26 (1978) 726-36.

7198. Hull, David L. Darwin and His Critics: The Reception of Darwin's Theory of Evolution by the Scientific Community. Cambridge, MA: Harvard University Press, 1973. xii + 473 pp.

7199. _____. "The Herd as a Means." PSA 1980. II, 73-92.

7200. Huxley, Julian. Evolution: The Modern Synthesis. London: Allen and Unwin, 1942. Reprint, with new Introduction: New York: John Wiley, 1964. 645 pp.

7201. _____. Evolution in Action. New York: Harper & Row,
1953. 141 pp.

7202. _____. New Bottles for New Wine. New York: Harper &
Brothers, 1957. Reprinted under title Knowledge, Morality, and
Destiny. New York: New American Library, 1960. 287 pp.

7203. Huxley, Julian, A. C. Hardy, and E. B. Ford, eds. Evolution
as a Process. London: George Allen & Unwin, 1954. Second edition,
1958. Reprint: New York: Collier Books, 1963. 416 pp.

7204. Hyndman, Olan R. The Origin of Life and the Evolution of
Living Things. An Environmental Theory. New York: Philosophical
Library, 1952. xxii + 648 pp.

7205. Johann, Robert O. "The Logic of Evolution." Thought. 36
(1961) 595-612.

7206. Kavanau, J. Lee. "A Theory on Causal Factors in the Origin
of Life." PS 12 (1945) 190-93.

7207. Keosian, John. The Origin of Life. New York: Reinhold,
1964. ix + 118 pp.

7208. Kimura, Motoo. "The Neutral Theory of Molecular Evolution
and Polymorphism." Scientia. 112 (1977) 687-707. Italian Tr.:
709-21.

7209. Koninck, Charles de. "Darwin's Dilemma." The Thomist. 24
(1961) 365-82. Reprinted in The Dignity of Science. James A. Weisheipl, ed. Baltimore: Thomist Press, 1961. 231-46.

7210. Laguna, Grace A. de. "The Role of Teleonomy in Evolution."
PS 29 (1962) 117-31.

7211. Lamotte, M. "La théorie actuelle des mécanismes de l'évolution." Archives de Philosophie. 23 (1960) 8-58.

7212. Lehman, Hugh. "On the Form of Explanation in Evolutionary
Theory." Theoria. 32 (1966) 14-24.

7213. Leitch, Addison H. "Evolution as an Easygoing Theory."
Pacific Philosophy Forum. 6/3 (1968) 69-78.

7214. Leonardi, Piero. "Finalistic Evolution or 'Teleogenesis'."
Laval Théologique et Philosophique. 8 (1952) 169-82.

7215. Lessertisseur, Jacques. "L'aspect causal et l'aspect formel
de la pensée évolutionniste." Les Etudes philosophiques. 15 (1960)
353-64.

7216. Lewontin, R. C. The Genetic Basis of Evolutionary Change.
New York: Columbia University Press, 1974. 346 pp.

7217. _____. "Evolution and the Theory of Games." BSPS 27 (1976) 286-311.

7218. Løvtrup, Søren. "Variation, Selection, Isolation, Environment: An Analysis of Darwin's Theory." Theoria. 43 (1977) 65-83.

7219. Luyten, Norbert. "Philosophical Implications of Evolution." New Scholasticism. 25 (1951) 290-312.

7220. Malkani, G. R. "Concept of Evolution." Philosophical Quarterly (India). 34 (1961) 103-10.

7221. Mandelbaum, Maurice. "The Scientific Background of Evolutionary Theory in Biology." Journal of the History of Ideas. 18 (1957) 342-61.

7222. Manier, Edward. "The Theory of Evolution as Personal Knowledge." PS 32 (1965) 244-52.

     a. "Mr. Manier's 'Theory of Evolution as Personal Knowledge': A Quasi-Reply," by Marjorie Grene. 33 (1966) 163-64.

7223. _____. "Methodology, Aesthetics, and Devolution." Pacific Philosophy Forum. 6/3 (1968) 79-85.

7224. _____. "Darwin's Language and Logic." SHPS 11 (1980) 305-23.

7225. Manser, A. R. "The Concept of Evolution." Philosophy. 40 (1965) 18-34.

     a. "The Concept of Evolution: A Comment," by Antony Flew. 41 (1966) 70-75.

     b. "The Concept of Evolution: A Comment on Papers by Mr. Manser and Professor Flew," by Kevin Connolly. 356-57.

7226. Maslov, S. Yu. "Macroevolution as Deduction Process." Synthese. 39 (1978) 417-34.

7227. Matisse, G. "Les déficiences de monde animé." Revue Phil. 137 (1947) 131-55; 138 (1948) 307-17.

7228. Mayr, Ernst, and William B. Provine, eds. The Evolutionary Synthesis: Perspectives on the Unification of Biology. Cambridge: Harvard University Press, 1980. xi + 487 pp.

     a. Review essay by Bentley Glass. Isis. 72 (1981). 642-47.

7229. Melsen, Andrew G. van. Evolution and Philosophy. Pittsburgh: Duquesne University Press, 1965. 208 pp.

7230. Meurers, Joseph. "Die wissenschaftstheoretische Position einer evolutiven Welterklärung." PN 8 (1964) 9-21.

7231. Meyer, Francois. Problématique de l'évolution. Paris:
Presses Universitaires de France, 1954.

   a. Review essay by Raymond Ruyer. Les Etudes philosophiques.
   10 (1955) 271-82.

7232. _____. "Le temps de l'évolution." Les Etudes philosophiques. 17 (1962) 13-18.

7233. Michod, Richard E. "Positive Heuristics in Evolutionary
Biology." BJPS 32 (1981) 1-36.

7234. Miklin, A. M. "Toward a Definition of the Concept of Pro-
gressive Evolution in Biological Phenomena." Soviet Studies in
Philosophy. 6, No. 4 (1968) 32-39. [English reprint from Vestnik
Leningradskogo universiteta. (1967, No. 5).]

7235. Milcu, St.-M. "The Biological Evolution as an Abstract Entity." Noesis. 2 (1974) 35-38.

7236. Miller, III, John F. "The Logic of Evolution." Southwestern
Journal of Philosophy. 3, No. 1 (1972) 147-56.

7237. Mills, Susan K., and John J. Beatty. "The Propensity Interpretation of Fitness." PS 46 (1979) 263-86.

7238. Monod, Jacques. "L'evolution microscopique." Theoria to
Theory. 10 (1977) 303-11.

7239. Montalenti, Giuseppe. "Evoluzione e genetica." Scientia.
110 (1975) 13-33. English Tr.: 35-50.

7240. Moreno, O.P., Antonio. "Some Philosophical Considerations on
Biological Evolution." Thomist. 37 (1973) 417-54.

7241. Morrison, Paul G. "On Evolution." Tulane Studies in Philosophy. 8 (1959) 15-26.

7242. _____. "Evolution-Revolution and the Cosmos." The
Philosophy Forum. 11 (1972) 71-98.

7243. Munson, Ronald. "Biological Adaptation." PS 38 (1971)
200-15. Reprinted in BSPS 27 (1976) 330-50.

   a. "Biological Adaptation," by Michael Ruse. PS 39 (1972)
   525-28.

   b. "Biological Adaptation: A Reply," by Ronald Munson.
   529-32.

7244. Nogar, O.P., Raymond J. "The Darwin Centennial: A Philosophical Intrusion." New Scholasticism. 33 (1959) 411-45.

7245. _____. "From the Fact of Evolution to the Philosophy of
Evolutionism." The Thomist. 24 (1961) 463-501. Reprinted in The
Dignity of Science. James A. Weisheipl, ed.

7246. _____. "The Lord of the Absurd. New York: Herder and Herder, 1966. 157 pp.

7247. Noüy, Lecomte du. Human Destiny. New York: Longsmans, Green & Co., 1947. Reprint: New York: New American Library. 189 pp.

7248. Onicescu, O. "Remarques sur l'evolution." Noesis. 4 (1978) 9-13.

7249. Oparin, A. I. The Origin of Life. Sergius Morgulis, tr. New York: MacMillan, 1938. Reprint: New York: Dover, 1953. xxiv + 270 pp.

7250. _____. "Les vues modernes sur l'origine de la vie." Scientia. 95 (1960) 322-26.

7251. _____. Life: Its Nature, Origin and Development. Ann Synge, tr. Edinburgh: Oliver and Boyd, 1962. Reprint: New York and London: Academic Press, 1964. xi + 207 pp.

7252. _____. "L'état actuel du problème de l'origine de la vie et ses perspectives." Scientia. 102 (1967) 318-23.

7253. _____. "Modern Aspects of the Problem of the Origin of Life." Scientia. 106 (1971) 195-206.

7254. _____. "Modern Concepts of the Origin of Life on the Earth." Scientia. 113 (1978) 7-16. Italian Tr.: 17-25.

7255. _____. "On the Essence of Life." Soviet Studies in Philosophy. 18 (1979-80) 19-39. [English reprint from Voprosy filosofii. (1979, No. 4).]

7256. Osche, G. "Mechanismen der Evolution und die Mannigfaltigkeit der Organismen." Studium Generale. 24 (1971) 191-201.

7257. Pawelzig, G. "Ueber den Charakter des Determinismus in der Ontogenese." DZP 9 (1961) 811-27.

7258. Peacocke, A. R. "Chance and the Life Game." Zygon. 14 (1979) 301-22.

7259. Pechhacker, S.J., P. Anton. "Der Evolutionismus in philosophischer Sicht." Salzburger Jahrbuch für Philosophie. 12-13. (1968/69) 323-56.

7260. Piaget, Jean. Le comportement: Moteur de l'évolution. Paris: Gallimard, 1976. 190 pp. Translated by Donald Nicholson-Smith under title Behavior and Evolution. New York: Pantheon Books, 1978. 165 pp.

7261. Pontet, Maurice. "Evolution according to Teilhard de Chardin." Thought. 36 (1961) 167-89.

7262. Popper, Karl. "Darwinism as a Metaphysical Research Program." Methodoogy and Science. 9 (1976) 103-19.

7263. Portmann, A. "L'ontogenèse et le problème de l'origine." Dialectica. 14 (1960) 37-52.

7264. Puligandla, R. "The Concept of Evolution and Revolution." The Philosophy Forum. 11 (1972) 41-70.

7265. Ramirez, J. Roland E. "The Ultimate Why of Evolution." New Scholasticism. 33 (1959) 446-92.

7266. Rensch, G. "Evolution als Eigenschaft des Lebendigen." Studium Generale. 12 (1959) 153-59.

7267. _____. "Die Evolutionsgesetze der Organismen in natur-philosophischer Sicht." PN 6 (1960-61) 288-326.

7268. Rivier, W. "L'apparition de la vie dans l'univers serait-elle compatible avec le calcul des probabilités?" Dialectica. 4 (1950) 158-62.

7269. Roldán, S.J., Alejandro. "Epistemología de la evolución bio-lógica." Pensamiento. 7 (1951) 583-601.

7270. Rosenberg, Alexander. "Ruse's Treatment of the Evidence for Evolution: A Reconsideration." PSA 1980. I, 83-93.

7271. Rudwick, M. J. S. "The Inference of Function from Structure in Fossils." BJPS 15 (1964) 27-40.

7272. Ruse, M. "Confirmation and Falsification of Theories of Evolution." Scientia. 104 (1969) 329-57. French Tr.: Supplement, 179-204.

7273. _____. "Natural Selection in The Origin of Species." SHPS 1 (1970-71) 311-51.

7274. _____. "Is the Theory of Evolution Different?" Scientia. 106 (1971) 765-83, 1069-93.

7275. _____. "Cultural Evolution." Theory and Decision. 5 (1974) 413-40.

7276. _____. The Darwinian Revolution: Science Red in Tooth and Claw. Chicago: University of Chicago Press, 1979. xiv + 320 pp.

7277. _____. Darwinism Defended: A Guide to the Evolution Controversies. Reading, MA: Addison-Wesley, 1982. 275 pp.

7278. Russell, John L. "The Theory of Evolution." Philosophy Today. 1 (1957) 63-64.

7279. Ruyer, R. "Les postulats du sélectionnisme." Revue Phil. 146 (1956) 318-53.

7280. Schrader, Malcolm E. "Is Life an Accident?" Zygon. 14 (1979) 323-28.

7281. Simpson, George Gaylord. The Meaning of Evolution: A Study of the History of Life and of Its Significance for Man. New Haven: Yale University Press, 1949. xv + 364 pp.

7282. _____. The Major Features of Evolution. New York: Columbia University Press, 1953. 434 pp.

7283. _____. This View of Life: The World of an Evolutionist. New York: Harcourt, Brace & World, 1964. ix + 308 pp.

7284. Skagestad, Peter. "Taking Evolution Seriously: Critical Comments on D. T. Campbell's Evolutionary Epistemology." Monist. 61 (1978) 611-20.

7285. Slobodkin, Lawrence B. "The Strategy of Evolution." American Scientist. 52 (1964) 342-57. Reprinted in BSPS (1976) 267-85.

7286. Smith, J. Maynard. "Time in the Evolutionary Process." Studium Generale. 23 (1970) 266-72.

7287. Smith, Vincent E. "Evolution and Entropy." Thomist. 24 (1961) 441-62. Reprinted in The Dignity of Science. James A. Weisheipl, ed. Baltimore: Thomist Press, 1961. 305-26.

7288. Sober, Elliot. "Evolution, Population Thinking, and Essentialism." PS 47 (1980) 350-83.

7289. _____. "Holism, Individualism, and the Units of Selection." PSA 1980. II, 93-121.

7290. Spilsbury, Richard. Providence Lost: A Critique of Darwinism. London: Oxford University Press, 1975. 133 pp.

7291. Stebbins, George Ledyard. "The Evolutionary Significance of Biological Templates," in Biology, History, and Natural Philosophy. A. D. Breck and W. Yourgrau, eds. New York: Plenum, 1972. 79-102.

7292. _____. "Adaptive Shifts and Evolutionary Novelty: A Compositionist Approach," in Studies in the Philosophy of Biology. F. J. Ayala and T. Dobzhansky, eds. Berkeley: University of California Press, 1974. 285-306.

7293. Stock, O.P., Michael. "Scientific vs. Phenomenological Evolution: A Critique of Teilhard de Chardin." New Scholasticism. 36 (1962) 368-80.

7294. Stöhr, H.-J. "Dialektischer Widerspruch und Faktoren der biologischen Evolution." DZP 25 (1977) 831-36.

7295. Sumner, L. W., John G. Slater, and Fred Wilson, eds. Prag-matism and Purpose: Essays Presented to Thomas A. Goudge. Toronto: University of Toronto Press, 1981. 362 pp.

7296. Szarski, H. "The Concept of Progress in Evolution." Scientia. 103 (1968) 152-59. French Tr.: Supplement, 94-101.

7297. Teilhard de Chardin, Pierre. Le phénomene humain. Paris: Editions du Seuil, 1955. English Tr.: The Phenomenon of Man. With an Introduction by Sir Julian Huxley. Bernard Wall tr. London: William Collins Sons; New York: Harper & Row, 1959. 318 pp.

   a. Review essay by Alan L. Stuart. BJPS 12 (1961) 235-45.

   b. Review essay by John L. Russell, S.J. Heythrop Journal. 1 (1960) 271-84; 2 (1961) 3-13.

   c. Review essay by P. B. Medawar. Mind. 70 (1961) 99-106.

   d. "Salvaging the 'Noosphere'," by T. A. Goudge. Mind. 71 (1962) 543-44.

7298. _____. L'Avenir de l'homme. Paris: Editions du Seuil, 1959. English Translation: The Future of Man. Norman Denny, tr. London: William Collins Sons; New York: Harper & Row, 1964. 332 pp.

7299. Teissier, G. "Enchaînement des générations et évolution." Scientia. 97 (1962) 247-53.

7300. Templeton, Alan R., and Edward D. Rothman. "Evolution and Fine-Grained Environmental Runs." WOSPS 13-II (1978) 131-83.

7301. Thompson, W. R. "The Status of Evolutionary Theory." Laval Théologique et Philosophique. 8 (1952) 196-202.

7302. Travis, Janet L. "A Criticism of the Use of the Concept of 'Dominant Group' in Arguments for Evolutionary Progressivism." PS 31 (1971) 369-75.

   a. "Biological Progress and Dominance: A Reply to Janet L. Travis," by Maurice J. A. Glickman. 39 (1972) 383-87.

7303. Turchin, Valentin F. The Phenomenon of Science. B. Frantz, tr. Forward by Loren Graham. New York: Columbia University Press, 1977. 344 pp.

7304. Versfeld, Martin. "Reflections on Evolutionary Knowledge." IPQ 5 (1975) 221-47.

7305. Waddington, C. H. The Ethical Animal. London: George Allen & Unwin, 1960. Reprint: Chicago: University of Chicago Press, 1967. 231 pp.

7306. _____. "The Theory of Evolution Today," in Beyond Reductionism. A. Koestler and J. R. Symthies, eds. New York: Macmillan, 1970. 357-95.

7307. Wallace, O.P., William A. "The Cosmogony of Teilhard de Chardin." New Scholasticism. 36 (1962) 353-67.

7308. Wassermann, Gerhard. "Testability of the Role of Natural Selection within Theories of Population Genetics and Evolution." BJPS 29 (1978) 223-42.

7309. _____. "On the Nature of the Theory of Evolution." PS 48 (1981) 416-37.

7310. Wicken, Jeffrey S. "Chance, Necessity, and Purpose: Toward a Philosophy of Evolution." Zygon. 16 (1981) 303-22.

7311. Williams, George C. Adaptation and Natural Selection: A Critique of Some Current Evolutionary Thought. Princeton, NJ: Princeton University Press, 1966. 307 pp.

7312. Williams, Mary B. "Falsifiable Predictions of Evolutionary Theory." PS 40 (1973) 518-37.

7313. _____. "The Logical Status of the Theory of Natural Selection and Other Evolutionary Controversies," in The Methodological Unity of Science. M. Bunge, ed. Dordrecht: Reidel, 1973. 84-102.

7314. _____. "Similarities and Differences between Evolutionary Theory and the Theories of Physics." PSA 1980. II, 385-96.

7315. Williams, Thomas Rhys. "The Evolution of a Human Nature." PS 26 (1959) 1-13.

7316. Wilson, Peter J. Man, The Promising Primate: The Conditions of Human Evolution. New Haven, CT: Yale University Press, 1980. 200 pp.

7317. Wimsatt, William C. "The Units of Selection and the Structure of the Multi-Level Genome." PSA 1980. II, 122-83.

7318. _____. "Randomness and Perceived-Randomness in Evolutionary Biology." Synthese. 43 (1980) 287-330.

7319. Witzemann, Edgar J. "Chemistry and Evolution." PS 12 (1945) 179-89.

7320. Woodfield, Andrew. "Darwin, Teleology and Taxonomy." Philosophy. 48 (1973) 35-50.

7321. Young, Robert M. "Darwin's Metaphor: Does Nature Select?" Monist. 55 (1971) 442-503.

7322. Zimmermann, Walter. Evolution und Naturphilosophie. Berlin: Duncker & Humblot, 1968. 313 pp.

7.4 Genetics

7323.   Abeloos, M.   "La régulation dans le développement des formes vivantes."   Revue Phil. 155 (1965) 265-84.

7324.   Aron, J.-P.   "Introduction à une problématique de l'hérédité." Revue Phil. 150 (1960) 65-83.

7325.   Berlinski, David.   "Philosophical Aspects of Molecular Biology." Journal of Philosophy.   69 (1972) 319-35.

  a.   "More Philosophical Aspects of Molecular Biology," by S. Wendell-Waechtler and E. Levy.   PS 42 (1975) 180-86.

7326.   Darden, Lindley.   "Theory Construction in Genetics."   BSPS 60 (1980) 151-70.

7327.   Dobzhansky, Theodosius.   Genetics and the Origin of the Species.   New York: Columbia University Press, 1951.   Third edition, revised.   364 pp.

7328.   _____.   Heredity and the Nature of Man.   New York: Harcourt, Brace and World, 1964.   Reprint: New York: New American Library, 1966.   175 pp.

7329.   Dubinin, N. P.   "Contemporary Natural Sciences and a Scientific World View."   Soviet Studies in Philosophy.   11 (1972-73) 248-69. [English reprint from Voprosy filosofii.   (1972, No. 3).]

7330.   Geissler, E.   "Die Gentheorie ist mit der materialistischen Dialektik vereinbar."   DZP 26 (1978) 765-70.

7331.   Glass, Bentley.   "The Establishment of Modern Genetical Theory as an Example of the Interaction of Different Models, Techniques and Inferences," in Scientific Change.   A. C. Crombie, ed. London: Heinemann, 1963.   521-41.

7332.   Goosens, William K.   "Reduction by Molecular Genetics."   PS 45 (1978) 73-95.

  a.   "Reduction in Genetics," by David L. Hull.   46 (1979) 316-20.

7333.   Hull, David L.   "Reduction in Genetics - Doing the Impossible." LMPS-4 (1971) 619-35.

7334.   _____.   "Reduction in Genetics - Biology or Philosophy?" PS 39 (1972) 491-99.

7335. Katz, Michael J., and William Goffman. "Preformation of On-togenetic Patterns." PS 48 (1981) 438-53.

7336. Kedrov, B. M. "The Road to Truth. (Some Reflections on the Problems of Natural Science)." Soviet Studies in Philosophy. 4, No. 2 (1965) 3-24. [English reprint from Novyi mir. (1965, No. 1).]

7337. Kimbrough, Steven Orla. "On the Reduction of Genetics to Molecular Biology." PS 46 (1979) 389-406.

7338. Łastowski, K. "The Method of Idealization in the Populational Genetics." Poznań Studies. 3 (1977) 199-212.

7339. Lenartowicz, Piotr. Phenotype-Genotype Dichotomy: An Essay in Theoretical Biology. Roma, 1975. 233 pp.

7340. L'Heritier, Ph. "Des lois de Mendel à l'autoreproduction du matériel génétique et à l'évolution." Les Etudes philosophiques. 15 (1960) 333-40.

7341. Lindenmayer, A., and N. Simon. "The Formal Structure of Genetics and the Reduction Problem." PSA 1980. I, 160-70.

7342. Mainx, Felix. "Ergebnisse und Probleme der modernen Genetik." Philosophisches Jahrbuch. 61 (1951) 132-45.

7343. _____. "Der Organismus als genetisches System." Studium Generale. 12 (1959) 147-53.

7344. _____. "Vererbung und Entwicklung als Grundlage für unser Verständnis des lebenden Organismus." Studium Generale. 12 (1959) 189-94.

7345. Manier, Edward (with Henry Bender). "Genetics and the Philosophy of Biology." Proc. Am. Cath. Phil. Assn. 39 (1965) 124-33.

7346. _____. "Genetics and the Future of Man: Scientific and Ethical Possibilities." Proc. Am. Cath. Phil. Assn. 42 (1968) 183-92.

7347. _____. "The Experimental Method in Biology: T. H. Morgan and the Theory of the Gene." Synthese. 20 (1969) 185-205.

7348. Mikulak, Maxim W. "Darwinism, Soviet Genetics, and Marxism-Leninism." Journal of the History of Ideas. 31 (1970) 359-76.

7349. Mortimer, H. "Probabilistic Definition on the Example of the Definition of Genotype." SL 87 (1977) 433-56. [Expanded version of a paper first published in Studia Logica. 15 (1964).]

7350. Przełęcki, Marian. "On the Concept of Genotype," in Form and Strategy in Science. J. R. Gregg and F. T. C. Harris, eds. Dordrecht: Reidel, 1964. 315-29.

7351. Rochhausen, R., and G. Ludwig. "Einige philosophische Probleme der modernen Genetik." DZP 11 (1963) 171-83.

7352. Rosenberg, Alexander. "Genetics and the Theory of Natural Selection: Synthesis or Sustinance?" Nature and System. 1 (1979) 3-15.

7353. Souriau, É. "Réflexions sur la notion d'hérédité." Revue Phil. 142 (1952) 165-86.

7354. Stent, G. S. "Explicit and Implicit Semantic Content of the Genetic Information." WOSPS 10 (1977) 131-49.

7355. Ueberschär, K. "Der Einfluss von reduktiver und integrativer Methode auf die Veränderung der methodologischen Struktur der modernen Genetik." DZP 26 (1978) 750-64.

7356. Woodger, J. H. "What Do We Mean by 'Inborn'?" BJPS 3 (1953) 319-26.

   a. "A Logical Basis for Genetics?" by J. B. S. Haldane. BJPS 6 (1955) 245-48.

   b. "A Reply to Professor Haldane," by J. H. Woodger. BJPS 7 (1956) 149-55.

APPENDIX 1
Boston Studies in the Philosophy
of Science

1. Marx W. Wartofsky (ed.). Proceedings of the Boston Colloquium
   for the Philosophy of Science 1961-1962. 1963. SL 6.

2. Robert S. Cohen and Marx W. Wartofsky (eds.). In Honor of
   Philipp Frank. 1965. SL 10.

3. Robert S. Cohen and Marx W. Wartofsky (eds.). Proceedings of
   the Boston Colloquium for the Philosophy of Science 1964-1966.
   In Memory of Norwood Russell Hanson. 1967. SL 14.

4. Robert S. Cohen and Marx W. Wartofsky (eds.). Proceedings of
   the Boston Colloquium for the Philosophy of Science 1966-1968.
   1969. SL 18.

5. Robert S. Cohen and Marx W. Wartofsky (eds.). Proceedings of
   the Boston Colloquium for the Philosophy of Science 1966-1968.
   1969. SL 19.

6. Robert S. Cohen and Raymond J. Seeger (eds.). Ernst Mach:
   Physicist and Philosopher. 1970. SL 27.

7. Milič Čapek. Bergson and Modern Physics. 1971. SL 37.

8. Roger C. Buck and Robert S. Cohen (eds.). PSA 1970. In
   Memory of Rudolf Carnap. 1971. SL 39.

9. A. A. Zinov'ev. Foundations of the Logical Theory of Scientific Knowledge (Complex Logic). (Revised and enlarged English
   edition with an appendix by G. A. Smirnov, E. A. Sidorenka, A.
   M. Fedina, and L. A. Bobrova.) 1973. SL 46.

10. Ladislav Tondl. Scientific Procedures. 1973. SL 47.

11. R. J. Seeger and Robert S. Cohen (eds.). Philosophical Foundations of Science. 1974. SL 58.

12. Adolf Grünbaum. Philosophical Problems of Space and Time. (Second, enlarged edition.) 1973. SL 55.

13. Robert S. Cohen and Marx W. Wartofsky (eds.). Logical and Epistemological Studies in Contemporary Physics. 1973. SL 59.

14. Robert S. Cohen and Marx W. Wartofsky (eds.). Methodological and Historical Essays in the Natural and Social Sciences. Proceedings of the Boston Colloquium for the Philosophy of Science 1969-1972. 1974. SL 60.

15. Robert S. Cohen, J. J. Stachel and Marx W. Wartofsky (eds.). For Dirk Struik. Scientific, Historical and Political Essays in Honor of Dirk Struik. 1974. SL 61.

16. Norman Geschwind. Selected Papers on Language and the Brain. 1974. SL 68.

17. Kuznetsov, B. G. Reason and Being: Studies in Classical Rationalism and Non-classical Science. In preparation 1984.

18. Peter Mittelstaedt. Philosophical Problems of Modern Physics. 1976. SL 95.

19. Henry Mehlberg. Time, Causality, and the Quantum Theory (2 vols.). 1980.

20. Kenneth F. Schaffner and Robert S. Cohen (eds.). Proceedings of the 1972 Biennial Meeting, Philosophy of Science Association. 1974. SL 64.

21. R. S. Cohen and J. J. Stachel (eds.). Selected Papers of Léon Rosenfeld. 1978. SL 100.

22. Milič Čapek (ed.). The Concepts of Space and Time. Their Structure and Their Development. 1976. SL 74.

23. Marjorie Grene. The Understanding of Nature. Essays in the Philosophy of Biology. 1974. SL 66.

24. Don Ihde. Technics and Praxis. A Philosophy of Technology. 1978. SL 130.

25. Jaakko Hintikka and Unto Remes. The Method of Analysis. Its Geometrical Origin and Its General Significance. 1974. SL 75.

26. John Emery Murdoch and Edith Dudley Sylla. The Cultural Context of Medieval Learning. 1975. SL 76.

27. Marjorie Grene and Everett Mendelsohn (eds.). Topics in the Philosophy of Biology. 1976. SL 84.

28. Joseph Agassi. Science in Flux. 1975. SL 80.

29. Jerzy J. Wiatr (ed.). Polish Essays in the Methodology of the Social Sciences. 1979. SL 131.

30. P. Janich. Protophysics of Time. In preparation 1984.

31. R. S. Cohen and M. W. Wartofsky (eds.). Language, Logic and Method. Including Selected Papers from the Boston Colloquia for the Philosophy of Science, 1973-1980. 1982.

32. R. S. Cohen, C. A. Hooker, A. C. Michalos, and J. W. van Evra (eds.). PSA 1974: Proceedings of the 1974 Biennial Meeting of the Philosophy of Science Association. 1976. SL 101.

33. Gerald Holton and William Blanpied (eds.). Science and Its Public: The Changing Relationship. 1976. SL 96.

34. Mirko D. Grmek (ed.). On Scientific Discovery. 1980.

35. Stefan Amsterdamski. Between Experience and Metaphysics. Philosophical Problems of the Evolution of Science. 1975. SL 77.

36. Mihailo Marković and Gajo Petrović (eds.). Praxis. Yugoslav Essays in the Philosophy and Methodology of the Social Sciences. 1979. SL 134.

37. Hermann von Helmholtz: Epistemological Writings. The Paul Hertz/Moritz Schlick Centenary Edition of 1921 with Notes and Commentary by the Editors. (Newly translated by Malcolm F. Lowe. Edited, with an Introduction and Bibliography, by Robert S. Cohen and Yehuda Elkana.). 1977. SL 79.

38. R. M. Martin. Pragmatics, Truth, and Language. 1979.

39. R. S. Cohen, P. K. Feyerabend, and M. W. Wartofsky (eds.). Essays in Memory of Imre Lakatos. 1976. SL 99.

40. B. M. Kedrov and V. Sadovsky. Current Soviet Studies in the Philosophy of Science. In preparation 1983.

41. M. Raphael. Theorie des Geistigen Schaffens auf Marxistischer Grundlage. In preparation 1984.

42. Humberto R. Maturana and Francisco J. Varela. Autopoiesis and Cognition. The Realization of the Living. 1980.

43. A. Kasher (ed.). Language in Focus: Foundations, Methods and Systems. Essays Dedicated to Yehoshua Bar-Hillel. 1976. SL 89.

44. T. D. Thao. Investigations into the Origin of Language and Consciousness. In preparation 1983.

45. A. Ishmimoto. Japanese Studies in the History and Philosophy of Science. In preparation 1984.

46. Peter L. Kapitza. Experiment, Theory, Practice. 1980.

47. Maria L. Dalla Chiara (ed.). Italian Studies in the Philosophy of Science. 1980.

48. Marx W. Wartofsky. Models: Representation and the Scientific Understanding. 1979. SL 129.

50. Yehuda Fried and Joseph Agassi. Paranoia: A Study in Diagnosis. 1976. SL 102.

51. Kurt H. Wolff. Surrender and Catch: Experience and Inquiry Today. 1976. SL 105.

52. Karel Kosík. Dialectics of the Concrete. 1976. SL 106.

53. Nelson Goodman. The Structure of Appearance. (Third Edition.) 1977. SL 107.

54. Herbert A. Simon. Models of Discovery and Other Topics in the Methods of Science. 1977. SL 114.

55. Morris Lazerowitz. The Language of Philosophy. Freud and Wittgenstein. 1977. SL 117.

56. Thomas Nickles (ed.). Scientific Discovery, Logic, and Rationality. 1980.

57. Joseph Margolis. Persons and Minds. The Prospects of Nonreductive Materialsm. 1977. SL 121.

58. Gerald Radnitzky and Gunnar Andersson (eds.). Progress and Rationality in Science. 1978. SL 125.

59. Gerard Radnitzky and Gunnar Andersson (eds.). The Structure and Development of Science. 1979. SL 136.

60. Thomas Nickles (ed.). Scientific Discovery: Case Studies. 1980.

61. Maurice A. Finocchiaro. Galileo and the Art of Reasoning. 1980.

62. William A. Wallace. Prelude to Galileo. 1981.

63. Friedrich Rapp. Analytical Philosophy of Technology. 1981.

64. Robert S. Cohen and Marx W. Wartofsky (eds.). Hegel and the Sciences. (Forthcoming).

65. Joseph Agassi. Science and Society. 1981.

66. Ladislav Tondl. Problems of Semantics. 1981.

67. Joseph Agassi and Robert S. Cohen (eds.). Scientific Philosophy Today. 1981.

68. Władysław Krajewski (ed.). Polish Essays in the Philosophy of Natural Sciences. 1982.

69. James H. Fetzer. Scientific Knowledge. 1981.

70. Stephen Grossberg. Studies of Mind and Brain. 1982.

71. R. S. Cohen and M. W. Wartofsky (eds.). Epistemology, Methodology, and the Social Sciences. Including Selected Papers from the Boston Colloquia for the Philosophy of Science, 1973-1980. In preparation 1983.

72. K. Berka. Measurement. Its Concepts, Theories and Problems. 1982.

73. G. L. Pandit. The Structure and Growth of Scientific Knowledge. A Study in the Methodology of Epistemic Appraisal. 1982.

74. A. A. Zinov'ev. Logical Physics. In preparation 1983.

75. G. G. Granger. Formal Thought and the Sciences of Man. In preparation 1983.

APPENDIX 2
Synthese Library

1. J. M. Bocheński, A Precis of Mathematical Logic. 1959.

2. P. L. Guiraud, Problèmes et méthodes de la statistique linguistique. 1960.

3. Hans Freudenthal (ed.). The Concept and the Role of the Model in Mathematics and Natural and Social Sciences. 1961.

4. Evert W. Beth. Formal Methods. An Introduction to Symbolic Logic and the Study of Effective Operations in Arithemetic and Logic. 1962.

5. B. H. Kazemier and D. Vuysje (eds.). Logic and Language. Studies Dedicated to Professor Rudolf Carnap on the Occasion of His Seventieth Birthday. 1962.

6. Marx W. Wartofsky (ed.). Proceedings of the Boston Colloquium for the Philosophy of Science 1961-1962. 1963. BSPS 1.

7. A. A. Zinov'ev. Philosophical Problems of Many-Valued Logic. 1963.

8. Georges Gurvitch. The Spectrum of Social Time. 1964.

9. Paul Lorenzen. Formal Logic. 1965.

10. Robert S. Cohen and Marx W. Wartofsky (eds.). In Honor of Philipp Frank. 1965. BSPS 2.

11. Evert W. Beth. Mathematical Thought. An Introduction to the Philosophy of Mathematics. 1965.

12. Evert W. Beth and Jean Piaget. Mathematical Epistemology and Psychology. 1966.

13. Guido Küng. Ontology and the Logistic Analysis of Language. An Enquiry into the Contemporary Views on Universals. 1967.

14. Robert S. Cohen and Marx W. Wartofsky (eds.). Proceedings of the Boston Colloquium for the Philosophy of Science 1964-1966. In Memory of Norwood Russell Hanson. 1967. BSPS 3.

15. C. D. Broad. Induction, Probability, and Causation. Selected Papers. 1968.

16. Günther Patzig. Aristotle's Theory of the Syllogism. A Logical-Philosophical Study of Book A of the Prior Analytics. 1968.

17. Nicholas Rescher. Topics in Philosophical Logic. 1968.

18. Robert S. Cohen and Marx W. Wartofsky (eds). Proceedings of the Boston Colloquium for the Philosophy of Science 1966-1968. 1969. BSPS 4.

19. Robert S. Cohen and Marx W. Wartofsky (eds.). Proceedings of the Boston Colloquium for the Philosophy of Science 1966-1968. 1969. BSPS 5.

20. J. W. Davis, D. J. Hockney, and W. K. Wilson (eds.). Philosophical Logic. 1969.

21. D. Davidson and J. Hintikka (eds.). Words and Objections. Essays on the Work of W. V. Quine. 1969.

22. Patrick Suppes. Studies in the Methodology and Foundations of Science. Selected Papers from 1911 to 1969. 1969.

23. Jaakka Hintikka. Models for Modalities. Selected Essays. 1969.

24. Nicholas Rescher et al. (eds.). Essays in Honor of Carl G. Hempel. A Tribute on the Occasion of His Sixty-Fifth Birthday. 1969.

25. P. V. Tavanec (ed.). Problems of the Logic of Scientific Knowledge. 1969.

26. Marshall Swain (ed.). Induction, Acceptance, and Rational Belief. 1970.

27. Robert S. Cohen and Raymond J. Seeger (eds.). Ernst Mach: Physicist and Philosopher. 1970. BSPS 6.

28. Jaakko Hintikka and Patrick Suppes. Information and Inference. 1970.

29. Karel Lambert. Philosophical Problems in Logic. Some Recent
     Developments. 1970.

30. Rolf A. Eberle. Nominalistic Systems. 1970.

31. Paul Weingartner and Gerhard Zecha (eds.). Induction, Physics,
     and Ethics. 1970.

32. Evert W. Beth. Aspects of Modern Logic. 1970.

33. Risto Hilpinen (ed.). Deontic Logic: Introductory and Syste-
     matic Readings. 1971.

34. Jean-Louis Krivine. Introduction to Axiomatic Set Theory.
     1971.

35. Joseph D. Sneed. The Logical Structure of Mathematical
     Physics. 1971.

36. Carl R. Kordig. The Justification of Scientific Change. 1971.

37. Milič Čapek. Bergson and Modern Physics. 1971. BSPS 7.

38. Norwood Russell Hanson. What I Do Not Believe, and Other Essays
     (ed. by Stephen Toulmin and Harry Woolf). 1971.

39. Roger C. Buck and Robert S. Cohen (eds.). PSA 1970. In Memory
     of Rudolf Carnap. 1971. BSPS 8.

40. Donald Davidson and Gilbert Harman (eds.). Semantics of Natural
     Language. 1972.

41. Yehoshua Bar-Hillel (ed.). Pragmatics of Natural Languages.
     1971.

42. Sören Stenlund. Combinators, Lambda-Terms, and Proof Theory.
     1972.

43. Martin Strauss. Modern Physics and Its Philosophy. Selected
     Papers in the Logic, History, and Philosophy of Science. 1972.

44. Mario Bunge. Method, Model and Matter. 1973.

45. Mario Bunge. Philosophy of Physics. 1973.

46. A. A. Zinov'ev. Foundations of the Logical Theory of Scienti-
     fic Knowledge (Complex Logic). (Revised and enlarged English
     edition with an appendix by G. A. Smirnov, E. A. Sidorenka, A.
     M. Fedina, and L. A. Bobrova.) 1973. BSPS 9.

47. Ladislav Tondl. Scientific Procedures. 1973. BSPS 10.

48. Norwood Russell Hanson. Constellations and Conjectures (ed. by
     Willard C. Humphreys, Jr.). 1973.

49. K. J. J. Hintikka, J. M. E. Moravcsik, and P. Suppes (eds.).
    Approaches to Natural Language. 1973.

50. Mario Bunge (ed.). Exact Philosophy - Problems, Tools and
    Goals. 1973.

51. Radu J. Bogdan and Ilkka Niiniluoto (eds.). Logic, Language,
    and Probability. 1973.

52. Glenn Pearce and Patrick Maynard (eds.). Conceptual Change.
    1973.

53. Ilkka Niiniluoto and Raimo Tuomela. Theoretical Concepts and
    Hypothetico-Inductive Inference. 1973.

54. Roland Fraissé. Course of Mathematical Logic - Volume 1:
    Relation and Logical Formula. 1973.

55. Adolf Grünbaum. Philosophical Problems of Space and Time.
    (Second, enlaged edition.) 1973. BSPS 12.

56. Patrick Suppes (ed.). Space, Time, and Geometry. 1973.

57. Hans Kelsen. Essays in Legal and Moral Philosophy (selected
    and introduced by Ota Weinberger). 1973.

58. R. J. Seeger and Robert S. Cohen (eds.). Philosophical Founda-
    tions of Science. 1974. BSPS 11.

59. Robert S. Cohen and Marx W. Wartofsky (eds.). Logical and
    Epistemological Studies in Contemporary Physics. 1973. BSPS
    13.

60. Robert S. Cohen and Marx W. Wartofsky (eds.). Methodological
    and Historical Essays in the Natural and Social Sciences.
    Proceedings of the Boston Colloquium for the Philosophy of
    Science 1969-1972. 1974. BSPS 14.

61. Robert S. Cohen, J. J. Stachel, and Marx W. Wartofsky (eds.).
    For Dirk Struik. Scientific, Historical and Political Essays
    in Honor of Dirk J. Struik. 1974. BSPS 15.

62. Kazimierz Ajdukiewicz. Pragmatic Logic (transl. from the
    Polish by Olgierd Wojtasiewicz). 1974.

63. Sören Stenlund (ed.). Logical Theory and Semantic Analysis.
    Essays Dedicated to Stig Kanger on His Fiftieth Birthday.
    1974.

64. Kenneth F. Schaffner and Robert S. Cohen (eds.). Proceedings
    of the 1972 Biennial Meeting, Philosophy of Science Association.
    1974. BSPS 20.

65. Henry E. Kyburg, Jr. The Logical Foundations of Statistical
    Inference. 1974.

66. Marjorie Grene. The Understanding of Nature. Essays in the Philosophy of Biology. 1974. BSPS 23.

67. Jan M. Brockman. Structuralism: Moscow, Prague, Paris. 1974.

68. Norman Geschwind. Selected Papers on Language and the Brain. 1974. BSPS 16.

69. Roland Fraissé. Course of Mathematical Logic - Volume 2: Model Theory. 1974.

70. Andrzej Grzegorczyk. An Outline of Mathematical Logic. Fundamental Results and Notions Explained with All Details. 1974.

71. Franz von Kutschera. Philosophy of Language. 1975.

72. Juha Manninen and Raimo Tuomela (eds.). Essays on Explanation and Understanding. Studies in the Foundations of Humanities and Social Sciences. 1976.

73. Jaakko Hintikka (ed.). Rudolf Carnap, Logical Empiricist. Materials and Perspectivies. 1975.

74. Milič Čapek (ed.). The Concepts of Space and Time. Their Structure and Their Development. 1976. BSPS 22.

75. Jaakko Hintikka and Unto Remes. The Method of Analysis. Its Geometrical Origin and Its General Significance. 1974. BSPS 25.

76. John Emery Murdoch and Edith Dudley Sylla. The Cultural Context of Medieval Learning. 1975. BSPS 26.

77. Stefan Amsterdamski. Between Experience and Metaphysics. Philosophical Problems of the Evolution of Science. 1975. BSPS 35.

78. Patrick Suppes (ed.). Logic and Probability in Quantum Mechanics. 1976.

79. Hermann von Helmholtz: Epistemological Writings. The Paul Hertz/Moritz Schlick Centenary Edition of 1921 with Notes and Commentary by the Editors. (Newly translated by Malcolm F. Lowe. Edited, with an Introduction and Bibliography, by Robert S. Cohen and Yehuda Elkana.) 1977. BSPS 37.

80. Joseph Agassi. Science in Flux. 1975. BSPS 28.

81. Sandra G. Harding (ed.). Can Theories Be Refuted? Essays on the Duhem-Quine Thesis. 1976.

82. Stefan Nowak. Methodology of Sociological Research. General Problems. 1977.

83. Jean Piaget, Jean-Blaise Grize, Alina Szeminska, and Vinh Bang. Epistemology and Psychology of Functions. 1977.

84. Majorie Grene and Everett Mendelsohn (eds.). Topics in the Philosophy of Biology. 1976. BSPS 27.

85. E. Fischbein. The Intuitive Sources of Probabilistic Thinking in Children. 1975.

86. Ernest W. Adams. The Logic of Conditionals. An Application of Probability to Deductive Logic. 1975.

87. Marian Przełęcki and Ryszard Wójcicki (eds.). Twenty-Five Years of Logical Methodology in Poland. 1977.

88. J. Topolski. The Methodology of History. 1976.

89. A. Kasher (ed.). Language in Focus: Foundations, Methods and Systems. Essays Dedicated to Yehoshua Bar-Hillel. 1976. BSPS 43.

90. Jaakko Hintikka. The Intentions of Intentionality and Other New Models of Modalities. 1975.

91. Wolfgang Stegmüller. Collected Papers on Epistemology, Philosophy of Science and History of Philosophy. 2 Volumes. 1977.

92. Dov M. Gabbay. Investigations in Modal and Tense Logics with Applications to Problems in Philosophy and Linguistics. 1976.

93. Radu J. Bogdan. Local Induction. 1976.

94. Stefan Nowak. Understanding and Prediction. Essays in the Methodology of Social and Behavorial Theories. 1976.

95. Peter Mittelstaedt. Philosophical Problems of Modern Physics. 1976. BSPS 18.

96. Gerald Holton and William Blanpied (eds.). Science and Its Public: The Changing Relationship. 1976. BSPS 33.

97. Myles Brand and Douglas Walton (eds.). Action Theory. 1976.

98. Paul Gochet. Outline of a Nominalist Theory of Proposition. An Essay in the Theory of Meaning. 1980.

99. R. S. Cohen, P. K. Feyerabend, and M. W. Wartofsky (eds.). Essays in Memory of Imre Lakatos. 1976. BSPS 39.

100. R. S. Cohen and J. J. Stachel (eds.). Selected Papers of Léon Rosenfeld. 1978. BSPS 21.

101. R. S. Cohen, C. A. Hooker, A. C. Michalos, and J. W. van Evra (eds.). PSA 1974: Proceedings of the 1974 Biennial Meeting of the Philosophy of Science Association. 1976. BSPS 32.

102. Yehuda Fried and Joseph Agassi. Paranoia: A Study in Diag-
     nosis. 1976. BSPS 50.

103. Marian Przełęcki, Klemens Szaniawski, and Ryszard Wójcicki
     (eds.). Formal Methods in the Methodology of Empirical
     Sciences. 1976.

104. John M. Vickers. Belief and Probability. 1976.

105. Kurt H. Wolff. Surrender and Catch: Experience and Inquiry
     Today. 1976. BSPS 51.

106. Karel Kosík. Dialectics of the Concrete. 1976. BSPS 52.

107. Nelson Goodman. The Structure of Appearance. (Third edition.)
     1977. BSPS 53.

108. Jerzy Giedymin (ed.). Kazimierz Ajdukiewicz: The Scientific
     World-Perspective and Other Essays 1931-1963. 1978.

109. Robert L. Causey. Unity of Science. 1977.

110. Richard E. Grandy. Advanced Logic for Applications. 1977.

111. Robert P. McArthur. Tense Logic. 1976.

112. Lars Lindahl. Position and Change. A Study in Law and Logic.
     1977.

113. Raimo Tuomela. Dispositions. 1978.

114. Herbert A. Simon. Models of Discovery and Other Topics in
     the Methods of Science. 1977. BSPS 54.

115. Roger D. Rosenkrantz. Inference, Method and Decision. 1977.

116. Raimo Tumela. Human Action and its Explanation. A Study on
     the Philosophical Foundations of Psychology. 1977.

117. Morris Lazerowitz. The Language of Philosophy. Freud and
     Wittgenstein. 1977. BSPS 55.

118. Stanislaw Leśniewski. Collected Works (ed. by S. J. Surma, J.
     T. J. Srzednicki, and D. I. Barnett, and an annotated biblio-
     graphy by V. Frederick Rickey). 1982. (Forthcoming.)

119. Jerzy Pelc. Semiotics in Poland, 1894-1969. 1978.

120. Ingmar Pörn. Action Theory and Social Science. Some Formal
     Models. 1977.

121. Joseph Margolis. Persons and Minds. The Prospects of Non-
     reductive Materialsm. 1977. BSPS 57.

122. Jaakko Hintikka, Ilkka Niiniluoto, and Esa Saarinen (eds.). Essays on Mathematical and Philosophical Logic. 1978.

123. Theo A. F. Kuipers. Studies in Inductive Probability and Rational Expectation. 1978.

124. Esa Saarinen, Risto Hilpinen, Ilkka Niiniluoto, and Merrill Provence Hintikka (eds.). Essays in Honour of Jaakko Hintikka on the Occasion of His Fiftieth Birthday. 1978.

125. Gerard Radnitzky and Gunnar Andersson (eds.). Progress and Rationality in Science. 1978. BSPS 58.

126. Peter Mittelstaedt. Quantum Logic. 1978.

127. Kenneth A Bowen. Model Theory for Modal Logic. Kripke Models for Modal Predicate Calculi. 1978.

128. Howard Alexander Bursen. Dismantling the Memory Machine. A Philosophical Investigation of Machine Theories of Memory. 1978.

129. Marx W. Wartofsky. Models: Representation and the Scientific Understanding. 1979. BSPS 48.

130. Don Ihde. Technics and Praxis. A Philosophy of Technology. 1978. BSPS 24.

131. Jerzy J. Wiatr (ed.). Polish Essays in the Methodology of the Social Sciences. 1979. BSPS 29.

132. Wesley C. Salmon (ed.). Hans Reichenbach: Logical Empiricist. 1979.

133. Peter Bieri, Rolf-P. Horstmann, and Lorenz Krüger (eds.). Transcendental Arguments in Science. Essays in Epistermology. 1979.

134. Mihalilo Marković and Gajo Petrović (eds.). Praxis. Yugoslav Essays in the Philosophy and Methodology of the Social Sciences. 1979. BSPS 36.

135. Ryszard Wójcicki. Topics in the Formal Methodology of Empirical Sciences. 1979.

136. Gerard Radnitzky and Gunnar Andersson (eds.). The Structure and Development of Science. 1979. BSPS 59.

137. Judson Chambers Webb. Mechanism, Mentalism, and Metamathematics. An Essay on Finitism. 1980.

138. D. F. Gustafson and B. L. Tapscott (eds.). Body, Mind, and Method. Essays in Honor of Virgil C. Aldrich. 1979.

139. Leszek Nowak. The Structure of Idealization. Towards a Systematic Interpretation of the Marxian Idea of Science. 1979.

140. Chaim Perelman. The New Rhetoric and the Humanities. Essays on Rhetoric and Its Applications. 1979.

141. Wlodzimierz Rabinowicz. Universalizability. A Study in Morals and Metaphysics. 1979.

142. Chaim Perelman. Justice, Law, and Argument. Essays on Moral and Legal Reasoning. 1980.

143. Stig Kanger and Sven Öhman (eds.). Philosophy and Grammar. Papers on the Occasion of the Quincentennial of Uppsala University. 1981.

144. Tadeusz Pawlowski. Concept Formation in the Humanities and the Social Sciences. 1980.

145. Jaakko Hintikka, David Gruender, and Evandro Agazzi (eds.). Theory Change, Ancient Axiomatics, and Galileo's Methodology. Proceedings of the 1978 Pisa Conference on the History and Philosophy of Science, Volume 1. 1981.

146. Jaakko Hintikka, David Gruender, and Evandro Agazzi (eds.). Probabilistic Thinking, Thermodynamics, and the Interaction of the History and Philosophy of Science. Proceedings of the 1978 Pisa Conference on the History and Philosophy of Science, Volume II. 1981.

147. Uwe Mönnich (ed.). Aspects of Philosophical Logic. Some Logical Forays into Central Notions of Linguistics and Philosophy. 1981.

148. Dov M. Gabbay. Semantical Investigations in Heyting's Intuitionistic Logic. 1981.

149. Evandro Agazzi (ed.). Modern Logic - A Survey. Historical, Philosophical, and Mathematical Aspects of Modern Logic and its Applications. 1981.

150. A. F. Parker-Rhodes. The Theory of Indistinguishables. A Search for Explanatory Principles below the Level of Physics. 1981.

151. J. C. Pitt. Pictures, Images, and Conceptual Change. An Analysis of Wilfrid Sellars' Philosophy of Science. 1981.

152. R. Hilpinen (ed.). New Studies in Deontic Logic. 1981.

153. C. Dilworth. Scientific Progress. A Study Concerning the Nature of the Relation Between Successive Scientific Theories. 1981.

APPENDIX 3

The University of Western Ontario
Series in Philosophy of Science

1. J. Leach, R. Butts, and G. Pearce (eds.). Science, Decision
    and Value. Proceedings of the Fifth University of Western
    Ontario Philosophy Colloquium. 1969. 1973. vii+213 pp.

2. C. A. Hooker (ed.). Contemporary Research in the Foundations
    and Philosophy of Quantum Theory. Proceedings of a Conference
    held at the University of Western Ontario, London, Canada.
    1973. xx+385 pp.

3. J. Bub. The Interpretation of Quantum Mechanics. 1974. ix+
    155 pp.

4. D. Hockney, W. Harper, and B. Freed (eds.). Contemporary Re-
    search in Philosophical Logic and Linguistic Semantics. Pro-
    ceedings of a Conference held at the University of Western
    Ontario, London, Canada. 1975. vii+332 pp.

5. C. A. Hooker (ed.). The Logico-Algebraic Approach to Quantum
    Mechanics. 1975. xv+607 pp.

6. W. L. Harper and C. A. Hooker (eds.). Foundations of Proba-
    bility Theory, Statistical Inference, and Statistical Theories
    of Science. 3 Volumes. Vol. I: Foundations and Philosophy
    of Epistemic Applications of Probability Theory. 1976. xi+308
    pp. Vol II: Foundations and Philosophy of Statistical Infer-
    ence. 1976. xi+455 pp. Vol. III: Foundations and Philosophy
    of Statistical Theories in the Physical Sciences. 1976. xii +
    241 pp.

7. C. A. Hooker (ed.). Physical Theory as Logico-Operational
    Structure. 1978. 334 pp.

8. J. M. Nicholas (ed.). Images, Perception, and Knowledge.
    Papers deriving from and related to the Philosophy of Science
    Workshop in Ontario, Canada, May 1974. 1977. ix+309 pp.

9. R. E. Butts and J. Hintikka (eds.). Logic, Foundations of
    Mathematics, and Computability Theory. Part One of the Pro-
    ceedings of the Fifth International Congress of Logic, Metho-
    dology and Philosophy of Science, London, Ontario, Canada,
    1975. 1977. x+406 pp.

10. R. E. Butts and J. Hintikka (eds.). Foundational Problems in
    the Special Sciences. Part Two of the Proceedings of the Fifth
    International Congress of Logic, Methodology and Philosophy of
    Science, London, Ontario, Canada, 1975. 1977. x+427 pp.

11. R. E. Butts and J. Hintikka (eds.). Basic Problems in Metho-
    dology and Linguistics. Part Three of the Proceedings of the
    Fifth International Congress of Logic, Methodology and Philo-
    sophy of Science, London, Ontario, Canada, 1975. 1977.
    x+321 pp.

12. R. E. Butts and J. Hintikka (eds.). Historical and Philoso-
    phical Dimensions of Logic, Methodology and Philosophy of
    Science. Part Four of the Proceedings of the Fifth Interna-
    tional Congress of Logic, Methodology and Philosophy of
    Science, London, Ontario, Canada, 1975. 1977. x + 336 pp.

13. C. A. Hooker (ed.). Foundations and Applications of Decision
    Theory. 2 volumes. Vol I: Theoretical Foundations. 1978.
    xxiii+442 pp. Vol. II: Epistemic and Social Applications.
    1978. xxiii+206 pp.

14. R. E. Butts and J. C. Pitt (eds.). New Perspectives on Galileo.
    Papers deriving from and related to a workshop on Galileo held
    at Virginia Polytechnic Institute and State University, 1975.
    1978. xvi+262 pp.

Titles of Periodicals Cited

Acta Philosophica Fennica

Actualités scientifiques et
industrielles

Ajatus

Algemeen Nederlands Tijd-
schrift voor Wijsbe-
geerete

Allgemeine Zeitschrift für
Philosophie

Analysis

Annales de l'Institut de
Philosophie (Brussels)

Annali di mathematica, pura
e applicata

Annals of Mathematics

Annals of the Japan Association for Philosophy
of Science (Kagaku
Kisoron Gakkai)

American Journal of Physics

American Philosophical
Quarterly

American Political Science
Review

American Scientist

Archiv für mathematische
Logik und Grundlagen-
forschung

Archiv für Philosophie

Archives de philosophie

Australasian Journal of
Philosophy

Australian Journal of
Science

Beihefte zur Zeitschrift
für philosophische
Forschung

British Journal for the
Philosophy of Science

Bulletin de la Société
Française de Philosophie

Bulletin of the American
Academy of Arts and
Sciences

Bulletin of the British
    Society for the His-
    tory of Science

Cambridge Journal

Cambridge Review

Canadian Journal of
    Philosophy

Centaurus

Communications in Pure and
    Applied Mathematics

Communications on Mathema-
    tical Physics

Comparative Studies in
    Society and History

Conceptus: Zeitschrift
    für Philosophie

Continuum

Czechoslovakian Journal
    of Theoretical Physics

Daedelus

Danish Yearbook of Phi-
    losophy

Deutsche Zeitschrift für
    Philosophie

Dialectica

Dialectics and Humanism

Dialogos

Dialogue: Canadian Philo-
    sophical Review - Re-
    vue Canadienne de
    Philosophie

Diánoia

Diogenes

Discovery

Duke Mathematical Journal

Duquesne Studies: Philo-
    sophical Series

Encounter

Epistemologia

Erkenntnis

Ethics

Etudes d'épistémologie
    génétique

Les Etudes philosophiques

Experientia

Foundations of Physics

Fragmenty Filozoficzne

Granta

Grazer philosophische
    Studien

Gregorianum

Helvetica Physica Acta

The Heythrop Journal

History and Theory

History of Science

Idealistic Studies

The Independent Journal
    of Philosophy

The Indian Journal of
    Philosophy

Inquiry

International Journal of
    Theoretical Physics

International Logic Review

International Philosophical
   Quarterly

International Studies in
   Philosophy

Isis

Iyyun

Journal of Chemical Edu-
   cation

Journal of Critical Analysis

The Journal of Interdis-
   ciplinary History

Journal of Mathematics and
   Mechanics

Journal of Philosophical
   Logic

Journal of Philosophy

The Journal of Symbolic
   Logic

Journal of the History of
   Ideas

Journal of Unified Science

The Journal of Value Inquiry

Kinesis

Laval Théologique et Philosophique

The Listener

Logique et analyse

Man and World

Metaphilosophy

Methodology and Science

Methodos

Midwest Studies in
Philosophy

Mind

Minerva

The Modern Schoolman

The Monist

Nature

Nature and System

Neue Hefte für Philosophie

New Literary History

The New Scholasticism

Noesis

Notre Dame Journal of Formal Logic

Noûs

Organon

Pacific Journal of Mathe-
   matics

Pacific Philosophical Quarterly

Pacific Philosophy Forum

Pensamiento

The Personalist

Perspektiven der Philosophie

Philosophia: Philosophical
   Quarterly of Israel

Philosophia Mathematica

Philosophia Naturalis

Philosophic Exchange

Philosophica

Philosophical Forum

Philosophical Inquiry

Philosophical Investigations

The Philosophical Journal

Philosophical Papers

The Philosophical Quarterly
(Scotland)

The Philosophical Quarterly
(India)

The Philosophical Review

Philosophical Studies
(Holland)

Philosophical Studies
(Ireland)

Philosophical Topics

Philosophische Perspektiven

Philosophische Rundschau

Philosophisches Jahrbuch

Philosophy

Philosophy and Literature

Philosophy and Phenomeno-
logical Research

Philosophy and Rhetoric

The Philosophy Forum

Philosophy of Science

Philosophy of the Social
Sciences

Philosophy Today

Poznań Studies in the Philosophy of the Sciences
and the Humanities

Proceedings and Addresses of
the American Philoso-
phical Association

Proceedings of the American
Academy of Arts and
Sciences

Proceedings of the American
Catholic Philosophical
Association

Proceedings of the American
Philosophical Society

Proceedings of the Aristote-
lian Society (New
Series & Supplementary
Volumes)

Proceedings of the British
Academy

Process Studies

Proteus

Przegląd Filozoficzny

Quality and Quantity

Rassegne de filosofia

Ratio

Reports on Philosophy

The Review of Metaphysics

Reviews of Modern Physics

Revue de l'Université
d'Ottawa

Revue de métaphysique et
de morale

Revue des sciences philosophiques et théologiques

Revue internationale de
philosophie

Revue philosophique de la
    France et de l'Etranger

Revue philosophique de
    Louvain

Salzburger Jahrbuch für Phi-
    losophie

Science

Science et esprit

Sciences ecclesiastiques

Scientia

Scientific Monthly

The Southern Journal of Phi-
    losophy

Southwestern Journal of Phi-
    losophy

Soviet Studies in Philosophy

Studia Filozoficzne

Studia Logica

Studia Philosophica

Studies in History and Phi-
    losophy of Science

Studies in Soviet Thought

Studium Generale

Synthese

Synthesis

Systematic Zoology

Telos

Theoria

Theoria to Theory

Theory and Decision

The Thomist

Thought

Tijdschrift voor Filosofie

Transactions of the Charles
    S. Peirce Society

Tulane Studies in Philosophy

Universities Quarterly

University of California
    Publications in Philo-
    sophy

Vistas in Astronomy

Voprosy filosofii

Wiener Jahrbuch für Philo-
    sophie

Zeitschrift für allgemeine
    Wissenschaftstheorie

Zeitschrift für philosophi-
    sche Forschung

Życie nauki

Zygon

Index of Personal Names

The numbers given below are entry numbers, not page numbers.
Underlined numbers indicate names within titles.

Abelé, J. 5209-10, 6188
Abell, M. A. 6823
Abeloos, M. 7096, 7323
Abercrombie, M. L. J. 1164
Abramenko, B. 5491, 6458
Abruzzi, A. 1165
Achinstein, P. 73, 474,
    525-26, 948, 1108,
    1211, 1384, 1523-24,
    1685-86, 1849-51, 1990,
    2137, 2254, 2471-73,
    2473-44, 2753, 3051,
    3491-92, 3627, 3902,
    4303, 4322, 6985
Ackermann, R. 1133, 1194-96,
    1825, 2255, 2474, 2904,
    3493, 3749, 4304-05,
    4411, 4501
Ackoff, R. L. 475, 1194,
    4306
Adam, E. 6189
Adams, E. P. 6245
Adams, E. W. 1037, 3105-06,
    3993-95, 4061
Adams, J. Q. 6459
Adams, Jr., J. S. 56
Addis, L. 527, 2475
Addison, J. 1687, 5936,
    6138

Adler, C. G. 5594
Adler, J. 2454, 2745, 2768
Adler, N. 5432
Adolphe, L. 5108
Adrian Marie, Sr. 7097
Aeschlimann, F. 1385
Afanasjew, W. G. 6797
Agassi, J. 1, 57, 163,
    313-14, 469, 528-31,
    580, 627, 950, 1008,
    1166-68, 1215, 1386-88,
    1525-28, 1688, 2109,
    2435-37, 2455, 2476,
    2746, 3107, 3494-96,
    3498, 3750-56, 3865,
    3900, 3907, 3996, 4098,
    4502-03, 4673-76, 4709,
    4921-24, 5036, 5120,
    5239, 5595, 5720, 6870
Agazzi, E. 164, 315, 532,
    1010-11, 4191, 4412,
    4925-27, 5211
Agar, W. E. 6798
Ageno, M. 6799
Ager, T. A. 2256
Ahundov, M. D. 6460
Ajdukiewicz, K. 2, 996,
    1169-70, 2712, 2816,

4099, 4928
Akchurin, J. A.   470, 504,
    1171, 1689, 5212-13
Albert, H.   5039
Albertson, J.   5655
Albrecht-Buehler, G.   3497
Albritton, Jr., C. G.   1797,
    5166
Alder, M. D.   2257
Aldrich, V. C.   1105
Aleksandrov, A. D.   6461
Alexander, H.   5100
Alexander, H. G.   2713,
    2747, 3498
Alexander, P.   533, 1529,
    2110, 2258-59, 2477
Alexander, R. D.   7098
Aliotta, A.   534, 6190
Allan, D. M.   4307
Allen, E. H.   3108
Almeder, R. F.   4677
Almog, J.   5656
Alston, W. P.   2478, 4100
Altham, J. E. J.   3067
Altschul, E.   1690
Altwegg, M.   2260
Ambacher, M.   165, 5433
Ambartsumian, V. A.   4678,
    5492
Ambrose, A.   1852, 2748
Amit, D.   4295
Amsterdamski, S.   58, 535,
    4679
Anderson, A. R.   1853, 4353
Anderson, O.   3109
Anderson, Jr., R. M.   1057,
    1172, 1389
Andersson, G.   166, 3757,
    3923, 4680, 4855-56
André, H.   7057
Angel, R. B.   2479, 6191
Angelidis, Th. D.   5657
Anguera-Argilaga, T.   1390
Annis, D. B.   3499
Anokhin, P. K.   6800
Anscombe, F. J.   3289
Anthony, G. F. P.   7099
Antiseri, D.   3758-59
Anton, P.   4101
Apostel, L.   1691, 2749
Aquinas, Thomas   5332
Arber, A.   6801-03

Archibald, G. C.   3760
Ardley, G. W. R.   59, 316,
    3706, 4799
Aristotle   202, 326, 399,
    643, 690, 2487, 2703,
    4373, 4375, 4468, 5484,
    5488, 6878, 7070
Armstrong, D. H.   4504
Armstrong, D. M.   1391, 5214
Aron, J. - P.   7100, 7324
Aron, R.   1173
Aronov, R. A.   5658
Aronson, J. L.   2256, 2480,
    4192
Arthur, R.   536, 1854
Ascheberg, R.   6625
Ashby, R. W.   1855
Ashley, B. M.   7101
Asquith, P. D.   92, 96, 108,
    110, 120, 143, 150,
    319, 351, 394, 951,
    1091-93, 1918, 2278,
    2553, 2947, 3206, 3839,
    3995, 4203, 4498, 5253,
    6893
Aster, E. von   4102
Asti Vera, A.   537
Atkinson, G.   2750
Atlan, H.   4308
Aubert, V.   2481
Auble, G. T.   6955
Audi, M.   5659
Audi, R.   2483
Audretsch, J.   4681
Auger, P.   1692, 5109
Augustynek, Z.   5215, 6192
    6462
Aune, B.   538, 4224
Auroux, S.   3707
Austin, J. H.   1530, 4193
Austin, W. H.   4682
Avaliani, S. S.   539
Avishai, Y.   4194
Axinn, S.   5040
Ayala, F. J.   317, 1549,
    4309, 4314, 4413, 4416,
    4423, 4437, 4466, 4468,
    4478, 4496, 5019,
    6804-05, 6853, 6944,
    6977, 6995, 7058,
    7102-03, 7166, 7150,
    7292

Ayer, A. J.   167, 2111,
    2751, 3110-12, 3500,
    3761, 4103
Ayers, M. R.   2112, 5110
Azar, T.,   7104

Bachelard, G.   165, 540-43,
    843, 957, 6193
Bachelerd, S.   5216
Bachem, A.   5331, 6806
Bacon, F.   2889
Bacon, J.   1392, 2113
Baekers, S. F.   318
Bafico, R.   6194
Bahm, A. J.   4414-15
Baigrie, B.   4810
Bailey, O. T.   6807
Baillie, P.   3501-02, 3708,
    4683
Baker, L. R.   6463-64
Balan, S.   5596
Baldamus, W.   4684
Baldini, M.   544
Balestra, D. J.   3709
Balinfante, F. J.   5664-65
Ball, T.   3762
Ballard, E. G.   1174, 1531,
    3965, 3997, 4195
Baltas, A.   3763
Balz, A. G. A.   5434
Balzer, W.   1090, 1393,
    2261-62, 2343, 4685,
    5597, 6056
Banigan, S.   650
Bantz, D. A.   1532
Baptist, J. H.   3113
Baran, B.   4929
Barashenkov, V. S.   5435,
    6195
Barber, B.   319
Barbour, I.   320-21, 5111
Bar-Hillel, M.   3114
Bar-Hillel, U.   5937
Bar-Hillel, Y.   3, 1038,
    1384, 1856, 2752, 2789,
    2925, 3019, 3195, 3389,
    3627, 3746, 4505, 4930
Barker, A. D.   7105
Barker, J. A.   4204
Barker, P.   4686
Barker, S. F.   948, 1798-99,
    273, 3014, 3022, 3051,

3503, 3765
Bärmark, J.   952
Barnard, G. A.   3158, 3228
Barnes, J. A.   7173
Barnett, S. A.   7106
Barnette, R. L.   5740
Barr, H. J.   4196
Barr, W. F.   1693-94
Barraud, H. - J.   545-46
Barreau, H.   5660, 6196,
    6465-66
Barrett, P. H.   1576
Barrett, R. B.   2029, 3710
Barricelli, N. A.   7107
Barter, E. G.   6197
Barthélemy-Madaule, M.   6808
Bartlett, M. S.   3115
Bartlett, S. J.   1533,
    1857-58
Bartley, S. H.   4065
Bartley, III, W. W.   2754,
    3765-66, 3935, 4931
    4941, 4979
Barzin, M.   4104
Basri, S. A.   6467
Bass, R. E.   2114
Bastin, T.   5661-62
Batens, D.   2755, 2939,
    3504, 4932
Bateson, G.   547
Baublys, K. K.   4310
Baum, R. J.   60
Baumer, W. H.   3351, 3480,
    3506-08
Baumgaertner, W.   2756
Baumrin, B.   42, 1071-72
Bavink, B.   548, 5041, 5167
Bawden, H. H.   6809
Bayertz, K.   549
Bayes, T.   851, 2442, 2819,
    3033, 3136, 3167, 3202,
    3213, 3218, 3246-47,
    3255, 3261-62, 3265,
    3347, 3360, 3423, 3443,
    3452, 3487, 3567, 3579,
    3639, 3669, 3811, 3818,
    5008, 5258
Bazenov, L. B.   1800
Bealer, G.   5436
Beard, R. W.   2490
Beardsley, E. L.   2115
Beatty, J. H.   1695, 7108,
    7237

Beauchamp, T. L.   2116,
    4216, 4260
Beauregard, L. A.   1859,
    6198-99
Bechert, K.   1175
Beckler, Z.   4687
Bechtel, P. W.   1394
Beck, G.   2263
Beck, L. W.   1176-77, 5331
Beck, W. S.   6810
Becker, G.   550
Becker, O.   5217
Beckermann, A.   4197
Beckman, T. A.   4098
Beckner, M.   4311, 4416,
    6811-13
Bedau, H.   5663
Beigbeder, M.   6814
Belis, M.   3116
Bell, C. G.   4312
Bell, John   6782
Bell, J. S.   5712, 5715,
    5880, 5926, 6112, 6234
Bellmann, R.   1271
Bellone, E.   4806, 5246
Belnap, N.   1917
Beltrametti, E. G.   5666-67
Benardete, J. A.   2757, 3998
Bendall, K.   5112
Bender, H.   7345
Bendmann, A.   6797
Benioff, P.   3117
Benjamin, A. C.   61, 551-52,
    3966-67, 4688
Benl, G.   7109
Bennett, J.   5113, 5493
Bennett, J. F.   3118
Bennett, J. G.   553
Bennett, H. S.   554
Benoist, R. W.   5668
Bense, M.   7020
Benson, A. J.   8
Benton, E.   4313
Benvenuto, E.   555
Berenda, C. W.   1860, 2117,
    4105, 4417, 5218,
    5494-95, 5669, 6200,
    6405, 6468
Berent, P.   2264, 3509
Berg, J.   1861-64
Berg, M.   2455
Berger, G.   6274, 6469-71,
    6529
Berghuys, J. J. W.   5437

Bergman, H.   4198
Bergmann, G.   5, 556-57,
    998, 1865-66, 1945,
    2758, 3487, 4106-08,
    4506, 5219, 6472
Bergmann, P. G.   168, 5496,
    6201, 6473
Bergson, H.   211, 1002,
    5249, 5251, 6344, 6765
Bergstein, T.   5670
Berk, E.   4507
Berka, K.   4000
Berkeley, G.   3893, 4503
Berkowitz, L. J.   1867
Berkson, W.   4675, 4689
Berlinski, D.   1696, 7325
Bernal, J. D.   2250
Bernard, C.   3854
Bernardini, C.   5220
Bernardini, S.   3767
Bernays, P.   62, 1178, 1395,
    2390, 3220, 4933, 5671
Bernier, R.   6815, 6964
Bernoulli, J.   3109
Bernsen, N. O.   3768
Bernstein, J.   3968
Bernstein, R. J.   1585
Berofsky, B.   2118, 4260
Berry, G.   1868
Berstein, R. J.   4508
Bertalanffy, L. von   761,
    1697, 2119, 4509, 6797,
    6816-24, 7110
Berteval, W.   4199
Bertholet, E.   322
Bertolet, R. J.   2759
Bertiau, F. C.   5497
Beth, E. W.   6, 29, 63, 64,
    169, 323, 558-61, 1080,
    869-71, 1916, 5672
Betz, F.   3866
Beurton, P.   7111-13
Beveridge, W. I. B.   1179
Bhaskar, R.   4510-11
Bhattacharya, N.   170, 3769
    4200
Bianca, M.   1698
Bibler, V. S.   4690
Biem, W.   5687
Bigelow, J. C.   3119, 5673
Bigelow, Julian   4377
Binkley, R. W.   4691
Binns, P.   3711

Birch, C. 4314
Bird, J. H. 5498
Bird, O. 171
Birjukov, B. V. 1180
Birkhoff, G. 5674, 6063,
    6200
Birnbaum, A. 1396
Birnbaum, I. 4201
Biser, E. 1690, 2120, 4512,
    4692, 6474
Bitsakis, E. 6202
Bjørnestad, Ø. 5675
Blachowicz, J. A. 4693
Black, M. 1181, 1699, 1872,
    1877, 2744, 2760-64,
    2899, 3510, 6475
Blackburn, S. 2765-66
Blackmore, J. T. 172, 562,
    4513, 4694
Blackwell, R. J. 65, 173,
    563, 912, 1534-38,
    2265, 4934, 6565
Blair, D. G. 3120
Blair, G. A. 324
Blair, G. W. S. 5168,
    5180, 6523
Blake, R. M. 174
Blakeley, T. J. 1355
Blanc, A. C. 7114
Blanché, R. 325, 564-65,
    1182, 2767
Blandino, G. 6825
Blanpied, W. 1014
Blanshard, B. 692
Bloch, K. 5042
Bloch, W. 5169
Block, N. J. 1183
Blokhintsev, D. I. 5676
Blom, S. 3121
Blondel, S. 5114
Bloomfield, L. 566
Bloor, D. 567, 1245, 3801
Blum, A. 2121, 3493
Blum, H. F. 6826, 7115
Boch, W. J. 7186
Bochenski, J. M. 1184
Bochner, S. 326, 5221-22
Bode, R. R. 5223
Boden, M. 2483, 6827
Boesiger, E. 7116
Bogaard, P. A. 4418
Bogdan, R. J. 53, 953,
    1041, 2768, 3401, 4053,
    5346

Boh, I. 227
Böhler, D. 282, 4695
Bohm, D. 568, 1185, 2266,
    3224, 3677-81, 5794,
    6875
Böhm, W. 569, 5174, 5515,
    6203, 6828, 6854
Böhme, G. 4001, 4696-97,
    6625
Bohnen, A. 4109
Bohnert, H. G. 175, 1882,
    2267, 2401, 5347
Bohr, N. 198, 259, 471,
    494, 5225-27, 5482,
    5669, 5682, 5788, 5795,
    5828, 5844, 5903, 5939,
    5943, 5972, 6028, 6085,
    6151
Boirel, R. 570
Boldrini, M. 5043
Bolker, E. D. 3122
Boltzmann, L. 954
Bona, E. 23, 176
Bondi, H. 3900, 5170, 5455,
    5499, 5591, 6203
Bone, F. 7117
Bonevac, D. 4419
Bonhoeffer, K. F. 6829
Bonin, G. von 6830
Bonjour, L. A. 5044
Bonnet de Viller, A. 327
Bonner, J. T. 1700
Bonnor, W. B. 3770
Bonsack, F. 328, 3123,
    5683, 6249
Boole, G. 5139, 5735,
    5758, 6185
Boon, L. 3771
Boorse, C. 7050
Bopp, E. 5598
Bopp, F. 6584
Borch, K. 3357
Boring, E. G. 955
Bork, A. M. 1186
Born, M. 3125, 3717,
    4514-15, 5007, 5228-31,
    5685-87, 6170
Bornemisza, S. T. 571
Borsari, R. 6476
Bory, C. 4002
Borzeszkowski, H.-H. von
    5500, 6204, 6477
Boudot, M. 2769
Bouligand, G. 957, 1539,

1801
Bouveresse, R. 3772
Bowdery, G. J. 5232
Bowie, G. L. 2122
Bowler, P. J. 7118
Bowman, P. A. 1523, 2400,
    6205, 6248, 6478-79
Boyd, R. N. 4516-17, 4546,
    5115
Boyle, R. 1527
Bradie, M. P. 177, 1540,
    4518-19, 7119
Bradley, F. 648
Bradley, J. 178, 3969, 5194
Bradley, M. C. 4656
Bradley, R. D. 5116
Bradshaw, G. L. 1672
Brain, W. R. 1397-99
Brainerd, B. 7059
Braithwaite, R. B. 771,
    1701, 2268, 2484-85,
    2800, 3126-27, 3296,
    7120
Brand, M. 4202-04
Brandon, R. N. 4315,
    7121-23
Brannigan, A. 1541
Branskij, V. P. 6022
Bratoev, G. 4205
Braun, E. 66
Braun, G. E. 3773
Breck, A. D. 131, 317, 350,
    389, 908, 958, 1132,
    1950, 3905, 4026, 4609,
    4635, 4900, 5530, 5611,
    5959, 6160, 6203, 6435,
    6824, 7015, 7103, 7131,
    7291
Bressan, A. 2078, 2123,
    5233
Bretzel, P. von 4206
Bridgman, P. W. 3965, 3966,
    3968, 3970-78, 3982,
    3989, 5234, 6206
Brier, B. 4207
Bright, L. 5235
Brillouin, L. 572-73, 5236
Briskman, L. 912, 1542,
    4935
Britton, Jr., G. G. 1150,
    4420
Broad, C. D. 574, 959,
    1107, 3101, 3289

Brodbeck, M. 931, 1134,
    1141, 2486, 2770, 2965,
    4421, 4520
Brody, B. 1135-36, 2487,
    3511, 4422
Broglie, Louis de 627,
    1187, 2288, 5237-38,
    5438, 5669, 5677,
    5688-89, 5944, 6207,
    6480
Bromberger, S. 818, 1543,
    2269-70, 2360
Bronowski, J. 575-77, 2862,
    4698, 6831
Brotman, H. 6481
Brown, G. B. 1188
Brown, G. S. 3128-29
Brown, H. 960
Brown, H. I. 1137, 1241,
    2271, 2771, 3774,
    4699-4702, 4936-37
Brown, H. R. 5699
Brown, J. R. 67, 2438
Brown, P. M. 3130
Brown, R. 1107, 2124, 4316
Brown, R. H. 578
Brühlmann, O. 6208
Bruna, M. 7124
Brüning, W. 179, 329
Brunner, Karl 2277
Brunner, O. 3709
Brush, S. G. 5117, 5439,
    5599, 5646
Bruzzaniti, G. 6482
Bub, J. 2266, 3371, 5661,
    5700-12, 5739
Bubner, R. 114
Buchanan, B. G. 2797
Buchanan, S. 5045
Buchdahl, G. 68, 180, 181,
    330, 579, 1065, 1544,
    2125, 2273, 2772-74,
    4938, 5239, 5600
Büchel, W. 3131, 5240,
    5440-41, 5713-15,
    6209-13, 6483, 7125
Buchler, J. 1189
Buck, R. C. 1088, 2137,
    2488
Bugajski, S. 5716-17
Buican, D. 6832
Builder, G. 6214
Bunch, B. L. 3006
Bunge, M. 69, 288, 580-83,
    950, 961-64, 1190,

1197, 1545, 1702,
1802–04, 1873–76, 2090,
2126–28, 2274–75, 2775,
2789, 3132–33, 3512,
3775, 4208, 4521–22,
5159, 5241–45, 5501,
5601, 5718–21, 6191,
6484–86, 6964, 7313
Bünning, E.   1400, 6833
Burch, G. B.   6834
Burch, R. W.   2489
Burgers, J. M.   584, 5629
6104
Burhenn, H.   2667
Burhoe, R. W.   761, 6831,
7126
Burian, R. M.   70, 1546,
4523, 4703
Burke, T. E.   3776
Burks, A. W.   1547, 2129,
2776–78, 3134–35, 3141,
3513, 4209
Buskovitch, A. V.   71,
1703–04
Butler, J. A. V.   6835
Butts, R. E.   7, 182, 331,
965–68, 1034, 1401,
2779, 4687, 4810
Buxton, R.   3136
Byerly, H. C.   1705, 2276,
4003–04, 4210, 7127
Byrne, Patrick H.   6215
Byrne, Peter   2130

Cahalan, J. C.   7144
Caldin, E. F.   332, 585,
5171–72
Caldirola, P.   5118,
5246–47, 5722–23, 6216, 6487,
6836
Callen, H.   5248
Callot, E.   586
Caloi, P.   1548
Campbell, C. A.   7053
Campbell, D. T.   1549,
3777, 4423, 4860, 7284
Campbell, K.   2780
Campbell, M.   3137, 7128
Campbell, N. R.   2273
Cancienne, D.   5502
Canfield, J.   2490, 2563

4317–18, 4377, 4379,
4390
Canguilhem, G.   183, 587,
1706, 6837
Cannavo, S.   1191, 3138,
3749
Cantore, E.   333–34, 588–89,
5724–25
Capaldi, N.   1136, 2781
Capek, M.   184, 185, 335,
1002, 4524, 5119, 5173,
5249–51, 6217–19,
6488–91
Capitan, W. H.   2730
Caplan, A.   4424, 6838,
7060, 7076, 7129–30
Cardwell, C. E.   2782
Carella, M. J.   5442
Cargile, J.   2783
Carles, J.   6839–40
Carleton, L. R.   1402
Carlo, W. E.   4319, 4425
Carloye, J. C.   1403,
1707–08, 2277
Carlson, R.   3442
Carlsson, G.   3139
Carmichael, P. A.   590
Carnap, R.   8, 64, 148, 175,
186, 189, 213, 214,
226, 247, 269, 287,
296, 471–72, 476, 478,
501, 507, 566, 591–92,
1008, 1065, 1088, 1096,
1108, 1278, 1856,
1877–84, 1896, 1913,
1948, 1951, 1990, 1998,
2009, 2758, 2778,
2784–91, 2823, 2852,
2878, 2882, 2883, 2896,
2908, 2919, 2925, 2970,
3018, 3023, 3050, 3052,
3140–43, 3154, 3241–42,
3256, 3281, 3327, 3375,
3389, 3348, 3487, 3492,
3514–15, 3577–78, 3590,
3609, 3627, 3629, 3664,
4112, 4134, 4137, 4139,
4159–61, 4190, 4646,
5049, 5092, 5098, 5103,
5252, 5291, 5514, 6492,
6590
Carnot, S.   5258
Carr, B.   3778
Carr, D.   717

Carrelli, A. 593
Carrier, L. S. 1404
Carter, G. S. 2018, 7186
Cartwright, H. M. 4005
Cartwright, N. D. 336,
    2132-33, 2491, 4211,
    4426, 5253, 5726-29,
    5822, 6071, 6107
Cartwright, R. 1885
Casagrande, F. 6493
Caspari, E. 6841
Casserly, J. V. L. 7131
Cassidy, J. 7132-33
Cassinelli, G. 5666-67
Cassirer, Ernest 5120
Cassirer, Eva 949
Castaneda, H. N. 2007
Catel, W. 594
Caton, H. 186
Causey, R. L. 473-74, 495,
    2256, 2266, 2278, 4006,
    4427-29
Cavaillès, J. 595
Caws, P. 596, 1138, 1184,
    1550, 1805, 2279, 2492,
    2792, 4212, 6494
Ceccato, S. 3965
Cedarbaum, D. 4704
Centore, F. F. 7134-35
Cerf, W. 844, 4110
Chaisson, E. J. 7136
Chalmers, A. F. 597, 1020,
    2439, 3779, 4705
Chandra, S. 2134
Changeaux, J.-P. 4706
Chapman, T. 5730, 6220
Chari, C. T. K. 5254, 5503,
    5731-34, 6221, 6495-99
Charpa, U. 101
Chatalian, G. 2793, 3144
Chatham, R. E. 6318
Chauchard, P. 337
Chaudhury, P. J. 598-600,
    1888, 2493, 4213, 5504
Chauviré, C. 1551, 3780
Chedd, G. 6842
Chernavska, A. 5735
Chernoff, P. 5736
Cherry, C. 2494
Chihara, C. S. 1552, 3102,
    3516, 4320
Child, J. 1889
Chisholm, R. 1405, 1880,
    1890, 2135-36, 3517
    5061

Chissick, S. 6065
Chomsky, N. 1406, 1891,
    3940
Chistensen, F. 5602, 6222,
    6500
Christian, C. 4007, 5046
Chudinov, E. M. 6223
Chung-Ying Cheng 2794-95
Church, A. 4385
Churchman, C. W. 475, 881,
    1192-94, 1369, 1892,
    2784, 2796-97, 3145-47,
    3518, 3866, 4008-09,
    4113, 4539
Churchland, P. M. 1407,
    3781, 4525
Chwistek, L. 601
Cimutta, J. 7137-38
Clagett, M. 1614
Clark, D. M. 4321
Clark, J. T. 187, 338-39,
    2714, 5255
Clark, R. 2137, 5047
Clarke, C. J. S. 5256,
    5505, 6224
Clarke, J. R. 7139
Clatterbaugh, K. C. 6501
Claude, W. N. 5900
Clavelin, M. 188
Clay, J. 2138, 2280, 3519,
    4526, 4707, 7140
Clement, W. C. 4527
Clendinnen, F. J. 2798-99,
    2905
Cleobury, F. H. 6225
Clifford, W. K. 602, 6600
Coburn, R. C. 2800
Cocchiarella, N. 7061
Cochran, A. 6843
Coder, D. 4353
Coe, G. 2177
Coffa, J. A. 189, 2495-98,
    2557, 3148, 3524, 6226
Coffey, B. 72, 5506
Cohen, B. 690
Cohen, I. B. 73, 603, 4708,
    5603
Cohen, Leon 5737-38
Cohen, L. J. 912, 1886-87,
    2499, 2801-05, 2871,
    3520-23, 3782-83, 4111,
    5048, 7072
Cohen, M. R. 1195

Cohen, R. S.   49, 190-91,
     340, 950, 956, 970-72,
     1006, 1023, 1064,
     1075-79, 1088-90, 1098,
     1110, 1196, 3719, 4112,
     4709, 4939, 5257, 6227
Cohen, Y.   2806
Cole, R.   2139, 2500, 3744,
     4214, 4710, 6502
Collingwood, R. G.   604,
     3939, 4905
Collins, A. W.   2501-02
     4322
Colodny, R. G.   973-77,
     5739, passim
Colonius, H.   4010
Commoner, B.   4430
Compton, J. J.   74, 605,
     4174, 5399, 6000
Comte, A.   232
Conant, J. B.   606-08
Conger, G. P.   609
Connell, R. J.   75
Connolly, F. G.   341
Connolly, K.   7225
Conrad-Martinus, H.   5174
Conradt, R.   610
Conway, P. H.   5332
Cook, K. C.   5443
Cooke, R.   2503, 2569
Cooley, J. C.   2841
Coombs, C. H.   1509
Cooper, N.   611, 3149
Copeland, A. H.   2807, 3525
Copernicus, N.   2609, 3704,
     6439
Copi, I. M.   2663
Corcoran, J.   1197
Cornforth, M.   612
Cornman, J. W.   1403,
     1408-10, 2281, 3526-27,
     4528-31
Corsiglia, L.   5604
Cortes, A.   5740
Corvez, M.   5741
Costa de Beauregard, O.
     4940, 5258, 5742-47,
     6228-31, 6503-10
Costantini, D.   2808-09,
     3150
Coulson, C. A.   342-44
Cousin, D. R.   3141, 5049
Cowan, T. A.   345, 1411,
     2796, 4113
Cox, L. H.   2810

Crahay, F.   4114
Craig, E. J.   6511
Craig, W.   1893, 1900,
     2027, 2330, 4529, 4573,
     4586
Craig, W. L.   6782
Cramer, F.   6844
Crawshay-Williams, R.   3528
Creary, L. G.   2133, 2811,
     4115
Creath, R.   2812
Creed, W.   346
Cresini, A.   1198
Cresswell, M. J.   2282
Crick, F.   4487
Crittenden, P. J.   3784
Crockett, C.   4116
Crombie, A. C.   1706, 3209,
     4711, 4800, 7331
Crosson, F. J.   4353
Croteau, J.   3785
Crow, C.   2813
Cuenod, M.   1709
Cullwick, E. G.   6232
Cummins, R.   2504, 4323
Cunningham, F.   4712
Cupples, B.   2505, 3612
Curd, M.   1553
Curi, U.   3979
Currie, G.   2453-54, 3786
Curry, H. B.   2715
Curthoys, J.   1199
Cyranski, J. F.   4011, 5748
Czerwiński, Z.   2814-16,
     3676
Czeżowski, T.   230, 1894,
     3529

Dacey, R.   2716
Dagognet, F.   5175
D'Agostino, F. B.   2395
Dako, M.   6512
Dale, A. I.   3151-52, 3398
Dalla Chiara, M. L.   2283,
     5121, 5749-51
D'Amour, G.   2440
Dandenault, G.   6943
Daniels, C. B.   1404
Danto, A.   1139
Dantzig, D. van   1200, 3154
Darachschani, H.   5752
Darden, L.   1554, 2284,
     4117, 4713, 7326

Darlington, J. 3530
D'Armagnac, C. 7141
Darmstadter, H. 2285
Darwin, Charles 195, 1412,
    1576-77, 1607, 2332,
    3847, 4468, 4713, 4748,
    6846, 7096-7322
    passim, 7348
Das, R. 2817
Dascal, M. 613
Daub, E. E. 5259
Davidson, D. 1895, 2842
Davidson, M. 5238
Davidson, W. 5509
Davies, J. T. 614, 5507-08
Davies, P. C. W. 6513-14
Davies, P. M. C. 835
Davis, P. B. 2506
Davis, W. A. 2166
Davison, J. 6845
Davison, R. M. 3787
Day, J. P. 3153
Day, M. 5260
Day, P. 2140
Deakin, M. A. B. 4714
Dear, G. F. 5122
Deely, J. N. 7142-44
Delaney, C. F. 192, 347
Delattre, P. 616-16, 5444
Deledalle, G. 6515
Del-Negro, W. 3155,
    5261-62, 5753
Democritus 4468
Demopoulos, W. 5711, 5739,
    5754-58, 6516
Demos, R. 4118
Denbigh, K. G. 5510,
    6517-18
Denecke, H.-M. 5759
Dennett, D. 2066
Denny, N. 7298
Denonn, L. E. 50
Deppert, W. 6519
Derden, Jr., J. K. 1896
Derksen, A. A. 4941
Derr, P. G. 2441
Desanti, J. T. 618
Descartes, R. 180, 1544,
    6846
Desmonde, W. H. 2717
d'Espagnat, B. 5263, 5660,
    5760-61, 6233-34
Dessauer, F. 4324
Destouches, J.-L. 615,
    1201-03, 1710, 2286,

3980, 4532, 4942,
    5264-67, 5762-69
Destouches, P. 5770
Destouches-Févier, P. 2287,
    4943, 5268-69, 5771-72
Detel, W. 3788, 4715
Dettering, R. W. 4119
Deutsch, K. 1711-12
Deutsch, M. 5270
Deutscher, M. 3789
Devitt, M. 4716
Dewey, J. 170, 264, 476,
    619, 1016, 1354, 2770,
    3823, 4614
d'Haëne, R. 5271
Diamond, C. 2141
Diederich, W. 894, 2412,
    5272
Diéguez, M. de 620
Dieks, D. 76, 4219, 5140
Diemer, A. 621, 4717
Dietl, P. 2507
Dietz, S. M. 2508
Diez Blanco, A. 622
Diggs, B. J. 2142
Dijk, H. van 1555
Dijk, R. J. A. van 276
Dilworth, C. 4718-19
Dima, T. 193
Dineen, J. A. 4121
Dingle, H. 194, 348, 373,
    623-23, 1221, 3717,
    4012, 5273-74, 5511,
    5773, 6235-39, 6317,
    6520-21
Dingler, H. 10, 77, 3981,
    4120, 5275, 6451, 6522
Dirac, P. A. M. 4234
Dishkant, H. 5774
Dobbs, H. A. C. 4325, 6484,
    6523-24
Dobrov, G. M. 625
Dobzhansky, T. 1549, 4314,
    4316, 4423, 4437, 4466,
    4468, 4478, 4496, 5019,
    6804-05, 6846, 6853,
    6944, 6977, 6995, 7062,
    7116, 7145-50, 7292,
    7327-28
Doherty, M. E. 1160
Dolby, R. G. A. 4720
Dommeyer, F. C. 1852
Domotor, Z. 1413, 4013,
    4215, 4721, 5775, 6525

Donagan, A. 3790
Donceel, J. 7151
Dongorozi, C.-S. 6526
Doppelt, G. 1416, 4722, 4810
Doring, J. 1245, 2442,
    2509, 2818-19, 5276,
    5605, 5661, 5974, 6240,
    6317
Dorolle, M. 1414
Doty, P. J. 5776
Dougherty, J. P. 2510
Doumit, E. 3791
Dowdeswell, W. H. 7152
Dretske, F. I. 1415-16,
    2143, 3156, 3712, 4216,
    5065, 6527
Driesch, H. 7006
Drieschner, M. 626, 5777
Drummond, I. 5661
Dubarle, D. 4533
Dubarle, R. P. 2718
Dubinin, N. P. 7329
Dubois, G. 6847, 7063
Dubouchet, J. 6848
Dubuisson, M. 6849
Ducasse, C. J. 174, 1556,
    1852, 3144, 4326, 4534
Duchesne, J. 7153
Duchesneau, F. 4327, 6850-51
Dudman, W. H. 3709
Dufault, L. 7154
Duhem, P. 220, 281, 627,
    843, 2288, 2442, 2469,
    3714, 3716, 3718-19,
    3723-24, 3736, 3745-48,
    4502, 4580
Duijn, P. van 1557
Duistermaat, J. J. 4014
Dumitriu, A. 628, 5123
Dummett, M. E. 670, 1897,
    2719, 5050-51
Dunham, B. 7155
Dunn, J. M. 5778
Dunne, J. W. 6495
Dupré, J. 7064
Durand, III, L. 4015
Durbin, P. R. 1140, 1536,
    1558-59
Dürr, K. 13, 1204
Düsberg, K. J. 3157, 4723,
    6528
Dussen, W. J. van der 3792
Dynine, B. S. 2303

Earle, W. 4944
Earman, J. 1059, 2144,
    4217, 4443, 4535, 4726,
    5124, 5277, 5512, 5606,
    6241-44, 6418, 6529-37,
    6723
Eberle, R. A. 2511, 4122,
    4431
Ebersole, F. B. 7156
Ebert, T. 3793
Eccles, J. C. 3794, 7157
Echarri, J. 2289, 4945,
    5278, 6852
Economos, J. J. 2512
Eddington, A. 194, 304,
    308, 4530, 4603, 5420
Edelman, G. M. 6853
Edidin, A. 3543
Edlin, G. 5779
Edwards, A. F. 3158
Edwards, D. A. 5780
Edwards, J. 3159
Edwards, J. S. 2766
Edwards, P. 2820
Edwards, R. B. 5052
Egami, F. 7158
Egerton, F. N. 7175
Egidi, R. 282, 629, 2290
Ehrenberg, R. 6854
Ehrenberg, W. 4218, 5781
Ehrlich, W. 6538
Eichinger, B. E. 6539
Eichner, K. 3737
Eigen, M. 630
Eilstein, H. 1424, 1967, 2291
Einstein, A. 11, 209-10,
    218, 236, 268, 301,
    709, 1106, 1300, 1686,
    1817, 1848, 2376, 3125,
    3893, 3975, 4133, 4599,
    5225, 5279, 5304, 5398,
    5482, 5613, 5650, 5743,
    5746-47, 5782 5843,
    5903, 5919, 5943, 6028,
    6085, 6188-6457 passim,
    6624, 6744, 6755
Einstein-Podolsky-Rosen
    5745-46, 5768, 5824,
    5919, 5932, 5945, 6020,
    6025, 6047, 6103, 6144,
    6148, 6154
Eisenstein, I. 7159
Eklund, H. 1417

Ekstein, H. 4194
Elgin, C. Z. 1898, 2454,
    4536
Elkana, Y. 349, 631, 1006,
    1526, 2447, 4016
Ellegård, A. 195
Elliott, F. 7160
Ellis, B. 2292, 2513-14,
    2821, 3160, 4017-21,
    5513, 5607-08, 6198,
    6247-48, 6540
Ellson, D. G. 1418
Elsasser, W. M. 350, 4452,
    5783-84, 6855-56
Emberger, L. 7161
Emch, G. 5785
Emerson, A. E. 7162
Emery, E. 1233
Emery, F. E. 4306
Emmerich, D. S. 3531
Emmet, D. M. 477
Enç, B. 1899, 2293, 4328,
    4432
Engelhardt, Jr., H. T. 2515
Engels, E.-M. 4329
Engfer, H. J. 2726
Englebretsen, G. 3532
English, J. 1900-01
Ennis, R. H. 351, 2856
Erbrich, P. 7163
Erismann, T. 3161
Erlichson, H. 5945, 6250,
    6326
Ernest, P. 1902
Erpenbeck, J. 4724
Erwin, E. 3533, 5786
Escat, G. 5280
Escobar, E. F. 196
'Espinasse, P. G. 7164,
    7178
Esposito, J. L. 197, 632,
    4330, 4946-47
Esser, P. H. 3534
Essler, W. K. 633, 1903-07,
    2516, 2822-23, 2883,
    6056
Euclid 6654, 6674, 6695,
    6749
Evans, M. G. 6249-50
Evans-Pritchard, E. E. 2476
Even-Granboulan, G. 941
Evra, J. W. van 1090
Ewing, A. C. 1107, 2824
Ewing, E. J. 7165

Eyk, B. J. van 1909

Faber, R. J. 3162
Fabian, A. P. 7008
Fabrizi, E. 7166
Faggiani, D. 634-35, 2294,
    4537, 5281-82, 5787
Fain, H. 2517, 2825
Fair, D. 4219
Fairchild, D. 2518
Fales, E. 1806
Fales, W. 4220
Falk, A. E. 1419, 1560,
    4331
Falmagne, J.-C. 4022-23
Fang, J. 1561
Fantoli, A. 6857
Fantomas 1215
Farber, E. 1562
Farber, M. 16
Farkas, J. 23, 176
Farre, G. L. 1563, 1777,
    2826, 5283
Farrell, R. 1094, 3535
Favarger, C. 6858, 7065
Favrholdt, D. 198, 5788
Faye, J. 5788
Feenberg, E. 6251-52
Feibleman, J. K. 14, 297,
    636-37, 1205-09, 2827,
    4123-24, 4538, 5789,
    7167
Feigl, H. 12, 38, 78, 199,
    200, 478, 989-91,
    1052-54, 1141-42, 1210,
    1582, 1799, 1910, 1992,
    2082, 2156, 2225, 2295,
    2451, 2828, 3014, 3364,
    3394, 3705, 4061,
    4125-27, 4539, 4645,
    5430, 5447, 5795, 5867,
    5901, 5963, 6281, 6663
Feinberg, G. 4540, 5284,
    5790
Feldman, F. 3623
Feleppa, R. 2868
Felt, J. W. 2849, 2991,
    6253
Fenstad, J. E. 3163
Féraud, L. 2829, 3164
Ferré, F. 6524, 6541
Fertig, H. 5285
Fetzer, J. H. 597, 638,

1911, 1993, 2145-46,
2232, 2472, 2519-22,
3165-66, 4221
Feuer, L. S. 201, 479, 612,
1807, 1912, 4334
Feuillée, P. 5176
Févier, P. 639, 5125,
5791
Feyerabend, P. K. 70,
640-46, 682, 695, 893,
971, 991, 1199,
1211-18, 1401, 1445-46,
1450, 1467, 1512, 1582,
1799, 1850, 1913, 1970,
2156, 2296-98, 2307,
2332, 2339, 2347, 2495,
2631, 2684, 2830-31,
3907, 3957, 4024,
4128-29, 4208, 4541-42,
4645, 4725, 4808, 4821,
4948, 4987, 5286-87,
5677, 5792-96, 5901,
6455-56
Feynman, R. 2147, 6468
Fiala, F. 4130
Fiala, S. 1420
Fiedler, F. 80, 480-81
Field, H. 4543, 4726,
6542
Fierz, M. 5177, 5288
Filiasi, C. 36
Filippi, U. 5797
Finch, H. A. 2148, 3167,
3536
Finch, P. D. 3168-71,
5798-99
Fine, A. I. 991, 1914,
2299, 2720, 4025,
4726-27, 5800-08, 6124,
6543, 6586, 6594
Fine, K. 2180
Fine. T. L. 3172-73,
3482, 5809
Finetti, B. de 1219,
3174-81, 3242, 3247
Finkelstein, D. 3182,
5289, 5810-11
Finlay-Freundlich, E. 5514
Findley, K. T. 124
Finocchiaro, M. A. 81,
82, 1564, 1652, 2455,
2523-23, 4949
Firestone, J. M. 352
Firth, R. 5061
Fischer, H. 6859

Fischer, K. 1215
Fischer, R. 6544
Fisher, A. L. 992
Fisher, R. A. 3106, 7168
Fisk, M. 2832, 3153, 3795,
4223-24, 5290, 6545,
6594
Fitzgerald, G. F. 3717,
6250, 6317
Fitzgerald, J. J. 647
Fitzgerald, P. 1915,
4225-26, 4810, 6254,
6546-47, 6565, 6751
Flach, W. 83, 84, 1220,
3911
Flaschner, L. 5609
Fleck, G. 2227
Fleck, L. 294, 648, 4684
Fleming, J. J. 5812
Fletcher, W. W. 7169
Fleury, N. 4227
Flew, A. 772, 7225
Fliedner, D. 6548
Fliegel, G. 5178
Fochen, C. M. 1221
Fock, V. 5813-14, 5816
Fodor, J. A. 482
Foerster, H. von 6860
Fogarasi, B. 5815
Foley, L. A. 1106
Foley, R. 3537
Folse, H. J. 4544, 5817-20
Ford, E. B. 7203
Forest, H. S. 7170
Forge, J. 2525
Förster, H. 6861
Foss, J. 4858
Foss, L. 353, 1421
Foster, J. A. 4260
Foster, L. 2831, 3538
Foster, M. H. 2833
Foster, S. 4728
Fothergill, P. G. 6862
Foucault, J. B. L. 7087
Fouchet, J. 6863
Foulis, D. J. 3001, 5821
Foulkes, P. 954, 1022,
1043, 1064, 1104
Fourastié, J. 649
Fourier, J. 5117
Fowler, D. R. 6255
Fowler, W. S. 1222
Fox, J. F. 2454
Fox, R. 650
Fraassen, B. C. van 202,

1422, 1916-18, 1975,
  2078, 2149, 2429, 2526,
  2689, 3184-87, 3327,
  4525, 4545-49, 5822-28
Fraïssé, R. 5829
Francis, R. 6256
Frank, P. 17, 203, 229,
  354-55, 483-84, 651-54,
  993, 1075, 1919,
  4131-34, 4149, 4539,
  5053, 5291, 6258
Frankel, H. 4228, 4729-31
Frankfurt, H. G. 1476,
  1565, 4317
Frankl, V. E. 4433
Franklin, A. 1423, 1566
Franquiz, J. A. 4135
Frantz, B. 7303
Franzwa, G. 2150
Fraser, D. A. S. 3188
Fraser, J. T. 6186, 6218,
  6230, 6239, 6544,
  6551-54, 6629, 6722,
  6766, 6779
Frazer, W. R. 485
Fréchet, M. 3189
Frede, D. 98
Freed, B. 1012
Freedman, P. 1223
Freeman, E. 4411, 4950-51,
  6555, 6599, 6646, 6664,
  6740
Freistadt, H. 655, 5830
Freud, S. 3805, 3807
Freudenberg, G. 3190
Freudenthal, G. 567
Freudenthal, H. 1713-14,
  3191-95, 5241
Freund, J. E. 2796,
  3539-40, 6711
Freundlich, Y. 356, 3541,
  3626, 5831-34, 6556
Freundlieb, D. 5054
Frey, G. 27, 357, 656-57,
  1022, 1224-25, 1715-16,
  1920, 2834, 5179, 5835,
  6557-58
Fridshal, D. 7155
Fridshal, R. 7155
Friedlander, M. 658, 1226
Friedman, K. S. 1808,
  2835-36, 3196, 4229,
  5644
Friedman, M. 2527, 5055,
  5606, 5836-37, 6259,
  6559-60

Friedmann, H. 5515
Friedrich, L. W. 994
Friedrich-Freska, H. 6945
Friedrichs, K. O. 5838
Friend, J. 1209
Fries, H. S. 659, 1809,
  5839
Frings, M. S. 5335
Frisch, J. E. 7171
Fritz, Jr., C. A. 995,
  2837
Froda, A. 5292
Frolov, I. T. 1567, 4333,
  6864-66
Frye, R. M. 6561
Fuchs, K. 5056
Fuchs, W. R. 5840
Fuchs-Kittowski, K. 4434,
  6867
Fujiwara, I. 5841
Fumerton, R. A. 2151, 2838
Funke, G. 660
Fürth, R. 1717

Gaa, J. 4732, 4952
Gabor, D. 1568
Gagnebin, E. 4334, 7066
Gagnebin, H.-S. 85
Gagnebin, S. 2300, 4953,
  6562
Gagnon, M. 86, 4733
Gaifman, H. 3197, 4550
Galanti, E. 3982
Galavotti, M. C. 3198
Gale, G. 1143, 1453
Gale, R. 2528, 6563-65,
  6751
Galeazzi, U. 3796
Galgani, L. 5610
Galileo 254, 627, 1010,
  2332, 2347, 2524, 4137,
  4269
Galinaitis, J. 7
Gallie, W. B. 661, 2529-30
Gal-Or, B. 5516
Galperin, F. 5445
Gatling, J. 358, 4136
Gamow, G. 5611
Garaëts, T. F. 4954
Gårdenfors, P. 2531-32,
  3199, 3200
Gardiner, P. 3895

Gardner, M. 2533
Gardner, M. R. 1921, 2152,
    4551, 5842-44, 6180,
    6260, 6566
Garrett, A. B. 1569
Garstens, M. A. 4026
Gaukroger, S. 2534
Gauss, K. F. 6674
Gaussen, H. 7172
Gauthier, Y. 662, 4549,
    5845
Gavin, W. J. 204
Gavroglu, K. 3763, 5293
Gay, H. 1570-71, 3797
Geatch, P. T. 1922
Gehlhar, F. 6267, 6567
Geiringer, H. 3201, 3373,
    3713
Geissler, E. 7330
Geldsetzer, L. 205
Gellner, E. 1215, 1227,
    7173
Gendron, B. 4649
Gent, W. 5294
George, F. H. 1923
George, P. M. 1228
Georgescu-Roegen, N. 3144
Gérard, R. 663, 4027,
    6268
Gerhard, W. A. 664
Gerharz, R. 5846-47
Gerholm, T. R. 4955
Gerjuoy, E. 5848
Gerlach, W. 5989
Gerlich, G. 5849
Germain, P. 4734
Geroch, R. 6269
Gex, M. 359
Geymonat, L. 665, 4552,
    4956
Ghigi, A. 7067
Ghiselin, M. T. 7174-76
Ghitǎ, S. 666
Ghose, A. 4028
Giannoni, C. 3714, 6261-66,
    6304
Gibbins, P. 5850, 6069
Gibbons, P. C. 3900
Gibbs, J. W. 5369, 5619,
    5638
Gibson, J. J. 4553
Gibson, L. 3571
Giedymin, J. 996, 1424,
    1924, 2712, 3676, 3715,
    4554, 5612, 6568-69
Giere, R. N. 87, 233, 997,

    1056, 1093, 1144, 1591,
    2779, 2839, 3202-06,
    3542, 3818, 4008
Gierer, A. 6868
Giesen, B. 2443
Gilbert, E. J. 667
Giles, R. 5295-96
Gillies, D. A. 3177,
    3207-08, 3983
Gillispie, C. C. 3209
Gilson, E. 360
Gini, C. 3210
Ginsberg, A. 5740
Ginzburg, V. L. 4735
Giorello, G. 1572
Giray, E. 7068
Girill, T. R. 1229, 1685,
    2017, 2455, 2535-38,
    4435
Glas, E. 361, 4736
Glass, B. 362, 7228, 7331
Glickman, M. J. A. 7302
Gluck, S. E. 2539
Glucksmann, A. 6869
Glymour, C. 1059, 1925,
    2540, 2583, 3543-45,
    3604, 4436, 4546, 4555,
    5126, 5702, 5836, 5851,
    6243-44, 6570-73
Gochet, P. 4, 206
Godambe, V. P. 3211
Goddard, L. 3546
Gödel, K. 2717, 2719, 2734,
    2741, 4320, 4338, 4353,
    4385, 6270, 6417
Godfrey-Smith, W. 6271
Godlovitch, S. 3798
Goe, G. 6674
Goenner, H. 6272
Goffman, W. 7335
Goguen, J. A. 1926
Goh, S. T. 2541, 2633
Gold, T. 6574-75
Goldberg, B. 2153
Goldblatt, R. 6576
Goldman, A. H. 4556
Goldman, A. I. 4335
Goldman, E. 2840
Goldman, S. 668
Goldman, S. L. 747
Goldstein, L. J. 4336
Gomez, B. 622
Gomperz, H. 486
Gonseth, F. 20, 88, 363-65,
    669, 823, 1206,

1230-33, <u>1238</u>, 1718,
2301, 3191, 3219-20,
4029, 4138, 4557-58,
5852, 6577-79, 6858
Gonzales-Gascon, F.   6273
Good, I. J.   1234, 2625,
3107, 3176, 3212-18,
3371, 3547, 3627, 3799,
3818, 3865, 4230-31,
4285, 5517, 6458
Goode, T. M.   3818
Goodfield, J.   1573, 4437,
5489, 6760, 6870
Goodman, N.   <u>213</u>, 670-71,
1425, 1810-16, <u>1842</u>,
<u>2141</u>, 2154, <u>2201</u>,
<u>2753-54</u>, <u>2765</u>, <u>2783</u>,
<u>2831</u>, 2841-42, <u>2887-88</u>,
<u>2895</u>, <u>2904</u>, <u>2912</u>, <u>2917</u>,
<u>2921</u>, <u>3006</u>, <u>3032</u>, <u>3045</u>,
<u>3049</u>, <u>3051</u>, <u>3057</u>, <u>3060</u>,
<u>3067</u>, <u>3072</u>, <u>3088</u>, <u>3095</u>,
<u>3102</u>, <u>3514</u>, <u>3538</u>, <u>3548</u>,
<u>3569-70</u>, <u>3586</u>, <u>3597</u>,
<u>3614</u>, <u>3635</u>, <u>3666</u>, <u>3696</u>,
4559
Goodstein, R. L.   2720
Goosens, W. K.   3716, 7332
Gora, G. K.   <u>4664</u>
Gordesch, J.   3221, 3288
Gordon, T. J.   1235
Gorgé, V.   5297
Gorostieta, M. G.   196
Gorovitz, S.   2542-43,
4232
Gorski, D. P.   1927-28,
2302, 4560
Gössmann, G.   6580
Got, E.   7177
Götlind, E.   1719, 1929
Götschl, J.   892
Gott, V. S.   487
Gottinger, H. W.   3222
Gottlieb, D.   2843, 4438
Goudge, T. A.   2544, 2721,
4737, 4878, 5519, 6819,
6897, 7178-81, <u>7295</u>,
7297
Gould, J. A.   207
Gould, R. P.   6871
Gould, S. J.   6872, 7182
Gower, B.   3511, 5298
Grabau, R. F.   147
Graham, A.   5180

Graham, L. R.   208, 4738
Gram, M. S.   5, 998, 1478
Grandy, R. E.   999, 3549
3616
Granger, G. G.   2722, 3550
Grant, C. K.   5574
Grant, J.   3551
Grassi, E.   365, 672-73
Graves, J. C.   2844, 4030
6274, 6418
Graves, S.   3223
Gray, B.   3800
Greechie, R. J.   5853-54
Greenaway, F.   6581
Greenberg, J. H.   7183
Greene, J. C.   7184
Greeno, J. G.   2545, 3411,
3531, 3552
Greenstein, H.   2546
Greenwood, D. C.   674
Gregg, J. R.   1000, 1700,
1737, 1871, 2259, 2382,
3363, 3390, 6812, 6826,
6845, 6884, 6887, 6923,
6973, 7042, 7069, 7090,
7092, 7155, 7164, 7350
Grégoire, F.   2845, 6973
7185
Gregory, R.   6809
Grene, M.   89, 675, 832-34,
836, 1196, 1426, 1479,
2155, 4233, 4321, 4337,
4439-41, 4455, 4480,
6874-80, 6966, 6998,
7070, 7186-90, 7222
Grewendorf, G.   90
Griaznov, B. S.   4690
Gribanov, D. P.   209
Grice, H. P.   1930
Grieder, A.   6275-76
Grier, B.   3553
Griese, A.   6277, 6582-83
Griffin, N.   210, 1817,
2846, 3554
Grigoryan, A. T.   5613
Grim, P.   4407
Grimal, E.   4957
Grize, J.-B.   2304
Grmek, M. D.   1574-75
Gröbner, W.   3224, 5299
Grobstein, C.   6881
Groen, G. J.   2410
Groenewold, H. J.   1720,
4739, 5855-57
Gromko, M.   7119

Grøn, Ø. 6318
Gross, E. 1931
Gross, L. 2563
Grossman, N. 5858-60
Grossman, R. 4442
Grove, J. W. 3801
Gruber, H. E. 1576-77
Gruender, D. 1010-11, 4561
Gruender, C. D. 1932, 2547
Grünbaum, A. 91, 272, 366,
    2156, 2243, 2360, 2488,
    2521, 2548, 2631, 3714,
    3717-21, 3738, 3747,
    3802-07, 4217, 4234,
    5206, 5518, 5861, 5997,
    6278-85, 6401, 6459,
    6523-24, 6541, 6559,
    6584-6602, 6613, 6624,
    6641, 6659, 6683, 6695,
    6705, 6723, 6745, 6794
Grunberg, E. 2488
Gruner, R. 676, 2549, 3555,
    7191
Grünfeld, J. 1236, 1933,
    2847, 3808-09, 4740,
    4958
Grunstra, B. R. 2848, 4031
Gryasnov, B. 677, 2303
Grzegorczyk, A. 1237, 1934,
    3556
Guccione, S. 3225, 5862
Gudder, S. 4032-33, 5854
Guérard des Lauriers, M. L.
    39
Guerlac, H. 4034
Guggenheimer, E. H. 1233
Guggenheimer, H. 1238, 6286
Gullvåg, I. 1935
Gunderson, K. 1058
Guntau, M. 1001
Gunter, P. A. Y. 211, 1002,
    5251, 6882
Gurwitsch, A. 778
Guttenberg, A. C. de 6287
Gutting, G. 92, 1027,
    1578-80, 2849, 4562-63,
    4741-42, 4810, 6288
Guzzo, A. 93, 678

Haack, S. 3810, 5057-58
Haberlin, P. 679, 5181
Habermas, J. 680-81
Habermehl, W. 3737

Hacking, I. 1092, 1145,
    2454, 2850-52, 3015,
    3019, 3158, 3226-34,
    3319, 3412, 3416-17,
    4564, 4806, 6603
Hadley, H. G. 6289
Haefner, J. W. 6883
Haering, T. 1936, 7055
Hager, N. 1427, 1721, 5614
Hägerström, A. 6290
Hagstroem, K. G. 3235
Hahn, L. E. 7192
Hainard, R. 367
Haldane, J. B. S. 7356
Halfpenny, P. 2550
Hall, E. W. 3557
Hall, R. J. 2451
Hall, T. S. 212
Haller, R. 2305
Hallie, P. P. 2157
Halsbury, Lord 5455
Hamann, J. R. 2306, 3236
Hamilton, H. J. 7193
Hamilton, W. R. 5736
Hammerton, M. 3811
Hampshire, S. 1239
Hanen, M. 3088, 3558
Hanna, J. F. 1937, 2551-54
Hansen, T. E. 45, 3911,
    4959
Hanson, E. D. 7194
Hanson, N. R. 21, 94, 95,
    368-69, 682-84,
    1003-04, 1076, 1240,
    1428-33, 1467, 1498,
    1510, 1581-87, 1938-40,
    2017, 2307-08, 2555,
    2723-25, 2853, 3284,
    4235, 4645, 4743, 5059,
    5300, 5446-49, 5520-21,
    5607, 5615-17, 5795,
    5863-69, 6604
Hanson, W. H. 3238, 4338
Hansson, B. 3237
Hao Wang 2854-55
Hardegree, G. M. 5870-77
Harding, S. G. 1434, 3722,
    4565
Hardy, A. C. 7203
Harman, G. 1897, 1941,
    2556, 2856-57, 2900,
    3239, 4649
Harnatt, J. 2557
Harper, W. L. 1012, 3240,
    3332, 3799, 4744-46
Harrah, D. 661, 1942

Harré, R.   41, 685-91, 1722,
     1818, 2158-59, 2609,
     2858-59, 2889, 3503,
     3554, 3812, 4228,
     4236-37, 4543, 4747,
     5522, 5878, 6605
Harris, E. E.   370, 692,
     1241-42, 3813, 4339,
     4960, 6706, 6812, 6826,
     6845, 6884, 6887, 6923,
     6973, 7042, 7090, 7092,
     7155, 7164, 7350
Harris, F. T. C.   1000,
     1700, 1737, 1871, 2259,
     2382, 3363, 3390
Harris, J. F.   2860
Harris, J. H.   2309-11,
     3814
Harrison, C.   6606
Harrod, R. F.   2861-63
Harsanyi, J. C.   3815, 4961
Hartley, R. V. L.   5301
Hartman, R. S.   371
Hartmann, M.   18, 693-94,
     4238, 5182, 6945
Hartmann, N.   695
Hartshorne, C.   1435,
     5879-80
Hartung, F. E.   372, 696
Harvey, D.   5183
Hasker, W.   5127
Hassenstein, B.   4340
Hattiangadi, J. N.   1215,
     1588-89, 4748, 4810,
     4962-63
Hauptli, B. W.   4566
Hausman, A.   213, 2864
Hausman, C. R.   1610
Hausman, D. B.   2558
Havas, P.   6291-92
Havermann, R.   5302
Hawkins, D.   697, 1436,
     4341, 5569
Hay, W. H.   2865, 3241, 5618
Hayek, F. A.   698, 2559
Healey, R.   1005, 5881-82
Heath, A. F.   2560
Heberer, G.   6945
Heckmann, O.   5523-24
Hedman, C. G.   4239
Heelan, P. A.   96, 699, 700,
     4567-68, 4749-50,
     4964-65, 5883-84
Heerden, P. J. van   5303,
     5450, 6236
Hegel, G. W. F.   181, 5120,

     6057
Hegenberg, L.   1146
Heidegger, M.   846, 4641
Heidelberger, M.   4751
Heidmann, J.   5525
Heidsieck, F.   5251
Heilig, K.   3242
Heim, K.   373-74
Hein, C. A.   3243
Hein, H.   4342-44
Heisenberg, W.   701-02,
     2312, 4279, 4664, 4749,
     4752, 5117, 5442, 5657,
     5869, 5885-93, 5909,
     5952, 5972, 6114, 6158,
     6806
Heitler, W.   703, 5060,
     5304
Hélal, G.   1243
Hellman, G.   1215, 1943,
     3244, 4443, 5894-96
Hellpach, W.   6607
Hellwig, K.-E.   6033
Helm, E.   704
Helmer, O.   934, 3559
Helmholtz, H. von.   1006
Hempel, C. G.   22, 214,
     670-71, 705, 1100,
     1147, 1244, 1944-46,
     2313-18, 2475, 2476,
     2492, 2497, 2508-11,
     2561-70, 2617, 2645,
     2647, 2670, 2866-68,
     3197, 3524, 3538, 3547,
     3554, 3560-62, 3569-70,
     3605, 3626, 3642, 3657,
     3666, 3668, 3691, 3984,
     4139, 4444, 4539, 4966,
     5061
Henderson, G. P.   2841
Hendrick, R. E.   4753
Hendry, H. E.   2264, 3563
Henkin, L.   1040, 1687,
     5062-63, 5936, 6138,
     6586
Henle, P.   1947-48, 3054
Henle, R. J.   375
Hennueller, F.   6608
Hennemann, G.   4140, 4240,
     4569, 5305-06, 5897-99
Henning, M. D.   1742
Henry, J.   3503
Henry-Hermann, G.   4241
Henson, R. G.   2555
Heraclitus   3832

Herburt, G. K.   3723
Herivel, J. W.   1590
Hermann, A.   5900
Hermes, H.   3245
Herrgott, G.   6625
Herrick, C. J.   1437
Herrmann, H.   6885-87
Hertz, P.   1006, 5619
Hertzberg, L.   2869, 4754
Herzberger, H. G.   5064
Herzfeld, K.   376
Heese, M.   97, 686, 706,
    930, 991, 1245-46,
    1438-39, 1591, 1723-27,
    1914, 2160-61, 2319,
    2571, 2631, 2804,
    2870-74, 2882, 2932,
    3246, 3564-67, 4474,
    4570-71, 4613, 4755,
    5065
Hesslow, C.   4244
Heuss, E.   7195
Heuts, M.-J.   7196
Heyde, J. E.   1247
Hiebert, E. N.   215-16
Hilbert, D.   1884
Hildebrand, J. H.   707
Hiley, B. J.   5681
Hiley, D. R.   4967
Hill, E. L.   5451, 5867,
    5901
Hillinger, C.   4244
Hillman, D. J.   1819, 2875
Hilpinen, R.   1102, 1440,
    2876-79, 2885, 4968
Hinckfuss, I.   6609
Hinshaw, Jr., V. G.   5066,
    5128
Hintikka, J.   15, 98, 217,
    965-68, 1007-11, 1102,
    1122, 1248-49, 1441-43,
    1592, 1949-52, 2320,
    2726, 2789, 2823,
    2876-77, 2880-86, 2916,
    2925, 3078, 3080, 3087,
    3105, 3247, 3419,
    3452-53, 3472, 3510,
    3568, 3669
Hintikka, M. P.   1102
Hirschmann, D.   2572
Hirsh, G.   3248-49
Hirst, R. J.   2485
Hjorth, S.   3250
Hobson, A.   5620

Hocevar, H. E.   6888
Hochberg, H.   1953, 2162,
    4572
Hoche, H.-U.   1954, 2887
Hockney, D.   1012, 1955,
    4756
Hodgson, P.   708
Hoering, W.   2321, 4969,
    5129, 6071
Hof, W.   6293
Høffding, H.   5788
Hofstadter, A.   2243, 2573
Hogben, L.   6889
Hollak, J. H. A.   1080
Hölling, J.   5307, 6294,
    6746
Hollinger, R.   1149, 3724
Holman, E. W.   4035
Holmes, S. J.   6890
Holt, D. C.   6610
Holton, G. J.   218, 709,
    1013-15, 1250, 1593,
    1956, 4430, 4757, 6295
Holz, H.   5130
Holzkamp, K.   710
Hönigswald, R.   6611
Hook, S.   264, 711, 1016-17,
    1939
Hooker, C. A.   712, 1018-19,
    1055, 1090, 1251,
    2322-23, 2574, 3240,
    3569-71, 3722, 3816,
    4445, 4500, 4573-75,
    4758, 5452, 5621, 5828,
    5902-06, 5945, 6103,
    6612, 6691
Hoover, K.   2860
Hooykaas, R.   5184
Hoppe, H.   2888
Hoppe, H.-H.   4036
Hopson, R. C.   2479, 2575
Horgan, T.   4446-47
Horigan, J. E.   713
Horn, J. C.   99
Horovitz, J.   2576
Horstmann, R.-P.   2727
Horton, D.   3865
Horton, M.   2889
Horwich, P.   3543, 3545,
    3572, 6613-14
Hörz, H.   714-15, 1721,
    4245, 5131, 5909-11,
    6296, 7197
Hosiasson-Lindenbaum, J.
    4759

Hosinski, T. E. 5526
Hospers, J. 2577
Hovard, R. B. 4448
Howard, T. 5367
Howell, R. 1444
Howson, C. 161, 219, 1020,
    1252, 2451, 2459, 2790,
    3208, 3232, 3251-54,
    3352, 4948
Hoy, R. C. 353, 1957, 6615
Hoyer, U. 4760, 5622, 5912,
    6297
Hsu, J. P. 6298, 6616
Hubbeling, H. G. 4141
Hübner, A. 4576
Hübner, K. 100, 220, 377,
    716, 844, 1253, 3141,
    3817, 3957, 4761, 4799,
    5067, 5527, 5913-15,
    6617
Hübner, R. 2324
Hucklenbroich, P. 4762
Hudgin, R. H. 6299
Hudimoto, H. 3250
Hudson, R. L. 5710
Huggett, W. J. 3691
Hughes, M. 6618
Hughes, R. I. G. 5916
Hull, D. L. 912, 2549,
    4449, 4810, 6891-93,
    6985, 7071-77, 7089,
    7198-99, 7332-34
Hull, L. W. H. 1254
Hull, R. 1445
Hullett, J. N. 2754, 3075
Humberto, M. 4393
Humburg, J. 3256
Hume, D. 111, 2218, 2955,
    3003, 3065, 3850, 4192,
    4279
Humphreys, P. 2578, 4246,
    4285, 5132
Humphreys, Jr., W. C. 1004,
    1240, 2325, 2579
Hund, F. 6300, 6619-20
Hung, H.-C. 2163, 2580
Hunt, G. M. K. 2872, 2890
Hunt, I. E. 5623
Huron, R. 6621
Husserl, E. 717, 4563
Hutchings, Jr., E. 1021
Hutten, E. H. 718-21, 1256,
    1583, 1728, 2326, 2581,
    2891, 3100, 3183, 3257,

3395, 4247, 5308,
    5528-29, 5624, 5694-95,
    6894, 7048
Huxley, J. 722, 7200-03,
    7297
Hwang, S.-T. 6622
Hyndman, O. R. 7204

Iaroshevski, M. G. 1628
Idan, A. 613
Ihde, A. J. 1594
Illy, J. 6301
Infeld, L. 5279, 6302-03
Ingle, D. J. 6895
Inhelder, B. 4172
Innis, R. E. 680, 4921
Inwagen, P. van 2164
Ionescu-Pallas, N. 1255
Irani, K. D. 1595
Isaye, R. P. G. 378
Issman, S. 2165, 2892-93,
    3258-59, 3573
Ito, S. 379
Itzkoff, S. W. 723
Ivanenko, D. D. 5530

Jackson, F. 2166, 2798,
    2894, 3260, 3574, 6304
Jaeglé, P. 6623
Jaki, S. 724-25, 1596,
    5068, 5309
James, W. 204
Jamison, Dean 3261
Jammer, M. 1729, 5310,
    5453, 5625, 5917-18,
    6624
Janich, P. 726, 4577, 5311,
    5626, 6370, 6625
Janis, A. I. 4217, 4234,
    6274, 6285, 6305
Janssen, P. 727
Janssen, P. A. E. M. 6024
Jardine, N. 6896
Jardine, R. 3575
Järnefelt, G. 5531
Jarvie, I. C. 4810, 4970,
    5036
Jason, G. J. 1597
Jasselette, P. 5919-20
Jauch, J. M. 5133, 5785,
    5889, 5921-24

Jaynes, E. T. 3262
Jecklin, H. 3220, 3263
Jeffrey, R. C. 1037, 2148,
    2545, 2582, 2790,
    2895-97, 2945, 3265-70,
    3389, 3411, 3431,
    3576-77, 3818, 4971
Jeffreys, J. 1256, 3264
Jellinghaus, K. T. 4345
Jelsukov, A. N. 1351
Jessup, J. A. 2898
Joad, C. E. M. 4142
Jobe, E. K. 2167-68, 2212,
    2583, 6626
Joerges, B. 1598
Joga, A. 1040
Johann, R. O. 7205
Johansson, I. 3819-20
Johnsen, B. C. 2899-2901
Johnson, G. 223
Johnson, M. C. 5069
Johnson, O. A. 4143
Johnston, W. H. 5438
Jöhr, W. A. 728
Joja, C. 4578
Jonas, H. 6897
Jones, G. A. 1446
Jones, G. E. 2798, 3821
    4763
Jones, K. E. 3822
Jones, R. 5359, 5925,
    6306
Jones, R. M. 3578
Jones, R. V. 4037
Jones, W. B. 1447, 5926,
    6111
Jones, W. T. 380
Jordan, P. 4038, 5532
Jørgensen, J. 25, 305, 488,
    4144-46, 5134
Joseph, G. 729, 4579, 6307,
    6627
Joussain, A. 1820
Joy, G. G. 4580
Jubien, M. 4450
Juffras, A. 3823
Juhos, B. 26, 221, 730-31,
    1022, 1257-59, 1730,
    2169-70, 2902-03,
    3271-77, 3725, 4147,
    4581, 5312-13, 5627,
    5927, 6308-11, 6358,
    6628
Jung, C. G. 732

Jürgen, K. 3824

Kading, D. 2828
Kadish, M. R. 1450
Kaeser, E. 4582
Kägi-Romano, U. 5928
Kahane, E. 6898
Kahane, H. 2904, 3032,
    3507, 7091
Kahl, R. 1148
Kahneman, D. 3466
Kaila, E. S. 1023, 2728,
    4248, 5929, 6312
Kaiser, C. H. 1260
Kaiser, H. 6899
Kaiser, W. 2327
Kalbfleisch, J. G. 3278
Kalckar, J. 4972
Kalikow, T. J. 4764
Kalish, D. 1958
Kallen, H. M. 489
Kalmus, H. 6629
Kambartel, F. 726, 733
Kamber, F. 2328
Kamino, K. 3825
Kaminski, S. 222
Kaminsky, J. 1959-60, 2171
Kamlah, A. 101, 1097, 1961,
    2172, 2329, 5930, 6143,
    6625, 6630
Kamp, H. 6631
Kane, R. H. 1108
Kane, W. H. 1127
Kanger, S. 4039
Kanitscheider, B. 734,
    2173, 3907, 5314,
    6632-34
Kannegiesser, K. 4765, 5628
Kant, I. 180, 1627, 2384,
    4840, 5331, 5470, 5551,
    5997, 6297, 6534
Kanthack, L. 5135, 5931
Kantor, J. R. 735
Kantorovich, A. 2444, 4766,
    5454
Kapitza, P. L. 736
Kaplan, D. 1962, 2511,
    2584
Kaplan, M. 3579, 5070
Kapp, R. O. 381, 1821,
    4325, 5315, 5455, 6635,
    6901

Kar, R. 2174, 6313-16
Karpinskaia, R. S. 4451
Käsbauer, M. 2585
Kasher, A. 1963
Kassiola, J. 2455, 3826
Kattsoff, L. O. 737, 1261,
    1449
Katz, A. 6943
Katz, J. J. 2905, 4767
Katz, M. 4040, 7335
Kauffman, S. A. 4454,
    6902-03
Kaufman, A. S. 738
Kaufman, F. 3487
Kaufman, S. A. 2330
Kaulbach, F. 5316
Kavanau, J. L. 7206
Kazemier, B. H. 4148
Kearney, H. 4768
Kedrov, B. M. 382, 490,
    739, 1262, 1599-1602,
    4769-70, 5136, 5456,
    5533, 7336
Keene, G. B. 3279, 3827
Kegley, C. W. 4149
Kekes, J. 1603, 3935, 4453,
    4675, 4973, 5036
Kelle, V. Z. 491
Kellett, B. H. 5932
Kelley, M. H. 2906
Kelly, D. A. 3828
Kelly, J. P. 383
Kelvin, Lord 4748
Kemble, E. C. 5933
Kemeny, J. G. 740, 1822-23,
    2586, 2907-08, 3141,
    3280-81, 3580, 4454
Kempermann, C. T. 6904
Kempthorne, O. 3262, 3282
Kennedy, R. 3102
Kennedy, R. J. 6326
Kenny, A. J. P. 4455
Kent, W. 1964
Keosian, J. 7207
Kepler, J. 709, 2358
Kerr, S. 5505
Kerridge, D. 3417
Keswani, G. H. 3717, 6317
Keuth, H. 3283, 3603,
    3829-31, 4977, 5071
Keynes, J. M. 3284
Khatchadourian, H. 741,
    4346
Kiefer, H. E. 1024
Kielkopf, C. F. 2909

Kiester, A. R. 7078
Kilmister, C. W. 3900
Kim, J. 2587-89, 4249-50,
    4456
Kimbrough, S. O. 1824,
    3581, 7337
Kimura, M. 7208
King, J. L. 2175, 2590
King, M. D. 4771
King-Farlow, J. 3285, 5072
Kingsley, J. M. 6318
Kirk, G. S. 3832, 3851
Kirschenmann, P. 1130,
    3286-87, 5137, 5934-35
Kisiel, T. 223, 700, 1027,
    1604-05, 5457
Kitchener, R. F. 102, 2591,
    4772
Kitcher, P. 2527, 2592,
    4481, 4773
Kiteley, M. J. 1987
Kitts, D. B. 2593, 5185,
    7076
Kitts, D. J. 7076
Klatt, B. 6905
Klaus, G. 1025, 4774
Klawiter, A. 1450
Klee, J. B. 1606
Kleene, S. C. 5073
Klein, B. 2594
Klein, M. J. 5629, 6906
Kleiner, S. A. 1607,
    1965-66, 2331-32,
    4775-76, 6907
Klemke, E. D. 5, 998,
    1149, 3833-34
Klibansky, R. 1026
Kline, A. D. 1149, 4251
Kling, F. R. 818
Klohr, O. 6908
Klotz, A. H. 6319
Klowski, J. 3835-36
Klüver, J. 3985, 4776
Kmita, J. 106, 1967, 4777
Knapp, H. G. 3221, 3288
Knauss, G. 5317
Kneale, W. C. 1451, 1968,
    2203, 2871, 3019, 3118,
    3289, 3486, 3837, 4778
Kneebone, G. T. 2861, 2910
Kneller, G. F. 742
Knorr, K. D. 743, 4779
Knorr-Cetina, K. D. 384
Koch, R. A. 5186-88
Kochanski, Z. 2595

Kochen, S. 5843, 5881,
    5895, 5936-38, 5978
Kockelmans, J. J. 224,
    744-45, 892-94, 1027,
    2445, 4780, 4978
Koehn, D. R. 2911
Koertge, N. 1055, 1263,
    1608-09, 2451, 2596,
    3726, 3838-39, 4781,
    4810, 4979
Koestler, A. 1610, 2119,
    4172, 4433, 4457-58,
    7030, 7306
Koethe, J. 4583
Koeze, J. 4782
Kogan, Z. 1611
Kohak, E. V. 385
Köhler, E. 2912
Köhler, G. 746
Kojève, A. 747
Kokoszyńska, M. 386, 1969,
    2913
Kolb, D. 4584
Kolman, E. 5318
Komar, A. 5939-40
Komarýt, J. 6900
Konczewska, H. 5941
Konecsni, J. 6909
König, G. 103, 5319
König, H. 1264, 4041, 5320
Konigsveld, H. 2333, 2446
Koninck, C. de 7209
Konrad, W. 5942
Konyndyk, Jr., K. 3060
Koopman, B. O. 3578
Kopnin, P. V. 104, 225,
    748, 1028
Koppelberg, D. 2463
Korch, H. 749, 5138, 6320
Kordig, C. R. 1452-53,
    1612, 1970, 3727,
    4783-86, 4809
Korff, F.-W. 716
Korn, J. 750
Körner, S. 387-89, 751,
    830, 1029, 1108, 1412
    1451, 2046, 2176, 2530,
    2597, 2667, 2729-30,
    3110, 3126, 4024, 4042,
    4089, 4787-88, 5177,
    5322, 5678, 5684, 5856,
    6062, 6084, 6159
Körner, U. 6910
Kosing, A. 105
Koslow, A. 1523, 2400,
    5321, 5458, 5630
Kossovsky, N. K. 3290
Kostiouk, V. N. 2914
Kotarbińska, J. 1971, 2731
Kotarbiński, T. 1265
Kottinger, W. 1266
Kournay, J. A. 4789
Koyré, A. 747, 1456
Kracklauer, A. F. 5944
Krah, W. 2598, 3841,
    4790-91
Krajewski, W. 1030, 1268,
    4500, 4792, 5074, 5323
Kraft, V. 27, 227, 270,
    492, 752, 1267, 2915,
    3840, 4120, 4150-51
Krampf, W. 753
Krantz, D. H. 4043-44
Krapiec, A. M. 5459
Kratzer, A. 1731, 5324
Krause, M. 3728
Krauss, P. 3419
Krausser, P. 4793, 6625
Krausz, M. 3907, 4980
Krauth, L. 226
Krbek, F. von 5631, 6636
Krebs, H. A. 1613
Kreisel, G. 4347
Kremiansky, V. I. 6911
Kremmeter, A.-F. 1269
Kretzmann, N. 4152
Kreyche, G. F. 4252
Krige, J. 4794
Krikorian, Y. H. 881, 4348
Krimerman, L. I. 2177
Krimsky, S. 1454, 1465,
    2599, 6637
Krips, H. 4810, 5945-47,
    6531
Kröber, G. 2178, 2600
Kroger, J. 1270
Kron, A. 4253
Kröner, F. 623, 1106, 5384
Kronfli, N. S. 1455, 5139,
    5632
Kronthaler, E. 2916-17
Kropp, G. 228, 5948
Kroy, M. 754
Krüger, L. 390, 1031, 2334,
    2601, 2727, 4523,
    4795-97, 4807
Kruseman, W. M. 6912-13
Kuchowicz, B. 6638
Kuhn, H. 421

Kuhn, T. S. 107, 391-94,
    648, 1020, 1450, 1456,
    1510, 1614-16, 2311,
    2397, 2451, 2455, 4045,
    4254, 4641, 4683, 4687,
    4704, 4712, 4722,
    4728-29, 4733, 4742-43,
    4753, 4763, 4782, 4789,
    4798-4808, 4822, 4826,
    4829, 4848-49, 4869,
    4871, 4874, 4881, 4888,
    4892, 4917, 4981, 4991,
    4996, 5086, 7010
Kuipers, T. A. F. 1732,
    2918-20, 3291-92
Kukla, A. 5140
Kulakov, Y. I. 5325
Kulka, T. 1215, 2447
Kumakura, K. 5949
Kundt, U. 5950, 6167
Kunsemüller, H. 5951
Kursanov, G. A. 229, 6176
Kurth, R. 3293, 5633, 5952,
    6639-40
Kutschera, F. von 755,
    2921, 3294-95, 4153
Kutschmann, W. 4715
Küttner, M. 2335, 2602
Kuznetsov, B. 5075, 5943,
    5953, 6321-22
Kuznetsov, I. V. 2336
Kuznicki, L. 6914
Kwassow, A. A. 6323
Kwiatkowski, T. 230
Kyburg, Jr. H. E. 92, 96,
    108, 110, 120, 143,
    150, 319, 351, 394,
    756, 951, 1108, 1825,
    1918, 2278, 2337, 2553,
    2569, 2732, 2922-27,
    2947, 3019, 3022, 3206,
    3232, 3296-3319, 3332,
    3410, 3414, 3421, 3839,
    3995, 3998, 4046, 4203,
    4498, 4982, 5253, 6707,
    6893

Lacey, H. M. 6641-42
Lacharité, N. 1131, 1972,
    4808
Lachman, S. J. 757
Laddaga, R. 3348
Ladrière, J. 231, 395,
    1973-74, 2733, 3842,
    4349, 5076
Laer, P. H. van 396,
    758-59, 2928, 5141
Lagerspetz, K. Y. H. 6915
Laguna, G. A. de 7210
Laitko, H. 1271, 5198, 5954
Lakatos, I. 912, 971, 1032,
    2435, 2437-39, 2445-47,
    2448-55, 2464-66, 2789,
    2871, 2882, 2925,
    2929-30, 3019, 3195,
    3268, 3797, 3843-44,
    4725, 4730, 4801, 4803,
    4825, 4850, 4903, 4913,
    4915
Lakoff, G. 1975
Lamaître, G. E. 6328
Lamarck, C. de 7116
Lamb, M. 493
Lambert, K. 1150, 5024
Lambros, C. H. 4154
Lamotte, M. 7211
Lamouche, A. 1826
Lanczos, C. 4983
Landé, A. 5142, 5369,
    5687, 5955-75, 6180
Landesman, C. 3894
Landsberg, P. T. 5976, 6643
Lane, G. 1976
Lane, N. R. 1457
Langevin, P. 6644
Langley, P. W. 1672
Langtry, B. 3845
Lanius, K. 5950
Lanz, L. 5977
Laplace, Marquis de 3203,
    3389, 3487, 3916, 5112,
    5124
Laptschinsky, D. 6358
Largeault, J. 1033,
    3320-21, 5143-44
Läsker, L. 6916-17
Łastowski, K. 7338
Lastrucci, C. 1272
Laszlo, E. 109, 760-62,
    2338, 4812-13
Latzer, R. W. 5978, 6645
Laucks, I. F. 6324
Laudan, L. 30, 110, 152,
    232-33, 1273, 1617,
    2931-32, 3232, 3729,
    4585, 4809-10
Laudan, R. 1618, 4811
Laue, M. von 6325

Lauer, Q. 763
Lauener, H. 54, 234
Lauter, H. A. 2179
Lauwers, S. 4216
Lavis, D. 5634
Lavoisier, A. L. <u>4256</u>
Lawden, D. F. 764
Ławniczak, W. 1733, 3846
Lawrence, N. 235, 6553-54,
    6646
Lay, R. 765
Laycock, H. 5460
Laymon, R. 1458, 2339,
    2603, 6326
Layzer, D. 5534, 6008
Lazara, V. A. 4004
Lázaro, J. 2933
Lazerowitz, M. 111
Leach, J. J. 1018, 1034,
    2604, 2868, 4984
Leaf, B. 6279, 6327
Lean, M. E. 5326
Leatherdale, W. H. 1734
Lebedev, S. A. 2934
Leblanc, H. 3322-27, 3530,
    3582-87, 3627
Leblanc, S. A. 5205
Leclerc, I. 112
Leclercq, R. 1274-75
Lecomte de Noüy, M. P.
    <u>4346</u>, 7247
LeCoq, J. P. 766
Lee, D. S. 1276, 1735, 5077
Lee, H. N. 2935
Lee, K. K. 3847
Leeds, S. 3307, 4586
LeGrand, Y. 113
Lehman, H. 2605, 4317,
    4350, 7079, 7212
Lehman, R. S. 3328
Lehmann, F. E. 6918
Lehrer, K. 1459, 1977,
    2340, 2490, 2634,
    2936-42, 2945, 2977,
    3329-30, 3588-89, 4647,
    4814-15
Leibniz, G. W. <u>2852</u>, <u>5740</u>,
    <u>5811</u>
Leinfellner, W. 767, 2341,
    4155
Leininger, C. W. 5979
Leist, A. 5078
Leitch, A. H. 7213
Leite-Lopes, J. 4227, 6647

Lejewski, G. 3848
Lektorsky, V. A. 104
Lelande, A. 16
Lelionnais, F. 1277
Lenartowicz, P. 7339
Lenin, V. I. <u>4769</u>, <u>5351</u>,
    <u>5435</u>, <u>6039</u>, <u>6461</u>, <u>7348</u>
Lenk, H. 1035, 2606, 4985,
    5980
Lenneberg, E. H. 1978
Lennox, J. 4328
Lenz, J. W. 2934, 3590
Lenzen, V. 236, 397, 1278,
    3591, 4255, 4587, 5981
Lenzen, W. 3592
Leonardi, P. 7214
Leplin, J. 1736, 1914,
    3730-31, 4588, 4816,
    6282
Lerner, D. 1036, 1460
Lesher, J. H. 1619
Lessertisseur, J. 7215
Lestienne, R. 5982, 6648-49
Leverett, H. M. 4047
Levi, I. 1279, 1461, 2607,
    2803, 2850-51, 2937,
    2944-47, 3313, 3331-43,
    3434, 3593-96, 3849,
    4008
Levin, M. E. 1620, 1914,
    2608, 4256, 4817, 6329
Levin, M. R. 2608, 4256
Levine, V. C. 4589
Levins, R. 6998
Levinson, H. B. 6317
Levison, A. B. 3732, 3850
Levy, E. 6330, 7325
Levy, I. 2884
Levy, M. 4459-60
Lévy, P. 3344-45, 6504
Lévy-Leblond, J.-M. 5983
Lewis, C. I. 1462, 2001
Lewis, D. 1979, 2180-82,
    2948, 3346, <u>4281</u>, 4353
Lewis, H. D. 3894
Lewontin, R. C. 1737, 6919,
    6998, 7216-17
Ley, H. 1738, 2183, 6920
Leyden, W. von 579
L'Heritier, P. 7340
Lieb, I. C. 2034
Lillie, R. S. 6809,
    6921-22, 7032
Lin Chao-Tien 3597
Lindemann, H. A. 7055

Lindenberg, S. 5984
Lindenmayer, A. 6923-24, 7341
Linder, A. 1463
Lindholm, L. M. 768
Lindley, D. V. 3347
Lindsay, R. B. 5327-28
Lineback, R. H. 33
Linhart, H. 3530
Lins, M. 1280-84, 2184
Linsky, B. 1980
Litt, T. 4986
Llewelyn, J. E. 2974
Lloyd, A. C. 1981, 2185
Lloyd, G. E. R. 3851
Lockemann, G. 1621
Loeb, J. 1739, 4351
Loewer, B. 3107, 3348, 5079
Loinger, A. 5247, 6331
Lonergan, B. 1270
Long, P. 2186
Longino, H. E. 3598
Long, J. W. 6744
Lopes, J. L. 5985
Lorch, J. 7080
Lorente, M. 5986-87
Lorentz, H. A. 3717, 6250, 6261, 6317, 6384, 6396, 6428, 6455
Lorenz, K. 114, 1042, 4764, 4818
Lorenzen, H.-P. 5080
Lorenzen, P. 769-70, 3349, 4048, 6332-33, 6650
Łos, J. 2949, 3350
LoSardo, P. 3225
Lossee, J. 912, 1151, 4810, 5988
Löther, R. 715, 6925
Lotz, J. 1982
Lovell, B. 5535
Lövestad, L. 2187, 3617
Løvtrup, S. 7218
Lowe, E. J. 2188
Lowe, M. F. 1006
Lowenthal, D. 4352
Lowry, A. 4461
Lucas, J. R. 2882, 3351-52, 4353, 6651
Lucca, J. de 398
Luce, R. D. 3353, 4049-54, 4077, 4078, 6334
Luchins, A. S. 771, 850
Luchins, E. H. 771

Ludwig, G. 2342-43, 5989, 6164, 7351
Ludwig, W. 6926
Lugg, A. 1285, 4810, 4819, 4987-88
Łukasiewicz, J. 4040
Lukes, S. 2344
Lungarzo, C. 2950
Łuszczewska-Romahnowa, S. 2951, 7081
Luyten, N. A. 5461, 7219
Lycan, W. 1827
Lyon, A. 2189, 4257

MacCormac, E. R. 1983
MacDonald, D. K. C. 5171
MacIntyre, A. 4820
MacIver, R. M. 6927
MacKay, D. 772
MacKinnon, D. M. 1984
MacKinnon, E. 115, 237-38, 399, 494, 773-74, 1622, 2609, 4590-94, 4649, 5081-82, 5702, 5990-91, 6652
McCarthy, T. 2487
MaCauley, D. E. 6940
McCauley, R. N. 495
McClennen, E. F. 1018
McConnell, J. R. 2345
McCrae, W. H. 5495, 5536-37, 5635, 6335
McDonagh, F. 421
McDowell, M. A. 6926
McEvoy, J. G. 3609, 4821
McGechie, J. E. 4021
McGilvray, J. A. 1409, 6653
McGonigle, T. G. 6654
McGowan, R. S. 2952-53
McGrath, J. H. 5992-93
McGuinness, B. F. 239, 954, 1123
McGuire, J. E. 5329
McIntosh, R. P. 6927
McKeon, R. 400
McKinney, J. C. 240
McKinney, J. P. 4509, 4655, 5933
McKinsey, J. C. C. 5636
McKnight, J. L. 4055-56
McLaughlin, A. 1286, 1464
McLaughlin, P. J. 775
McLaughlin, R. N. 2112

McMahon, W. E.   1985
McMorris, M. N.   6336
McMullin, E.   74, 116-20,
     241-43, 401, 785, 1740,
     2340, 2450, 2610, 4156,
     4224, 4462, 4595, 4810,
     4989-90, 5449, 5462-64,
     5538, 5637, 5885
McMurrin, S. M.   402
McNicholl, A.   121, 403, 776
McRae, R.   180
McShane, P.   3354, 6928
McTaggart, J. E. M.   6563,
     6725
McVittie, G. C.   6637
McWilliams, J. A.   4258

Mace, C. A.   3896
Macfie, A. L.   404
Mach, E.   30, 168, 172, 178,
     185, 190-91, 203,
     215-16, 218, 249, 253,
     484, 1043, 1465, 1482,
     3893, 4133, 5439, 5458,
     5601, 6253, 6272, 6297,
     6322, 6356, 6454, 6456,
     6717, 6755
Machamer, P. K.   1408, 1466,
     2347, 4784, 6655
Machan, T. R.   4822, 4870,
     4991
Machlup, F.   5465
Mackay, D. M.   5994
Mackay, J.   3188
Mackey, G. W.   5995
Mackie, J. L.   2871, 3371,
     3599, 3600, 3935,
     4259-61, 6987
Macklin, R.   2611, 4354
Macovschi, E.   6929
Madden, E. H.   122, 174,
     690, 955, 1044,
     1986-87, 2612, 2631,
     2954-55, 4236-37, 4262
Madsen, K. B.   2348
Magala, S.   3907
Magee, B.   3852
Magnini, C.   777
Magnus, A.   5189
Magyar, G.   1287, 5330
Mahalanobis, P. C.   3355

Mainx, F.   6930, 7342-44
Majone, G.   4263
Makai, M.   4463
Malament, D. B.   5638,
     6573, 6656, 6734
Malcolm, N.   4355
Malek, I.   1377
Malherbe, J.-F.   244, 1988,
     3853-54
Malicot, G.   6943
Malisoff, W. M.   123
Malkani, G. R.   7220
Mamchur, E. A.   4823
Manara, C. F.   1288
Mandelbaum, M.   4596-97,
     4799, 7221
Mandelstam, S.   6179
Manders, K. L.   1808
Manier, E.   405, 6931-32,
     7222-24, 7345-47
Manninen, J.   1045
Mannoia, Jr., V. J.   1152
Mannoury, G.   6933
Manser, A. R.   2613, 7225
March, A.   3356, 5996
Marchand, J.-P.   5668
Marchi, P.   1623
Marcus, H.   4264
Marcus, S.   1289, 4598
Marcuse, H.   778
Mardiros, A. M.   3855
Mare, C.   4824
Maré, J. de   3207
Marek, J. C.   892
Margáin, H.   2614
Margalit, A.   1384, 2805
Margenau, H.   109, 245,
     2615, 3487, 3856,
     4057-58, 4599, 4600,
     5145, 5331-34, 5997-
     6001, 6103, 6338, 6657
Margolis, J.   1467, 2956,
     4601-02, 4656
Marhenke, P.   1989
Marí, E. E.   2349
María de Alejandro, J.
     6934
Marioni, C.   6002
Maritain, J.   291
Markarian, E. S.   491
Markova, I. A.   315
Marković, M.   1046, 2190,
     4992, 5083

Marley, A. A. J. 4054,
    4059-60
Marlow, A. 6003
Marquardt, H. 4265
Marquis, D. B. 4603
Marshak, J. 3357
Marsden, J. E. 5736
Martin, N. M. 3358
Martin, M. 124-25, 1990,
    2350, 2833, 3511, 3601,
    4355, 4993-95
Martin, R. M. 779, 1290,
    1991, 2351-52, 2616,
    5084
Martin-Löf, P. 3359
Maruyama, M. 4266
Marx, K. 490, 811-13,
    5335, 7348
Masani, A. 6658
Mascall, E. L. 406
Maschler, C. 824
Maslov, S. Y. 7226
Maslow, A. H. 780, 1291
Mason, R. O. 3866
Massey, G. J. 496, 2617,
    6659-60
Masterman, M. 4825
Matchinski, M. 6339
Matczak, S. A. 532
Matisse, G. 7227
Matson, F. W. 4464
Matson, W. I. 2957
Matteuzzi, M. L. M. 2353
Matthew, A. 2618
Matthews, G. 6661
Matthews, R. J. 2619
Mattick, P. 5335
Maugé, F. 497
Maull, N. L. 126, 498,
    1654, 2284
Maund, B. 6340-41
Maurin, K. 5336
Mauskopf, S. H. 1047
Maxfield, M. W. 3262
Maxwell, G. 990-91, 1053-54,
    1057, 1582, 1799, 1992,
    2156, 2225, 2307, 2698,
    2958, 3014, 3360, 3364,
    3394, 3602, 3875, 4061,
    4125, 4604-07, 4645,
    5430, 5795, 5867, 5901,
    5963, 6281, 6663
Maxwell, J. C. 1591, 3095
Maxwell, N. 1624, 1828,

2959, 3858, 4267, 5337,
    6004-05
May, E. 781-82, 1292,
    2960, 6935
Mayants, L. S. 3361
Maynard, P. 4842
Mayo, B. 3469, 5146, 6726,
    6739
Mayo, D. 2961, 3362
Mayr, D. 4465, 6071
Mayr, E. 4356, 6870, 6936,
    6938, 7082, 7228
Mays, W. 246-48, 825, 3363
Maziarz, E. A. 407, 783,
    854
Mead, G. H. 240, 6662
Meana, L. 3859
Medawar, P. B. 1625, 2962,
    3860, 4466, 6937, 7297
Meehl, P. E. 2620, 3603,
    4467, 5147
Mehlberg, H. 408, 784,
    1294-95, 2225, 2354,
    5963, 6006, 6342,
    6663-65
Mehlberg, J. J. 3364-65
Mehra, J. 5338, 6007
Meigne, M. 5466
Meiland, J. W. 3861, 4996,
    6666
Meister, A. 7022
Meixner, J. 2621
Mejbaum, W. 1468, 2191,
    5339
Mellor, D. H. 1048, 1469,
    1741, 1993, 2922-24,
    3366-68, 3935, 4608,
    4810, 5085
Mellor, D. P. 5467
Mellor, H. 6692
Melsen, A. G. van 785-87,
    5468, 7229
Meltzer, B. 2625
Mendel, G. 7178, 7340
Mendelsohn, E. 6880,
    6938-39
Mendelson, E. 1037
Menger, K. 249, 1049, 4061,
    6343, 6608
Menges, G. 3369
Menzel, D. H. 6008
Mercier, A. 250, 409, 1293,
    2192, 4268, 4609-10,
    5340, 6009, 6667-68

Merkens, H. 3771
Merlan, P. 410
Merleau-Ponty, J. 5525,
    5539-43, 6594, 6669-70
Mcrmin, N. D. 6010
Merrill, D. D. 2730
Merrill, G. H. 127, 2963,
    3604, 4611-12
Merton, R. K. 648, 1626
Mertz, D. B. 6940
Mertz, D. W. 4269, 4826
Merwe, A. van der 6180
Mesarovic, E. 1742
Metz, A. 128, 3370, 4270,
    4997, 6011, 6344-46,
    6671-72
Meurers, J. 788-90, 4271,
    5544-46, 7230
Meyard, L. 1153
Meyer, F. 7231-32
Meyer, H. 1743
Meyer, M. 1627, 2626,
    4998
Meyer, S. L. 3605
Meyer, W. 1296
Meyer-Abich, A. 1051
Meyers, R. G. 1994, 3862
Meyerson, E. 1683
Meynell, H. 791, 1215,
    5086
Michalos, A. C. 124, 282,
    1090, 1154, 2877, 3520,
    3606-10, 3849, 4827,
    4999, 5000
Michalski, K. 5336
Michelson, A. A. 6295,
    6326, 6434
Michod, R. E. 7233
Miescher, K. 1297
Miettinen, S. K. 2641
Miklin, A. M. 7234
Mikulak, M. W. 251, 7348
Milcu, St.-M. 7235
Miles, T. R. 411
Mill, J. S. 68, 2949
Millard, R. M. 6673
Miller, A. I. 6347, 6455,
    6589, 6674
Miller, D. L. 690, 965,
    1298, 1995, 2562,
    2627-28, 2819, 2964,
    3254, 3371, 3818,
    3863-65, 3948, 4157,
    5087, 5341, 6675-76,
    7001

Miller, D. S. 3093
Miller, D. W. 3019
Miller, G. 6348
Miller, H. 792
Miller, III, J. F. 7236
Miller, M. B. 4664
Miller, P. 6677
Miller, R. W. 3372
Millman, A. B. 2457
Milmed, B. K. 2193
Mills, S. K. 7237
Milne, E. A. 5547, 6227,
    6349, 6425, 6523, 6678
Miloš, P. 6941
Mink, L. O. 4902
Minkowski, H. 6312, 6470-71,
    6576, 6744
Minoque, B. P. 4613
Minton, A. J. 4403
Mirabelli, A. 3611
Mirman, R. 6679-80
Mirski, E. M. 252, 412
Mischel, T. 2629
Misiek, J. 1299, 5342
Misis, R. von 253, 3270,
    3359, 3373, 3487
Misner, C. W. 5548, 6350
Mitchell, E. T. 413
Mitroff, I. I. 129-30,
    3733, 3866
Mittasch, P. A. 4273-74
Mittelstaedt, P. 930, 2355,
    5343, 6012-19, 6351
Mittelstrass, J. 254, 726,
    733, 793-95, 4828
Moberg, D. W. 2356
Mocek, R. 6942
Moch, F. 3374
Modigliani, F. 2488
Mohorvičic, S. 6352
Mohr, H. 796
Moisil, G. C. 1040
Molchanov, I. B. 6353
Moldauer, P. A. 6020
Moles, A. 797
Molina, A. M. 1300
Monk, R. 1629
Monod, J. 4394, 6814,
    6852, 6857, 6869, 6920,
    6943-44, 6949, 7238
Monson, T. C. 3603
Montague, R. 2511
Montaldi, E. 6002

Montalenti, G. 7083, 7239
Montefiore, A. 2529, 4357,
    5148
Montgomery, H. 6681
Mooij, J. J. A. 6682
Moon, P. 6354-55, 6683
Moor, J. H. 3612
Moore, A. 2965
Moore, G. E. 111
Moore, H. 4614-15
Moran, B. T. 1630
Morchio, R. 7084
Moreau, J. 5149
Moreland, J. 2966
Moreno, A. 6356, 6684,
    7240
Moretti, J. 5469
Morgan, C. G. 1631, 1744,
    2474, 2587, 2641, 2693
Moregenbesser, S. 556, 808,
    1061-62, 1139, 1239,
    1290, 1342, 1354, 1396,
    1432, 1470, 1816, 2423,
    2568, 2630, 2763-64,
    2946, 3334, 3565,
    3613-14, 3630, 3867,
    4054, 4444, 4540, 4616,
    5062-63, 5073, 5465,
    5630, 6593, 6813, 6846
Morgulis, S. 7249
Morick, H. 4158
Morin, E. 798
Morin, H. 4358
Morison, R. S. 1015
Morley, E. W. 6326, 6434
Morrill, T. 7170
Morris, C. E. 38, 255,
    471-72, 476, 489, 499,
    501, 507, 566, 818,
    1065, 1278, 1996-98,
    3375, 4159-60, 5291,
    5514
Morris, J. M. 2967
Morrison, P. 5549-50
Morrison, P. G. 4272,
    7241-42
Mortensen, C. 4173, 6609,
    6685, 6691
Mortimer, H. 1999, 2000,
    2968, 7349
Morton, A. 2194
Mosedale, F. E. 799
Moser, M. 6357-58
Moser, S. 927, 3892, 6945

Mostepanenko, A. M., and
    V. M. 4617
Mostowski, A. 800
Mott, P. L. 3868
Motycka, A. 4829
Mould, R. A. 6338
Moule, C. F. D. 2161
Moulines, C. U. 2261,
    2357-58, 2412, 4830,
    5344-46
Mouloud, N. 1301-02
Moulyn, A. C. 4062, 4390
Moyal, J. E. 4275
Mshvénieradzé, V. V. 1303
Muguerza, J. 3869
Mühlhölzer, F. 6686
Mukherji, S. R. 2969
Mulckhuyse, J. J. 5190
Mulder, H. L. 1063
Mulkay, M. 801, 3807
Mullen, J. D. 4999
Müller, A. 4260
Müller, G. 6687
Müller, G. H. 670, 1201
Müller, H. 500, 2195
Müller, W. 4776
Müller-Markus, S. 256-59,
    802, 6021-23, 6359-63
Mundle, C. W. K. 1471,
    3128, 6523
Munévar, G. 803
Munitz, M. K. 1024, 5551-57
Munson, R. 6946-47, 6985,
    7079, 7243
Murakami, Y. 6688
Murdoch, J. E. 414
Murphy, A. 4753
Murray, G. B. 992
Musgrave, A. 1032, 1304,
    2450, 2455, 2458-60,
    3615, 3708, 3871-73,
    4618-19, 4725, 4799,
    4801, 4803, 4810, 4825,
    4831, 4850, 4903, 4913,
    4915
Mussachia, M. M. 3874
Muynck, W. M. de 6024
Myers, C. M. 1472
Myhill, J. 5347
Mynatt, C. R. 1160
Myro, G. 3616

Naess, A. 31, 131, 640,
  804, 2359, 3617, 4161
Nagai, H. 260-62
Nagasaka, G. 1745, 6025
Nagel, E. 263-64, 805-08,
  910, 1037, 1062, 1354,
  1543, 1877, 2027, 2177,
  2276, 2360-61, 2485,
  2503, 2510, 2631, 2734,
  2927, 2970, 3093, 3375,
  3487, 3721, 4162,
  4359-62, 4366, 4381,
  4430, 4469, 4539, 6026,
  6079
Nalimov, V. V. 809, 1305,
  3618
Namer, E. 3619
Narens, L. 3353, 4063-64,
  6334
Narlikar, J. 5558, 6689
Narskii, I. S. 3875
Nartonis, D. K. 6027
Nash, L. K. 810
Naslin, P. 1746
Nathanson, S. L. 4975
Naumann, H. 6028
Needham, J. 6948
Needham, R. 7173
Neeman, U. 6690
Nejedlý, R. 3376
Nell, E. J. 2567
Nelson, A. F. 2632, 3620,
  4470-71
Nelson, B. 265
Nelson, E. J. 2001,
  2971-73
Nelson, G. 7085
Nelson, J. O. 2974-75,
  3621, 4620
Nelson, R. J. 1829, 1960,
  4363-64
Nelson, T. M. 4065
Nemetz, T. 3622
Nerlich, G. 3699, 3876,
  4276, 6685, 6691-93
Neuhäusler, A. 5470
Neumann, J. von 5674, 5708,
  5930, 6029, 6063, 6115,
  6157
Neumann, M. 4621
Neurath, M. 1064
Neurath, O. 266, 471-72,
  476, 489, 501, 507,
  566, 1064-65, 1278,
  1998, 3375, 4159, 4161,

5291, 5514
Neville, R. C. 282
Newburgh, R. G. 6454
Newman, F. 2633-34
Newman, J. R. 602, 2734
Newton, I. 212, 1458, 1564,
  1589, 2358, 2703, 4044,
  4234, 4830, 5594-5654
  passim, 6324, 6333,
  6420, 6471, 6501, 6560,
  6642, 6761, 6792
Newton, R. G. 5642
Newton-Smith, W. H. 2362,
  4832, 5001, 6694
Neyman, J. 2961, 3207,
  3223-24, 3441
Nicholas, J. M. 1066, 5003
Nicholson-Smith, D. 7260
Nicklas, U. 44
Nickles, T. 502, 1632-38,
  2406, 2523, 2599, 2635,
  4472, 5002
Nicod, J. 3572
Nicola, M. 6318
Nidditch, P. H. 1155
Nielsen, H. A. 1306, 2976
Nielsen, K. 2002, 5150
Niessen, M. 2636
Niggli, P. 5191
Niiniluoto, I. 1041, 1102,
  2143, 2330, 2363, 2637,
  2926, 2977-82, 3377,
  3609, 3877, 4687,
  4833-36, 5088
Nikitina, A. G. 2638
Nikitine, E. P. 2303
Nilson, D. R. 47, 6030
Nisbett, R. E. 3064
Nissen, L. 4317, 4365-66,
  4398
Nissim-Sabat, C. 4234
Nitschke, A. 4837
Njegovan, V. 5471
Noble, D. 4367
Nogar, R. J. 2365, 5559,
  7144, 7244-46
Nola, R. 2003, 4622
Noland, A. 1129
Nolfi, P. 3161, 3220,
  3378
Noonan, H. W. 6364
Norburn, G. 831
Nordberg, R. 6474
Nordenson, H. 6365

Nordenstam, T.   267
Nordin, I.   6031
Noren, S. J.   415, 2017,
    4530, 4623
Norris, L. W.   2639
North, J. D.   1747, 5560,
    6366
North, J. H.   7155
Northrop, F. S. C.   268,
    416-18, 5887
Norton, B. G.   269
Novakovic, S.   4838
Novik, I. B.   132
Novikov, I. D.   5593
Nowaczyk, A.   2366
Nowak, I.   5348-49
Nowak, L.   811, 1830, 2631,
    2640, 5089
Nowak, S.   2983-84, 4277
Nowakowa, I.   4839
Nowiński, C.   6949-50
Nuchelmans, G.   2196
Núñez de Castro, I.   6951
Nute, D. E.   2146, 2197,
    4221
Nuzzaci, F.   3878
Nyasani, J. M.   814
Nyberg, T.   52
Nye, M. J.   5350
Nyman, A.   1473, 2985

O'Brien, J. F.   5192, 5472
Occam, W.   1827
Ochs, W.   6032-33
O'Connor, D. J.   2198, 5151
O'Connor, J.   3045
Oddie, G.   3254, 3879-80
Odegard, D.   1699, 3379
Oeser, E.   133, 270, 815-16
Oetjens, H.   3881
Oeumov, A. I.   1308
Ofierska, U.   43
Öfsti, A.   2986
O'Gorman, F. P.   6695
O'Hair, S. G.   2069
Ohe, S.   271
O'Hear, A.   3882-83
Ohm, G. S.   4751
Øhmori, S.   817
Øhrstrøm, P.   6367
Olding, A.   3884
Oliver, W. D.   2987

Olmstead, P. S.   419
Olsen, J. H.   2199
Olsoufieff, D.   6825
O'Malley, J. B.   282
Omelianovsky, M. E.   1474,
    4624, 5193, 5351-52, 6034-39
Omer, I. A.   2641-42
O'Neil, W. M.   2367
Onicescu, O.   1307, 3380,
    7248
Oparin, A. I.   7249-55
Öpik, E. J.   5561
Oppenheim, P.   503, 818,
    2511, 2562, 2586, 3559, 3562,
    3580, 4454, 5663, 5984
Oppenheimer, J. R.   5353
Orens, I. P.   420
Orlicki, W.   5562
Os, C. H. van   4625
Osborne, D. K.   4066-67
Osche, G.   7256
Ostien, P. A.   1408, 1475
O'Toole, E. J.   3381
Otte, R.   4278
Ovchinnikov, N. F.   504
Overington, M. A.   819

Pagano, F.   4163
Pahler, K.   3885, 3935
Pais, A.   5473
Pakswer, S.   2988
Pal, R.   2989
Palcos, A.   820
Palmieri, L. E.   2004, 2200,
    2862, 4164
Palter, R.   134, 272-73,
    2368, 3986, 4527, 6696
Pandit, G. L.   3886
Paneth, E. and F. A.   5194
Pannenberg, W.   421
Pantin, C. F. A.   422, 7048
Pagels, H. R.   6040
Pap, A.   227, 821, 2005-10,
    2841, 3382, 4165-69, 4473, 5090
Papepetrou, A.   6368
Papineau, D.   2011-13, 2369,
    3383
Pappas, G. S.   1068
Parain-Vial, J.   6943
Pargetter, R.   3260, 3574,
    6304
Parijs, P. van   3853

Paris, C. 274, 2370
Park, D. 6041, 6554,
    6697-6700
Park, J. L. 4068, 6001,
    6042-44
Parker-Rhodes, A. F. 5354
Parlavecchia, P. 2014
Parry, W. T. 2201
Parsegian, V. L. 6952
Parsons, C. 1897, 4523
Parsons, K. P. 2015-16,
    4786
Partashnikov, A. 6953
Pârvu, I. 2371, 4840
Pasch, A. 1476, 4170-71
Pasquinelli, A. 2643
Passmore, J. A. 423, 2644,
    3887
Pastin, M. 3623
Paszkiewicz, E. 1639
Paterson, A. M. 4841
Patin, H. A. 275
Patryas, W. 2202, 3624,
    4839
Pattee, H. H. 6954
Patten, B. C. 6955
Patten, S. C. 2645
Paty, M. 5985, 6045
Paul, A. M. 2017
Paul, S. 1748
Pauli, W. 732, 2372, 3384,
    4626, 6046-47, 6079
Paulos, J. 3625
Pawelzig, G. 7257
Pawłowski, T. 2021
Pawson, R. 4069
Peacocke, A. R. 424, 4474,
    7258
Peacocke, C. A. B. 2646
Pearce, D. 2373-74
Pearce, G. 1034, 4842
Pears, D. 2203
Pearson, C. I. 4627
Pearson, E. 2961, 3223-24,
    3441
Pearson, K. 602
Peat, F. D. 6048
Pechhacker, P. A. 7259
Pedersen, S. A. 4843
Peetz, D. W. 3734
Peierls, R. 5355
Peirce, C. S. 171, 192,
    233, 285, 1547, 1551,
    1565, 1585, 1744, 2407,
    2794-95, 2898, 2911,

2963, 3006, 3060, 3372,
3468, 3780, 3877, 4737,
4885, 4951, 5879, 6488
Pelč, J. 1069
Pepper, S. C. 761
Pera, M. 1640, 3888
Percy, W. 1309
Perelman, C. 425, 1749
Peres, A. 2204
Perrin, F. 5152
Perry, C. 3889
Peschke, J. von 6049
Peters, H. M. 1750, 6956
Peters, R. H. 6957
Petersen, A. 4475, 6050-51
Petersen-Falshoft, G. 1215
Petrie, H. G. 1310, 5004
Petrović, G. 1046
Petry, G. 6052, 6369
Pettijohn, W. C. 3408
Peursen, C. A. van 276
Pfannenstill, B. 3987
Pfanzagl, J. 4070
Pfarr, J. 6370, 6633
Pfister, F. 1311
Pham Xuân Yêm 5356, 5474
Phelps, E. R. 426
Phipps, Jr., T. E. 4071,
    6053
Piaget, J. 427, 822-25,
    1576, 4172, 4733, 4772,
    6958, 7260
Piattelli-Palmarini, M.
    6959
Piccone, P. 4368
Pietarinen, J. 2205, 2886,
    2948, 2990
Pietruska-Madej, E. 4844-45
Pietschmann, H. 826, 1641,
    5357
Piguet, J.-C. 359
Pihl, M. 5005
Pike, M. 682
Pikler, A. G. 5358
Pilet, P.-E. 428, 1751,
    6960-62
Pilot, H. 1312, 3907
Pinkham, G. N. 1477
Pinto, A. V. 827
Pirie, N. W. 2018
Pirlot, P. 6815, 6963-64
Piron, C. 5924, 6054-56
Pitcher, G. 1478
Pitt, A. 6057

Pitt, D. 37
Pitt, J. 37, 277-78,
    2647-48, 4846-47
Pittioni, V. 4848
Plamondon, A. L. 279, 2991,
    4369
Planck, M. 228, 4280, 4694,
    5359-60
Platnick, N. I. 7085
Plato 30, 1853, 1868
Plato, J. von 2992, 3385
Platt, J. R. 4430
Plaut, H. C. 4288
Plochmann, G. K. 4072, 6965
Pluhar, E. B. 4476
Podlaha, M. F. 6317, 6370,
    6455
Poincaré, H. 207, 280,
    3912, 6261, 6568-69,
    6682, 6695, 6764
Poirer, R. 280-81
Polanyi, M. 828-36, 1113,
    1270, 1479, 1540,
    1642-43, 2019, 4427,
    4430, 4477, 4628, 6281,
    6966
Poldrack, H. 4849
Pole, N. 2993
Poli, E. 4370
Polikarov, A. 505, 1313-14,
    1644, 2376, 5153, 5563,
    6058
Politis, C. 135
Pollard, E. C. 6967
Pollock, J. L. 837, 2735,
    2994-96, 3626
Pollock, S. 661
Pols, E. 4237
Pólya, G. 1645-47
Pomerans, A. J. 5891,
    6982
Pomerantz, I 4371
Ponce, M. 2649, 4372-73
Poncelet, G. 4279
Pontet, M. 7261
Pool, J. C. T. 6059-60
Poole, B. 4317
Poole, H. W. 5564
Popovich, M. V. 2020
Popowitsch, M. W. 1028
Popper, K. R. 45, 136, 138,
    284, 1551, 2206, 2455,
    2468, 2533, 2789, 2862,
    2925, 2997, 3283, 3327,
    3371, 3372, 3386-90,

    3484, 3609, 3627, 3664,
    3691, 3717, 3749-3964
    passim, 4478, 4575,
    4580, 4698, 4744, 4763,
    4803, 4808, 4850, 4933,
    4951, 5006, 5091, 5147,
    5154, 5361, 5703, 5834,
    6061-63, 6140, 6164,
    6317, 6601, 6782, 6794,
    7262
Portmann, A. 6968, 7263
Pos, H. J. 1080
Post, E. J. 5362, 5826,
    6064, 6701-03
Post, H. R. 1831, 2461,
    2650, 2998, 3913, 5475,
    5739
Potter, O. 652
Powers, W. T. 6969
Prado, C. G. 6970
Prat, N. 6971
Pratt, V. 2455, 4479,
    6985, 7086-87
Presley, C. F. 2022, 2207
Preston, M. G. 429
Price, J. F. 3251
Price, W. O. 1465, 6065
Priest, G. 2999, 4173
Prigogine, I. 838, 1648,
    4480, 5195-96, 5364
Primeaux, D. 1495
Prior, A. N. 6565
Proctor, G. L. 2208
Prokhovnik, S. J. 5565,
    6317, 6371-72, 6455,
    6704, 6718
Prokop, W. 6373
Protasiewicz, T. I. 5325
Provençal, Y. 839
Provine, W. B. 7228
Prugovečki, E. 4073
Przełęcki, M. 1086-87,
    1480, 2023-26, 2377-79,
    4074, 4851, 7350
Pugliese, O. 840
Puigrefagut, R. 4280,
    6066-67, 6374-75
Puligandla, R. 1481, 5007,
    7264
Pun, L. 1752
Purtill, R. L. 2209, 2651,
    4799
Purton, A. C. 6972
Putnam, H. 503, 671, 1094,

1753, <u>1980</u>, 2027-31,
2066, <u>2210</u>, 2380, 2684,
<u>2689</u>, 3628-30, 3914,
<u>3932</u>, 4583, 4629-31,
<u>4662</u>, 5008, 5837,
6068-71, 6103, 6705-07
Putnam, R. A.   1679, 4174

Quay, P. M.   430, 2381,
    4852, 5639
Quine, W. V.   48, <u>292</u>, 670,
    1156, <u>1394</u>, <u>1838</u>, <u>1891</u>,
    <u>1897-98</u>, <u>1921-22</u>,
    2032-36, <u>2093</u>, 2382-83,
    3000, 3516, <u>3714</u>, <u>3719</u>,
    3745-47, 3915, <u>4173</u>,
    <u>4175</u>, <u>4566</u>, 4632-36,
    5092
Quinn, J. M.   6708
Quinn, P. L.   2452, 2462,
    3735-36, 6287, 6597
Quintelier, G.   5009
Quinton, A.   6709

Rabb, J. D.   4853
Rabbeno, G.   6376
Rabinowicz, W.   3391
Radermacher, H.   66, 2384
Radner, M.   1055, 2385,
    3371, 3916
Radnitzky, G.   137-39, 166,
    282-84, 1315-24, 1363,
    3917-23, 4854-56
Rădulet, R.   4637
Radžabov, U. A.   6072
Raffensperger, M. J.   1235
Raggio, A. R.   2037
Railton, P.   2652-53
Rakitov, A. N.   3392
Ramirez, J. R. E.   7265
Ramsay, J. O.   4075
Ramsey, F. P.   <u>1900</u>, <u>2267</u>,
    <u>2401</u>, <u>2410</u>, <u>2424</u>, <u>4535</u>,
    <u>4744</u>
Ramsperger, A. G.   4539
Randall, C. H.   3001, 5821
Randall, J. A.   2038
Rankin, K. W.   1107, 3002
Rantala, V.   140, 841, 2039,
    2386, 2654, 5365

Rapaport, A.   2040, 2387,
    2655, 5640
Rapaport, S. M.   6867
Rapp, F.   3737, 4857
Rapport, S.   1095
Rasch, G.   5010
Rashevsky, N.   1325, 6973-74
Rat, P.   5197
Ratliff, F.   1482
Ratoosh, P.   4009
Raub, W.   2463
Raven, C. P.   1326, 6975-76
Ravetz, J. R.   842
Raymond, A.   506
Rayski, J.   6073-76
Recami   4227, 6216, 6487
Redhead, M. L. G.   1754,
    2388, 2442, 3207, 5659,
    5699, 5922, 6077-78,
    6093, 6455
Redman, C. L.   5205
Redondi, P.   843
Reenpää, Y.   1483, 2211,
    2656
Reeve, E. G.   4143
Rehder, W.   1216, 1484
Reichenbach, H.   <u>49</u>, <u>184</u>,
    <u>197</u>, <u>290</u>, 844, <u>1096-97</u>,
    <u>1098</u>, 1103, 1405, 1485,
    <u>1859</u>, <u>1925</u>, <u>1977</u>, 2041,
    <u>2168</u>, <u>2179</u>, <u>2212</u>, 3003,
    <u>3043</u>, <u>3134</u>, <u>3167</u>, 3395,
    3487, 4176, <u>5792</u>, <u>5844</u>,
    5874-75, 5930, <u>6030</u>,
    6079-80, <u>6226</u>, <u>6377-79</u>,
    <u>6437</u>, <u>6630</u>, 6710-11
Reichenbach, M.   49, 1096-98,
    3395, 6378, 6711
Reidemeister, K.   6712
Reignier, J.   5476
Reinberg, A.   6713
Reiser, O. L.   5477
Reitzer, A.   1327, 1755
Remes, U.   1249, 2726
Rensch, B.   4374, 6977,
    7266-67
Rescher, N.   285, 845, 1099,
    1100, <u>1116</u>, 1328-29,
    2042-43, 2213-15, 2274,
    2340, 2471, 2657-62,
    2765, 2848, 3004-07,
    3393-94, 3440, 3631,
    3753, 4076, 4290-91,
    4608, <u>4737</u>, 4858-59,
    5093-94, 6732

Resnick, L. 3632
Reutterer, A. 6978
Reyna, R. 5478
Riaza, F. 6943
Richards, R. J. 4375,
    4860
Richardson, R. C. 4481
Richardson, W. J. 846
Richfield, J. 2663
Richman, R. J. 2044, 4177
Richmond, S. 5036
Richter, E. 6081
Richter, G. 2216
Richter, H. 3396
Richter, K. 5198
Richter, M. N. 847
Ricoeur, P. 671
Riedel, M. 2664
Riemann, B. 4840, 6627
Rietdijk, C. W. 5155, 6380
Rietzler, K. 5366
Rifkin, J. 5367
Risse, W. 28
Ritchie, A. D. 1330, 5479
Rivier, W. 7268
Robbins, J. C. 3924
Robert, J.-D. 35, 431-32,
    848-49, 4638, 5095
Robert, S. 1649
Roberts, F. S. 4077-78
Roberts, G. W. 6714
Robertson, H. P. 6381
Robinson, A. 1884, 4376
Robinson, G. S. 3925
Robinson, R. 4079
Robles, J. A. 2649
Rochhausen, R. 433, 4765,
    4861-62, 6979, 7351
Rodni, N. I. 433
Rodrigues, J. R. 3926
Rody, P. J. 3633
Roelofs, R. 3589
Rogers, B. 2665, 3397,
    3634, 6715
Rogers, R. 141, 2389
Röhler, G. 5199
Rohmer, J. 5566
Rohs, P. 2915
Roig Gironella, J. 6082
Roldán, A. 286, 7269
Roll-Hansen, N. 282, 4482,
    6985
Rolston, III, H. L. 1331,
    1832

Roman, P. 5368
Romanos, G. D. 2488
Rombach, H. 1101
Rootselaar, B. van 1039
Roper, J. E. 3563, 4030
Rorty, A. 4455
Rorty, R. 2045, 4649, 4863,
    7070
Rose, L. E. 435, 3398
Rose, M. C. 4178
Röseberg, U. 4724, 5614,
    6083
Rosen, D. 4242
Rosen, P. 6382
Rosen, R. 5369
Rosen, S. M. 6716
Rosenberg, A. 2046, 4260,
    4281, 6980-81, 7088,
    7270, 7352
Rosenberg, J. F. 4639, 4864
Rosenblatt, L. D. 1486
Rosenblueth, A. 850, 1756,
    4377, 4390
Rosenfeld, L. 972, 6084-85
Rosenkrantz, B. 648
Rosenkrantz, R. D. 851,
    1487, 1833, 2666, 3008,
    3348, 3399-3401, 3423,
    3635, 3865, 5096
Rosenthal, H.-A. 6867
Rosenthal, S. B. 2390
Rosenthal-Schnieder, I.
    1488, 2217, 6383
Ross, A. 9
Ross, D. J. 6086
Ross, J. J. 5011
Ross, S. D. 852, 5097
Rossel, J. 2391, 5370
Rosser, W. G. V. 6384
Rossi, A. 3927
Rossi, P. 853
Rossi, R. J. 3009
Rossteutscher, H. 6087
Rostand, J. 6982
Roth, P. A. 2392
Rothbart, D. 3928
Rothman, E. D. 7300
Rothstein, J. 2393, 3988,
    5371-73
Rottschaefer, W. A. 1489-90,
    4640
Rouse, J. 4641
Rousseau, G. S. 436
Rowan, M. 3941

Rowe, W. L.   4355
Roxburgh, I. W.   6282, 6385,
    6500
Rozeboom, W. W.   1881, 1884,
    2047-50, 3010, 3371,
    3394, 3577, 3636-37,
    4080
Ruben, P.   1757
Ruddick, C. T.   2218
Ruddick, S.   680
Ruddick, W.   4282
Rudich, N.   1211
Rudner, R. S.   1799, 1834,
    2051, 3014, 4645, 5012
Rudwick, M. J. S.   7271
Runggaldier, E.   287
Rupke, N. A.   1650
Ruse, M.   1651, 1758-59,
    2549, 2667, 3929, 4378,
    4400, 4406, 4483-84,
    4747, 4858, 4865-66,
    6485, 6880, 6983-87,
    7089, 7243, 7270,
    7272-77
Rusk, G. Y.   4081
Russell, B.   50, 177, 263,
    507, 602, 854-55, 995,
    1105, 2820, 2865, 2901,
    3003, 4179, 4492, 4527,
    5374, 6386
Russell, E. S.   6988
Russell, J. L.   856, 7278,
    7297
Russell, L. J.   854, 1105,
    3402
Russo, F.   288, 4082, 5567,
    6088
Rutherford, F. J.   4149
Rutte, H.   289
Ruttiman, G.   6009
Ruytinx, J.   508-10
Ruyer, R.   359, 437-39,
    5375, 6989-90, 7231,
    7279
Ruzavin, G. I.   3403
Rybicki, P.   1332
Ryder, J. M.   3404
Ryle, G.   2052, 2667, 2743
Rynin, D.   1491, 2053, 3405,
    4180, 4183

Saarinen, E.   1102, 2054

Sachkov, I. V.   857, 1689,
    6091
Sachs, M.   511, 690, 3907,
    3930, 4083, 5376,
    5480-82, 6089-90,
    6387-89, 6454, 6717
Sachsse, H.   858
Sachsteder, W.   1333, 2394
Sadovsky, V. N.   2055, 2736,
    4687, 4867
Sagal, P. T.   142, 3638,
    4868
Sagoroff, S.   3406
Săhleanu, V.   6991
Saini, H.   2219, 5483
Saint-Sernin, B.   1147
Sakai, H.   859
Salmon, M. H.   2737
Salmon, W. C.   143, 290,
    1103, 1334, 1799, 2057,
    2212, 2220, 2497, 2540,
    2545, 2596, 2605,
    2668-74, 2850, 2925,
    3011-22, 3369, 3383,
    3407-11, 3639-40, 3924,
    4283-85, 6390-91,
    6718-19
Salt, D.   5241
Sambursky, S.   1492
Sampson, G.   2395, 3940
Samuel, H. L.   5377
Samuel, V.   860-61, 5378-79
Sandgathe, F.   6392
Santirocco, R. A.   5641
Santman, A.   6024
Sapper, K.   5380, 6393
Sarkar, H.   1215, 1304,
    1335, 1562, 2464,
    391-33
Sarndahl, C.-E.   3023
Sarton, G.   862
Sastri, P. S.   512, 2056
Saunders, L. R.   2738
Saur, K.   6358
Savage, C. W.   1060
Savage, L. J.   3024, 3412-13
Savary, C.   4869
Savigny, E. V.   591, 910
Savitt, S. F.   2421
Sawyer, C.   2437
Sayre, K. M.   4286, 4353
Schächter, J.   1104, 4181
Schaerer, R.   440
Schäfer, K.   6720

Schäfer, L. 2396-97, 3934
Schaffner, K. F. 513, 1089,
    1653-54, 2398-2400,
    4487-88, 6455
Schagrin, M. L. 3025, 4870
Schapp, W. 863
Schatzmann, E. 6092
Scheffler, I. 671, 1336,
    1864, 2401, 2543, 2675,
    3032, 3538, 3691, 3732,
    4182, 4379, 4871, 4996,
    5013
Scheichert, H. 892
Scheibe, E. 46, 2402,
    2676-77, 4872, 6063,
    6093
Schermer, J. 2600
Scheurer, P. 4873
Schick, F. 3414
Schields, M. C. 11
Schievella, P. S. 4485
Schild, A. 6394
Schiller, R. 6094-95
Schilling, K. 5381, 6992
Schilpp, P. A. 1105-08,
    3935, passim
Schlanger, J. 3641
Schlaretzki, W. E. 3026
Schlegel, R. 864, 2739,
    4208, 4642, 5290, 5382,
    5568-69, 6096-99,
    6395-97, 6721-22
Schleichert, H. 2221, 4084,
    5200, 6238, 6358,
    6398-99
Schlesinger, G. N. 441,
    1337-38, 1807, 1821,
    1835-36, 1987, 2057-58,
    3027-28, 3600, 3612,
    3642-46, 3738, 3936,
    3989, 4486, 6723-27
Schlessinger, M. 6185
Schlick, M. 289, 865-66,
    1006, 1063, 4154,
    4183-84, 4287-88
Schlossberger, E. 5484
Schmid, M. 2442, 3725
Schmidt, H.-J. 5989
Schmidt, P. F. 1339-40
Schmutzer, E. 2222, 6400-01
Schnädelbach, H. 2059
Schneider, E. F. 2060
Schneider, F. 867, 5485
Schnell, W. 868

Schnorr, C. P. 3415
Schock, R. 869, 2061, 2403,
    3029, 5156, 6402
Schoenberg, J. 3647
Schon, D. 1581, 1655
Schopman, J. 4874
Schorsch, A. 3030
Schrader, M. E. 7280
Schramm, A. 971, 2454,
    2465, 4875
Schreiber, A. 1493, 3031,
    3648
Schrödinger, E. 870-72,
    927, 2223, 5570, 5709,
    5743, 5869, 5993,
    6100-01, 6169
Schroeder-Heister, P. 10
Schubert-Soldern, R. von
    6993
Schüler, W. 5201
Schulman, L. S. 5642
Schultz, J. 4384
Schultzer, B. 4185
Schulz, D. J. 2224
Schumacher, D. L. 6575
Schuntermann, M. F. 3416-17
Schurig, V. 6994
Schurupow, A. 6373
Schütt, K.-P. 3418
Schwartz, H. T. 4380
Schwartz, J. 1341
Schwartz, R. J. 1693,
    3032, 3075, 3649-50
Sciacca, M. F. 442
Sciama, D. W. 5571-73
Scoledes, G. M. 6102
Scott, D. 3419, 4085
Scott, W. T. 1666
Scozzafava, R. 3033
Scriven, M. 443, 1052-53,
    1342, 1837, 2225, 2548,
    2631, 2678-81, 4289,
    5157, 5574
Seager, W. 2404
Seall, R. E. 2226
Seaman, F. 3718, 6403
Searles, H. L. 1343, 4186
Seeger, R. J. 191, 873,
    1110
Seelig, W. 5383
Seeliger, R. 1760
Seely, C. S. 4643
Segeth, W. 1344
Seidenfeld, T. 3307,

3420-24
Seiffert, A. 1345
Seiffert, H. 874
Seifullaev, R. S. 5158
Seigfried, H. 4644
Sellars, R. W. 444, 6404-09
Sellars, W. 51, 277-78,
    1142, 1992, 2062-66,
    2405, 2647-48, 2682,
    2684, 2941, 3034-35,
    4187-88, 4411, 4419,
    4467, 4501, 4508, 4523,
    4528, 4531, 4545, 4562,
    4584, 4589, 4640,
    4645-51, 4876, 5044,
    5061, 6555, 6599, 6646,
    6664, 6728, 6740
Selvaggi, F. 875, 5384
Sen, P. K. 3651
Senechal, M. 2227
Senior, J. K. 144
Sepetys, J. 6414
Serrano, J. A. 876
Servien, P. 1494, 3425
Settle, T. 2831, 3372,
    3426-27, 3485, 3501,
    3789, 3937-38, 5014,
    5036, 5159
Severi, F. 6410
Shackle, G. L. S. 3342,
    6729
Shafer, G. 3342, 3428-32
Shames, M. L. 1495
Shapere, D. 1157, 1496,
    2406, 4652, 4799,
    4877-80, 5015, 5065,
    5651, 6730-31, 6995
Shapiro, J. 680
Sharp, D. H. 6103
Sharp, R. A. 2228
Sharpe, R. 3036, 3652,
    3739
Sharżyński, E. 2379
Shea, W. R. 145, 1112
Shelanski, V. B. 4381
Sheldrake, R. 6996
Shelley, J. H. 1613
Sherrington, C. 6997
Sherry, T. N. 6298
Shibata, S. 1346
Shils, E. 1113
Shimony, A. 1497, 2485,
    3037, 3412, 3653, 3677,
    4881, 5252, 5646, 5807,
    6104-06 6108

Shirley, E. S. 2753
Shiva, V. 5712, 6107
Shoemaker, S. 3038
Shope, K. 4232
Short, T. L. 2407, 4887
Shorter, J. M. 6726
Shrader, Jr., D. W. 2605,
    4286, 4883
Shrader-Frechette, K.
    5385-86, 5486
Shrewsbury, M. M. 7156
Shtokhamer, R. 5642
Shvyrev, V. S. 1356, 2067,
    2080
Sibley, W. M. 4653
Siciński, M. 5387
Siegel, A. 2308, 6094,
    6108
Siegel, H. 1667, 4722,
    4806, 4996
Siemens, Jr., D. F. 3039
Siemens, W. D. 4884
Silva, H. 5016
Silvers, S. 1838
Sikora, J. J. 146, 445-46,
    3040
Sikora, R. I. 3674
Simard, E. 1347-48
Simberloff, D. 6998
Siminel, G. 6411
Simon, H. A. 877, 1540,
    1668-72, 1770, 1839,
    2068, 2408-10, 3654-55,
    4290-91, 5017, 5643
Simon, M. A. 6999
Simon, N. 7341
Simon, R. 5202
Simon, T. W. 4382, 5008
Simon, Y. 291, 878
Simons, J. H. 879
Simons, L. 2631, 4489
Simpson, G. G. 880, 7000,
    7281-83
Singer, Jr., E. A. 881,
    4189, 4348, 4383
Singerman, O. 447
Sinha, A. K. 5575
Sinks, J. D. 4654
Sinnott, E. W. 7001-02
Sipfle, D. A. 5251
Sivadjian, J. 6412-13
Siwek, P. 4384
Skagestad, P. 2421, 3939,
    4885, 7284

Skarsgård, L. 2683
Skiebe, K. 7003
Skjervheim, H. 267
Sklar, A. 7090
Sklar, L. 514, 2072,
    3433-35, 3656, 3741,
    4490-91, 4886, 5070,
    5277, 5644-45, 6732-37,
    6786
Skolimowski, H. 4951,
    5018-20
Skotnicky, J. 4292-93
Skyrms, B. 2229-32, 2548,
    2662, 3041-43, 3657,
    3740
Slaatte, H. A. 1158
Slaght, R. L. 24
Slater, J. G. 7295
Sleinis, E. E. 1498
Slichter, C. S. 882
Slobodkin, L. B. 7285
Sloman, A. 3044
Slote, M. A. 3045-46, 3658
Smart, J. J. C. 292, 474,
    883, 1108, 1409, 1499,
    2032, 2411, 2455, 2631,
    2684, 3047, 4086, 4385,
    4655-56, 6709-10,
    6738-40, 7004
Smart, W. M. 5576
Smirnov, S. N. 4887
Smirnov, V. A. 1673, 2055
Smith, C. S. 1674
Smith, C. U. M. 7005
Smith, D. W. 1500
Smith, E. T. 7006
Smith, J. E. 5021
Smith, J. M. 7286
Smith, G. J. 5702
Smith, M. 1349, 3048
Smith, P. 4657
Smith, V. E. 1115, 1350,
    1501, 5388-90, 6114,
    7007, 7287
Smithson, A. 3941
Smokler, H. 2069, 3049-50,
    3319, 3414, 3436-38,
    3659-60
Smolicz, J. J. 4888
Smullyan, A. 3661-62
Smythies, J. R. 1399, 2119,
    4172, 4433, 4457-58,
    7030, 7306
Sneed, J. D. 2262, 2311,

    2397, 2412-15, 3439,
    4687, 4804, 4889-90,
    4892, 5346, 6115-16
Snell, B. 884
Snider, C. W. 6741
Snyder, A. 4616
Snyder, D. P. 515
Sober, E. 1840-41, 5022,
    7288-89
Sofonea, L. 1255, 6109
Solla Price, D. de 1675
Sollazzo, G. 3051
Somenzi, V. 293, 1676,
    3540, 3965, 3990,
    5391-93
Somerville, J. 448, 885
Sommers, F. 3532, 3663
Sontag, F. 147
Sosa, E. 1116, 2233, 4289,
    4294
Soran, V. 7008
Souriau, E. 7353
Southwood, T. R. E. 7009
Sparkes, J. J. 4891
Speakman, J. C. 5487
Specker, E. P. 5843, 5895,
    5936-38, 5978, 6117
Spector, M. 1062, 1771,
    2416, 4492-93, 6415
Spencer, D. E. 6354-55,
    6683
Spencer, H. 7118
Sperber, G. 6118
Spicker, S. F. 2515
Spielman, S. 851, 2948,
    3052, 3107, 3207, 3421,
    3440-44
Spilsbury, R. J. 3053,
    7178, 7290
Spinner, H. F. 886, 4494
Spinney, G. H. 449
Spirito, U. 450, 887-88
Spisani, F. 6742, 6743
Spohn, W. 2982
Sprigge, T. L. S. 4386
Sprott, D. A. 3278
Srzednicki, J. 3445
Staal, J. F. 6, 1039
Stace, W. T. 3054
Stachel, J. J. 970, 972,
    1059, 5826, 5990, 6119,
    6292, 6416
Stachow, E. W. 2070,
    6018-19, 6120-23, 6162

Stachowiak, H. 1772-74, 2685
Stairs, A. 6124
Stakhanov, I. P. 6125
Stalnaker, R. C. 3184, 3446, 4744, 5871
Stanciu, L. 6743
Stancovici, V. 6743
Standen, A. 889
Stapp, H. P. 5880, 5926, 6110-13
Starostin, B. A. 890
Stauffer, R. C. 1117
Stebbins, G. L. 7291-92
Stefansen, N. C. 5023
Stegmüller, W. 670, 891-97, 1118, 2032, 2234, 2311, 2349, 2397, 2686-87, 2791, 3416, 3447-49, 4167, 4387, 4676, 4723, 4892-95, 5098
Stehr, G. 6135
Stein, H. 2400, 6126-27, 6417-19, 6744-45
Steinberg, D. 9, 2635
Stelzl, I. 892, 1775, 3664
Stemmer, N. 2487, 3055-62, 3665
Stengers, I. 838, 1648, 5196, 7010
Stenius, E. 52, 1119, 2071, 2740
Stenner, A. 2474, 3063
Stent, G. S. 7011, 7354
Stepnin, W. S. 1351
Stern, A. 898, 5099
Stern, C. 4256
Stern, H. S. 451
Stern, O. 5989
Stewart, J. P. 2841
Stich, S. P. 2072, 3064, 3940
Stiernotte, A. P. 7001
Stiffler, E. 1394
Stock, M. 7293
Stock, W. G. 294, 899
Stocker, O. 4388
Stöhr, H.-J. 7294
Stonert, A. 2073
Stopes-Row, H. V. 2074, 2743
Storer, T. 4072
Stove, D. C. 2235, 3065, 3571, 3666-68, 3691,

3900, 3941
Strasser, H. 4779
Straubinger, H. 6128
Strauss, M. 2417-18, 3450, 4896, 5241, 5394-95, 6129-33, 6401, 6420
Strauss und Torney, L. von 5396, 6134
Strawson, P. F. 148, 2032, 3013, 3028
Strohal, R. 3451
Ströker, E. 516, 892-93, 4897, 6746
Stroll, A. 5100
Strombach, W. 900, 4658, 5203, 5397, 5577
Strong, Jr. D. R. 7012
Stroud, B. 2075
Struik, D. 970
Stuart, A. L. 7297
Stuewer, R. H. 1056, 5398-99
Subbotin, A. L. 1352
Such, J. 3742, 5400
Suchting, W. A. 1199, 2236, 2688, 3876, 5623
Sudbury, A. W. 2237-38
Sumner, L. W. 7295
Supek, I. 901
Suppe, F. 73, 295, 1091, 1502, 1504, 2076, 2266, 2318, 2406, 2419-22, 2425, 2689, 4802
Suppes, P. 53 149-50, 517, 953, 1037, 1040, 1062, 1120, 1248, 1353-53, 1503-04, 1776, 1842-43, 2077-78, 2230, 2423, 2690, 2876, 2881, 2884-86, 2925, 3078, 3080, 3087, 3105, 3195, 3268, 3401, 3419, 3452-55, 3510, 3669-70, 3702, 4053, 4085-88, 4197, 4278, 4295, 5024, 5346, 5488, 5605, 5636, 5807, 6136-46, 6586 6747-49
Sussman, H. J. 7013
Süssmann, G. 4089
Svenonius, L. 1844, 2079 4495
Swain, M. 1068, 3066, 3589, 4204, 5025

Swanson, J. W.   1771, 3743,
    4454
Swartz, N.   3456
Sweigart, J.   296, 2841
Swijtink, Z. G.   2424
Swinburne, R. G.   2081,
    2780, 2810, 2820, 2943,
    2959, 3045, 3067-69,
    3084, 3457, 3671-75,
    3744, 3942, 5578-79,
    6750-52
Swinton, W. E.   452
Swobada, W.   172
Swoyer, C.   3603
Synge, A.   7251
Synge, J. L.   902
Szaniawski, K.   151, 1087,
    1505, 3070, 3458-60,
    3676
Szarski, H.   7014, 7296
Székely, D. L.   518-20, 3991
Szent-Györgyi, A.   7015
Szumilewicz, I.   1945, 4898

Tagliagambe, S.   903
Takeda, K.   297
Talbott, G. R.   904
Tallet, J.   3461
Tarozzi, G.   6147-48
Tarski, A.   1037, 1687,
    5071, 5098, 5936, 6138,
    6586
Tassi, A.   453
Tatarkiewicz, K.   1778
Tatarkiewicz, W.   454
Tati, T.   6753
Taton, R.   455, 1677
Tavakol, R. K.   6385
Tavanec, P. V.   1355-56
Tavel, M.   4847
Taylor, C.   4367, 4389,
    4403
Taylor, C. W.   1615
Taylor, F. S.   905
Taylor, H. A.   456
Taylor, J. G.   5401, 6754
Taylor, R.   4390
Teensma, E.   3071
Teilhard de Chardin, P.
    7099, 7149, 7261, 7293,
    7297-98, 7307
Teissier, G.   7299

Teller, E.   1846
Teller, P.   1506, 2980,
    3072, 3646, 3677, 5808,
    6149-51
Temple, D.   2239, 3678
Templeton, A. R.   7300
Tennessen, H.   3073
Teranaka, H.   298
Terletzkis, J. P.   6161
Tetze, H.   5402
Teune, H.   1507
Thackray, A.   152
Thagard, P.   906, 3074,
    4899, 6987
Thakur, S. C.   3943
Thalberg, I.   2528
Thales,   5284
Thayer, L.   761
Theobald, D. W.   1159,
    1779, 4659
Thienemann, A.   1508
Thom, A.   4765
Thom, R.   907, 1780, 6755
Thomas, E.   7016
Thomas, J.-P.   2590
Thomas, L.   7017
Thomasma, D. C.   5403-04
Thomason, R. H.   6756, 7091
Thompson, D.   6872, 6965
Thompson, F. W.   4443
Thompson, G.   1357
Thompson, J. W.   1509, 4090
Thompson, M.   2240
Thompson, M. E.   3211
Thompson, R. P.   7018
Thompson, W. R.   457, 7301
Thomson, J. J.   3075, 3751
Thorndike, E. M.   6326
Thornton, J. B.   4660
Thornton, M.   4545
Thorpe, D. A.   2691
Thorpe, W. H.   4391, 4496,
    7033
Thum, P. B.   7019
Thüring, B.   5405-08, 5580,
    5648-49
Thurnher, R.   2692
Thyssen-Rutten, N.   3944-45
Tibbetts, P.   1215, 1510,
    1678, 2082, 3946, 4661
Tichý, P.   3947-49
Tiercy, G.   1511
Timoféeff-Ressovsky, N. W.
    7020

Tinnon, J. 2072
Tisza, L. 3201, 5409,
    5621, 5646-47
Titiev, R. J. 4091
Titze, H. 458, 5160, 6421,
    6757-58
Todd, W. 3076, 3679
Tomlin, E. W. F. 7021
Tondl, L. 1358, 1377,
    2083-84, 4092
Tonini, V. 299, 459, 5161,
    5410, 6422
Tonnetat, M.-A. 5411-12,
    5650, 6423
Tooley, M. 2241
Topitsch, E. 460
Toraldo de Francia, G.
    5121, 5413
Törnebohm, H. 139, 908-09,
    1359-63, 3077, 3680-81,
    4900-01, 6152, 6279,
    6424-34, 6759
Toro, T. 6153-54
Torretti, R. 3019, 6304,
    6589
Tosi, P. 3079
Toulmin, S. 300, <u>771</u>,
    910-12, 1003, <u>2209</u>,
    2425, 2466, <u>2703</u> 3141,
    <u>3285</u>, 3462, <u>4392</u>, 4878,
    <u>4902</u>-04, <u>4989</u>, 5018,
    5027-30, <u>5414</u>, 5489,
    5581, 5651, 5679, 5794,
    5865, 6760
Tournier, F. 3745
Townsend, B. 1512, 6304
Trainor, P. 4905
Tranøy, K. E. 5026
Trapp, R. 1907, 2085
Trautman, A. 6761
Trautteur, G. 3463
Travis, J. L. 7302
Treder, H.-J. 6435-36, 6762
Tredwell, R. F. 2242
Trenholme, R. 3464
Trenn, T. J. 648
Troll, W. 1513, 7022
Truesdell, C. 2962
Trusted, J. 1364
Tuana, N. 3736
Tucci, P. 6437
Tuchańska, B. 4093-94
Tucker, J. 1365
Tugendhat, E. 5098

Tummers, J. H. 913, 6763
Tuomela, R. 1045, 1409,
    1781, 1952, 2086-92,
    2364, 2426, 2693-97,
    3080, 3950, 4662, 4836,
    4906-07
Turbayne, C. M. 2427
Turchin, V. F. 914, 7303
Turner, D. 915
Turner, J. 1782, 2698,
    3966
Turnbull, R. G. 6655
Turner, S. 2428
Tversky, A. 3465-66
Tweeney, R. D. 1160

Ubbink, J. B. 1783
Uchii, S. 1382, <u>2980</u>,
    3081-83, 3467
Ueberschär, K. 7355
Uemov, A. 917, 1784
Ueno, Y. 6438
Uexküll, T. von 672-73
Uhlmann, A. 6155
Ujemow, A. I. 6439
Ukolova, V. I. 5158
Ullian, J. S. 1156, 2753,
    2843, 3102, 3468, 3682,
    4788
Ullmann, C. 2699
Ullmo, J. 916, 5031, 5204,
    5415, 6156-57
Ulmer, K. 5101
Urbach, P. 2467, 3683,
    3951
Urbani, E. 7023
Urmson, J. O. 3084, 3469
Ushenko, A. P. 301,
    2243-44, 5102
Utz, S. 4403

Vácha, J. 7024
Valcke, L. 4394
Valen, L. van 7092-93,
    7186
Vallée, R. 1514
Vandamme, F. J. 4296, 4908
Vandel, A. 918
Varadarajan, V. S. 3470
Varela, F. 4393

Vayonis, G. C. 5016
Veatch, H. 461
Vedenov, M. F. 1689
Velde-Schlick, B. F. B. van
    de 1063
Velikovsky, I. 1226, 4841,
    5582
Verbruggen, F. 5032
Verde, M 6158
Vermeersch, E. 5033
Vermeulen, R. 3648
Versfeld, M. 7304
Vet, P. van der 1285
Vetter, H. 3471-72,
    3952-53
Vetterling, M. K. 462, 2488
Vickers, J. M. 3085,
    3473-76, 4909
Vietoris, L. 3477-79
Vigier, J. P. 5162, 6159-60
Vincent, R. H. 2154, 3480,
    3685-88, 3954-55
Vogt, H. 5583
Vogtherr, K. 6440-43
Voigt, W. 6899
Vossius, G. 7025
Vries, H. de 4713
Vries, J. de 3086
Vuillemin, J. 2093, 3746,
    4095, 6764

Waddington, C. H. 919-20,
    6875, 7026-27, 7305-06
Waelhens, A. de 463, 921
Waerden, B. L. van der
    3481, 6674
Wagner, M 4910
Wahl, D. 1366
Wahl, J. 5887
Wahlert, G. von 7186
Wahsner, R. 5500, 6161,
    6204, 6477, 6583
Waismann, F. 1123, 2074
Walcott, G. D. 153, 1161
Walentynowicz, B. 302
Walk, K. 3087
Walker, E. R. 3689
Walker, M. 922
Wall, B. 7297
Wallace, J. R. 3088
Wallace, R. 6162

Wallace, W. A. 464, 4096,
    4297, 4663-65, 5490,
    7307
Wallén, G. 139
Wallenmaier, T. E. 4395
Walley, P. 3482
Walsh, B. 6958
Walter, E. J. 923, 1453
Walton, G. 3690
Walton, K. L. 2095
Warnock, G. J. 2096
Wartofsky, M. W. 154-55,
    924, 956, 970-71,
    1074-79, 1211, 1679-80,
    1785, 5257, 5463
Wassermann, G. D. 3910,
    4400, 7308-09
Wasserstrom, R. 111
Watanabe, S. 4497, 4911,
    5034, 6765-67
Watkins, J. W. N. 1847,
    2700, 2789, 3019, 3498,
    3691, 3956-59, 4597,
    4912-13, 5035-36
Watling, J. 2124, 2841,
    3483, 3692
Watson, J. 4487
Watson, P. J. 5205
Watson, R. A. 5166
Watson, W. H. 5416
Watts, F. 5006
Weaver, W. 521
Webb, C. W. 6768
Weber, H. 7028
Wechsler, J. 1681
Wedberg, A. 1878, 4190,
    5103
Wedeking, G. 3747
Wegener, U. 3484, 5931,
    6163-64
Weiler, G. 5037
Weill, R. 7029
Weimer, W. B. 1367, 1682
Wein, H. 465, 927
Weinberg, J. R. 925
Weinberger, C. 2245
Weinberger, O. 2246, 5163
Weiner, J. 2247
Weingard, R. 2256, 5584,
    6444-46, 6529, 6603,
    6734, 6769-71, 6782
Weingartner, P. 1124-26,
    1162, 1786, 2701, 5104

Weisheipl, J. A. 121,
    1127, 6926, 7209, 7245,
    7287
Weiss, P. A. 926, 1515,
    5061, 5206, 7030
Weisskopf, V. F. 5417, 6165
Weizsäcker, C. F. von
    522-23, 927-30, 1516
    4298, 6166
Wells, R. 2034
Wendell-Waechtler, S. 7325
Wendt, H. 100, 2248
Wenzl, A. 6447
Wenzlaff, B. 2249, 6167,
    6772
Werkmeister, W. H. 931-33,
    4666, 5331, 6168, 6921,
    7031
Werner, C. 466
Werth, R. 1517
Wesley, P. 4914
Wessel, H. 1383
Wessels, L. 2429, 6169-70
Westfall, R. S. 233, 997,
    1591, 2779
Westland, G. 156
Westman, R. S. 4810
Wetter, G. 303
Wettersten, J. R. 3799,
    3960, 4676
Weyl, H. 934, 2250, 6226,
    6773
Wheatley, J. 3089
Wheeler, J. A. 5418-19,
    5585, 6448-49, 6468,
    6600, 6774
Whewell, Wm. 68, 1556,
    2779, 2932, 3877
Whitbeck, C. 157, 6450
White, A. R. 3485
White, M. 1062, 1354, 2097,
    3513
White, M. J. D. 7094
Whitehead, A. N. 235, 248,
    279, 5235, 5817, 6104,
    6112, 6255, 6403
Whiteley, C. H. 467, 3469,
    3561, 3693
Whiteman, M. 6775
Whitmore, C. E. 1518, 4396
Whitrow, G. J. 158, 935,
    2098, 5574, 5586-91,
    5652, 6523, 6776-82

Whittaker, E. 5420, 6100
Whyte, L. L. 936, 5105,
    5421-26, 5652, 6171,
    6783, 7032-35
Wiatr, J. J. 1128
Wicken, J. S. 4397, 5653,
    7310
Wiebe, D. 421, 3961
Wiegand, F. 6172
Wiener, N. 1756, 4377,
    4390
Wiener, P. P. 627, 1129,
    1163, 2288
Wigner, E. P. 1130, 4667,
    5427, 6103, 6173-74
Wilhelm IV of Hesse-Kassel
    1630
Wilkerson, T. E. 3090
Wilkie, J. S. 4299, 7036
Wilkinson, J. 524
Wilks, S. S. 4097
Will, F. L. 1192, 2135,
    2855, 3091-93, 3486,
    3694-95, 4668
Willer, J. 1787, 6451
Williams, D. C. 2243, 2793,
    3091, 3093-94, 3141,
    3144, 3487, 6489, 6784
Williams, D. H. 5359
Williams, G. 5106, 7311
Williams, L. P. 4915, 6452
Williams, M. B. 2702,
    7312-14
Williams, P. M. 2430, 3428,
    3488, 3696-97
Williams, R. J. 7037
Williams, T. R. 7315
Williamson, R. B. 1848
Willing, A. 2194
Wilmsen, A. 6175
Wilson, Jr., E. B. 1368
Wilson, E. O. 7038-39
Wilson, F. 213, 525,
    2099-2101, 2703, 3232,
    3992, 7295
Wilson, M. 2251, 2431,
    2852, 3095
Wilson, N. L. 1108, 2102
Wilson, P. J. 7316
Wilson, P. R. 1788, 3698-99
Wimsatt, W. C. 2704, 4398,
    4498-99, 7040-41,
    7317-18

Winch, P. 941, <u>3889</u>, 3962
Winkler, R. 630
Winnie, J. A. 2103–04,
   6453, 6785
Winokur, S. 1055
Wintgen, G. 6867
Wisdom, J. O. 937, 1107,
   1369, 1519, 2432, 2817,
   3963–64, 4300, 4916,
   7045
Wit, H. F. de 3700–01
Withers, R. F. J. 1370,
   5180, 7042
Witkowski, L. 4839
Witmer, E. E. 4692
Wittenberg, A. 159
Wittich, D. 4917
Wittgenstein, L. <u>138</u>, <u>284</u>,
   <u>1009</u>, <u>1872</u>, <u>3490</u>, <u>3834</u>,
   <u>3919</u>, <u>4541</u>, <u>4619</u>
Witt-Hansen, J. 304–06
Wittrock, B. 3737
Witzemann, E. J. 938, 7319
Wohlfahrt, T. A. 1789
Wohlgenannt, R. 939
Wohlhueter, W. 894
Wohlrapp, H. 160
Wójcicki, R. 1086–87,
   1371–74, 2105–06, 2379,
   2433, 5164
Wojciechowski, J. 509, 940,
   1131, 4301, 4669–70,
   5038
Wojtasiewicz, O. 1170
Wolf, R. 5654
Wolff, E. 1520, 4399, 7043
Wolfson, P. 6786
Wollgast, S. 715
Wolter, H. 1757
Woltjer, H. R. 5428
Wolvekamp, H. P. 7044
Wonsowski, G. W. 6176
Woodfield, A. 4400–01,
   7320
Woodger, J. H. 831, <u>1000</u>,
   1375–76, 2107, <u>7045</u>–49,
   7356
Woods, J. 7095
Woodson, M. I. C. E. 3603
Woodward, J. 2705–06, 6454,
   6786
Woolf, H. 1003
Woolhouse, R. S. 2252, 4401

Workman, R. W. 2707,
   3096–97, 6177
Works, C. 2108
Worrall, J. 161, 219, 1215,
   2434, 2453–54, 2468–69
Wright, G. H. von <u>941</u>,
   <u>1009</u>, 1045, 3098–3101,
   <u>3141</u>, 3489–90, <u>3506</u>,
   3702–02, 4302, <u>5165</u>,
   6787
Wright, H. 1095
Wright, L. 2708, 4378,
   4400, 4402–07, 7051
Wright, S. 7051
Wrightsman, B. 3704
Wuketits, F. M. 4408, 7052
Wüstneck, K. D. 1790–91
Wykstra, S. J. 162, 6455

Yaker, H. M. 307
Yamamoto, M. 4409
Yanase, M. M. 4671, 5338,
   5429, 6174, 6178,
   6788–89
Yildirim, C. 942
Yilmaz, H. 1521
Yolton, J. W. 308, 2709–10
Yoshida, R. M. 3748, 4493,
   4500
Yost, R. M. 1107
Young, J. Z. 5107, 7053
Young, R. M. 991, 7321
Young, T. <u>161</u>
Yourgrau, W. 111, 131, 317,
   350, 389, 908, 958,
   1132, 1950, 2108, 2741,
   3905, 4026, 4410, 4609,
   4635, 4900, 5430, 5530,
   5592, 5611, 5668, 5959,
   5962, 6160, 6179–80,
   6203, 6435, 6454, 6824,
   7015, 7103, 7131, 7291
Yudin, B. G. 4918
Yukawa, H. 943, 6790

Zabell, S. L. 5638
Zabludowski, A. 3102
Zaffron, R. 2711
Zahar, E. 309, 1683, 2453,

      2470, 6455-56
Zahler, R. S. 7013
Zanotti, M. 2690, 4088,
      6145-46
Zanstra, H. 3103
Zaragüeta Bengoechea, J. 944
Zaret, D. 6791-92
Zaslawsky, D. 310, 428, 468
Zaw, S. K. 7054
Zecha, G. 19, 311, 1126
Zeh, H. D. 6181-82
Zeigler, B. P. 1792
Zelbstein, U. 1793
Zeldovich, I. B. 5593, 6457
Zeman, J. 6793
Zeman, J. J. 6183
Zeman, V. 312
Zeno, 5212, 6459, 6595
Zenzen, M. 6794
Zeppelin, A. von 4184
Zich, O. 1377
Ziedins, R. 1522
Ziemba, Z. 3104
Ziemski, S. 945, 1378-80
Zierler, N. 6184-85
Ziff, P. 4919
Zilian, H. G. 4778
Zilsel, E. 4920
Ziman, J. 946-47
Zimmer, K. G. 7020
Zimmerman, E. J. 6186
Zimmermann, W. 7322
Zinkernagel, P. 4672
Zinoviev, A. A. 1381-83,
      2742, 6187
Zinzen, A. 5207-08
Zöllner, W. 7055
Zoubek, G. 6056
Zucker, F. J. 523, 928
Zukav, G. 5431
Zupan, M. L. 4870
Zwart, P. J. 2253, 6795-96,
      7056
Zweig, A. 3705
Zwicky, F. 1684